INTERNATIONAL MATHEMATICAL SERIES

Series Editor: **Tamara Rozhkovskaya**
Novosibirsk, Russia

This series was founded in 2002 and is a joint publication of Springer and "Tamara Rozhkovskaya Publisher." Each volume presents contributions from the Volume Editors and Authors exclusively invited by the Series Editor Tamara Rozhkovskaya who also prepares the Camera Ready Manuscript. This volume is distributed by "Tamara Rozhkovskaya Publisher" (tamara@mathbooks.ru) in Russia and by Springer over all the world.

For other titles published in this series, go to
www.springer.com/series/6117

AROUND THE RESEARCH OF VLADIMIR MAZ'YA II

Partial Differential Equations

Editor: **Ari Laptev**

Imperial College London, UK
Royal Institute of Technology, Sweden

SPRINGER
TAMARA ROZHKOVSKAYA PUBLISHER

Editor
Ari Laptev
Department of Mathematics
Imperial College London
Huxley Building, 180 Queen's Gate
London SW7 2AZ
United Kingdom
a.laptev@imperial.ac.uk

ISSN 1571-5485 e-ISSN 1574-8944
ISBN 978-1-4419-1342-5 e-ISBN 978-1-4419-1343-2
ISBN 978-5-9018-7342-7 (Tamara Rozhkovskaya Publisher)
DOI 10.1007/978-1-4419-1343-2
Springer New York Dordrecht Heidelberg London

Library of Congress Control Number: 2009941304

Printed on acid-free paper

Springer is part of Springer Science+Business Media (www.springer.com)

1975. V. Maz'ya lectures at the Winter Voronezh Mathematical School, an annual meeting of Soviet mathematicians.

Vladimir Maz'ya was born on December 31, 1937, in Leningrad (present day, St. Petersburg) in the former USSR. His first mathematical article was published in Doklady Akad. Nauk SSSR when he was a fourth-year student of the Leningrad State University. From 1961 till 1986 V. Maz'ya held a senior research fellow position at the Research Institute of Mathematics and Mechanics of LSU, and then, during 4 years, he headed the Laboratory of Mathematical Models in Mechanics at the Institute of Engineering Studies of the Academy of Sciences of the USSR. Since 1990, V. Maz'ya lives in Sweden. At present, Vladimir Maz'ya is a Professor Emeritus at Linköping University and Professor at Liverpool University. He was elected a Member of Royal Swedish Academy of Sciences in 2002. The list of publications of V. Maz'ya contains 20 books and more than 450 research articles covering diverse areas in Analysis and containing numerous fundamental results and fruitful techniques. Research activities of Vladimir Maz'ya have strongly influenced the development of many branches in Analysis and Partial Differential Equations, which are clearly highlighted by the contributions to this collection of 3 volumes, where the world-recognized specialists present recent advantages in the areas I. *Function Spaces* (Sobolev type spaces, isoperimetric and capacitary inequalities in different contexts etc.) II. *Partial Differential Equations* (asymptotic analysis, boundary value problems etc.) III. *Analysis and Applications* (the oblique derivative problem, ill-posed problems etc.)

The quartet of Swedish mathematicians. Lund, 1991
Left to right: Lars Hörmander, Vladimir Maz'ya, Lars Gårding, Anders Melin.

I. Function Spaces

Ari Laptev Ed.

Hardy Inequalities for Nonconvex Domains ... 1
Farit Avkhadiev and Ari Laptev

Distributions with Slow Tails and Ergodicity of Markov Semigroups in Infinite Dimensions ... 13
Sergey Bobkov and Boguslaw Zegarlinski

On Some Aspects of the Theory of Orlicz–Sobolev Spaces ... 81
Andrea Cianchi

Mellin Analysis of Weighted Sobolev Spaces with Nonhomogeneous Norms on Cones ... 105
Martin Costabel, Monique Dauge, and Serge Nicaise

Optimal Hardy–Sobolev–Maz'ya Inequalities with Multiple Interior Singularities ... 137
Stathis Filippas, Achilles Tertikas, and Jesper Tidblom

Sharp Fractional Hardy Inequalities in Half-Spaces ... 161
Rupert L. Frank and Robert Seiringer

Collapsing Riemannian Metrics to Sub-Riemannian and the Geometry of Hypersurfaces in Carnot Groups ... 169
Nicola Garofalo and Christina Selby

Sobolev Homeomorphisms and Composition Operators ... 207
Vladimir Gol'dshtein and Aleksandr Ukhlov

Extended L^p Dirichlet Spaces ... 221
Niels Jacob and René L. Schilling

Characterizations for the Hardy Inequality ... 239
Juha Kinnunen and Riikka Korte

Geometric Properties of Planar BV-Extension Domains ... 255
Pekka Koskela, Michele Miranda Jr., and Nageswari Shanmugalingam

On a New Characterization of Besov Spaces with Negative Exponents . 273
Moshe Marcus and Laurent Véron

Isoperimetric Hardy Type and Poincaré Inequalities on Metric Spaces .. 285
Joaquim Martín and Mario Milman

Gauge Functions and Sobolev Inequalities on Fluctuating Domains 299
Eric Mbakop and Umberto Mosco

A Converse to the Maz'ya Inequality for Capacities under Curvature Lower Bound ... 321
Emanuel Milman

Pseudo-Poincaré Inequalities and Applications to Sobolev Inequalities . 349
Laurent Saloff-Coste

The p-Faber-Krahn Inequality Noted ... 373
Jie Xiao

Index ... 391

References to Maz'ya's Publications Made in Volume I ... 393

II. Partial Differential Equations

Ari Laptev Ed.

Large Solutions to Semilinear Elliptic Equations with Hardy Potential and Exponential Nonlinearity ... 1
Catherine Bandle, Vitaly Moroz, and Wolfgang Reichel

Stability Estimates for Resolvents, Eigenvalues, and Eigenfunctions of Elliptic Operators on Variable Domains ... 23
Gerassimos Barbatis, Victor I. Burenkov, and Pier Domenico Lamberti

Operator Pencil in a Domain with Concentrated Masses. A Scalar Analog of Linear Hydrodynamics ... 61
Gregory Chechkin

Selfsimilar Perturbation near a Corner: Matching Versus Multiscale Expansions for a Model Problem ... 95
Monique Dauge, Sébastien Tordeux, and Grégory Vial

Stationary Navier–Stokes Equation on Lipschitz Domains in Riemannian Manifolds with Nonvanishing Boundary Conditions ... 135
Martin Dindoš

On the Regularity of Nonlinear Subelliptic Equations ... 145
András Domokos and Juan J. Manfredi

Rigorous and Heuristic Treatment of Sensitive Singular Perturbations Arising in Elliptic Shells ... 159
Yuri V. Egorov, Nicolas Meunier, and Evariste Sanchez-Palencia

On the Existence of Positive Solutions of Semilinear Elliptic Inequalities on Riemannian Manifolds ... 203
Alexander Grigor'yan and Vladimir A. Kondratiev

Recurrence Relations for Orthogonal Polynomials and Algebraicity of Solutions of the Dirichlet Problem ... 219
Dmitry Khavinson and Nikos Stylianopoulos

On First Neumann Eigenvalue Bounds for Conformal Metrics ... 229
Gerasim Kokarev and Nikolai Nadirashvili

Necessary Condition for the Regularity of a Boundary Point for Porous Medium Equations with Coefficients of Kato Class ... 239
Vitali Liskevich and Igor I. Skrypnik

The Problem of Steady Flow over a Two-Dimensional Bottom Obstacle ... 253
Oleg Motygin and Nikolay Kuznetsov

Well Posedness and Asymptotic Expansion of Solution of Stokes Equation Set in a Thin Cylindrical Elastic Tube ... 275
Grigory P. Panasenko and Ruxandra Stavre

On Solvability of Integral Equations for Harmonic Single Layer Potential on the Boundary of a Domain with Cusp ... 303
Sergei V. Poborchi

Hölder Estimates for Green's Matrix of the Stokes System in Convex Polyhedra ... 315
Jürgen Roßmann

Boundary Integral Methods for Periodic Scattering Problems 337
Gunther Schmidt

Boundary Coerciveness and the Neumann Problem for 4th Order Linear Partial Differential Operators 365
Gregory C. Verchota

Index 379

References to Maz'ya's Publications Made in Volume II 381

III. Analysis and Applications

Ari Laptev Ed.

Optimal Control of a Biharmonic Obstacle Problem 1
David R. Adams, Volodymyr Hrynkiv, and Suzanne Lenhart

Minimal Thinness and the Beurling Minimum Principle 25
Hiroaki Aikawa

Progress in the Problem of the L^p-Contractivity of Semigroups for Partial Differential Operators 47
Alberto Cialdea

Uniqueness and Nonuniqueness in Inverse Hyperbolic Problems and the Black Hole Phenomenon 77
Gregory Eskin

Global Green's Function Estimates 105
Michael W. Frazier and Igor E. Verbitsky

On Spectral Minimal Partitions: the Case of the Sphere 153
Bernard Helffer, Thomas Hoffmann-Ostenhof, and Susanna Terracini

Weighted Sobolev Space Estimates for a Class of Singular Integral Operators 179
Dorina Mitrea, Marius Mitrea, and Sylvie Monniaux

On general Cwikel–Lieb–Rozenblum and Lieb–Thirring Inequalities 201
Stanislav Molchanov and Boris Vainberg

Estimates for the Counting Function of the Laplace Operator on Domains with Rough Boundaries 247
Yuri Netrusov and Yuri Safarov

$W^{2,p}$-Theory of the Poincaré Problem 259
Dian K. Palagachev

Weighted Inequalities for Integral and Supremum Operators 279
Luboš Pick

Finite Rank Toeplitz Operators in the Bergman Space 331
Grigori Rozenblum

Resolvent Estimates for Non-Selfadjoint Operators via Semigroups 359
Johannes Sjöstrand

Index 385

References to Maz'ya's Publications Made in Volume III 387

Contributors

Editor

Ari Laptev

President
The European Mathematical Society

Professor
Head of Department
Department of Mathematics
Imperial College London
Huxley Building, 180 Queen's Gate
London SW7 2AZ, UK
a.laptev@imperial.ac.uk

Professor
Department of Mathematics
Royal Institute of Technology
100 44 Stockholm, Sweden
laptev@math.kth.se

Ari Laptev is a world-recognized specialist in Spectral Theory of Differential Operators. He discovered a number of sharp spectral and functional inequalities. In particular, jointly with his former student T. Weidl, A. Laptev proved sharp Lieb–Thirring inequalities for the negative spectrum of multidimensional Schrödinger operators, a problem that was open for more than twenty five years.

A. Laptev was brought up in Leningrad (Russia). In 1971, he graduated from the Leningrad State University and was appointed as a researcher and then as an Assistant Professor at the Mathematics and Mechanics Department of LSU. In 1982, he was dismissed from his position at LSU due to his marriage to a British subject. Only after his emigration from the USSR in 1987 he was able to continue his career as a mathematician. Then A. Laptev was employed in Sweden, first as a lecturer at Linköping University and then from 1992 at the Royal Institute of Technology (KTH). In 1999, he became a professor at KTH and also Vice Chairman of its Department of Mathematics. From January 2007 he is employed by Imperial College London where from September 2008 he is the Head of Department of Mathematics.

A. Laptev was the Chairman of the Steering Committee of the five years long ESF Programme SPECT, the President of the Swedish Mathematical Society from 2001 to 2003, and the President of the Organizing Committee of the Fourth European Congress of Mathematics in Stockholm in 2004. He is now the President of the European Mathematical Society for the period January 2007–December 2010.

Authors

David R. Adams Vol. III
University of Kentucky
Lexington, KY 40506-0027
USA
dave@ms.uky.edu

Hiroaki Aikawa Vol. III
Hokkaido University
Sapporo 060-0810
JAPAN
aik@math.sci.hokudai.ac.jp

Farit Avkhadiev Vol. I
Kazan State University
420008 Kazan
RUSSIA
favhadiev@ksu.ru

Catherine Bandle Vol. II
Mathematisches Institut
Universität Basel
Rheinsprung 21, CH-4051 Basel
SWITZERLAND
catherine.bandle@unibas.ch

Gerassimos Barbatis Vol. II
University of Athens
157 84 Athens
GREECE
gbarbatis@math.uoa.gr

Sergey Bobkov Vol. I
University of Minnesota
Minneapolis, MN 55455
USA
bobkov@math.umn.edu

Victor I. Burenkov Vol. II
Università degli Studi di Padova
63 Via Trieste, 35121 Padova
ITALY
burenkov@math.unipd.it

Grigori Chechkin Vol. II
Lomonosov Moscow State University
Vorob'evy Gory, Moscow
RUSSIA
chechkin@mech.math.msu.su

Alberto Cialdea Vol. III
Università della Basilicata
Viale dell'Ateneo Lucano 10,
85100, Potenza
ITALY
cialdea@email.it

Andrea Cianchi Vol. I
Università di Firenze
Piazza Ghiberti 27, 50122 Firenze
ITALY
cianchi@unifi.it

Martin Costabel Vol. I
Université de Rennes 1
Campus de Beaulieu
35042 Rennes
FRANCE
martin.costabel@univ-rennes1.fr

Monique Dauge Vols. I, II
Université de Rennes 1
Campus de Beaulieu
35042 Rennes
FRANCE
monique.dauge@univ-rennes1.fr

Martin Dindoš Vol. II
Maxwell Institute of
Mathematics Sciences
University of Edinburgh
JCMB King's buildings Mayfield Rd
Edinbugh EH9 3JZ
UK
m.dindos@ed.ac.uk

András Domokos Vol. II
California State
University Sacramento
Sacramento 95819
USA
domokos@csus.edu

Yuri V. Egorov Vol. II
Université Paul Sabatier
118 route de Narbonne
31062 Toulouse Cedex 9
FRANCE
egorov@mip.ups-tlse.fr

Gregory Eskin Vol. III
University of California
Los Angeles, CA 90095-1555
USA
eskin@math.ucla.edu

Nicola Garofalo Vol. I
Purdue University
West Lafayette, IN 47906
USA
garofalo@math.purdue.edu
and

Università di Padova
35131 Padova
ITALY
garofalo@dmsa.unipd.it

Vladimir Gol'dshtein Vol. I
Ben Gurion University of the Negev
P.O.B. 653, Beer Sheva 84105
ISRAEL
vladimir@bgu.ac.il

Alexander Grigor'yan Vol. II
Bielefeld University
Bielefeld 33501
GERMANY
grigor@math.uni-bielefeld.de

Stathis Filippas Vol. I
University of Crete
71409 Heraklion
Institute of Applied and
Computational Mathematics
71110 Heraklion
GREECE
filippas@tem.uoc.gr

Rupert L. Frank Vol. I
Princeton University
Washington Road
Princeton, NJ 08544
USA
rlfrank@math.princeton.edu

Michael W. Frazier Vol. III
University of Tennessee
Knoxville, Tennessee 37922
USA
frazier@math.utk.edu

Bernard Helffer Vol. III
Université Paris-Sud
91 405 Orsay Cedex
FRANCE
Bernard.Helffer@math.u-psud.fr

Thomas Hoffmann-Ostenhof Vol. III
Institut für Theoretische Chemie
Universität Wien
Währinger Strasse 17, and
International Erwin
Schrödinger Institute
for Mathematical Physics
Boltzmanngasse 9
A-1090 Wien
AUSTRIA
thoffman@esi.ac.at

Volodymyr Hrynkiv Vol. III
University of Houston-Downtown
Houston, TX 77002-1014
USA
HrynkivV@uhd.edu

Niels Jacob Vol. I
Swansea University
Singleton Park
Swansea SA2 8PP
UK
n.jacob@swansea.ac.uk

Dmitry Khavinson Vol. II
University of South Florida
4202 E. Fowler Avenue, PHY114
Tampa, FL 33620-5700
USA
dkhavins@cas.usf.edu

Juha Kinnunen Vol. I
Institute of Mathematics
Helsinki University of Technology
P.O. Box 1100, FI-02015
FINLAND
juha.kinnunen@tkk.fi

Gerasim Kokarev Vol. II
University of Edinburgh
King's Buildings, Mayfield Road
Edinburgh EH9 3JZ
UK
G.Kokarev@ed.ac.uk

Vladimir A. Kondratiev Vol. II
Moscow State University
119992 Moscow
RUSSIA
vla-kondratiev@yandex.ru

Riikka Korte Vol. I
University of Helsinki
P.O. Box 68
Gustaf Hällströmin katu 2 b
FI-00014
FINLAND
riikka.korte@helsinki.fi

Pekka Koskela Vol. I
University of Jyväskylä
P.O. Box 35 (MaD), FIN–40014
FINLAND
pkoskela@maths.jyu.fi

Nikolay Kuznetsov Vol. II
Institute for Problems
in Mechanical Engineering
Russian Academy of Sciences
V.O., Bol'shoy pr. 61
199178 St. Petersburg
RUSSIA
nikolay.g.kuznetsov@gmail.com

Pier Domenico Lamberti Vol. II
Universitá degli Studi di Padova
63 Via Trieste, 35121 Padova
ITALY
lamberti@math.unipd.it

Ari Laptev Vol. I
Imperial College London
Huxley Building, 180 Queen's Gate
London SW7 2AZ
UK
a.laptev@imperial.ac.uk
and
Royal Institute of Technology
100 44 Stockholm
SWEDEN
laptev@math.kth.se

Suzanne Lenhart Vol. III
University of Tennessee
Knoxville, TN 37996-1300
USA
lenhart@math.utk.edu

Vitali Liskevich Vol. II
Swansea University
Singleton Park
Swansea SA2 8PP
UK
v.a.liskevich@swansea.ac.uk

Juan J. Manfredi Vol. II
University of Pittsburgh
Pittsburgh, PA 15260
USA
manfredi@pitt.edu

Moshe Marcus Vol. I
Israel Institute of Technology-Technion
33000 Haifa
ISRAEL
marcusm@math.technion.ac.il

Joaquim Martín Vol. I
Universitat Autònoma de Barcelona
Bellaterra, 08193 Barcelona
SPAIN
jmartin@mat.uab.cat

Eric Mbakop Vol. I
Worcester Polytechnic Institute
100 Institute Road
Worcester, MA 01609
USA
steve055@WPI.EDU

Nicolas Meunier Vol. II
Université Paris Descartes (Paris V)
45 Rue des Saints Pères
75006 Paris
FRANCE
nicolas.meunier@math-info.univ-paris5.fr

Emanuel Milman Vol. I
Institute for Advanced Study
Einstein Drive, Simonyi Hall
Princeton, NJ 08540
USA
emilman@math.ias.edu

Mario Milman Vol. I
Florida Atlantic University
Boca Raton, Fl. 33431
USA
extrapol@bellsouth.net

Michele Miranda Jr. Vol. I
University of Ferrara
via Machiavelli 35
44100, Ferrara
ITALY
michele.miranda@unife.it

Dorina Mitrea Vol. III
University of Missouri at Columbia
Columbia, MO 65211
USA
mitread@missouri.edu

Marius Mitrea Vol. III
University of Missouri
at Columbia
Columbia, MO 65211
USA
mitream@missouri.edu

Stanislav Molchanov Vol. III
University of North Carolina
at Charlotte
Charlotte, NC 28223
USA
smolchan@uncc.edu

Sylvie Monniaux Vol. III
Université Aix-Marseille 3
F-13397 Marseille Cédex 20
FRANCE
sylvie.monniaux@univ.u-3mrs.fr

Vitaly Moroz Vol. II
Swansea University
Singleton Park
Swansea SA2 8PP
UK
v.moroz@swansea.ac.uk

Umberto Mosco Vol. I
Worcester Polytechnic Institute
100 Institute Road
Worcester, MA 01609
USA
mosco@WPI.EDU

Oleg Motygin Vol. II
Institute for Problems
in Mechanical Engineering
Russian Academy of Sciences
V.O., Bol'shoy pr. 61
199178 St. Petersburg
RUSSIA
o.v.motygin@gmail.com

Nikolai Nadirashvili Vol. II
Centre de Mathématiques
et Informatique
Université de Provence
39 rue F. Joliot-Curie
13453 Marseille Cedex 13
FRANCE
nicolas@cmi.univ-mrs.fr

Yuri Netrusov Vol. III
University of Bristol
University Walk
Bristol BS8 1TW
UK
y.netrusov@bristol.ac.uk

Serge Nicaise Vol. I
Université Lille Nord de France
UVHC, 59313
Valenciennes Cedex 9
FRANCE
snicaise@univ-valenciennes.fr

Dian K. Palagachev Vol. III
Technical University of Bari
Via E. Orabona 4
70125 Bari
ITALY
palaga@poliba.it

Grigory P. Panasenko Vol. II
University Jean Monnet
23, rue Dr Paul Michelon
42023 Saint-Etienne
FRANCE
Grigory.Panasenko@univ-st-etienne.fr

Luboš Pick Vol. III
Charles University
Sokolovská 83, 186 75 Praha 8
CZECH REPUBLIC
pick@karlin.mff.cuni.cz

Sergei V. Poborchi Vol. II
St. Petersburg State University
28, Universitetskii pr., Petrodvorets
St. Petersburg 198504
RUSSIA
poborchi@mail.ru

Wolfgang Reichel Vol. II
Universität Karlsruhe (TH)
D-76128 Karlsruhe
GERMANY
wolfgang.reichel@math.uni-karlsruhe.de

Jürgen Roßmann Vol. II
University of Rostock
Institute of Mathematics
D-18051 Rostock
GERMANY
juergen.rossmann@uni-rostock.de

Grigori Rozenblum Vol. III
Chalmers University of Technology
University of Gothenburg
S-412 96, Gothenburg
SWEDEN
grigori@math.chalmers.se

Yuri Safarov Vol. III
King's College London
Strand, London WC2R 2LS
UK
yuri.safarov@kcl.ac.uk

Laurent Saloff-Coste Vol. I
Cornell University
Mallot Hall, Ithaca, NY 14853
USA
lsc@math.cornell.edu

Evariste Sanchez-Palencia Vol. II
Université Pierre et Marie Curie
4 place Jussieu
75252 Paris
FRANCE
sanchez@lmm.jussieu.fr

René L. Schilling Vol. I
Technische Universität Dresden
Institut für Stochastik
D-01062 Dresden
GERMANY
rene.schilling@tu-dresden.de

Gunther Schmidt Vol. II
Weierstrass Institute of
Applied Analysis and Stochastics
Mohrenstr. 39, 10117 Berlin
GERMANY
schmidt@wias-berlin.de

Robert Seiringer Vol. I
Princeton University
P. O. Box 708
Princeton, NJ 08544
USA
rseiring@princeton.edu

Christina Selby Vol. I
The Johns Hopkins University
11100 Johns Hopkins Road
Laurel, MD 20723.
USA
Christina.Selby@jhuapl.edu

Nageswari Shanmugalingam Vol. I
University of Cincinnati
Cincinnati, OH 45221-0025
USA
nages@math.uc.edu

Johannes Sjöstrand Vol. III
Université de Bourgogne
9, Av. A. Savary, BP 47870
FR-21078 Dijon Cédex
and UMR 5584, CNRS
FRANCE
johannes@u-bourgogne.fr

Igor I. Skrypnik Vol. II
Institute of Applied
Mathematics and Mechanics
Donetsk
UKRAINE
iskrypnik@iamm.donbass.com

Ruxandra Stavre Vol. II
Institute of Mathematics
"Simion Stoilow" Romanian Academy
P.O. Box 1-764
014700 Bucharest
ROMANIA
Ruxandra.Stavre@imar.ro

Nikos Stylianopoulos Vol. II
University of Cyprus
P.O. Box 20537
1678 Nicosia
CYPRUS
nikos@ucy.ac.cy

Susanna Terracini Vol. III
Università di Milano Bicocca
Via Cozzi, 53 20125 Milano
ITALY
susanna.terracini@unimib.it

Achilles Tertikas Vol. I
University of Crete and
Institute of Applied and
Computational Mathematics
71110 Heraklion
GREECE
tertikas@math.uoc.gr

Jesper Tidblom Vol. I
The Erwin Schrödinger Institute
Boltzmanngasse 9
A-1090 Vienna
AUSTRIA
Jesper.Tidblom@esi.ac.at

Sébastien Tordeux Vol. II
Institut de Mathématiques de Toulouse
31077 Toulouse
FRANCE
sebastien.tordeux@insa-toulouse.fr

Alexander Ukhlov Vol. I
Ben Gurion University of the Negev
P.O.B. 653, Beer Sheva 84105
ISRAEL
ukhlov@math.bgu.ac.il

Boris Vainberg Vol. III
University of North Carolina
at Charlotte
Charlotte, NC 28223
USA
brvainbe@uncc.edu

Igor E. Verbitsky Vol. III
University of Missouri
Columbia, Missouri 65211
USA
verbitskyi@missouri.edu

Gregory C. Verchota Vol. II
Syracuse University
215 Carnegie
Syracuse NY 13244
USA
gverchot@syr.edu

Laurent Véron Vol. I
Université de Tours
Parc de Grandmont
37200 Tours
FRANCE
veronl@univ-tours.fr

Grégory Vial Vol. II
IRMAR, ENS Cachan Bretagne
CNRS, UEB
35170 Bruz
FRANCE
gvial@bretagne.ens-cachan.fr

Jie Xiao Vol. I
Memorial University
of Newfoundland
St. John's, NL A1C 5S7
CANADA
jxiao@mun.ca

Boguslaw Zegarlinski Vol. I
Imperial College London
Huxley Building
180 Queen's Gate
London SW7 2AZ
UK
b.zegarlinski@imperial.ac.uk

Partial Differential Equations

Contributions of Vladimir Maz'ya

Numerous fundamental contributions of V. Maz'ya to the theory of partial differential equations are related to diverse areas.

— In the 1970's and 1980's, V. Maz'ya worked with his colleagues (V. Kozlov, B. Plamenevskii, J. Rossmann et al.) to extend the results by V. Kondratiev of 1967 to general classes of higher order elliptic operators on bounded and unbounded domains with point singularities (cones, cusps etc.) and also higher dimensional singularities (edges, polyhedral vertices etc.). In 1979, V. Maz'ya with his colleagues started to study the case of domains with singularly perturbed boundaries such as "multistructures" which have different dimensions at different points, leading to compound asymptotic expansions; this work includes the solution of the 1D–3D junction problem.

— In 1981, V. Maz'ya proposed a new approach to boundary integral equations on nonsmooth surfaces, which was later applied to problems of elasticity and (in several papers with A. Solov'ev) to problems in planar domains with cuspidal boundary irregularities.

— V. Maz'ya obtained deep results on the regularity of a boundary point in the sense of Wiener for solutions of elliptic equations. In 1962, he derived an estimate of the continuity modulus of a harmonic function in terms of the Wiener integral. In 1972, he found a condition of the regularity of a boundary point for a class of quasilinear second order equations containing the p-Laplacian. In 2002, V. Maz'ya generalized the Wiener criterion for higher order elliptic equations.

— Starting in the 1970's, V. Maz'ya with B. Vainberg and N. Kuznetsov developed the mathematical theory of water waves. In 1977, V. Maz'ya found a solution to the problem on small oscillations of a fluid in the presence of an immersed body. Recently V. Maz'ya and J. Rossmann obtained results on the Navier–Stokes equations in polyhedral domains.

— Together with S. Mayboroda, V. Maz'ya obtained sharp regularity results for solutions of the polyharmonic equations in an arbitrary open set.

— Yu. Burago and V. Maz'ya developed a theory of harmonic single and double layer potentials in the spaces C and C^* for nonsmooth surfaces (1967).

Main Topics

In this volume, the following topics are discussed:

- Semilinear elliptic equation in a bounded smooth domain with an exponential nonlinearity and a Hardy potential depending on the distance to the boundary of the domain. /Bandle–Moroz–Reichel/
- Estimates for the variation of resolvents, eigenvalues, and eigenfunctions of general second order uniformly elliptic operators in perturbed domains. /Barbatis–Burenkov–Lamberti/

- Homogenization methods in the study of hydrodynamics problems. /Chechkin/
- Multiscale expansions versus matched asymptotic expansions in the problem for the Laplace–Dirichlet equation in a polygonal domain perturbed at the small scale near a vertex. /Dauge–Tordeux–Vial/
- The stationary Navier–Stokes equation on Lipschitz domains in Riemannian manifolds.The existence of a solution. /Dindoš/
- Nondegenerate quasilinear subelliptic equations of p-Laplacian type. /Domokos–Manfredi/
- Singular perturbations of elliptic systems depending on a parameter ε such that for $\varepsilon = 0$ the boundary conditions do not satisfy the Shapiro–Lopatinskii condition. /Egorov–Meunier–Sanchez-Palencia/
- Elliptic inequalities on geodesically complete Riemannian manifolds and sharp sufficient conditions in terms of capacities and volumes for the nonexistence of positive solutions. /Grigor'yan–Kondratiev/
- Recurrence relations for orthogonal polynomials and polynomial solutions to the Dirichlet problem. /Khavinson–Stylianopoulos/
- Estimate of the first Neumann eigenvalues for a conformal class of Riemannian metrics in terms of Hersch's isoperimetric inequality. /Kokarev–Nadirashvili/
- The boundary regularity problem for generalized porous medium type quasilinear equations with measurable coefficients. The necessity of the Wiener test. /Liskevich–Skrypnik/
- The existence and uniqueness of a solution to the linear boundary value problem describing a steady flow over a two-dimensional obstacle. /Motygin–Kuznetsov/
- The well posedness and asymptotic analysis for the Stokes equation describing the creeping flow of a viscous incompressible fluid through a long and narrow cylindrical elastic tube. /Panasenko–Stavre/
- Solvability of integral equations for harmonic single layer potential on the boundary of a domain with cusp /Poborchi/
- The Dirichlet problem for the Stokes system in a convex polyhedron. Hölder estimates for entries of the Green matrix. /Roßmann/
- Boundary integral methods for periodic scattering problems. /Schmidt/
- The Neumann problem for 4th order linear partial differential operators. Boundary coerciveness. /Verchota/

Contents

Large Solutions to Semilinear Elliptic Equations with Hardy Potential and Exponential Nonlinearity 1
Catherine Bandle, Vitaly Moroz, and Wolfgang Reichel
1 Introduction 1
2 Some Definitions and Tools 3
2.1 Sub- and super-harmonics 4
2.2 Hardy constant 5
2.3 Phragmen–Lindelöf alternative 5
2.4 Sub- and super-solutions 6
2.5 Solutions with zero boundary data 8
3 A Priori Upper Bounds 10
4 Nonexistence of Large Solutions if $\mu < 0$ 13
5 Asymptotic Behavior of Large Solutions near the Boundary. 14
5.1 Global sub-solutions 14
6 Uniqueness and Existence of Large Solutions 16
6.1 Uniqueness 16
6.2 Existence 17
7 Borderline Potentials. Summary and Open Problems 17
7.1 Summary and open problems 21
References 22

Stability Estimates for Resolvents, Eigenvalues, and Eigenfunctions of Elliptic Operators on Variable Domains 23
Gerassimos Barbatis, Victor I. Burenkov, and Pier Domenico Lamberti
1 Introduction 23
2 General Setting 27
3 Perturbation of ϕ 29
4 Stability Estimates for the Resolvents and Eigenvalues 31
5 Stability Estimates for Eigenfunctions 40
6 On the Regularity of Eigenfunctions 48
7 Estimates via Lebesgue Measure 52

7.1 Local perturbations ... 53
7.2 Global normal perturbations ... 56
8 Appendix ... 57
References ... 59

Operator Pencil in a Domain with Concentrated Masses. A Scalar Analog of Linear Hydrodynamics ... 61
Gregory Chechkin
1 Introduction ... 61
2 Statement of the Problem ... 62
3 Homogenization of Boundary Value Problems in Domains with Concentrated Masses Periodically Distributed along the Boundary ... 63
3.1 Notation and the main results ... 63
3.2 Preliminaries ... 66
3.3 Proof of the Estimate ... 68
3.4 Proof of the main assertions ... 76
4 Asymptotic Behavior of Eigenvalues and Eigenfunctions of the Operator Pencil ... 82
4.1 Reduction of the problem to the case of an operator pencil ... 82
4.2 Properties of the operator pencils ... 83
4.3 Remarks ... 84
4.4 Homogenization theorem ... 85
4.5 Small oscillations of a viscous inhomogeneous fluid in a stationary open vessel with a rigid net on the surface. Normal oscillations ... 88
References ... 91

Selfsimilar Perturbation near a Corner: Matching Versus Multiscale Expansions for a Model Problem ... 95
Monique Dauge, Sébastien Tordeux, and Grégory Vial
1 Introduction ... 95
2 Notation. Outline of Results ... 98
2.1 Selfsimilar perturbations ... 98
2.2 The Dirichlet problem and its singularities ... 99
2.3 Outline of results ... 100
3 Super-Variational Problems ... 102
3.1 Variational problems ... 102
3.2 Super-variational problems in ω. Behavior at the origin ... 103
3.3 Super-variational problems in Ω. Behavior at infinity ... 105
4 Multiscale Expansion ... 107
4.1 The construction of the first terms ... 108
4.2 The general construction ... 110

4.3 Optimal error estimate 111
5 Matching of Asymptotic Expansions 113
5.1 Formal derivation of the asymptotic expansions 113
5.2 Definition of the asymptotic terms 115
5.3 Global error estimates 116
5.4 Local error estimates 119
6 Comparison of the Two Expansions 121
7 Extensions and Generalizations 122
7.1 Smooth data without condition of support 123
7.2 Neumann boundary conditions 123
7.3 Small holes 125
7.4 Domains with multiply connected junction sectors 126
7.5 Nonhomogeneous operators. Helmholtz equation 127
8 Conclusion: A Practical Two-Term Expansion 130
8.1 Compound expansion 130
8.2 Application: asymptotics of coefficients of singularities 131
References 132

Stationary Navier–Stokes Equation on Lipschitz Domains in Riemannian Manifolds with Nonvanishing Boundary Conditions 135
Martin Dindoš
1 Introduction 135
2 Statements of the Main Results 137
3 Linear Theory Revisited 138
References 143

On the Regularity of Nonlinear Subelliptic Equations 145
András Domokos and Juan J. Manfredi
1 Introduction 146
2 Second Order Horizontal Differentiability of Weak Solutions 148
3 C^∞ Regularity 155
References 156

Rigorous and Heuristic Treatment of Sensitive Singular Perturbations Arising in Elliptic Shells 159
Yuri V. Egorov, Nicolas Meunier, and Evariste Sanchez-Palencia
1 Introduction 159
2 Generalities on the Koiter Shell Model 162
3 The Ellipticity of Systems and the Shapiro–Lopatinskii Condition 166
3.1 Elliptic systems in the sense of Douglis and Nirenberg 166
3.2 Shapiro–Lopatinskii conditions for elliptic systems in the sense of Douglis and Nirenberg 168

3.3 Some Results for "Well-Posed" Elliptic Systems 170
4 Study of Four Systems Involved in Shell Theory 172
4.1 The rigidity system 172
4.2 The system of *membrane tensions* 176
4.3 The *membrane* system 177
4.4 The *Koiter shell* system 180
5 A Sensitive Singular Perturbation Problem Arising in the Koiter Linear Shell Theory 181
5.1 Definition of the problem 181
5.2 Sensitive character 181
5.3 Abstract convergence results as $\varepsilon \to 0$ 183
6 Heuristic Asymptotics in the Previous Problem 185
6.1 Introduction to the heuristic asymptotic 187
6.2 The boundary layer on Γ_0 188
6.3 Formulation of the problem in the heuristic asymptotics 197
6.4 The formal asymptotics and its sensitive behavior .. 198
References 201

On the Existence of Positive Solutions of Semilinear Elliptic Inequalities on Riemannian Manifolds 203
Alexander Grigor'yan and Vladimir A. Kondratiev
1 Introduction and the Main Results 203
2 Proof of the Capacity Tests 207
3 Examples to the Capacity Test 210
4 Proof of the Volume Test 214
5 Examples to the Volume Test 216
References 218

Recurrence Relations for Orthogonal Polynomials and Algebraicity of Solutions of the Dirichlet Problem 219
Dmitry Khavinson and Nikos Stylianopoulos
1 Introduction 219
2 The Main Results 221
3 Proofs 223
4 Concluding Remarks 227
References 228

On First Neumann Eigenvalue Bounds for Conformal Metrics 229
Gerasim Kokarev and Nikolai Nadirashvili
1 Introduction 229
2 Preliminaries 230
2.1 Notation 230
2.2 Reformulation of the theorem 231
2.3 The Yang-Yau estimate revisited 233
3 The Proof 233

4 Appendix: Auxiliary Propositions 236
References 237

Necessary Condition for the Regularity of a Boundary Point for Porous Medium Equations with Coefficients of Kato Class 239
Vitali Liskevich and Igor I. Skrypnik
1 Introduction and the Main Results 239
2 Auxiliary Estimates 242
3 Proof of Theorem 1.1 248
References 251

The Problem of Steady Flow over a Two-Dimensional Bottom Obstacle 253
Oleg Motygin and Nikolay Kuznetsov
1 Introduction 253
1.1 Statement of the problem 254
1.2 The main result 256
2 Auxiliary Assertions 258
2.1 Asymptotics at infinity 258
2.2 Uniqueness theorem for the supercritical regime.... 259
2.3 The Fredholm alternative for the integral equation . 261
2.4 The limit problem of flow under rigid lid 262
2.5 Inequality 264
3 Proof of Theorem 1.1 265
3.1 The existence of a solution for $\nu > 1$ 265
3.2 The uniqueness of a solution for $\nu > 1$ 271
References 273

Well Posedness and Asymptotic Expansion of Solution of Stokes Equation Set in a Thin Cylindrical Elastic Tube 275
Grigory P. Panasenko and Ruxandra Stavre
1 Introduction 275
2 Notation and Problem Statement 277
3 Variational Analysis of the Problem 280
4 Asymptotic Analysis of the Problem 291
5 The Justification of the Asymptotic Solution 296
References 301

On Solvability of Integral Equations for Harmonic Single Layer Potential on the Boundary of a Domain with Cusp 303
Sergei V. Poborchi
1 Introduction 303
2 Domains and Function Spaces 304
3 Description of the Spaces $Tr(\Gamma)$ and $Tr(\Gamma)^*$ 305
4 The Single Layer Potential for Γ 309
5 The Dirichlet Problem for the Laplace Equation 311

6 The Neumann Problem for the Laplace Equation 312
References 313

Hölder Estimates for Green's Matrix of the Stokes System in Convex Polyhedra 315
Jürgen Roßmann
1 Introduction 315
2 The Green Matrix for the Stokes System 318
3 A Local Estimate for Solutions of the Stokes System 319
4 Point Estimates of the Green Matrix 323
5 Hölder Estimates for the Elements of the Green Matrix 330
6 Proof of Theorem 5.1 331
References 335

Boundary Integral Methods for Periodic Scattering Problems 337
Gunther Schmidt
1 Introduction 337
2 Conical Diffraction 340
2.1 Maxwell equations 340
2.2 Helmholtz equations 341
3 Potential Methods 344
3.1 Potentials of periodic diffraction 344
3.2 Boundary integrals for periodic diffraction 346
3.3 Boundary integrals for the Laplacian 348
4 Integral Formulation 349
4.1 Integral equation 350
4.2 Equivalence 352
5 Existence and Uniqueness of Solutions 353
5.1 Fredholmness 353
5.2 Uniqueness 356
5.3 Existence of solutions 360
References 363

Boundary Coerciveness and the Neumann Problem for 4th Order Linear Partial Differential Operators 365
Gregory C. Verchota
References 377

Index 379

References to Maz'ya's Publications Made in Volume II 381

Large Solutions to Semilinear Elliptic Equations with Hardy Potential and Exponential Nonlinearity

Catherine Bandle, Vitaly Moroz, and Wolfgang Reichel

Dedicated to Vladimir Maz'ya

Abstract On a bounded smooth domain $\Omega \subset \mathbb{R}^N$, we study solutions of a semilinear elliptic equation with an exponential nonlinearity and a Hardy potential depending on the distance to $\partial\Omega$. We derive global a priori bounds of the Keller–Osserman type. Using a Phragmen–Lindelöf alternative for generalized sub- and super-harmonic functions, we discuss the existence, nonexistence, and uniqueness of so-called *large* solutions, i.e., solutions which tend to infinity at $\partial\Omega$. The approach develops the one used by the same authors for a problem with a power nonlinearity instead of the exponential nonlinearity.

1 Introduction

Let $\Omega \subset \mathbb{R}^N$ be a bounded smooth domain (say, $\partial\Omega \in C^3$), and let $\delta(x)$ be the distance from a point $x \in \Omega$ to the boundary $\partial\Omega$. In this paper, we study semilinear problems of the form

Catherine Bandle
Mathematisches Institut, Universität Basel, Rheinsprung 21, CH-4051 Basel, Switzerland
e-mail: catherine.bandle@unibas.ch

Vitaly Moroz
Department of Mathematics, Swansea University, Singleton Park, Swansea SA2 8PP, Wales, UK
e-mail: v.moroz@swansea.ac.uk

Wolfgang Reichel
Fakultät für Mathematik, Universität Karlsruhe (TH), D-76128 Karlsruhe, Germany
e-mail: wolfgang.reichel@math.uni-karlsruhe.de

A. Laptev (ed.), *Around the Research of Vladimir Maz'ya II: Partial Differential Equations*, International Mathematical Series 12, DOI 10.1007/978-1-4419-1343-2_1,

$$-\Delta u - \frac{\mu}{\delta^2} u + e^u = 0 \text{ in } \Omega, \tag{1.1}$$

where $\mu \in \mathbb{R}$ is a given constant. The case without Hardy potential

$$-\Delta u + e^u = 0 \text{ in } \Omega \tag{1.2}$$

is well understood. In particular, for any continuous function $\varphi \in C(\partial\Omega)$ the boundary value problem (1.2) with $u = \varphi$ on $\partial\Omega$ has a unique classical solution. Moreover, there exists a unique solution of (1.2) (cf., for example, [3, 4]) with the property that

$$u(x) \to \infty \text{ as } x \to \partial\Omega. \tag{1.3}$$

This solution dominates all other solutions and is therefore commonly called *large*. Near the boundary, it behaves like [4]

$$u(x) = \log \frac{2}{\delta^2(x)} + (N-1)\mathcal{H}_0(\sigma(x))\delta(x) + o(\delta(x)) \text{ as } x \to \partial\Omega, \tag{1.4}$$

where $\sigma : \Omega \to \partial\Omega$ denotes the nearest-point projection of x onto the boundary and $\mathcal{H}_0(y)$ is the mean curvature of the boundary at $y \in \partial\Omega$.

The presence of a Hardy potential has a significant effect on the set of solutions of (1.1). Because of the singularity of the potential the boundary values φ in the problem

$$-\Delta u - \frac{\mu}{\delta^2} u + e^u = 0 \text{ in } \Omega, \quad u = \varphi \text{ on } \partial\Omega \tag{1.5}$$

cannot, in general, be prescribed arbitrarily. For instance, it is not difficult to show (cf. Theorem 2.2 below) that if $\varphi = 0$, then the problem (1.5) admits a unique solution for every $\mu < C_H(\Omega)$, where $C_H(\Omega) > 0$ is the optimal constant in the Hardy inequality

$$\int_\Omega |\nabla\phi|^2 \, dx \geqslant C_H(\Omega) \int_\Omega \frac{\phi^2}{\delta^2} \, dx, \qquad \forall \phi \in C_0^\infty(\Omega).$$

On the other hand, if $\varphi > 0$ is continuous, then the problem (1.5) has no solution unless $\mu = 0$. This can be seen as follows. Without loss of generality let us assume that u is positive in Ω (otherwise, replace Ω by a neighborhood of $\partial\Omega$). Suppose for contradiction that (1.5) has a $C^2(\Omega) \cap C(\overline{\Omega})$-solution. Then the problem

$$-\Delta v = \frac{u}{\delta^2} \text{ in } \Omega, \quad v = 0 \text{ on } \partial\Omega \tag{1.6}$$

has a $C^2(\Omega) \cap C(\overline{\Omega})$-solution, where $v = \frac{1}{\mu}(u + z - h)$, z is the Newtonian-potential of e^u, and h is the harmonic extension of $(\varphi + z)|_{\partial\Omega}$. Let $f_k(x) :=$

$\min\{\frac{u(x)}{\delta^2(x)}, k\}$ for $k \in \mathbb{N}$, and let v_k be the weak $H_0^1(\Omega)$-solution of $-\Delta v_k = f_k$ in Ω with $v_k = 0$ on $\partial\Omega$. Then $v_k \in C(\overline{\Omega})$ and

$$v_k(x) = \int_\Omega G(x,y) f_k(y)\,dy \quad \text{for all} \quad x \in \Omega,$$

where $G(x,y)$ is the Dirichlet Green-function of $-\Delta$ on Ω. The comparison principle yields $v_k(x) - \frac{1}{k} \leqslant u(x)$ for all $x \in \Omega$ and all $k \in \mathbb{N}$. However, by monotone convergence,

$$v_k(x) = \int_\Omega G(x,y) f_k(y)\,dy \to \int_\Omega G(x,y)\frac{u(y)}{\delta^2(y)}\,dy = \infty \quad \text{as} \quad k \to \infty$$

for all $x \in \Omega$. This is a contradiction.

The fact that no solutions exist with finite, nonzero boundary data motivated us to study solutions which are unbounded near the boundary. The goal of this paper is to study *large solutions* of (1.1), i.e., solutions which satisfy (1.3).

MAIN RESULT. (i) *If* $\mu < 0$, *then* (1.1) *has no large solutions.*

(ii) *If* $0 \leqslant \mu < C_H(\Omega)$, *then there exists a unique large solution of* (1.1). *It is pointwise larger than any other solution of* (1.1).

The paper is organized as follows. In Section 2, we set up the notation and introduce some basic definitions and tools. We also provide an existence proof for the solution of (1.1) vanishing on the boundary. In Section 3, we establish a Keller–Osserman type a priori upper bound on solutions of (1.1). In Section 4, we prove the nonexistence of large solutions in the case $\mu < 0$, while, in Sections 5 and 6, we establish the asymptotic behavior, existence, and uniqueness of large solutions of (1.1) when $0 \leqslant \mu < C_H(\Omega)$. Finally, in Section 7, we construct a borderline case of a function $\gamma > 0$ such that $0 < \gamma(\delta) \leqslant 1$ and $\gamma(\delta) = o(\delta)$ as $\delta \to 0$ and for which the problem

$$-\Delta u + \frac{\gamma(\delta)}{\delta^2} u + e^u = 0 \quad \text{in} \quad \Omega$$

has a large solution. We also discuss some open questions related to (1.1).

2 Some Definitions and Tools

For $\rho > 0$ and $\varepsilon \in (0, \rho)$ we use the notation

$$\Omega_\rho := \{x \in \Omega : \delta(x) < \rho\}, \qquad \Omega_{\varepsilon,\rho} := \{x \in \Omega : \varepsilon < \delta(x) < \rho\},$$
$$D_\rho := \{x \in \Omega : \delta(x) > \rho\}, \qquad \Gamma_\rho := \{x \in \Omega : \delta(x) = \rho\}.$$

2.1 Sub- and super-harmonics

For the sake of simplicity, we set

$$\mathscr{L}_\mu := -\Delta - \frac{\mu}{\delta^2}.$$

Let $G \subset \Omega$ be open. Following [2], we call solutions h of the equation

$$\mathscr{L}_\mu h = 0 \text{ in } G \tag{2.1}$$

harmonics of $\mathscr{L}_\mu$ in G. If $G = \Omega$, we often omit G and say that h is a *global harmonic* of $\mathscr{L}_\mu$. By interior regularity, weak solutions of (2.1) are classical, so in what follows we assume that all harmonics are of class $C^2(G)$.

We define *super-harmonics* in G as functions $\overline{h} \in H^1_{\text{loc}}(G) \cap C(G)$ which solve in the weak sense the differential inequality

$$\mathscr{L}_\mu \overline{h} \geqslant 0 \text{ in } G. \tag{2.2}$$

Similarly, $\underline{h} \in H^1_{\text{loc}}(G) \cap C(G)$ is called a *sub-harmonic* in G if the inequality sign is reversed.

If $\underline{h}$ and $\overline{h}$ satisfy (2.2) in Ω, then they are called *global sub-harmonic* and *global super-harmonic* respectively. If $\underline{h}$ and $\overline{h}$ satisfy (2.2) in a neighborhood of the boundary Ω_ε, then they are called *local sub-harmonic* and *local super-harmonic* respectively.

By the classical strong maximum principle for the Laplacian with potentials applied locally in small subdomains of Ω, any nontrivial super-harmonic $\overline{h} \gneq 0$ is strictly positive in Ω, while any sub-harmonic $\underline{h}$ in Ω is locally bounded above.

The following examples of explicit local sub- and super-harmonics will play an important role in our considerations.

Examples [2, Lemma 2.8]. Let $\mu < 1/4$, and let

$$\beta_\pm = \frac{1}{2} \pm \sqrt{\frac{1}{4} - \mu}.$$

The function δ^β is a local super-harmonic of $\mathscr{L}_\mu$ if $\beta \in (\beta_-, \beta_+)$. It is a local sub-harmonic if $\beta \notin [\beta_-, \beta_+]$. In the borderline cases $\beta = \beta_\pm$, for small $\varepsilon > 0$

$$\overline{h} = \delta^{\beta_+}(1 - \delta^\varepsilon), \quad \overline{H} = \delta^{\beta_-}(1 + \delta^\varepsilon)$$

are local super-harmonics and

$$\underline{h} = \delta^{\beta_+}(1 + \delta^\varepsilon), \quad \underline{H} = \delta^{\beta_-}(1 - \delta^\varepsilon)$$

are local sub-harmonics.

2.2 Hardy constant

The constant

$$C_H(\Omega) = \inf_{0\neq\phi\in W_0^{1,2}(\Omega)} \frac{\displaystyle\int_\Omega |\nabla\phi|^2\,dx}{\displaystyle\int_\Omega \delta^{-2}(x)\phi^2\,dx}$$

is called the *global Hardy constant*. It is well known that $0 < C_H(\Omega) \leqslant 1/4$. In general, $C_H(\Omega)$ varies with the domain. For convex domains $C_H(\Omega) = 1/4$, but there exist smooth domains for which $C_H(\Omega) < 1/4$. A review with an extensive bibliography and where, in particular, Maz'ya's relevant earlier contributions [9] are mentioned, can be found in [5]. Improvements of this inequality by adding an additional L^q norm were obtained by Filippas, Maz'ya, and Tertikas in a series of papers. The most recent results can be found in [6]. This paper contains also references to previous related works. It turns out (cf. [8]) that $C_H(\Omega)$ is attained if and only if $C_H(\Omega) < 1/4$. Note that $C_H(\Omega)$ is, in general, not monotone with respect to Ω.

The relation between the Hardy constant, existence of positive super-harmonics in Ω, and validity of a comparison principle for $\mathscr{L}_\mu$ is explained by the following classical result (cf. [1, Theorem 3.3]).

Lemma 2.1. *The following three statements are equivalent:*

(i) $\mu \leqslant C_H(\Omega)$.

(ii) $\mathscr{L}_\mu$ *admits a positive super-harmonic in* Ω.

(iii) *For any subdomain* G *with* $\overline{G} \subset \Omega$ *and any two sub- and super-harmonics* $\underline{h}$, $\overline{h}$ *of* $\mathscr{L}_\mu$ *in* G *with* $\underline{h} \leqslant \overline{h}$ *on* ∂G *it follows that* $\underline{h} \leqslant \overline{h}$ *a.e. in* G.

2.3 Phragmen–Lindelöf alternative

Observe that global positive super-harmonics of $\mathscr{L}_\mu$ exist for all $\mu \leqslant C_H(\Omega)$, while the existence of local positive super-harmonics of $\mathscr{L}_\mu$ is controlled by the *local Hardy constant*

$$C_H^{\rm loc}(\Omega_\rho) := \inf_{0\neq\phi\in W_0^{1,2}(\Omega_\rho)} \frac{\displaystyle\int_{\Omega_\rho} |\nabla\phi|^2\,dx}{\displaystyle\int_{\Omega_\rho} \delta^{-2}(x)\phi^2\,dx}.$$

Note that, in general, $C_H(\Omega_\rho) \neq C_H^{\rm loc}(\Omega_\rho)$ because $\delta(x) := \operatorname{dist}(x, \partial\Omega) \neq \operatorname{dist}(x, \partial\Omega_\rho)$. It is known [2, Lemma 2.5] that $C_H^{\rm loc}(\Omega_\rho) = 1/4$ if $\rho > 0$ is sufficiently small.

If $\mu \leqslant C_H^{\mathrm{loc}}(\Omega_\rho)$, then $\mathscr{L}_\mu$ admits positive local super-harmonics and satisfies the comparison principle between sub- and super-harmonics in Ω_ρ for all sufficiently small $\rho > 0$ (cf. [2]). Furthermore, the following *Phragmen–Lindelöf alternative* holds for $\mathscr{L}_\mu$. We repeat the statement and its proof from [2, Theorem 2.6] for the sake of convenience.

Theorem 2.1. *Let $\mu \leqslant 1/4$, and let $\underline{h}$ be a local positive sub-harmonic. Then the following alternative holds: either*

(i) *for every local super-harmonic $\overline{h} > 0$*

$$\limsup_{x \to \partial\Omega} \underline{h}/\overline{h} > 0 \tag{2.3}$$

or

(ii) *for every local super-harmonic $\overline{h} > 0$*

$$\limsup_{x \to \partial\Omega} \underline{h}/\overline{h} < \infty. \tag{2.4}$$

Proof. Assume that (i) does not hold, i.e., there exists a super-harmonic $\overline{h}_* > 0$ such that

$$\lim_{x \to \partial\Omega} \underline{h}/\overline{h}_* = 0. \tag{2.5}$$

Let $\overline{h} > 0$ be an arbitrary super-harmonic in Ω_ρ for some sufficiently small $\rho > 0$. Then there exists a constant $c > 0$ such that $\overline{h} \geqslant c\underline{h}$ on $\Gamma_{\rho/2}$. For $\tau > 0$ we define a comparison function

$$v_\tau := c\underline{h} - \tau\overline{h}_*.$$

Then (2.5) implies that for every $\tau > 0$ there exists $\varepsilon = \varepsilon(\tau) \in (0, \rho)$ such that $v_\tau \leqslant 0$ on Ω_ε. Applying the comparison principle in $\Omega_{\varepsilon/2,\rho/2}$, we conclude that $\overline{h} \geqslant v_\tau$ in $\Omega_{\varepsilon/2,\rho/2}$ and hence in $\Omega_{\rho/2}$. So, considering arbitrary small $\tau > 0$, we conclude that for every super-harmonic $\overline{h} > 0$ in Ω_ρ there exists $c > 0$ such that $\overline{h} \geqslant c\underline{h}$ holds in Ω_ρ. This implies (2.4). □

If we apply this alternative to the above-mentioned special super-harmonics, we get for sub-harmonics the following boundary behavior. If $\mu < 1/4$, then either

(i) $\limsup_{x \to \partial\Omega} \frac{\underline{h}(x)}{\delta(x)^{\beta_-}} > 0$

or

(ii) $\limsup_{x \to \partial\Omega} \frac{\underline{h}(x)}{\delta(x)^{\beta_+}} < \infty$.

2.4 Sub- and super-solutions

Let $G \subset \Omega$ be open. A function $\overline{u} \in H^1_{\mathrm{loc}}(G) \cap C(G)$ satisfying the inequality

$$\mathscr{L}_\mu \overline{u} + e^{\overline{u}} \geqslant 0 \text{ in } G$$

in the weak sense is called a *super-solution* of (1.1) on G. Similarly, $\underline{u} \in H^1_{\text{loc}}(G) \cap C(G)$ is called a *sub-solution* of (1.1) if the inequality sign is reversed. A function u is a *solution* of (1.1) in G if it is a sub-solution and a super-solution in G. By interior elliptic regularity, weak solutions of (1.1) are classical. Hence in what follows we assume that all solutions of (1.1) are of class $C^2(\Omega)$.

Observe that solutions and sub-solutions are sub-harmonics of $\mathscr{L}_\mu$.

The following comparison principle is based on an argument used in [2] and plays a crucial role in our estimates. Part (i) relies heavily on the fact that $\mu < C_H(\Omega)$. Part (ii) is an extension of (i) for arbitrary μ under an additional assumption.

Lemma 2.2 (comparison principle). *Let $G \subset \Omega$ be open, and let $\underline{u}, \overline{u} \in H^1_{\text{loc}}(G) \cap C(G)$ be a pair of sub-, super-solutions of (1.1) satisfying*

$$\limsup_{x \to \partial G} [\underline{u}(x) - \overline{u}(x)] < 0.$$

(i) *If $\mu < C_H(\Omega)$, then $\underline{u} \leqslant \overline{u}$ in G.*

(ii) *If $\mu \geqslant C_H(\Omega)$ and, in addition, $\overline{u} > 1$ in G, then $\underline{u} \leqslant \overline{u}$ in G.*

Proof. Let $G_+ := \{x \in G : \underline{u}(x) > \overline{u}(x)\}$. In view of the boundary conditions, we have $\overline{G}_+ \subset G$. In the weak formulation of the inequality

$$\mathscr{L}_\mu(\overline{u} - \underline{u}) \geqslant -(e^{\overline{u}} - e^{\underline{u}}) \text{ in } G, \tag{2.6}$$

we use the test function $(\underline{u} - \overline{u})_+ \in H^1_0(G)$ and obtain

$$\int_G |\nabla(\underline{u} - \overline{u})_+|^2 \, dx \leqslant \mu \int_G \delta^{-2} (\underline{u} - \overline{u})_+^2 \, dx.$$

Case (i). Unless $G_+ = \varnothing$, this implies

$$\mu \geqslant \inf_{0 \neq \phi \in W_0^{1,2}(G)} \frac{\int_G |\nabla \phi|^2 \, dx}{\int_G \delta^{-2}(x) \phi^2 \, dx} \geqslant \inf_{0 \neq \phi \in W_0^{1,2}(\Omega)} \frac{\int_\Omega |\nabla \phi|^2 \, dx}{\int_\Omega \delta^{-2}(x) \phi^2 \, dx} = C_H(\Omega),$$

which contradicts our assumption.

Case (ii). If $\mu \geqslant C_H(\Omega)$, we make use of the following argument. In the weak formulation (2.6), we again use the test function $(\underline{u} - \overline{u})_+ \in H^1_0(G)$ and obtain

$$\int_G |\nabla(\underline{u} - \overline{u})_+|^2 \, dx - \mu \int_G \delta^{-2} (\underline{u} - \overline{u})_+^2 \, dx \leqslant \int_{G_+} \frac{e^{\overline{u}} - e^{\underline{u}}}{\underline{u} - \overline{u}} (\underline{u} - \overline{u})_+^2 \, dx. \tag{2.7}$$

Since $\overline{u} > 1$ in G, we can write $(\underline{u} - \overline{u})_+ = \phi\overline{u}$, where $\phi \in W_0^{1,2}(G)$ and the support of ϕ lies in the closure of G_+. Then

$$\begin{aligned}\int_G |\nabla(\underline{u} - \overline{u})_+|^2 \, dx &= \int_{G_+} (\phi^2|\nabla\overline{u}|^2 + 2\phi\overline{u}\nabla\overline{u}\cdot\nabla\phi + \overline{u}^2|\nabla\phi|^2)\, dx \\ &= \int_{G_+} [\overline{u}^2|\nabla\phi|^2 + \nabla\overline{u}\cdot\nabla(\phi^2\overline{u})]\, dx.\end{aligned}$$

Recalling that $\overline{u}$ is a super-solution and $\phi^2\overline{u} \geqslant 0$, we conclude that

$$\int_{G_+} \nabla\overline{u}\cdot\nabla(\phi^2\overline{u})\, dx \geqslant \int_{G_+} \big[\frac{\mu}{\delta^2}\phi^2\overline{u}^2 - e^{\overline{u}}\phi^2\overline{u}\big]\, dx.$$

This leads to

$$\int_G |\nabla(\underline{u} - \overline{u})_+|^2 \, dx - \mu\int_{G_+} \delta^{-2}(\underline{u} - \overline{u})^2\, dx \geqslant -\int_{G_+} \frac{e^{\overline{u}}}{\overline{u}}(\underline{u} - \overline{u})^2\, dx. \tag{2.8}$$

Since

$$\frac{e^{\overline{u}} - e^{\underline{u}}}{\underline{u} - \overline{u}} \leqslant -e^{\overline{u}} \text{ whenever } \underline{u} \geqslant \overline{u}$$

by convexity and $\overline{u} > 1$ by assumption, we find that (2.8) contradicts (2.7) unless $G_+ = \varnothing$. □

2.5 Solutions with zero boundary data

We show that the problem

$$\mathscr{L}_\mu u + e^u = 0, \qquad u \in H_0^1(\Omega), \tag{2.9}$$

admits a solution for all $\mu < C_H(\Omega)$. For this purpose, we need the following lemma.

Lemma 2.3. *Let $\mu < C_H(\Omega)$. Then the boundary value problem*

$$\mathscr{L}_\mu \phi = -1, \qquad \phi \in H_0^1(\Omega), \tag{2.10}$$

admits a unique solution $\phi < 0$. In addition, ϕ is bounded in $\overline{\Omega}$.

Proof. Results of this type are standard (cf., for example, [8] and references therein). For the sake of completeness, we sketch the proof. Consider the quadratic form associated to $\mathscr{L}_\mu$:

$$\mathcal{E}_\mu(u) := \int_\Omega \Big(|\nabla u|^2 - \mu\frac{u^2}{\delta^2}\Big)\, dx.$$

By the definition of the Hardy constant $C_H(\Omega)$,

$$\mathcal{E}_\mu(u) \geqslant \Big(1 - \frac{\mu}{C_H(\Omega)}\Big) \int_\Omega |\nabla u|^2 dx. \tag{2.11}$$

We conclude that $\mathcal{E}_\mu$ is a coercive and continuous quadratic form on $H_0^1(\Omega)$. Since $-1 \in [H_0^1(\Omega)]^*$, the existence and uniqueness of a solution $\phi \in H_0^1(\Omega)$ follows by the Lax–Milgram theorem. Since $\mathscr{L}_\mu \phi < 0$ in Ω, the comparison principle of Lemma 2.1 implies that $\phi < 0$. By the classical regularity theory, ϕ is bounded in every compact subset of Ω. A straightforward computation (using formula (5.2)) shows that for large A and small ϵ

$$\underline{\phi} = -A\delta^\nu, \quad \nu = \min\{2, \beta_+ - \epsilon\}$$

is a sub-solution in Ω_{δ_0} for a small $\delta_0 > 0$. Choosing $A > 0$ so large that, in particular, $\underline{\phi} \leqslant \phi$ on Γ_{δ_0}, we can apply the comparison principle and conclude that ϕ is bounded in $\overline{\Omega}$. □

Theorem 2.2. *Let $\mu < C_H(\Omega)$. Then the problem* (2.9) *has a unique solution u_0. Moreover, $\phi < u_0 < 0$, where ϕ is defined in Lemma* 2.3.

Proof. Consider the energy functional corresponding to (2.9):

$$J(u) := \frac{1}{2}\mathcal{E}_\mu(u) + \int_\Omega e^u \, dx.$$

In view of (2.11), it is clear that $J : H_0^1(\Omega) \to \mathbb{R} \cup \{+\infty\}$ is coercive, convex, and weakly lower semicontinuous on $H_0^1(\Omega)$. Hence J admits a unique minimizer $u_0 \in H_0^1(\Omega)$. Note that $J(u^-) \leqslant J(u)$ for every $u \in H_0^1(\Omega)$. As a consequence, $u_0 \leqslant 0$. Hence e^{u_0} is bounded from above, and thus u_0 satisfies the Euler–Lagrange equation and solves (2.9). Further, since $u = 0$ is not a solution of (2.9), we conclude that $u_0 < 0$.

Let $\phi \in H_0^1(\Omega)$ be the same as in Lemma 2.3. Since $\phi < 0$ in Ω, we have $\mathscr{L}_\mu \phi + e^\phi \leqslant 0$ in Ω, so ϕ is a sub-solution of (2.9). From the comparison principle of Lemma 2.2 (i) it follows that $\phi < u_0$. □

Remark 2.1. Suppose that a domain Ω is such that $C_H(\Omega) < 1/4$. Then there exists a positive solution $\phi_1 \in H_0^1(\Omega)$ of $\mathscr{L}_{C_H(\Omega)}\phi_1 = 0$ in Ω (cf. [8]). We claim that if $\mu \geqslant C_H(\Omega)$, then (2.9) has no negative solution. Suppose that $u \in H_0^1(\Omega)$ is a negative solution of (2.9). Then we obtain the contradiction

$$0 \leqslant \int_\Omega \frac{(C_{H(\Omega)} - \mu)}{\delta^2} u\phi_1 \, dx = \int_\Omega \nabla u \cdot \nabla \phi_1 - \frac{\mu}{\delta^2} u\phi_1 \, dx = -\int_\Omega e^u \phi_1 \, dx < 0.$$

Hence, if for $C_H(\Omega) < 1/4$ and $\mu \geqslant C_H(\Omega)$ a solution of (2.9) exists, then it must be sign-changing (cf. Question 1 in Section 7). The same statement holds for solutions of (2.10).

3 A Priori Upper Bounds

In this section, we construct a universal upper bound for all solutions of (1.1) by means of a super-solution which tends to infinity at the boundary. The construction is inspired by the Keller–Osserman bound given in [2] for power nonlinearities. The terminology *Keller–Osserman bound* refers the universal upper bound of Lemma 3.1 and Lemma 3.2. Such upper bounds, which hold for all solutions of a nonlinear equation, were observed in the classical papers by Keller [7] and Osserman [10].

For our purpose, we need the *Whitney distance* $d : \Omega \to \mathbb{R}_+$ which is a $C^\infty(\Omega)$-function such that for all $x \in \Omega$

$$\begin{aligned} &c^{-1}\delta(x) \leqslant d(x) \leqslant c\delta(x), \\ &|\nabla d(x)| \leqslant c, \quad |\Delta d(x)| \leqslant c d^{-1}(x), \end{aligned}$$

with a constant $c > 0$ which is independent of x. These properties of the Whitney distance can be found in [11].

For $\varepsilon > 0$ we use the notation $\mathcal{D}_\varepsilon = \{x \in \Omega : d(x) > \varepsilon\}$.

Lemma 3.1. *Let $\mu \leqslant 0$. Then there exists a number $A > 0$ such that for every solution u of* (1.1) *we have*

$$u(x) \leqslant \log \frac{A}{d^2(x)} \quad \text{in } \Omega.$$

Proof. For small $\varepsilon > 0$ we consider the function

$$f_\varepsilon(x) = \log \frac{A}{(d(x)-\varepsilon)^2} \quad \text{in } \mathcal{D}_\varepsilon.$$

It satisfies the equation

$$\Delta f_\varepsilon = \frac{2}{(d-\varepsilon)^2}|\nabla d|^2 - \frac{2}{d-\varepsilon}\Delta d \qquad \text{in } \mathcal{D}_\varepsilon.$$

Thus, taking into account the properties of the Whitney distance and the fact that μ is nonpositive, we find

$$\Delta f_\varepsilon + \frac{\mu}{\delta^2} f_\varepsilon - e^{f_\varepsilon} \leqslant \frac{c_1 - A}{(d-\varepsilon)^2} \qquad \text{in } \mathcal{D}_\varepsilon.$$

For sufficiently large A the right-hand side of this inequality is negative. Hence f_ε is a super-solution satisfying $f_\varepsilon > u$ on $\partial\mathcal{D}_\varepsilon$. The comparison principle implies

$$u(x) \leqslant f_\varepsilon(x) \ \text{ in } \mathcal{D}_\varepsilon.$$

Since $\varepsilon > 0$ is arbitrary, the conclusion follows. □

If μ is positive, we proceed in a different way. For $A > 0$ the function $L_A(d(x))$, $d(x)$ =Whitney-distance, will play an essential role in the following construction of upper bounds for all solutions of (1.1). The definition of $L_A(t)$ is given implicitly by the formula

$$\frac{e^{L_A(t)}}{L_A(t)} = \frac{A}{t^2}, \quad A > 0, \quad L_A(t) > 1. \tag{3.1}$$

It is clear that the function $L_A(t)$ is monotone increasing in A and decreasing in t. Also, from the relation $L_A(t) = \log \frac{A}{t^2} + \log L_A(t)$ one finds successively

$$\begin{aligned}
L_A(t) &\geqslant \log \frac{A}{t^2}, \\
L_A(t) &\geqslant \log \frac{A}{t^2} + \log\log \frac{A}{t^2}, \\
L_A(t) &\geqslant \log \frac{A}{t^2} + \log\left(\log \frac{A}{t^2} + \log\log \frac{A}{t^2}\right), \\
L_A(t) &\geqslant \dots
\end{aligned}$$

Moreover,

$$\lim_{t\to 0+} \frac{L_A(t)}{\log(1/t^2)} = \lim_{t\to 0+} \frac{L_A'(t)}{-2/t} = \lim_{t\to 0+} \frac{L_A(t)}{L_A(t)-1} = 1 \tag{3.2}$$

since $L_A(0) = \infty$.

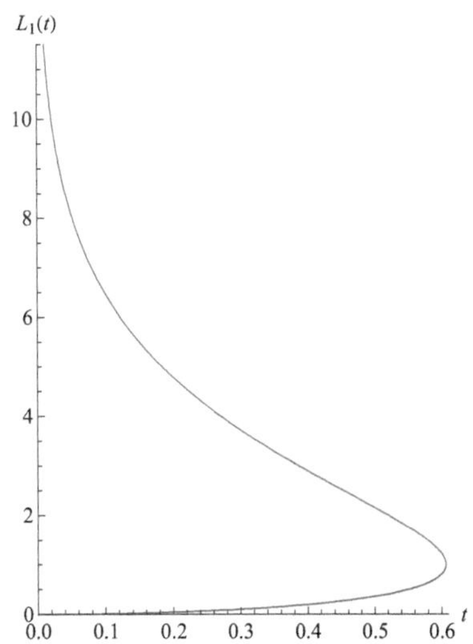

Fig. 1 Lambert function $L_1(t)$.

As a historical note, let us mention that the function $L_A(t)$ is related to the Lambert W-function which satisfies the equation

$$W(s)e^{W(s)} = s$$

and which has a long history starting with J.H. Lambert and L. Euler. Indeed we have

$$L_A(t) = -W\Big(-\frac{t^2}{A}\Big)$$

if one takes for W again the upper branch.

Next we show that $L_A(d(x))$ is indeed a universal upper bound for all solutions of (1.1) if $A > 0$ is sufficiently large. The estimate is based on the extended comparison principle of Lemma 2.2 (ii).

Lemma 3.2. *There exists $A > 0$ such that every solution of* (1.1) *satisfies the inequality*

$$u(x) \leqslant L_A(d(x)) \ \ in \ \ \Omega.$$

Proof. In order to define $L_A(d(x))$ with the property (3.1), we must take $A > 0$ so large that $\inf_\Omega \frac{A}{d^2(x)} > e$. A straightforward computation yields

$$\Delta L_A(d) = \frac{2L_A(d)}{d^2(L_A(d)-1)}|\nabla d|^2\left\{1-\frac{2}{(L_A(d)-1)^2}\right\} - \frac{2L_A(d)}{d(L_A(d)-1)}\Delta d. \tag{3.3}$$

For $\varepsilon \geqslant 0$, let $\overline{u}_\varepsilon : \mathcal{D}_\varepsilon \to \mathbb{R}$ be defined as

$$\overline{u}_\varepsilon(x) := L_A(d(x)-\varepsilon). \tag{3.4}$$

Then, by (3.1), (3.3), and the properties of the Whitney distance, we have

$$\begin{aligned}\Delta\overline{u}_\varepsilon + \frac{\mu}{\delta^2}\overline{u}_\varepsilon - e^{\overline{u}_\varepsilon} \leqslant & \frac{2L_A(d-\varepsilon)}{(d-\varepsilon)^2(L_A(d-\varepsilon)-1)}c^2\left\{1-\frac{2}{(L_A(d-\varepsilon)-1)^2}+\frac{2}{c}\right\}\\ &+\frac{L_A(d-\varepsilon)}{(d-\varepsilon)^2}\left\{c_\pm\mu - A\right\}, \ \text{ where } c_\pm = \begin{cases}0 & \text{if } \mu \leqslant 0\\ c^2 & \text{if } \mu > 0.\end{cases}\end{aligned}$$

Taking A sufficiently large, we can always achieve that the right-hand side is negative independently of ε. Consequently, $\overline{u}_\varepsilon$ is a super-solution of (1.1) in $\mathcal{D}_\varepsilon$ for all sufficiently small $\varepsilon > 0$.

Let u be an arbitrary solution of (1.1). It is clear that $u < \overline{u}_\varepsilon$ on $\partial\mathcal{D}_\varepsilon$. Moreover, by definition, $\overline{u}_\varepsilon(x) = L_A(d(x)-\varepsilon) > 1$. Thus, Lemma 2.2 (ii) applies and yields

$$u(x) \leqslant L_A(d(x)-\varepsilon) \ \text{ in } \ \mathcal{D}_\varepsilon.$$

Since $\varepsilon > 0$ is an arbitrary small number, this concludes the proof of the lemma. ◻

Remark 3.1. From the above proof it is clear that for sufficiently large $A > 0$ the function

$$\overline{u}(x) := L_A(d(x)) \tag{3.5}$$

is a super-solution of (1.1) in Ω.

Remark 3.2. Note that

$$\frac{e^{L_A(d(x))}}{L_A(d(x))} = \frac{A}{d^2(x)} \leqslant \frac{Ac^2}{\delta^2(x)} = \frac{e^{L_{c^2A}(\delta(x))}}{L_{c^2A}(\delta(x))}.$$

Replacing the Whitney distance by the standard distance, we obtain the universal a priori bound

$$u(x) \leqslant L_{c^2A}(\delta(x)),$$

and, by (3.2), we obtain

$$\limsup_{x\to\partial\Omega} \frac{u(x)}{\log \delta^{-2}(x)} \leqslant 1. \tag{3.6}$$

It should be pointed out that the above-constructed bound holds for every $\mu \in \mathbb{R}$.

4 Nonexistence of Large Solutions if $\mu < 0$

Lemma 3.2, together with the Phragmen–Lindelöf alternative, gives rise to a nonexistence result.

Theorem 4.1. *If* $\mu < 0$, *then* (1.1) *does not have large solutions.*

Proof. If a solution u of (1.1) exists with $u(x) \to \infty$ as $x \to \partial\Omega$, then, by the conclusion drawn from the Phragmen–Lindelöf alternative of Theorem 2.1, it must satisfy

$$\limsup_{x\to\partial\Omega} \frac{u(x)}{\delta(x)^{\beta_-}} > 0, \quad \text{where } \beta_- = \frac{1}{2} - \sqrt{\frac{1}{4} - \mu}.$$

On the other hand, (3.6) implies

$$\limsup_{x\to\partial\Omega} \frac{u(x)}{\delta(x)^{\beta_-}} \leqslant \limsup_{x\to\partial\Omega} \delta(x)^{-\beta_-} \log \frac{1}{\delta(x)^2} = 0.$$

This is impossible, and therefore u does not exist. □

This nonexistence result, together with the Phragmen–Lindelöf alternative, leads to the following conclusion.

Corollary 4.1. *If $\mu < 0$, then all solutions of* (1.1) *vanish on the boundary.*

5 Asymptotic Behavior of Large Solutions near the Boundary

5.1 Global sub-solutions

Since the case $\mu = 0$ is well known and no large solutions exist for negative μ, we assume throughout this section that $\mu > 0$.

Let L_A be defined as in (3.1). We construct local sub-solutions which have the same asymptotic behavior as the super-solution $L_A(d(x))$ from Lemma 3.2.

Proposition 5.1. *Let $0 < B \leqslant \mu$. Then there exists a small positive $\varepsilon_0 < \frac{1}{2}\sqrt{B/e}$ such that $\underline{u}_\varepsilon(x) := L_B(\delta(x)+\varepsilon)$ is a sub-solution of* (1.1) *in Ω_{ε_0} for any $\varepsilon \in [0, \varepsilon_0]$.*

Proof. Since $(\delta(x)+\varepsilon)^2 \leqslant 4\varepsilon_0^2 < B/e$, the function $\underline{u}_\varepsilon$ is well defined in Ω_{ε_0}. As in the proof of Lemma 3.2, we have

$$\begin{aligned}\Delta \underline{u}_\varepsilon + \frac{\mu}{\delta^2}\underline{u}_\varepsilon - e^{\underline{u}_\varepsilon} = {} & \frac{2\underline{u}_\varepsilon}{(\delta+\varepsilon)^2(\underline{u}_\varepsilon - 1)}\left\{1 - \frac{2}{(\underline{u}_\varepsilon - 1)^2}\right\} \\ & - \frac{2\underline{u}_\varepsilon}{(\delta+\varepsilon)(\underline{u}_\varepsilon - 1)}\Delta\delta + \frac{\mu}{\delta^2}\underline{u}_\varepsilon - \frac{B}{(\delta+\varepsilon)^2}\underline{u}_\varepsilon . \end{aligned} \tag{5.1}$$

In Ω_{ε_0} one has the expansion

$$\Delta\delta(x) = -(N-1)\mathcal{H}_0(\sigma(x)) + o(\delta(x)). \tag{5.2}$$

Hence $\Delta\delta \leqslant K$ in Ω_{ε_0} for some constant $K > 0$ independently of ε_0. Next we choose ε_0 so small that $1 - \frac{2}{(\underline{u}_\varepsilon - 1)^2} \geqslant \frac{1}{2}$ in Ω_{ε_0}. Since $0 < B \leqslant \mu$, we find

$$\Delta \underline{u}_\varepsilon + \frac{\mu}{\delta^2}\underline{u}_\varepsilon - e^{\underline{u}_\varepsilon} \geqslant \frac{\underline{u}_\varepsilon}{(\delta+\varepsilon)(\underline{u}_\varepsilon - 1)}\left(\frac{1}{\delta+\varepsilon} - 2K\right)$$

in Ω_{ε_0}. The right-hand side is positive provided that $\epsilon_0 < 1/(4K)$. Thus, $\underline{u}_\varepsilon$ is a sub-solution in Ω_{ε_0} for all $\varepsilon \in [0, \varepsilon_0]$. □

At the next step, we extend the local sub-solution $\underline{u}_\varepsilon$ to a global sub-solution $\underline{\mathcal{U}}_\varepsilon$ in the whole domain such that $\underline{\mathcal{U}}_\varepsilon = \underline{u}_\varepsilon$ near the boundary.

Proposition 5.2. *Assume that* $0 < \mu < C_H(\Omega)$. *Then there exists a global sub-solution* $\underline{\mathcal{U}}_0$ *with* $\underline{\mathcal{U}}_0(x) = L_\mu(\delta(x))\big(1-O(\delta(x)^{\beta-})\big)$. *Moreover, if* u *is any solution of* (1.1) *which tends to infinity at the boundary, then* $u \geqslant \underline{\mathcal{U}}_0$ *and, in particular,*

$$\liminf_{x\to\partial\Omega} \frac{u(x)}{\log \delta^{-2}(x)} \geqslant 1. \tag{5.3}$$

Proof. Let $\phi \in H_0^1(\Omega)$ be as defined in Lemma 2.3. Since ϕ is nonpositive, we have $\mathscr{L}_\mu\phi + e^\phi \leqslant 0$ in Ω. Therefore, ϕ is a sub-solution of (1.1). Let $\underline{u}_\varepsilon(x) = L_\mu(\delta(x)+\varepsilon)$ by the local sub-solution from Proposition 5.1. Consider the local super-harmonic (cf. Examples in Section 2)

$$\overline{H} = \delta^{\beta-}(1+\delta^\nu), \quad \text{where } \nu << 1.$$

It is clear that $\underline{u}_\varepsilon - C\overline{H}$ is also a local sub-solution of (1.1) in Ω_{ε_0}, where C is an arbitrary positive number. Choose $C > 0$ so large that $\underline{u}_\varepsilon - C\overline{H} < \varphi$ on Γ_{ε_0}, i.e.,

$$L_\mu(\varepsilon_0+\varepsilon) - C\varepsilon_0^{\beta-}(1+\varepsilon_0^\nu) < \min_{\Gamma_{\varepsilon_0}} \phi.$$

Because of the inequality $L_\mu(\varepsilon_0+\varepsilon) \leqslant L_\mu(\varepsilon_0)$ the value $C = C(\varepsilon_0)$ can be chosen independently of $\varepsilon \in [0, \epsilon_0]$. With this fixed C we now define the function

$$\underline{\mathcal{U}}_\epsilon = \begin{cases} \max\{\underline{u}_\varepsilon - C\overline{H}, \phi\} \text{ in } \Omega_{\varepsilon_0}, \\ \phi \text{ in } D_{\varepsilon_0}. \end{cases} \tag{5.4}$$

The function $\underline{\mathcal{U}}_\varepsilon$ is a global sub-solution for all $\epsilon \in [0, \epsilon_0]$. Moreover, since $\overline{H} = 0$ on $\partial\Omega$ and $\underline{u}_\varepsilon$ is positive in Ω_{ε_0}, we have $\underline{\mathcal{U}}_\varepsilon = \underline{u}_\varepsilon - C\overline{H}$ near $\partial\Omega$. Set $\omega_\varepsilon := \{x \in \Omega_{\varepsilon_0} : \underline{u}_\varepsilon - C\overline{H} > \phi\}$ and note that $\omega_\varepsilon \supset \omega_{\varepsilon_0}$ for all $\varepsilon \in [0, \varepsilon_0]$, so that each ω_ε contains a fixed neighborhood of the boundary $\partial\Omega$. Thus,

$$\underline{\mathcal{U}}_\varepsilon(x) = L_\mu(\delta(x)+\varepsilon) - C(\varepsilon_0)\delta^{\beta-}(x)(1+\delta^\nu(x)) \quad \text{for } x \in \omega_{\varepsilon_0} \text{ and all } \epsilon \in (0, \varepsilon_0).$$

If u is any solution of (1.1) which tends to infinity at the boundary, then the comparison principle of Lemma 2.2 implies that $u(x) \geqslant \underline{\mathcal{U}}_\epsilon(x)$ in Ω for all $\varepsilon \in (0, \epsilon_0]$. Letting $\epsilon \to 0$, we get $u(x) \geqslant \underline{\mathcal{U}}_0(x)$ in Ω and, in particular, near the boundary

$$u(x) \geqslant \underline{\mathcal{U}}_0(x) = L_\mu(\delta(x)) - C(\varepsilon_0)\delta^{\beta-}(x)(1+\delta^\nu(x))$$

This, together with (3.2), implies (5.3). □

Remark 5.1. If a domain Ω is small in the sense that its inradius ρ_0 satisfies the condition

$$\frac{\mu}{\rho_0^2} \geqslant e,$$

then $v = 1$ is a global sub-solution. If $\mu \geqslant C_H(\Omega)$, it is not clear whether we can deduce from this fact that $u > 1$ for large solutions u.

Proposition 5.2 and the inequality (3.6) imply the following assertion.

Theorem 5.1. *If* $0 < \mu < C_H(\Omega)$, *then every large solution of* (1.1) *satisfies*

$$\lim_{x \to \partial\Omega} \frac{u(x)}{\log \delta^{-2}(x)} = 1. \tag{5.5}$$

6 Uniqueness and Existence of Large Solutions

6.1 Uniqueness

Theorem 6.1. *Assume that* $0 < \mu < C_H(\Omega)$. *Then* (1.1) *has at most one large solution.*

Proof. Suppose that (1.1) has two large solutions U_1 and U_2. If the domain is large, the solutions can become negative. In this case, we add a sufficiently large negative multiple of the function $\phi \in H_0^1(\Omega)$ of Lemma 2.3 (recall that $\mathscr{L}_\mu \phi = -1$ and $\phi < 0$ in Ω) such that $w_i := U_i - t\phi > 1$ for $i = 1, 2$ and $t > 0$ is taken sufficiently large. Then

$$\mathscr{L}_\mu w_i = -\underbrace{a(x)}_{:=e^{t\phi(x)}} e^{w_i} + t \text{ in } \Omega, \quad w_i(x) \to \infty \text{ as } x \to \partial\Omega, \quad i = 1, 2.$$

Define a function $\sigma(x)$ by the equality $w_1(x) = \sigma(x) w_2(x)$. Because of the asymptotic behavior of U_1 and U_2 known from Theorem 5.1, we have $\sigma(x) = 1$ on $\partial\Omega$. Then

$$\begin{aligned} t = \mathscr{L}_\mu w_1 + a e^{w_1} &= -\sigma \Delta w_2 - w_2 \Delta\sigma - 2\nabla\sigma \cdot \nabla w_2 - \mu\delta^{-2}\sigma w_2 + a e^{\sigma w_2} \\ &= -w_2 \Delta\sigma - 2\nabla\sigma \cdot \nabla w_2 - \sigma a e^{w_2} + t\sigma + a e^{\sigma w_2}. \end{aligned}$$

Suppose that $w_1 > w_2$ (or, equivalently, $\sigma > 1$) in a subset Ω' of Ω. Since $w_1/w_2 \to 1$ as x approaches the boundary of Ω', we have $\sigma(x) = 1$ on $\partial\Omega'$. Using our assumption $w_2 > 1$, we conclude that $e^{\sigma(x) w_2} > \sigma(x) e^{w_2}$ in Ω'. Thus,

$$-w_2 \Delta\sigma - 2\nabla\sigma \cdot \nabla w_2 < t(1 - \sigma) < 0 \text{ in } \Omega',$$

and, by the maximum principle, $\sigma \leqslant 1$ in Ω'. This contradicts the fact that $w_1 > w_2$ in Ω'. Consequently, $w_1 \leqslant w_2$. Similarly, we show that $w_2 > w_1$ is impossible. This completes the proof. □

6.2 Existence

Theorem 6.2. *If $0 < \mu < C_H(\Omega)$, then* (1.1) *has a large solution.*

Proof. Let $\overline{u}$ be a super-solution of (1.1) which blows up at $\partial\Omega$, as constructed in (3.5). Let $\underline{u}_m$ be a sub-solution of (1.1) defined in (5.4) and chosen in such a way that $\underline{u}_m = m$ on $\partial\Omega$ for $m \in \mathbb{N}$. Let $\{M_n\}_{n\in\mathbb{N}}$ be a monotone increasing sequence of numbers such that

$$\underline{u}_m < M_n < \overline{u} \text{ on } \Gamma_{1/n}.$$

Let $u_{m,n}$ be a solution of the problem

$$\mathscr{L}_\mu u_{m,n} + e^{u_{m,n}} = 0 \text{ in } D_{1/n}, \quad u_{m,n} = M_n \text{ on } \partial D_{1/n}.$$

Such a solution could be constructed, for example, by minimizing the energy functional

$$J(u) = \frac{1}{2}\mathcal{E}_\mu(u) + \int_\Omega e^u \, dx,$$

which is coercive and weakly lower semicontinuous on the convex set

$$\mathcal{M}_n = \{u \in H^1(D_{1/n}),\ u = M_n \text{ on } \partial D_{1/n}\}.$$

From the comparison principle of Lemma 2.2 (i) it follows that

$$\underline{u}_m \leqslant u_{m,n} \leqslant \overline{u} \text{ in } D_{1/n}.$$

Thus, by standard compactness and diagonalization arguments, we conclude that there exists a subsequence $\{u_{m,n(m)}\}_{m\in\mathbb{N}}$ which converges as $m \to \infty$ to a large solution u of (1.1) in Ω. □

7 Borderline Potentials. Summary and Open Problems

By Theorem 4.1, no large solution of (1.1) exists if μ is negative. This is due to the fact that the corresponding large sub-harmonics which interact with the nonlinear regime are too large near the boundary and hence incompatible with the a priori bound constructed in Lemma 3.1. We construct a maximal (in a certain sense) positive perturbation of $-\Delta$ of the form

$$\mathscr{L}_{\gamma(\delta)} := -\Delta + \frac{\gamma(\delta)}{\delta^2},$$

where $\gamma(\delta) > 0$, $\gamma(\delta) = o(1)$ as $\delta \to 0$, and such that the semilinear problem

$$\mathscr{L}_{\gamma(\delta)}u + e^u = 0 \text{ in } \Omega \tag{7.1}$$

admits a large solution. Observe the different signs in the definition of $\mathscr{L}_{\gamma(\delta)}$ and $\mathscr{L}_\mu$. Lemma 3.1 and the Phragmen–Lindelöf alternative suggest that it is reasonable to look for a function γ for which the operator $\mathscr{L}_{\gamma(\delta)}$ admits large local sub-harmonics with the same or with a smaller order of magnitude as the Keller–Osserman bound near $\partial\Omega$.

The asymptotic behavior given in (1.4) suggests to use

$$h := \left(\log\frac{1}{\delta^2}\right)^m, \quad m > 0,$$

as a "prototype" family of sub- and super-harmonics in order to determine the borderline potential $\gamma(\delta)$. By direct computations, we have

$$\mathscr{L}_{\gamma(\delta)}h = -\Delta h + \frac{\gamma(\delta)}{\delta^2}h = -h'\Delta\delta - h''(\delta)|\nabla\delta|^2 + \frac{\gamma(\delta)}{\delta^2}h,$$

where $|\nabla\delta| = 1$ and $\Delta\delta = -(N-1)\mathcal{H}_0 + o(\delta)$. Therefore,

$$\mathscr{L}_{\gamma(\delta)}h = \left\{\frac{2m}{\delta}\left(\log\frac{1}{\delta^2}\right)^{m-1}\Delta\delta - \frac{4m(m-1)}{\delta^2}\left(\log\frac{1}{\delta^2}\right)^{m-2}|\nabla\delta|^2\right\}$$

$$-\frac{2m}{\delta^2}\left(\log\frac{1}{\delta^2}\right)^{m-1}|\nabla\delta|^2 + \frac{\gamma(\delta)}{\delta^2}\left(\log\frac{1}{\delta^2}\right)^m,$$

where the expression in brackets is of lower order as $\delta \to 0$. Now, we want to construct $\gamma(\delta)$ such that h is either a sub-harmonic or a super-harmonic, depending on the value of m. Set

$$\gamma(\delta) := \beta\min\left\{\left|\log\frac{1}{\delta^2}\right|^{-1}, 1\right\}$$

for some $\beta > 0$. With such a choice of γ we find that

$$\mathscr{L}_{\gamma(\delta)}h = \frac{\beta - 2m}{\delta^2}\left(\log\frac{1}{\delta^2}\right)^{m-1}(1 + o(1))$$

in a small parallel strip Ω_ρ. Therefore,

$$\overline{H} := \left(\log\frac{1}{\delta^2}\right)^m$$

is a local super-harmonic of $\mathscr{L}_{\gamma(\delta)}$ for all $0 < m < \beta/2$. Otherwise, for $m > \beta/2$, $\overline{H}$ is a local sub-harmonic of $\mathscr{L}_{\gamma(\delta)}$.

Further, a simple computation verifies that

$$\overline{h} = \delta^\alpha$$

is also a local super-harmonic of $\mathscr{L}_{\gamma(\delta)}$ for all $0 \leqslant \alpha < 1$. Thus, a Phragmen–Lindelöf type argument similar to the one used in Theorem 2.1, applied here to $\overline{H}$ and $\overline{h}$ defined above, shows that if $\underline{h} \geqslant 0$ is a local sub-harmonic of $\mathscr{L}_{\gamma(\delta)}$, then either

(i) $\limsup_{x \to \partial\Omega} \underline{h}(x)\Big(\log \frac{1}{\delta^2}\Big)^{-m} > 0$ for all $0 < m < \beta/2$

or

(ii) $\underline{h} = 0$ on Ω.

In particular, every large solution of (7.1) must satisfy (i).

Note that operator $\mathscr{L}_{\gamma(\delta)}$ is positive definite on Ω, simply because $\gamma(x) > 0$ in Ω. As a consequence, a comparison principle similar to Lemma 2.2 (i) is valid for the problem (7.1). Exactly the same arguments as in Lemma 3.1 imply that for large $A > 0$ every solution u of (7.1) satisfies a Keller–Osserman type bound

$$u(x) \leqslant \log \frac{A}{\delta^2(x)} \quad \text{in } \Omega. \tag{7.2}$$

Combining (7.2) with the Phragmen–Lindelöf bound (i) which holds for any $m < \beta/2$, we immediately obtain the following nonexistence result.

Theorem 7.1. *If* $\beta > 2$, *then* (7.1) *does not have large solutions.*

Next observe that if $0 < \beta < 2$, then for $0 < B < 2 - \beta$ the function

$$\underline{u} = \log \frac{B}{\delta^2} \tag{7.3}$$

is a local sub-solution of (7.1) with infinite boundary values. This local sub-solution can be extended to a global sub-solution in the same way as in (5.4). However, contrary to the construction in Proposition 5.2, this time we cannot construct sub-solutions with everywhere finite and nonzero boundary values, cf. (i) in the conclusion from the Phragmen–Lindelöf argument above.

In fact, we can prove the following existence and nonuniqueness result.

Theorem 7.2. *If* $0 < \beta < 2$, *then* (7.1) *has a large solution* u *such that*

$$\lim_{x \to \partial\Omega} \frac{u(x)}{\log \delta^{-2}(x)} = 1,$$

and for every $M > 0$ *there exists a large solution* v_M *such that*

$$\lim_{x \to \partial\Omega} \frac{v_M(x)}{\big(\log \delta^{-2}(x)\big)^{\beta/2}} = M.$$

Proof. Recall that in Theorem 6.2 the existence was based on a family of sub-solutions with finite boundary values and a super-solution with infinite boundary value. Since such sub-solutions are no longer available in the present case, we sketch a different argument for the proof of the above existence result. For $k \in \mathbb{N}$ let u_k be a large solution of the problem

$$\mathscr{L}_{\gamma(\delta)} u_k + e^{u_k} = 0 \text{ in } D_{1/k}, \quad u_k = \infty \text{ on } \partial D_{1/k}.$$

The sequence u_k is monotonically decreasing, and if $\underline{u}$ is the sub-solution from (7.3) extended to the whole of Ω, then $u_k \geqslant \underline{u}$ by the comparison principle. Therefore, $u_k \to u$ as $k \to \infty$ locally uniformly in Ω, where u is a large solution of (7.1) in Ω with $u \geqslant \underline{u}$. Hence

$$\lim_{x \to \partial\Omega} \frac{u(x)}{\log \delta^{-2}(x)} \geqslant 1.$$

Together with the Keller-Osserman upper bound from (7.2), this establishes the first claim of the theorem.

We now proceed to the construction of the large solution v_M. Let $M > 0$ be any given number, and let

$$\underline{H}_{M,k} := M\Big(\log \frac{1}{\delta^2}\Big)^{\beta/2} - k.$$

A straightforward computation yields for small $\delta(x)$

$$\begin{aligned}\mathscr{L}_{\gamma(\delta)} \underline{H}_{M,k} + e^{\underline{H}_{M,k}} &= \Big\{M\beta(2-\beta)\delta^{-2}(\log(\delta^{-2}))^{\frac{\beta}{2}-2}(1+o(1))\Big\} \\ &\quad - k\beta\delta^{-2}(\log(\delta^{-2}))^{-1} + e^{-k}e^{M(\log(\delta^{-2}))^{\beta/2}}.\end{aligned}$$

Since $\beta < 2$, the expression in the parenthesis $\{\dots\}$ is of lower order as $\delta \to 0$. Let $0 < \varepsilon < 1$, and let δ_0 be such that $M < (1-\epsilon)(\log(\delta_0^{-2}))^{1-\beta/2}$. Then for all $x \in \Omega$ with $\delta(x) \leqslant \delta_0$ one finds

$$\mathscr{L}_{\gamma(\delta)} \underline{H}_{M,k} + e^{\underline{H}_{M,k}} \leqslant -k\beta\delta^{-2}(\log(\delta^{-2}))^{-1}(1+o(1)) + e^{-k}\delta^{-2(1-\varepsilon)} \leqslant 0$$

provided that $k > 0$ is chosen sufficiently large. Hence $\underline{H}_{M,k}$ is a local sub-solution. Let $\phi \in H_0^1(\Omega)$ be a solution of $\mathscr{L}_{\gamma(\delta)}\phi = -1$ (cf. Lemma 2.3 with μ replaced by $-\gamma(\delta)$). Similarly to (5.4), one can choose $k > 0$ large enough so that $\underline{v}_M := \max\{\underline{H}_{M,k}, \phi\}$ is a global sub-solution of (7.1) in Ω.

To construct a super-solution, set

$$\overline{H}_{M,K} := M\Big(\log \frac{1}{\delta^2}\Big)^{\beta/2} + K,$$

which for large K and $\delta(x)$ small is a local super-solution. Let $A > 0$ be as in Lemma 3.2, so that $L_A(d(x))$ is a global super-solution of (7.1) in Ω. Then one can choose $K > 0$ so large that $\overline{v}_M := \min\{L_A(d(x)), \overline{H}_{M,K}\}$ is a global super-solution of (7.1) in Ω, which coincides with $\overline{H}_{M,K}$ near the boundary.

Since $\underline{v}_M < \overline{v}_M$ in Ω, a global large solution v_M of (7.1) with the required asymptotic can be constructed by using a diagonalization procedure similar to the one used in Theorem 6.2. We omit details. □

7.1 Summary and open problems

Our results are summarized as follows. The existence/nonexistence of large solutions of the problem

$$-\Delta u - V(x)u + e^u = 0 \text{ in } \Omega, \quad u = 0 \text{ on } \partial\Omega$$

can be read from the following table, where we use the notation

$$\gamma_0 = \min\left\{ \left| \log \frac{1}{\delta^2} \right|^{-1}, 1 \right\}.$$

	$V(x) = \frac{\mu}{\delta^2}$	$V(x) = \frac{-\beta\gamma_0(\delta)}{\delta^2}$	$V(x) = \frac{\mu-\beta\gamma_0(\delta)}{\delta^2}$
$\nexists$	$\mu < 0$	$\beta > 2$	$\mu < 0$ or $\mu = 0,\ \beta > 2$
$\exists$	$0 \leqslant \mu < C_H(\Omega)$	$0 < \beta < 2$	$0 < \mu < C_H(\Omega)$ or $\mu = 0$ and $0 < \beta < 2$.
critical borderline	no	$\beta = 2$	$\mu = 0,\ \beta = 2$

Except for $\mu = 0$, the assertions in the last row were not proved in the present paper, but they can be obtained with little changes since for $\mu \neq 0$ the perturbation $\frac{\gamma_0(\delta)}{\delta^2}$ is of lower order than the dominant term $\frac{\mu}{\delta^2}$.

We finish our discussion with the following open questions:

1. Does $\mathscr{L}_\mu u + e^u = 0$, $u \in H_0^1(\Omega)$, admit a solution for $\mu > C_H(\Omega)$ (cf. also Remark 2.1)?

2. Does (1.1) admit a large solution for $\mu \geqslant C_H(\Omega)$?

3. Does a solution of (1.1) exist with $u = \infty$ on Γ_∞ and $u = 0$ on Γ_0, where $\Gamma_\infty \cup \Gamma_0 = \partial\Omega$?

4. Does a large solution of (7.1) exist in the critical case $\beta = 2$?

Acknowledgments. Part of this work was supported by the Royal Society grant "Liouville theorems in nonlinear elliptic equations and systems." Part of the research was done while C.B. and V.M. were visiting the University of Karlsruhe. The authors would like to thank this institution.

References

1. Agmon, S.: On positivity and decay of solutions of second order elliptic equations on Riemannian manifolds. In: Methods of Functional Analysis and Theory of Elliptic Equations' (Naples, 1982), pp. 19–52. Liguori, Naples (1983)
2. Bandle, C., Moroz, V., Reichel, W.: "Boundary blowup" type sub-solutions to semilinear elliptic equations with Hardy potential. J. London Math. Soc. **77**, 503–523 (2008)
3. Bandle, C., Marcus, M.: Large solutions of semilinear elliptic equations: existence, uniqueness and asymptotic behaviors. J. Anal. Math. **58**, 9–24 (1992)
4. Bandle, C., Marcus, M.: Dependence of blowup rate of large solutions of semilinear elliptic equations, on the curvature of the boundary. Compl. Var. **49**, 555–570 (2004)
5. Davies, E.B.: A review of Hardy inequalities. In: The Maz'ya Anniversary Collection 2. Oper. Theory Adv. Appl. **110**, 55–67 (1999)
6. Filippas, S., Maz'ya, V., Tertikas A.: Critical Hardy-Sobolev inequalities. J. Math. Pures Appl. **87**, 37–56 (2007)
7. Keller, J.B.: On solutions of $\Delta u = f(u)$. Commun. Pure Appl. Math. **10**, 503–510 (1957)
8. Marcus, M., Mizel, V.J., Pinchover, Y.: On the best constant for Hardy's inequality in $\mathbb{R}^n$. Trans. Am. Math. Soc. **350**, 3237–3255 (1998)
9. Maz'ya, V.G.: Sobolev Spaces. Springer, Berlin etc. (1985)
10. Osserman, R.: On the inequality $\Delta u \geqslant f(u)$. Pacific J. Math. **7**, 1641–1647 (1957)
11. Stein, E.M.: Singular Integrals and Differentiability Properties of Functions. Princeton Univ. Press, Princeton, NJ (1970)

Stability Estimates for Resolvents, Eigenvalues, and Eigenfunctions of Elliptic Operators on Variable Domains

Gerassimos Barbatis, Victor I. Burenkov, and Pier Domenico Lamberti

Dedicated to Vladimir Maz'ya

Abstract We consider general second order uniformly elliptic operators subject to homogeneous boundary conditions on open sets $\phi(\Omega)$ parametrized by Lipschitz homeomorphisms ϕ defined on a fixed reference domain Ω. For two open sets $\phi(\Omega)$ and $\widetilde{\phi}(\Omega)$ we estimate the variation of resolvents, eigenvalues, and eigenfunctions via the Sobolev norm $\|\widetilde{\phi} - \phi\|_{W^{1,p}(\Omega)}$ for finite values of p, under natural summability conditions on eigenfunctions and their gradients. We prove that such conditions are satisfied for a wide class of operators and open sets, including open sets with Lipschitz continuous boundaries. We apply these estimates to control the variation of the eigenvalues and eigenfunctions via the measure of the symmetric difference of the open sets. We also discuss an application to the stability of solutions to the Poisson problem.

1 Introduction

This paper is devoted to the proof of stability estimates for the nonnegative selfadjoint operator

Gerassimos Barbatis
Department of Mathematics, University of Athens, 157 84 Athens, Greece e-mail: gbarbatis@math.uoa.gr

Victor I. Burenkov
Dipartimento di Matematica Pura ed Applicata, Università degli Studi di Padova, Via Trieste, 63, 35121 Padova, Italy e-mail: burenkov@math.unipd.it

Pier Domenico Lamberti
Dipartimento di Matematica Pura ed Applicata, Università degli Studi di Padova, Via Trieste, 63, 35121 Padova, Italy e-mail: lamberti@math.unipd.it

A. Laptev (ed.), *Around the Research of Vladimir Maz'ya II: Partial Differential Equations*, International Mathematical Series 12, DOI 10.1007/978-1-4419-1343-2_2,

$$Lu = -\sum_{i,j=1}^{N} \frac{\partial}{\partial x_i}\Big(A_{ij}(x)\frac{\partial u}{\partial x_j}\Big), \quad x \in \Omega, \tag{1.1}$$

subject to homogeneous boundary conditions, upon variation of the open set Ω in $\mathbf{R}^N$. Here, A_{ij} are fixed bounded measurable real-valued functions defined in $\mathbf{R}^N$ satisfying $A_{ij} = A_{ji}$ and a uniform ellipticity condition.

The focus is on explicit quantitative estimates for the variation of the resolvents, eigenvalues, and eigenfunctions of L on a class of open sets diffeomorphic to Ω.

In the first part of the paper, we consider two diffeomorphisms ϕ and $\widetilde{\phi}$ from Ω onto $\phi(\Omega)$ and $\widetilde{\phi}(\Omega)$ respectively, and we compare the resolvents, eigenvalues, and eigenfunctions of L on the open set $\widetilde{\phi}(\Omega)$ with those of L on $\phi(\Omega)$. To compare the operators, defined on different domains $\varphi(\Omega)$ and $\widetilde{\varphi}(\Omega)$, we compare their pull-backs to the same domain Ω (cf. Section 2). The main goal is to provide stability estimates via $\|\widetilde{\phi}-\phi\|_{W^{1,p}(\Omega)}$ for finite values of p. These estimates are applied in the last part of the paper where we take $\phi = Id$ and, given a deformation $\widetilde{\Omega}$ of Ω, construct a special diffeomorphism $\widetilde{\phi}$ representing $\widetilde{\Omega}$ in the form $\widetilde{\Omega} = \widetilde{\phi}(\Omega)$, and obtain stability estimates in terms of the Lebesgue measure $|\Omega \vartriangle \widetilde{\Omega}|$ of the symmetric difference of Ω and $\widetilde{\Omega}$.

Our method allows us to treat the general case of the mixed homogeneous Dirichlet–Neumann boundary conditions

$$u = 0 \quad \text{on } \Gamma \quad \text{and} \quad \sum_{i,j=1}^{N} A_{ij}\frac{\partial u}{\partial x_j}\nu_i = 0 \quad \text{on } \partial\Omega \setminus \Gamma, \tag{1.2}$$

where $\Gamma \subset \partial\Omega$ and ν denotes the exterior unit normal to $\partial\Omega$. To our knowledge, our results are new also for the Dirichlet and for Neumann boundary conditions.

There is vast literature concerning domain perturbation problems (cf., for example, the extensive monograph [14]). The problem of finding explicit quantitative estimates for the variation of the eigenvalues of elliptic operators has been considered in [3]–[6], [8]–[10], [16, 17, 21] (cf. [7] for a survey on the results of these papers). However, less attention has been devoted to the problem of finding explicit estimates for the variation of the eigenfunctions. With regard to this, we mention the estimate in [21] concerning the first eigenfunction of the Dirichlet Laplacian and the estimates in [16, 17] concerning the variation of the eigenprojections of the Dirichlet and Neumann Laplacian. In particular, in [16, 17], the variation of the eigenvalues and eigenprojections of the Laplace operator was estimated via $\|\nabla\widetilde{\phi} - \nabla\phi\|_{L^\infty(\Omega)}$ under minimal assumptions on the regularity of Ω, ϕ and $\widetilde{\phi}$.

In all the cited papers and in this paper, perturbations of domains may be considered as in some sense regular perturbations. There is also vast literature concerning a wide range of perturbation problems of different type which may be characterized as singular perturbations (which are out of scope of this paper). Typically, formulations of such problems involve a small parameter ε and the problem degenerates in that sense or other as $\varepsilon \to 0$. Say, the domain may contain small holes, or boundaries which may include blunted angles, cones and edges, narrow slits, thin bridges etc, or the limit region may consist of subsets of different dimension, or it could be a homogenization problem. V.G. Maz'ya and his co-authors V.A. Kozlov, A.B. Movchan, S.A. Nazarov, B.A. Plamenevskii and others developed the powerful asymptotic theory which allowed to find asymptotic expansions of solutions for all aforementioned problems and can be applied in many other cases (cf., for example, [15, 18]).

In this paper, we consider the same class of transformations $\phi, \widetilde{\phi}$ as in [16, 17] ($\phi, \widetilde{\phi}$ are bi-Lipschitz homeomorphisms) and, making stronger regularity assumptions on $\phi(\Omega)$ and $\widetilde{\phi}(\Omega)$, we estimate the variation of the resolvents, eigenvalues, eigenprojections, and eigenfunctions of L via the measure of vicinity

$$\delta_p(\phi, \widetilde{\phi}) := \|\nabla\widetilde{\phi} - \nabla\phi\|_{L^p(\Omega)} + \|A \circ \widetilde{\phi} - A \circ \phi\|_{L^p(\Omega)} \tag{1.3}$$

for any $p \in]p_0, \infty]$, where $A = (A_{ij})_{i,j=1,\dots,N}$ is the matrix of coefficients.

Here, $p_0 \geqslant 2$ is a constant depending on the regularity assumptions. The best p_0 that we obtain is $p_0 = N$ which corresponds to the highest degree of regularity (cf. Remark 4.8), while the case $p_0 = \infty$ corresponds to the lowest degree of regularity in which case only the exponent $p = \infty$ can be considered. The regularity assumptions are expressed in terms of summability properties of the eigenfunctions and their gradients, see Definition 4.2. Note that if the coefficients A_{ij} of the operator L are Lipschitz continuous, then $\delta_p(\phi, \widetilde{\phi})$ does not exceed a constant independent of $\phi, \widetilde{\phi}$ multiplied by the Sobolev norm $\|\phi - \widetilde{\phi}\|_{W^{1,p}(\Omega)}$. Moreover, if the coefficients A_{ij} are constant, then the second summand on the right-hand side of (1.3) vanishes.

More precisely, we prove stability estimates for the resolvents in the Schatten classes (Theorem 4.6), stability estimates for eigenvalues (Theorem 4.11), eigenprojections (Theorem 5.2), and eigenfunctions (Theorem 5.6). In Appendix, we also consider an application to the Poisson problem (we refer to [23] for stability estimates for the solutions to the Poisson problem in the case of the Dirichlet boundary conditions obtained by a different approach). To prove the resolvent stability estimates in the Schatten classes, we follow the method developed in [1, 2].

In Section 7, we apply our general results and, for a given deformation $\widetilde{\Omega}$ of Ω, we prove stability estimates in terms of $|\Omega \vartriangle \widetilde{\Omega}|$. This is done in two cases: the case where $\widetilde{\Omega}$ is obtained by a localized deformation of the boundary of Ω and the case where $\widetilde{\Omega}$ is a deformation of Ω along its normals. We also require that the deformation $\widetilde{\Gamma}$ of Γ is induced by the deformation of Ω (cf. conditions (7.3) and (7.14)). In these cases, similarly to [5], we can construct special bi-Lipschitz transformations $\widetilde{\phi} : \Omega \to \widetilde{\Omega}$ such that $\widetilde{\phi}(\Gamma) = \widetilde{\Gamma}$ and

$$\|\nabla\widetilde{\phi} - I\|_{L^p(\Omega)} \leqslant c|\Omega \vartriangle \widetilde{\Omega}|^{1/p}, \tag{1.4}$$

where $c > 0$ is independent of Ω and $\widetilde{\Omega}$. Observe that using finite values of p is essential, since in the case $p = \infty$ the exponent on the right-hand side of (1.4) vanishes.

"Let us describe these results in the regular case in which Ω and $\widetilde{\Omega}$ are of class $C^{1,1}$, Γ, $\widetilde{\Gamma}$ are connected components of the corresponding boundaries, and the coefficients A_{ij} are Lipschitz continuous. In Theorems 7.3 and 7.6, we prove that for any $r > N$ there exists a constant $c_1 > 0$ such that

$$\left(\sum_{n=1}^{\infty}\left|\frac{1}{\lambda_n+1} - \frac{1}{\widetilde{\lambda}_n+1}\right|^r\right)^{1/r} \leqslant c_1|\Omega \vartriangle \widetilde{\Omega}|^{\frac{1}{r}} \tag{1.5}$$

if $|\Omega \vartriangle \widetilde{\Omega}| < c_1^{-1}$. Here, λ_n, $\widetilde{\lambda}_n$ are the eigenvalues of the operators (1.1) corresponding to the domains Ω, $\widetilde{\Omega}$ and the associated portions of the boundaries Γ, $\widetilde{\Gamma}$ respectively. Moreover, for a fixed Ω and any $r > N$ there exists $c_2 > 0$ such that if $\lambda_n = \ldots = \lambda_{n+m-1}$ is an eigenvalue of multiplicity m, then for any choice of orthonormal eigenfunctions $\widetilde{\psi}_n, \ldots, \widetilde{\psi}_{n+m-1}$ corresponding to $\widetilde{\lambda}_n, \ldots, \widetilde{\lambda}_{n+m-1}$, there exist orthonormal eigenfunctions $\psi_n, \ldots, \psi_{n+m-1}$ corresponding to $\lambda_n, \ldots, \lambda_{n+m-1}$ such that[1]

$$\|\psi_k - \widetilde{\psi}_k\|_{L^2(\Omega\cup\widetilde{\Omega})} \leqslant c_2|\Omega \vartriangle \widetilde{\Omega}|^{\frac{1}{r}} \tag{1.6}$$

for all $k = n, \ldots, n+m-1$ provided that $|\Omega \vartriangle \widetilde{\Omega}| < c_2^{-1}$. Here, it is understood that the eigenfunctions are extended by zero outside their domains of definition.

In the general case of open sets Ω, $\widetilde{\Omega}$ with Lipschitz continuous boundaries and Γ, $\widetilde{\Gamma}$ with Lipschitz continuous boundaries in $\partial\Omega$, $\partial\widetilde{\Omega}$, our statements still hold for a possibly worse range of exponents (cf. Theorems 7.3 and 7.6).

We emphasize that, in the spirit of [16, 17], in this paper we never assume that the transformation ϕ belongs to a family of transformations ϕ_t depending analytically on one scalar parameter t, as often done in the literature (cf., for example, [14] for references). In that case, one can use proper methods of bifurcation theory in order to prove existence of branches of eigenvalues and eigenfunctions depending analytically on t. In this paper, $\widetilde{\phi}$ is an arbitrary perturbation of ϕ and this requires a totally different approach.

The paper is organized as follows. In Section 2, we describe the general setting. In Section 3, we describe our perturbation problem. In Section 4, we prove stability estimates for the resolvents and the eigenvalues. In Section 5, we prove stability estimates for the eigenprojections and eigenfunctions. In Section 6, we give sufficient conditions providing the required regularity of

---

[1] Note that, for a fixed Ω and variable $\widetilde{\Omega}$, one first chooses eigenfunctions in $\widetilde{\Omega}$ and then finds eigenfunctions in Ω, while the opposite is clearly not possible.

the eigenfunctions. In Section 7, we prove stability estimates via the Lebesgue measure of the symmetric difference of sets. In Appendix, we briefly discuss the Poisson problem.

2 General Setting

Let Ω be a domain, i.e., an open connected set, in $\mathbf{R}^N$ of finite measure. We consider a family of open sets $\phi(\Omega)$ in $\mathbf{R}^N$ parametrized by bi-Lipschitz homeomorphisms ϕ of Ω onto $\phi(\Omega)$. Namely, following [16], we consider the family of transformations

$$\Phi(\Omega) := \Big\{\phi \in \left(L^{1,\infty}(\Omega)\right)^N : \text{the continuous representative of } \phi \\ \text{is injective, } \operatorname*{ess\,inf}_{\Omega} |\det \nabla\phi| > 0\Big\}, \quad (2.1)$$

where $L^{1,\infty}(\Omega)$ denotes the space of the functions in $L^1_{loc}(\Omega)$ which have weak derivatives of first order in $L^\infty(\Omega)$. Note that if $\phi \in \Phi(\Omega)$, then ϕ is Lipschitz continuous with respect to the geodesic distance in Ω.

Note that if $\phi \in \Phi(\Omega)$, then $\phi(\Omega)$ is open, ϕ is a homeomorphism of Ω onto $\phi(\Omega)$, and the inverse vector-function $\phi^{(-1)}$ of ϕ belongs to $\Phi(\phi(\Omega))$. Moreover, any transformation $\phi \in \Phi(\Omega)$ allows changing variables in integrals. Accordingly, the operator C_ϕ from $L^2(\phi(\Omega))$ to $L^2(\Omega)$ defined by

$$C_\phi[v] := v \circ \phi, \quad v \in L^2(\phi(\Omega)),$$

is a linear homeomorphism which restricts to a linear homeomorphism of the space $W^{1,2}(\phi(\Omega))$ onto $W^{1,2}(\Omega)$, and of $W^{1,2}_0(\phi(\Omega))$ onto $W^{1,2}_0(\Omega)$, where $W^{1,2}(\Omega)$ denotes the standard Sobolev space and $W^{1,2}_0(\Omega)$ denotes the closure of $C^\infty_c(\Omega)$ in $W^{1,2}(\Omega)$. Furthermore, $\nabla(v \circ \phi) = \nabla v(\phi)\nabla\phi$ for all $v \in W^{1,2}(\phi(\Omega))$. Note that if $\phi \in \Phi(\Omega)$, then the measure of $\phi(\Omega)$ is finite (cf. [16] for details).

Let $A = (A_{ij})_{i,j=1,\dots,N}$ be a real symmetric matrix-valued measurable function defined on $\mathbf{R}^N$ such that for some $\theta > 0$

$$\theta^{-1}|\xi|^2 \leqslant \sum_{i,j=1}^N A_{ij}(x)\xi_i\xi_j \leqslant \theta|\xi|^2 \quad (2.2)$$

for all $x, \xi \in \mathbf{R}^N$. Note that (2.2) implies that $A_{ij} \in L^\infty(\mathbf{R}^N)$ for all $i, j = 1, \dots, N$.

Let $\phi \in \Phi(\Omega)$, and let $\mathcal{W}$ be a closed subspace of $W^{1,2}(\phi(\Omega))$ containing $W^{1,2}_0(\phi(\Omega))$. We consider the nonnegative selfadjoint operator L on $L^2(\phi(\Omega))$ canonically associated with the sesquilinear form Q_L given by

$$\mathrm{Dom}(Q_L) = \mathcal{W}\,, \quad Q_L(v_1, v_2) = \int_{\phi(\Omega)} \sum_{i,j=1}^{N} A_{ij} \frac{\partial v_1}{\partial y_i} \frac{\partial \overline{v}_2}{\partial y_j} dy\,, \quad v_1, v_2 \in \mathcal{W}. \tag{2.3}$$

Recall that $v \in \mathrm{Dom}(L)$ if and only if $v \in \mathcal{W}$ and there exists $f \in L^2(\phi(\Omega))$ such that

$$Q_L(v, \psi) = \langle f, \psi \rangle_{L^2(\phi(\Omega))} \tag{2.4}$$

for all $\psi \in \mathcal{W}$, in which case $Lv = f$ (cf., for example, [11]). The choice of the space $\mathcal{W}$ determines the boundary conditions. For example, if $\mathcal{W} = W_0^{1,2}(\phi(\Omega))$ (respectively, $\mathcal{W} = W^{1,2}(\phi(\Omega))$), then the operator L satisfies the homogeneous Dirichlet (respectively, Neumann) boundary conditions.

We also consider the operator H on $L^2(\Omega)$ obtained by pulling-back L to $L^2(\Omega)$ as follows. Let $v \in W^{1,2}(\phi(\Omega))$ be given, and let $u = v \circ \phi$. Note that

$$\int_{\phi(\Omega)} |v|^2 dy = \int_{\Omega} |u|^2 |\det \nabla \phi|\, dx\,.$$

Moreover, a simple computation shows that

$$\int_{\phi(\Omega)} \sum_{i,j=1}^{N} A_{ij} \frac{\partial v}{\partial y_i} \frac{\partial \overline{v}}{\partial y_j} dy = \int_{\Omega} \sum_{i,j=1}^{N} a_{ij} \frac{\partial u}{\partial x_i} \frac{\partial \overline{u}}{\partial x_j} |\det \nabla \phi|\, dx\,,$$

where $a = (a_{ij})_{i,j=1,\dots,N}$ is the symmetric matrix-valued function defined in Ω by

$$a_{ij} = \sum_{r,s=1}^{N} \left(A_{rs} \frac{\partial \phi_i^{(-1)}}{\partial y_r} \frac{\partial \phi_j^{(-1)}}{\partial y_s} \right) \circ \phi = ((\nabla \phi)^{-1} A(\phi) (\nabla \phi)^{-t})_{ij}\,. \tag{2.5}$$

The operator H is defined as the nonnegative selfadjoint operator on the Hilbert space $L^2(\Omega, |\det \nabla \phi|\, dx)$ associated with the sesquilinear form Q_H given by

$$\begin{aligned} \mathrm{Dom}(Q_H) &= C_\phi[\mathcal{W}],\ Q_H(u_1, u_2) \\ &= \int_{\Omega} \sum_{i,j=1}^{N} a_{ij} \frac{\partial u_1}{\partial x_i} \frac{\partial \overline{u}_2}{\partial x_j} |\det \nabla \phi|\, dx,\ u_1, u_2 \in C_\phi[\mathcal{W}]. \end{aligned}$$

Formally,

$$Hu = -\frac{1}{|\det \nabla \phi|} \sum_{i,j=1}^{N} \frac{\partial}{\partial x_j} \left(a_{ij} \frac{\partial u}{\partial x_i} |\det \nabla \phi| \right).$$

Alternatively, the operator H can be defined as

$$H = C_\phi L C_{\phi^{(-1)}}.$$

In particular, H and L are unitarily equivalent and the operator H has compact resolvent if and only if L has compact resolvent. (Note that the embedding $\mathcal{W} \subset L^2(\phi(\Omega))$ is compact if and only if the embedding $C_\phi[\mathcal{W}] \subset L^2(\Omega)$ is compact.)

We set $g(x) := |\det \nabla\phi(x)|$, $x \in \Omega$, and denote by $\langle\cdot,\cdot\rangle_g$ the inner product in $L^2(\Omega, g\,dx)$ and also in $(L^2(\Omega, g\,dx))^N$.

Let $T : L^2(\Omega, g\,dx) \to (L^2(\Omega, g\,dx))^N$ be the operator defined by

$$\mathrm{Dom}(T) = C_\phi[\mathcal{W}], \qquad Tu = a^{1/2}\nabla u, \quad u \in C_\phi[\mathcal{W}].$$

Then it is easy to see that

$$H = T^*T,$$

where the adjoint T^* of T is understood with respect to the inner products of $L^2(\Omega, g\,dx)$ and $(L^2(\Omega, g\,dx))^N$.

3 Perturbation of ϕ

In this paper, we study the variation of the operator L defined by (2.3) upon variation of ϕ. Our estimates depend on $\operatorname{ess\,inf}_\Omega |\det \nabla\phi|$ and $\|\nabla\phi\|_{L^\infty(\Omega)}$. Thus, in order to obtain uniform estimates it is convenient to consider the families of transformations

$$\Phi_\tau(\Omega) = \left\{\phi \in \Phi(\Omega) : \ \tau^{-1} \leqslant \operatorname*{ess\,inf}_{\Omega} |\det \nabla\phi| \ \text{ and } \ \|\nabla\phi\|_{L^\infty(\Omega)} \leqslant \tau\right\}$$

for all $\tau > 0$, as in [16]. Hereinafter, for a matrix-valued function $B(x)$, $x \in \Omega$, we set $\|B\|_{L^p(\Omega)} = \| \, |B| \, \|_{L^p(\Omega)}$, where $|B(x)|$ denotes the operator norm of the matrix $B(x)$.

Let $\phi, \widetilde{\phi} \in \Phi_\tau(\Omega)$. Let $\mathcal{W}$ and $\widetilde{\mathcal{W}}$ be closed subspaces of $W^{1,2}(\phi(\Omega))$, $W^{1,2}(\widetilde{\phi}(\Omega))$ respectively, containing $W_0^{1,2}(\phi(\Omega))$, $W_0^{1,2}(\widetilde{\phi}(\Omega))$ respectively. We use tildes to distinguish objects induced by $\widetilde{\phi}$, $\widetilde{\mathcal{W}}$ from those induced by ϕ, $\mathcal{W}$. We consider the operators L and $\widetilde{L}$ defined on $L^2(\phi(\Omega))$, $L^2(\widetilde{\phi}(\Omega))$ respectively, as in Section 2.

In order to compare L and $\widetilde{L}$, we make a "compatibility" assumption on the respective boundary conditions; namely, we assume that

$$C_\phi[\mathcal{W}] = C_{\widetilde{\phi}}[\widetilde{\mathcal{W}}]. \tag{3.1}$$

This means that $\mathrm{Dom}(Q_H) = \mathrm{Dom}(Q_{\widetilde{H}})$, a property which is important in what follows. It is clear that (3.1) holds if either L and $\widetilde{L}$ both satisfy the homogeneous Dirichlet boundary conditions or they both satisfy homogeneous Neumann boundary conditions.

We always assume that the spaces $\mathcal{W}$, $\widetilde{\mathcal{W}}$ are compactly embedded in $L^2(\phi(\Omega))$, $L^2(\widetilde{\phi}(\Omega))$ respectively, or equivalently that the space $\mathcal{V} := C_\phi[\mathcal{W}] = C_{\widetilde{\phi}}[\widetilde{\mathcal{W}}]$ is compactly embedded in $L^2(\Omega)$.

Moreover, we require that the nonzero eigenvalues λ_n of the Laplace operator associated in $L^2(\Omega)$ with the quadratic form $\int_\Omega |\nabla u|^2 dx$, $u \in \mathcal{V}$, defined on $\mathcal{V}$, satisfy the condition

$$c^* := \sum_{\lambda_n \neq 0} \lambda_n^{-\alpha} < \infty \tag{3.2}$$

for some fixed $\alpha > 0$. (This is in fact a very weak condition on the regularity of the set Ω and the associated boundary conditions.)

For brevity, we refer to assumption (A) as the following set of conditions which summarize the setting described above:

$$\text{(A)}: \quad \begin{cases} \phi, \widetilde{\phi} \in \Phi_\tau(\Omega), \\ \mathcal{V} := C_\phi[\mathcal{W}] = C_{\widetilde{\phi}}[\widetilde{\mathcal{W}}] \ \text{ is compactly embedded in } L^2(\Omega), \\ \text{condition (3.2) holds}\ . \end{cases}$$

Remark 3.1. We note that if Ω is a domain of class $C^{0,1}$, i.e., Ω is locally a subgraph of Lipschitz continuous functions, then the inequality (3.2) holds for any $\alpha > N/2$ (cf., for example, [3, 20] and also Remark 6.5 below). We also note that by the Min-Max Principle [11, p. 5] and by comparing with the Dirichlet Laplacian on a ball contained in Ω, the condition (3.2) does not hold for $\alpha \leqslant N/2$ (no matter whether Ω is regular or not).

To compare L and $\widetilde{L}$, we compare the respective pull-backs H and $\widetilde{H}$. Since they act on different Hilbert spaces – $L^2(\Omega, g\,dx)$ and $L^2(\Omega, \widetilde{g}\,dx)$ – we use the canonical unitary operator

$$w : L^2(\Omega, g\,dx) \longrightarrow L^2(\Omega, \widetilde{g}\,dx)\ , \qquad u \mapsto wu\,,$$

defined as the multiplication by the function $w := g^{1/2}\widetilde{g}^{-1/2}$. We also introduce the multiplication operator S on $(L^2(\Omega))^N$ by the symmetric matrix

$$w^{-2} a^{-1/2} \widetilde{a} a^{-1/2}\,, \tag{3.3}$$

where the matrix a is defined by (2.5) and the matrix $\widetilde{a}$ is defined in the same way with $\widetilde{\phi}$ replacing ϕ. If there is no ambiguity, we also denote by S the matrix (3.3).

As we see in the sequel, in order to compare H and $\widetilde{H}$, we need an auxiliary operator. Namely, we consider the operator T^*ST, which is the nonnegative selfadjoint operator in $L^2(\Omega, g\,dx)$ canonically associated with the sesquilinear form

$$\int_{\Omega} (\widetilde{a}\nabla u_1 \cdot \nabla \overline{u}_2)\widetilde{g}dx\,, \quad u_1, u_2 \in \mathcal{V}\,.$$

It is easily seen that the operator T^*ST is the pull-back to Ω via $\widetilde{\phi}$ of the operator

$$\widehat{L} := \frac{\widetilde{g} \circ \widetilde{\phi}^{(-1)}}{g \circ \widetilde{\phi}^{(-1)}} \widetilde{L} \tag{3.4}$$

defined on $L^2(\widetilde{\phi}(\Omega))$. Thus, in the sequel, we deal with the operators $L, \widetilde{L}$ and $\widehat{L}$ and the respective pull-backs H, $\widetilde{H}$, and T^*ST. We repeatedly use the fact that these are pairwise unitarily equivalent.

We denote by $\lambda_n[E]$, $n \in \mathbf{N}$, the eigenvalues of a nonnegative selfadjoint operator E with compact resolvent, arranged in nondecreasing order and repeated according to multiplicity, and by $\psi_n[E]$, $n \in \mathbf{N}$, a corresponding orthonormal sequence of eigenfunctions.

Lemma 3.2. *Let* (A) *be satisfied. Then the operators* L, $\widetilde{L}$, $\widehat{L}$, H, $\widetilde{H}$, *and* T^*ST *have compact resolvents and the corresponding nonzero eigenvalues satisfy the inequality*

$$\sum_{\lambda_n[E]\neq 0} \lambda_n[E]^{-\alpha} \leqslant cc^* \tag{3.5}$$

for $E = L$, $\widetilde{L}$, $\widehat{L}$, H, $\widetilde{H}$, T^*ST, *where* c *depends only on* N, τ, *and* θ.

Proof. We prove the inequality (3.5) only for $E = T^*ST$, the other cases being similar. Note that the Rayleigh quotient corresponding to T^*ST is given by

$$\frac{\langle T^*STu, u\rangle_g}{\langle u, u\rangle_g} = \frac{\langle STu, Tu\rangle_g}{\langle u, u\rangle_g} = \frac{\displaystyle\int_{\Omega} (\widetilde{a}\nabla u \cdot \nabla \overline{u})\widetilde{g}dx}{\displaystyle\int_{\Omega} |u|^2 g dx}, \qquad u \in \mathcal{V}.$$

Then the inequality (3.5) easily follows by observing that

$$\widetilde{a}\nabla u \cdot \nabla \overline{u} \geqslant \theta^{-1}|(\nabla\widetilde{\phi})^{-1}\nabla u|^2 \geqslant \theta^{-1}\tau^{-2}|\nabla u|^2,$$

$$|\mathrm{det}\nabla\phi| \leqslant N!|\nabla\phi|^N \tag{3.6}$$

and using the Min-Max Principle [11, p. 5]. □

4 Stability Estimates for the Resolvents and Eigenvalues

The following lemma is based on the well-known commutation formula (4.3) (cf. [12]). We denote by $\sigma(E)$ the spectrum of an operator E.

Lemma 4.1. *Let* (A) *be satisfied. Then for all* $\xi \in \mathbf{C} \setminus (\sigma(H) \cup \sigma(\widetilde{H}) \cup \sigma(T^*ST))$

$$(w^{-1}\widetilde{H}w - \xi)^{-1} - (H - \xi)^{-1} = A_1 + A_2 + A_3 + B\,, \tag{4.1}$$

where

$$A_1 = (1 - w)(wT^*STw - \xi)^{-1},$$

$$\begin{aligned} A_2 &= w(wT^*STw - \xi)^{-1}(1 - w), \\ A_3 &= -\xi(T^*ST - \xi)^{-1}(w - w^{-1})(wT^*STw - \xi)^{-1}w, \\ B &= T^*S^{1/2}(S^{1/2}TT^*S^{1/2} - \xi)^{-1}S^{1/2}(S^{-1} - I)(TT^* - \xi)^{-1}T\,. \end{aligned}$$

Proof. It suffices to prove (4.1) for $\xi \neq 0$ since the case $\xi = 0 \notin \sigma(H) \cup \sigma(\widetilde{H}) \cup \sigma(T^*ST)$ is then obtained by letting $\xi \to 0$.

Recall that $T^*T = H$. Similarly $\widetilde{T}^{\widetilde{*}}\widetilde{T} = \widetilde{H}$, where we have emphasized the dependence of the adjoint operation on the specific inner-product used. In this respect we note that the two adjoints of an operator E are related by the conjugation relation $E^{\widetilde{*}} = w^2E^*w^{-2}$. This allows us to use only $*$ and not $\widetilde{*}$.

Note that

$$\widetilde{H} = (\widetilde{a}^{1/2}\nabla)^{\widetilde{*}}\widetilde{a}^{1/2}\nabla = w^2(\widetilde{a}^{1/2}\nabla)^*w^{-2}\widetilde{a}^{1/2}\nabla = w^2T^*ST\,. \tag{4.2}$$

Therefore, by simple computations, we obtain

$$\begin{aligned} &(w^{-1}\widetilde{H}w - \xi)^{-1} - (H - \xi)^{-1} \\ &= w^{-1}(\widetilde{H} - \xi)^{-1}w - (H - \xi)^{-1} \\ &= w^{-1}(w^2T^*ST - \xi)^{-1}w - (T^*T - \xi)^{-1} \\ &= w^{-1}(w^2T^*ST - \xi)^{-1}w - (T^*ST - \xi w^{-2})^{-1}w^{-1} \\ &\quad +(T^*ST - \xi w^{-2})^{-1}w^{-1} - (T^*ST - \xi w^{-2})^{-1} \\ &\quad +(T^*ST - \xi w^{-2})^{-1} - (T^*ST - \xi)^{-1} \\ &\quad +(T^*ST - \xi)^{-1} - (T^*T - \xi)^{-1} \\ &= A_1 + A_2 + A_3 + ((T^*ST - \xi)^{-1} - (T^*T - \xi)^{-1})\,. \end{aligned}$$

To compute the last term we use the commutation formula

$$-\xi(E^*E - \xi)^{-1} + E^*(EE^* - \xi)^{-1}E = I \tag{4.3}$$

which holds for any closed and densely defined operator E (cf. [12]). We write (4.3) first for $E = T$, then for $E = S^{1/2}T$, and then we subtract the two relations. After some simple calculations we obtain $(T^*ST - \xi)^{-1} - (T^*T - \xi)^{-1} = B$, as required. □

We now introduce a regularity property which is important for our estimates. Sufficient conditions for its validity are given in Section 6.

Definition 4.2. Let U be an open set in $\mathbf{R}^N$, and let E be a nonnegative selfadjoint operator on $L^2(U)$ with compact resolvent and $\mathrm{Dom}(E) \subset W^{1,2}(U)$. We say that E satisfies property (P) if there exist $q_0 > 2$, $\gamma \geqslant 0$, $C > 0$ such that the eigenfunctions $\psi_n[E]$ of E satisfy the following conditions:

$$\|\psi_n[E]\|_{L^{q_0}(U)} \leqslant C\lambda_n[E]^{\gamma} \tag{P1}$$

and

$$\|\nabla\psi_n[E]\|_{L^{q_0}(U)} \leqslant C\lambda_n[E]^{\gamma+\frac{1}{2}} \tag{P2}$$

for all $n \in \mathbf{N}$ such that $\lambda_n[E] \neq 0$.

Remark 4.3. It is known that if Ω, A_{ij} and Γ are sufficiently smooth, then for the operator L in (1.1), subject to the boundary conditions (1.2), property (P) is satisfied with $q_0 = \infty$ and $\gamma = N/4$ (cf. Theorem 6.3 and the proof of Theorem 7.3).

By interpolation, it follows that if conditions (P1) and (P2) are satisfied, then

$$\begin{aligned} \|\psi_n[E]\|_{L^q(U)} &\leqslant C^{\frac{q_0(q-2)}{q(q_0-2)}}\lambda_n[E]^{\frac{q_0(q-2)\gamma}{q(q_0-2)}}, \\ \|\nabla\psi_n[E]\|_{L^q(U)} &\leqslant C^{\frac{q_0(q-2)}{q(q_0-2)}}\lambda_n[E]^{\frac{1}{2}+\frac{q_0(q-2)\gamma}{q(q_0-2)}} \end{aligned} \tag{4.4}$$

for all $q \in [2, q_0]$.

In the sequel, we require that property (P) is satisfied by the operators H, $\widetilde{H}$ and T^*ST which, according to the following lemma, is equivalent to requiring that property (P) is satisfied by the operators L, $\widetilde{L}$ and $\widehat{L}$ respectively.

Lemma 4.4. *Let* (A) *be satisfied. Then the operators H, $\widetilde{H}$, and T^*ST respectively, satisfy property* (P) *for some $q_0 > 2$ and $\gamma \geqslant 0$ if and only if the operators L, $\widetilde{L}$, and $\widehat{L}$ respectively, satisfy property* (P) *for the same q_0 and γ.*

Let E be a nonnegative selfadjoint operator on a Hilbert space the spectrum of which consists of isolated positive eigenvalues of finite multiplicity and may also contain zero as an eigenvalue of possibly infinite multiplicity. Let $s > 0$. For a function $g : \sigma(E) \to \mathbf{C}$ we define

$$\begin{aligned} |g(E)|_{p,s} &= \Big(\sum_{\lambda_n[E]\neq 0} |g(\lambda_n[E])|^p\lambda_n[E]^s\Big)^{1/p}, \ 1 \leqslant p < \infty, \\ |g(E)|_{\infty,s} &= \sup_{\lambda_n[E]\neq 0} |g(\lambda_n[E])|, \end{aligned}$$

where, as usual, each positive eigenvalue is repeated according to its multiplicity.

The next lemma involves the Schatten norms $\|\cdot\|_{\mathcal{C}^r}$, $1 \leqslant r \leqslant \infty$. For a compact operator E on a Hilbert space they are defined by $\|E\|_{\mathcal{C}^r} = (\sum_n \mu_n(E)^r)^{1/r}$, if $r < \infty$, and $\|E\|_{\mathcal{C}^\infty} = \|E\|$, where $\mu_n(E)$ are the singular values of E, i.e., the nonzero eigenvalues of $(E^*E)^{1/2}$; recall that the Schatten space $\mathcal{C}^r$, defined as the space of those compact operators for which the Schatten norm $\|\cdot\|_{\mathcal{C}^r}$ is finite, is a Banach space (cf. [22] or [24] for details).

Let $F := TT^*$. Recall that $\sigma(F) \setminus \{0\} = \sigma(H) \setminus \{0\}$ (cf. [12]). In the next lemma, $g(H)$ and $g(F)$ are operators defined in the standard way by functional calculus. The following lemma is a variant of Lemma 8 of [2].

Lemma 4.5. *Let $q_0 > 2$, $\gamma \geqslant 0$, $p \geqslant q_0/(q_0-2)$, $2 \leqslant r < \infty$ and $s = 2q_0\gamma/[p(q_0-2)]$. Then the following statements hold.*

(i) *If the eigenfunctions of H satisfy* (P1), *then for any measurable function $R : \Omega \to \mathbf{C}$ and function $g : \sigma(H) \to \mathbf{C}$ we have*

$$\|Rg(H)\|_{\mathcal{C}^r} \leqslant \|R\|_{L^{pr}(\Omega)} \Big(|\Omega|^{-\frac{1}{pr}} |g(0)| + C^{\frac{2q_0}{pr(q_0-2)}} |g(H)|_{r,s} \Big) . \tag{4.5}$$

(ii) *If the eigenfunctions of H satisfy* (P2), *then for any measurable matrix-valued function R in Ω and function $g : \sigma(F) \to \mathbf{C}$ such that if $0 \in \sigma(F)$, then $g(0) = 0$, we have*

$$\|Rg(F)\|_{\mathcal{C}^r} \leqslant C^{\frac{2q_0}{pr(q_0-2)}} \|a\|^{\frac{1}{r}}_{L^\infty(\Omega)} \|R\|_{L^{pr}(\Omega)} |g(F)|_{r,s} . \tag{4.6}$$

Proof. We only prove statement (ii) since the proof of (i) is simpler. It is enough to consider the case where R is bounded and g has finite support: the general case then follows by approximating R in $\|\cdot\|_{L^{pr}(\Omega)}$ by a sequence R_n, $n \in \mathbf{N}$, of bounded matrix-valued functions and g in $|\cdot|_{r,s}$ by a sequence g_n, $n \in \mathbf{N}$, of functions with finite support, and observing that, by (4.6), the sequence $R_n g_n(F)$, $n \in \mathbf{N}$, is then a Cauchy sequence in $\mathcal{C}^r$.

Since R is bounded and g has finite support, $Rg(F)$ is compact. Hence the inequality (4.6) is trivial for $r = \infty$. Thus, it is enough to prove (4.6) for $r = 2$ since the general case follows by interpolation (cf. [24]). It is easily seen that $z_n := T\psi_n[H]/\|T\psi_n[H]\| = \lambda_n[H]^{-1/2} T\psi_n[H]$, for all $n \in \mathbf{N}$ such that $\lambda_n[H] \neq 0$, are orthonormal eigenfunctions of F, $Fz_n = \lambda_n[H] z_n$, $n \in \mathbf{N}$, and $\operatorname{span}\{z_n\} = \operatorname{Ker}(F)^\perp$. Since $g(0) = 0$,

$$\|Rg(F)\|^2_{\mathcal{C}^2} = \sum_{n=1}^\infty \|Rg(F)z_n\|^2_{L^2(\Omega)} = \sum_{n=1}^\infty |g(\lambda_n[H])|^2 \|Rz_n\|^2_{L^2(\Omega)}$$

$$= \sum_{n=1}^\infty \lambda_n[H]^{-1} |g(\lambda_n[H])|^2 \|Ra^{1/2}\nabla\psi_n[H]\|^2_{L^2(\Omega)}$$

$$\leqslant \|a^{1/2}\|^2_{L^\infty(\Omega)}\|R\|^2_{L^{2p}(\Omega)}\sum_{n=1}^{\infty}\lambda_n[H]^{-1}|g(\lambda_n[H])|^2\|\nabla\psi_n[H]\|^2_{L^{2p/(p-1)}(\Omega)}$$

$$\leqslant C^{\frac{2q_0}{p(q_0-2)}}\|a^{1/2}\|^2_{L^\infty(\Omega)}\|R\|^2_{L^{2p}(\Omega)}\sum_{n=1}^{\infty}|g(\lambda_n[H])|^2\lambda_n[H]^{\frac{2q_0\gamma}{p(q_0-2)}}, \tag{4.7}$$

where for the last inequality we used (4.4). This proves (4.6) for $r = 2$, thus completing the proof of the lemma. □

Recall that $\delta_p(\phi, \widetilde{\phi})$, $1 \leqslant p \leqslant \infty$, is defined in (1.3).

Theorem 4.6 (stability of resolvents). *Let* (A) *be satisfied. Let* $\xi \in \mathbf{C} \setminus (\sigma(H) \cup \{0\})$. *Then the following statements hold.*

(i) *There exists* $c_1 > 0$ *depending only on* N, τ, θ, α, c^*, *and* ξ *such that if* $\delta_\infty(\phi, \widetilde{\phi}) \leqslant c_1^{-1}$, *then* $\xi \notin \sigma(\widetilde{H})$ *and*

$$\|(w^{-1}\widetilde{H}w - \xi)^{-1} - (H - \xi)^{-1}\|_{\mathcal{C}^\alpha} \leqslant c_1\delta_\infty(\phi, \widetilde{\phi}). \tag{4.8}$$

(ii) *Let, in addition,* (P) *be satisfied by the operators* H, $\widetilde{H}$, *and* T^*ST *for the same* q_0, γ, *and* C. *Let* $p \geqslant q_0/(q_0 - 2)$ *and* $r \geqslant \max\{2, \alpha + \frac{2q_0\gamma}{p(q_0-2)}\}$. *Then there exists* $c_2 > 0$ *depending only on* N, τ, θ, α, c^*, r, p, q_0, C, γ, $|\Omega|$, *and* ξ *such that if* $\delta_{pr}(\phi, \widetilde{\phi}) \leqslant c_2^{-1}$, *then* $\xi \notin \sigma(\widetilde{H})$ *and*

$$\|(w^{-1}\widetilde{H}w - \xi)^{-1} - (H - \xi)^{-1}\|_{\mathcal{C}^r} \leqslant c_2\delta_{pr}(\phi, \widetilde{\phi}). \tag{4.9}$$

Remark 4.7. Let $s = [q_0/(q_0 - 2)]\max\{2, \alpha + 2\gamma\}$. It follows by Theorem 4.6 (ii) (choosing $p = q_0/(q_0 - 2)$) that if $\delta_s(\phi, \widetilde{\phi}) \leqslant c_2^{-1}$, then $\xi \notin \sigma(\widetilde{H})$ and

$$\|(w^{-1}\widetilde{H}w - \xi)^{-1} - (H - \xi)^{-1}\| \leqslant c_2\delta_s(\phi, \widetilde{\phi}). \tag{4.10}$$

Remark 4.8. As we will see in Section 7 below, the best range for s in (4.10) used in our applications is $s > N$; this corresponds to the case where $q_0 = \infty$, $\gamma = N/4$, and $\alpha > N/2$ (cf. Remarks 3.1 and 4.3).

Proof of Theorem 4.6. In this proof, we denote by c a positive constant depending only on N, τ, θ, α, and c^* the value of which may change along the proof. When dealing with statement (ii) constant c may depend also on r, p, q_0, C, γ, and $|\Omega|$. We divide the proof into two steps.

Step 1. We assume first that $\xi \notin \sigma(\widetilde{H}) \cup \sigma(T^*ST)$ and set

$$d_\sigma(\xi) = \mathrm{dist}(\xi, \sigma(H) \cup \sigma(\widetilde{H}) \cup \sigma(T^*ST)).$$

At this first step, we prove (4.8) and (4.9) without any smallness assumptions on $\delta_\infty(\phi,\widetilde{\phi})$ and $\delta_{pr}(\phi,\widetilde{\phi})$ respectively.

We first prove (4.8). We use Lemma 4.1, and to do so we first estimate the terms A_1, A_2, A_3 in the identity (4.1). It is clear that

$$\frac{\lambda_n[\widetilde{H}]}{|\lambda_n[\widetilde{H}]-\xi|} \leqslant \left(1+\frac{|\xi|}{d(\xi,\sigma(\widetilde{H}))}\right). \tag{4.11}$$

Since the eigenvalues of the operator wT^*STw coincide with the eigenvalues of $\widetilde{H}$ (cf. (4.2)), it follows that

$$\begin{aligned}\|(wT^*STw-\xi)^{-1}\|_{\mathcal{C}^\alpha}^\alpha &= \sum_{n=1}^\infty \frac{1}{|\lambda_n[\widetilde{H}]-\xi|^\alpha}\\ \leqslant \frac{1}{|\xi|^\alpha} &+ \left(1+\frac{|\xi|}{d(\xi,\sigma(\widetilde{H}))}\right)^\alpha \sum_{\lambda_n[\widetilde{H}]\neq 0} \lambda_n[\widetilde{H}]^{-\alpha}\\ = \frac{1}{|\xi|^\alpha} &+ c\left(1+\frac{|\xi|}{d(\xi,\sigma(\widetilde{H}))}\right)^\alpha.\end{aligned} \tag{4.12}$$

Taking into account (3.6) and observing that

$$\left|\det\nabla\phi-\det\nabla\widetilde{\phi}\right| \leqslant N!N\left|\nabla\phi-\nabla\widetilde{\phi}\right| \max\left\{\left|\nabla\phi\right|,\left|\nabla\widetilde{\phi}\right|\right\}^{N-1}, \tag{4.13}$$

we find

$$|1-w|, |w-w^{-1}| \leqslant c|\nabla\phi-\nabla\widetilde{\phi}|. \tag{4.14}$$

Combining the inequalities (4.12) and (4.14), we obtain

$$\begin{aligned}\|A_1\|_{\mathcal{C}^\alpha}, \|A_2\|_{\mathcal{C}^\alpha} &\leqslant c\left(1+\frac{1}{|\xi|}+\frac{|\xi|}{d_\sigma(\xi)}\right)\|\nabla\phi-\nabla\widetilde{\phi}\|_{L^\infty(\Omega)},\\ \|A_3\|_{\mathcal{C}^\alpha} &\leqslant c\left(\frac{1+|\xi|}{d_\sigma(\xi)}+\frac{|\xi|^2}{d_\sigma(\xi)^2}\right)\|\nabla\phi-\nabla\widetilde{\phi}\|_{L^\infty(\Omega)}.\end{aligned} \tag{4.15}$$

We now estimate the term B in (4.1). We recall that $F=TT^*$ and set $F_S = S^{1/2}TT^*S^{1/2}$. Then, by polar decomposition, there exist partial isometries $Y, Y_S : L^2(\Omega, g\,dx) \to (L^2(\Omega, g\,dx))^N$ such that $T = F^{1/2}Y$ and $S^{1/2}T = F_S^{1/2}Y_S$. We have

$$B = Y_S^* F_S^{1/2}(F_S-\xi)^{-1}S^{1/2}(S^{-1}-I)(F-\xi)^{-1}F^{1/2}Y.$$

Hence, by the Hölder inequality for the Schatten norms (cf. [22, p. 41]), it follows that

$$\|B\|_{\mathcal{C}^\alpha} \leqslant \|F_S^{1/2}(F_S-\xi)^{-1}\|_{\mathcal{C}^{2\alpha}}\|S^{1/2}(S^{-1}-I)\|_{L^\infty(\Omega)}\|(F-\xi)^{-1}F^{1/2}\|_{\mathcal{C}^{2\alpha}}. \tag{4.16}$$

Since $\sigma(F) \setminus \{0\} = \sigma(H) \setminus \{0\}$, we may argue as before and obtain

$$\begin{aligned}
\|(F-\xi)^{-1}F^{1/2}\|_{\mathcal{C}^{2\alpha}}^{2\alpha} &\leqslant c\left(1+\frac{|\xi|}{d(\xi,\sigma(H))}\right)^{2\alpha}, \\
\|(F_S-\xi)^{-1}F_S^{1/2}\|_{\mathcal{C}^{2\alpha}}^{2\alpha} &\leqslant c\left(1+\frac{|\xi|}{d(\xi,\sigma(T^*ST))}\right)^{2\alpha}.
\end{aligned} \tag{4.17}$$

Now, it is easy to see that

$$\begin{aligned}
|S^{-1}-I| &\leqslant |(w^2-1)a^{1/2}\widetilde{a}^{-1}a^{1/2}| + |a^{1/2}(\widetilde{a}^{-1}-a^{-1})a^{1/2}| \\
&\leqslant c(|\nabla\phi - \nabla\widetilde{\phi}| + |A\circ\phi - A\circ\widetilde{\phi}|).
\end{aligned} \tag{4.18}$$

Combining (4.16), (4.17), and (4.18), we conclude that

$$\|B\|_{\mathcal{C}^\alpha} \leqslant c\left(1+\frac{|\xi|}{d_\sigma(\xi)}\right)^2 \delta_\infty(\phi,\widetilde{\phi}). \tag{4.19}$$

By Lemma 4.1 and the estimates (4.15) and (4.19), we deduce that

$$\|(w^{-1}\widetilde{H}w-\xi)^{-1}-(H-\xi)^{-1}\|_{\mathcal{C}^\alpha} \leqslant c_1\left(1+\frac{1}{|\xi|}+\frac{1}{d_\sigma(\xi)}+\frac{|\xi|^2}{d_\sigma(\xi)^2}\right)\delta_\infty(\phi,\widetilde{\phi}). \tag{4.20}$$

We now prove (4.9). In order to estimate A_1, A_2, and A_3, we use the estimate (4.5) and get

$$\|A_1\|_{\mathcal{C}^r}, \|A_2\|_{\mathcal{C}^r} \leqslant c\left(1+\frac{1}{|\xi|}+\frac{|\xi|}{d_\sigma(\xi)}\right)\|\nabla\phi-\nabla\widetilde{\phi}\|_{L^{pr}(\Omega)}, \tag{4.21}$$

$$\|A_3\|_{\mathcal{C}^r} \leqslant c\left(\frac{1+|\xi|}{d_\sigma(\xi)}+\frac{|\xi|^2}{d_\sigma(\xi)^2}\right)\|\nabla\phi-\nabla\widetilde{\phi}\|_{L^{pr}(\Omega)}. \tag{4.22}$$

We now estimate B. We assume without loss of generality that $S^{-1}-I \geqslant 0$. Thus, in order to estimate the $\mathcal{C}^r$ norm of B, we estimate the $\mathcal{C}^{2r}$ norms of $F_S^{1/2}(F_S-\xi)^{-1}S^{1/2}(S^{-1}-I)^{1/2}$ and $(S^{-1}-I)^{1/2}(F-\xi)^{-1}F^{1/2}$. By Lemma 4.5, it follows that

$$\begin{aligned}
&\|(S^{-1}-I)^{1/2}(F-\xi)^{-1}F^{1/2}\|_{\mathcal{C}^{2r}}^{2r} \\
&\quad\leqslant c\|(S^{-1}-I)^{1/2}\|_{L^{2pr}(\Omega)}^{2r}\sum_{n=1}^\infty\left|\frac{\lambda_n[H]}{\lambda_n[H]-\xi}\right|^{2r}\lambda_n[H]^{\frac{2q_0\gamma}{p(q_0-2)}-r} \\
&\quad\leqslant c\|S^{-1}-I\|_{L^{pr}(\Omega)}^r\left(1+\frac{|\xi|}{d_\sigma(\xi)}\right)^{2r}.
\end{aligned} \tag{4.23}$$

The same estimate holds also for the operator $F_S^{1/2}(F_S-\xi)^{-1}S^{1/2}(S^{-1}-I)^{1/2}$. Thus, by the Hölder inequality for the Schatten norms, it follows that

$$\|B\|_{\mathcal{C}^r} \leqslant c\left(1+\frac{|\xi|}{d_\sigma(\xi)}\right)^2 \|S^{-1}-I\|_{L^{pr}(\Omega)}. \tag{4.24}$$

Using Lemma 4.1 and combining the estimates (4.18), (4.22), and (4.24), we deduce that

$$\|(w^{-1}\widetilde{H}w-\xi)^{-1}-(H-\xi)^{-1}\|_{\mathcal{C}^r} \leqslant c_1\left(1+\frac{1}{|\xi|}+\frac{1}{d_\sigma(\xi)}+\frac{|\xi|^2}{d_\sigma(\xi)^2}\right)\delta_{pr}(\phi,\widetilde{\phi}). \tag{4.25}$$

Step 2. We prove statement (i). First of all, we prove that there exists $c>0$ such that if

$$\delta_\infty(\phi,\widetilde{\phi}) < \frac{d(\xi,\sigma(H))}{c(1+|\xi|^2+d(\xi,\sigma(H))^2)}, \tag{4.26}$$

then $\xi \notin \sigma(\widetilde{H})\cup\sigma(T^*ST)$ and

$$d(\xi,\sigma(\widetilde{H})),\ d(\xi,\sigma(T^*ST)) > \frac{d(\xi,\sigma(H))}{2}. \tag{4.27}$$

We begin with T^*ST. By recalling that $B=(T^*ST-\xi)^{-1}-(T^*T-\xi)^{-1}$ (cf. the proof of Lemma 4.1) and using the estimate (4.19) with $\xi=-1$ and the inequality (4.36), we find that there exists $C_1>0$ such that for all $n\in\mathbf{N}$

$$\left|\frac{1}{\lambda_n[T^*ST]+1}-\frac{1}{\lambda_n[H]+1}\right| \leqslant C_1\delta_\infty(\phi,\widetilde{\phi}). \tag{4.28}$$

Assume that $n\in\mathbf{N}$ is such that

$$\lambda_n[T^*ST] \leqslant |\xi|+d(\xi,\sigma(H)).$$

By (4.28), it follows that if

$$C_1(1+|\xi|+d(\xi,\sigma(H)))\delta_\infty(\phi,\widetilde{\phi}) < \frac{|\xi|+d(\xi,\sigma(H))}{2(|\xi|+d(\xi,\sigma(H)))+1},$$

then

$$\begin{aligned}\lambda_n[H] &\leqslant \frac{|\xi|+d(\xi,\sigma(H))+C_1[1+|\xi|+d(\xi,\sigma(H))]\delta_\infty(\phi,\widetilde{\phi})}{1-C_1[1+|\xi|+d(\xi,\sigma(H))]\delta_\infty(\phi,\widetilde{\phi})}\\ &\leqslant 2(|\xi|+d(\xi,\sigma(H))),\end{aligned} \tag{4.29}$$

(the elementary inequality $(A+t)(1-t)^{-1}<2A$ if $0<t<A(2A+1)^{-1}$ was used). Thus, by (4.28) and (4.29), it follows that if

$$\delta_\infty(\phi,\widetilde{\phi}) \leqslant \frac{d(\xi,\sigma(H))}{2C_1[1+|\xi|+d(\xi,\sigma(H))][1+2(|\xi|+d(\xi,\sigma(H)))]}$$

then

$$\begin{aligned}
|\xi-\lambda_n[T^*ST]| &\geqslant |\xi-\lambda_n[H]| - |\lambda_n[H]-\lambda_n[T^*ST]| \\
&\geqslant d(\xi,\sigma(H)) - C_1[1+|\xi|+d(\xi,\sigma(H))][1+2(|\xi|+d(\xi,\sigma(H)))]\delta_\infty(\phi,\widetilde{\phi}) \\
&\geqslant \frac{d(\xi,\sigma(H))}{2}
\end{aligned} \tag{4.30}$$

for all $n \in \mathbf{N}$ such that $\lambda_n[T^*ST] \leqslant |\xi| + d(\xi,\sigma(H))$. Thus, the inequality (4.27) for $d(\xi,\sigma(T^*ST))$ follows by (4.30) and by observing that if $n \in \mathbf{N}$ is such that $\lambda_n[T^*ST] > |\xi| + d(\xi,\sigma(H))$, then

$$|\xi-\lambda_n[T^*ST]| > d(\xi,\sigma(H)).$$

The inequality (4.27) for $d(\xi,\sigma(\widetilde{H}))$ can be proved in the same way. Indeed, it suffices to observe that, by Step 1, there exists $C_2 > 0$ such that

$$\left|\frac{1}{\lambda_n[\widetilde{H}]+1} - \frac{1}{\lambda_n[H]+1}\right| \leqslant C_2\delta_\infty(\phi,\widetilde{\phi}); \tag{4.31}$$

we then proceed exactly as above.

By (4.20) and (4.27), there exists $c > 0$ such that if

$$\delta_\infty(\phi,\widetilde{\phi}) \leqslant \frac{d(\xi,\sigma(H))}{c(1+|\xi|^2+d(\xi,\sigma(H))^2)}, \tag{4.32}$$

then $\xi \notin \sigma(\widetilde{H}) \cup \sigma(T^*ST)$ and

$$\begin{aligned}
&\|(w^{-1}\widetilde{H}w-\xi)^{-1} - (H-\xi)^{-1}\|_{C^\alpha} \\
&\leqslant c\Big(1+\frac{1}{|\xi|}+\frac{1}{d(\xi,\sigma(H))}+\frac{|\xi|^2}{d(\xi,\sigma(H))^2}\Big)\delta_\infty(\phi,\widetilde{\phi}).
\end{aligned} \tag{4.33}$$

This completes the proof of statement (i).

The argument above works word by word also for the proof of statement (ii) provided that $\delta_\infty(\phi,\widetilde{\phi})$ is replaced with $\delta_{pr}(\phi,\widetilde{\phi})$. □

Remark 4.9. The proof of Theorem 4.6 gives some information about the dependence of the constants c_1 and c_2 on ξ, which is useful in the sequel. For instance, in the case of statement (i), in fact it was proved that there exists c depending only on N, τ, θ, α, and c^* such that if (4.32) holds, then (4.33) holds. Exactly the same holds for statement (ii) where c depends also on r, p, q_0, C, γ, and $|\Omega|$. Moreover, for such ϕ and $\widetilde{\phi}$, if $0 \notin \sigma(H)$, then $0 \notin \sigma(\widetilde{H})$ and the summand $1/|\xi| + 1/d(\xi,\sigma(H)$ can be removed from the right-hand side of (4.33). Furthermore, in this case, statements (i) and (ii) also hold for $\xi = 0$. This can be easily seen by looking closely at the proofs of (4.15) and (4.21).

Remark 4.10. By the proof of Theorem 4.6, for a fixed $\xi \in \mathbf{C} \setminus [0, \infty[$ no smallness conditions on $\delta_\infty(\phi, \widetilde{\phi})$ and $\delta_{pr}(\phi, \widetilde{\phi})$ are required for the validity of (4.8) and (4.9) respectively.

Theorem 4.11 (stability of eigenvalues). *Let* (A) *be satisfied. Then the following statements hold.*

(i) *There exists* $c_1 > 0$ *depending only on* N, τ, θ, α, *and* c^* *such that if* $\delta_\infty(\phi, \widetilde{\phi}) \leqslant c_1^{-1}$, *then*

$$\left(\sum_{n=1}^{\infty}\left|\frac{1}{\lambda_n[\widetilde{L}]+1} - \frac{1}{\lambda_n[L]+1}\right|^{\alpha}\right)^{1/\alpha} \leqslant c_1 \delta_\infty(\phi, \widetilde{\phi}). \tag{4.34}$$

(ii) *Let, in addition,* (P) *be satisfied by the operators* L, $\widetilde{L}$, *and* $\widehat{L}$ *for the same* $q_0 > 2$, $\gamma \geqslant 0$, *and* $C > 0$. *Let* $p \geqslant q_0/(q_0 - 2)$ *and* $r \geqslant \max\{2, \alpha + \frac{2q_0\gamma}{p(q_0-2)}\}$. *Then there exists* $c_2 > 0$ *depending only on* N, τ, θ, α, c^*, r, p, q_0, C, γ, *and* $|\Omega|$ *such that if* $\delta_{pr}(\phi, \widetilde{\phi}) \leqslant c_2^{-1}$, *then*

$$\left(\sum_{n=1}^{\infty}\left|\frac{1}{\lambda_n[\widetilde{L}]+1} - \frac{1}{\lambda_n[L]+1}\right|^{r}\right)^{1/r} \leqslant c_2 \delta_{pr}(\phi, \widetilde{\phi}). \tag{4.35}$$

Proof. The theorem follows by Theorem 4.6 and by applying the inequality

$$\left(\sum_{n=1}^{\infty}\left|\frac{1}{\lambda_n[E_1]+1} - \frac{1}{\lambda_n[E_2]+1}\right|^{r}\right)^{1/r} \leqslant \|(E_1+1)^{-1} - (E_2+1)^{-1}\|_{\mathcal{C}^r}, \tag{4.36}$$

with $E_1 = w^{-1}\widetilde{H}w$, $E_2 = H$ (cf. [24, p. 20]). □

Remark 4.12. We note that, in the case of the Dirichlet boundary conditions, i.e., $\mathcal{V} = W_0^{1,2}(\Omega)$, the inequality (4.34) directly follows from Lemma 6.1 in [5] the proof of which is based on the Min-Max Principle.

5 Stability Estimates for Eigenfunctions

Definition 5.1. Let E be a nonnegative selfadoint operator with compact resolvent on a Hilbert space $\mathcal{H}$. For a finite subset G of $\mathbf{N}$ we denote by

$P_G(E)$ the orthogonal projection from $\mathcal{H}$ onto the linear space generated by all the eigenfunctions corresponding to the eigenvalues $\lambda_k[E]$ with $k \in G$.

Note that the dimension of the range of $P_G(E)$ coincides with the number of elements of G if and only if no eigenvalue with index in G coincides with an eigenvalue with index in $\mathbf{N} \setminus G$; this will always be the case in what follows.

In the following statements, it is understood that whenever $n = 1$ the term λ_{n-1} has to be dropped.

Theorem 5.2. *Let* (A) *be satisfied. Let* λ *be a nonzero eigenvalue of* H *of multiplicity* m*, let* $n \in \mathbf{N}$ *be such that* $\lambda = \lambda_n[H] = \ldots = \lambda_{n+m-1}[H]$*, and let* $G = \{n, n+1, \ldots, n+m-1\}$*. Then the following statements hold.*

(i) *There exists* $c_1 > 0$ *depending only on* N, τ, θ, α, c^*, $\lambda_{n-1}[H]$, λ*, and* $\lambda_{n+m}[H]$ *such that if* $\delta_\infty(\phi, \widetilde{\phi}) \leqslant c_1^{-1}$*, then* $\dim \operatorname{ran} P_G(w^{-1}\widetilde{H}w) = m$ *and*

$$\|P_G(H) - P_G(w^{-1}\widetilde{H}w)\| \leqslant c_1 \delta_\infty(\phi, \widetilde{\phi}). \tag{5.1}$$

(ii) *Let, in addition,* (P) *be satisfied by the operators* H, $\widetilde{H}$ *and* T^*ST *for the same* q_0, γ *and* C*. Let* $s = [q_0/(q_0-2)] \max\{2, \alpha + 2\gamma\}$*. Then there exists* $c_2 > 0$ *depending only on* N, τ, θ, α, c^*, q_0, C, γ, $|\Omega|$, $\lambda_{n-1}[H]$, λ*, and* $\lambda_{n+m}[H]$ *such that if* $\delta_s(\phi, \widetilde{\phi}) \leqslant c_2^{-1}$*, then* $\dim \operatorname{ran} P_G(w^{-1}\widetilde{H}w) = m$ *and*

$$\|P_G(H) - P_G(w^{-1}\widetilde{H}w)\| \leqslant c_2 \delta_s(\phi, \widetilde{\phi}). \tag{5.2}$$

Proof. We set $\rho = \frac{1}{2} \operatorname{dist}(\lambda, (\sigma(H) \cup \{0\}) \setminus \{\lambda\})$ and $\lambda^* = \lambda$ if λ is the first nonzero eigenvalue of H, and $\lambda^* = \lambda_{n-1}[H]$ otherwise.

By Theorem 4.11 (i), it follows that

$$|\lambda_k[H] - \lambda_k[\widetilde{H}]| \leqslant c(\lambda_k[H] + 1)(\lambda_k[\widetilde{H}] + 1)\delta_\infty(\phi, \widetilde{\phi}). \tag{5.3}$$

This implies that there exists $c > 0$ such that if

$$\delta_\infty(\phi, \widetilde{\phi}) < c^{-1}\lambda_k[H]/(\lambda_k[H] + 1)^2,$$

then

$$\lambda_k[\widetilde{H}] \leqslant 2\lambda_k[H].$$

This together with (5.3) implies the existence of $c > 0$ such that if

$$\delta_\infty(\phi, \widetilde{\phi}) < c^{-1} \min\{\rho, \lambda_k[H]\}/(\lambda_k[H] + 1)^2,$$

then

$$|\lambda_k[H] - \lambda_k[\widetilde{H}]| < \rho/2.$$

Applying this inequality for $k = n-1, \ldots, n+m$, we deduce that if

$$\delta_\infty(\phi,\widetilde{\phi}) < \frac{\min\{\rho,\lambda^*\}}{c(\lambda_{n+m}[H]+1)^2},$$

then

$$\begin{aligned}|\lambda_k[\widetilde{H}]-\lambda| &\leqslant \rho/2 \ \ \forall\, k\in G\,,\\ |\lambda_k[\widetilde{H}]-\lambda| &\geqslant 3\rho/2 \ \ \forall\, k\in \mathbf{N}\setminus G\,.\end{aligned} \tag{5.4}$$

Hence $\dim\operatorname{ran} P_G(w^{-1}\widetilde{H}w)=m$ and, by the well-known Riesz formula, we have

$$P_G[H] = -\frac{1}{2\pi i}\int_\Gamma (H-\xi)^{-1}d\xi\,, \tag{5.5}$$

$$P_G[w^{-1}\widetilde{H}w] = -\frac{1}{2\pi i}\int_\Gamma (w^{-1}\widetilde{H}w-\xi)^{-1}d\xi\,, \tag{5.6}$$

where $\Gamma(\theta)=\lambda+\rho e^{i\theta}$, $0\leqslant\theta<2\pi$. Hence

$$\|P_G[H]-P_G[w^{-1}\widetilde{H}w]\| \leqslant \rho\sup_{\xi\in\Gamma}\|(w^{-1}\widetilde{H}w-\xi)^{-1}-(H-\xi)^{-1}\|. \tag{5.7}$$

Let c_1 be the same as in Theorem 4.6 (i). Using Theorem 4.6 (i) and Remark 4.9 and observing that $\lambda-\rho\leqslant|\xi|\leqslant\lambda+\rho$ and $1/|\xi|\leqslant 1/\rho$ for all $\xi\in\Gamma$, we find that

$$\delta_\infty(\phi,\widetilde{\phi}) < \frac{\rho}{c_1(1+\lambda^2_{n+m}[H]+\rho^2)}$$

implies

$$\|(w^{-1}\widetilde{H}w-\xi)^{-1}-(H-\xi)^{-1}\| \leqslant c_1\left(1+\frac{1}{\rho}+\frac{\lambda^2}{\rho^2}\right)\delta_\infty(\phi,\widetilde{\phi}). \tag{5.8}$$

The proof of statement (i) then follows by combining (5.7) and (5.8). The proof of (ii) is similar. □

Remark 5.3. The proof of Theorem 5.2 gives some information about the dependence of the constants c_1,c_2 on $\lambda_{n-1}[H],\lambda,\lambda_{n+m}[H]$ which is useful in the sequel. For instance, in the case of statement (i), in fact we proved that there exists $c>0$ depending only on N, τ, θ, α, and c^* such that

$$\delta_\infty(\phi,\widetilde{\phi}) \leqslant \frac{\min\{\rho,\lambda^*\}}{c(1+\rho^2+\lambda_{n+m}[H]^2)}$$

implies

$$\|P_G(H)-P_G(w^{-1}\widetilde{H}w)\| \leqslant c\left(1+\rho+\frac{\lambda^2}{\rho}\right)\delta_\infty(\phi,\widetilde{\phi}). \tag{5.9}$$

Exactly the same is true for statement (ii), where c depends also on q_0, C, γ, and $|\Omega|$.

We are going to apply the stability estimates of Theorem 5.2 to obtain stability estimates for eigenfunctions. For this purpose, we need the following lemma.

Lemma 5.4 (selection lemma). *Let U and V be finite dimensional subspaces of a Hilbert space $\mathcal{H}$, $\dim U = \dim V = m$, and let $u_1, \dots, u_m$ be an orthonormal basis for U. Then there exists an orthonormal basis $v_1, \dots, v_m$ for V such that*

$$\|u_k - v_k\| \leqslant 5^k \|P_U - P_V\|, \quad k = 1, \dots, m, \tag{5.10}$$

where P_U and P_V are the orthogonal projections onto U and V respectively.

Proof. Step 1. It is clear that $\|P_U - P_V\| \leqslant 2$. If $1 \leqslant \|P_U - P_V\| \leqslant 2$, then the estimate (5.10) obviously holds for any choice of an orthonormal basis $v_1, \dots, v_m$ for V. So, we assume that $\|P_U - P_V\| < 1$. Let $u \in U$, $\|u\| = 1$. Then

$$\|P_V u\| = \|u + (P_V - P_U)u\| \geqslant 1 - \|P_U - P_V\| > 0. \tag{5.11}$$

Letting $z = P_V u / \|P_V u\|$, we have $\|z\| = 1$ and

$$\langle u, z\rangle = \frac{\langle u, P_V^2 u\rangle}{\|P_V u\|} = \|P_V u\|.$$

Hence

$$\|P_U - P_V\|^2 \geqslant \|(P_U - P_V)u\|^2 = \|u\|^2 - \|P_V u\|^2 \geqslant \frac{\|u - z\|^2}{2},$$

and therefore

$$\|u - z\| \leqslant \sqrt{2}\|P_U - P_V\|. \tag{5.12}$$

Step 2. Assume that $\|P_U - P_V\| \leqslant 1/6$. Let

$$z_k = \frac{P_V u_k}{\|P_V u_k\|}, \quad k = 1, \dots, m.$$

We prove that

$$|\langle z_k, z_l\rangle| \leqslant 3\|P_U - P_V\|, \quad k, l = 1, \dots, m, \ k \neq l. \tag{5.13}$$

Indeed, for $k \neq l$ we have

$$\begin{aligned} |\langle P_V u_k, P_V u_l\rangle| &= |\langle P_V u_k - u_k, P_V u_l\rangle + \langle u_k, u_l\rangle + \langle u_k, P_V u_l - u_l\rangle| \\ &= |\langle (P_V - P_U)u_k, P_V u_l\rangle + \langle u_k, (P_V - P_U)u_l\rangle| \\ &\leqslant 2\|P_U - P_V\|, \end{aligned}$$

and the claim is proved by recalling (5.11).

Step 3. It is easy to see that since $\|P_U - P_V\| < 1$, the vectors $z_1, \dots, z_m$ are linearly independent. Thus, we can apply the Gram–Schmidt orthogonalization procedure, i.e., define

$$v_1 = z_1 \ , \quad v_k = \frac{z_k - \sum_{l=1}^{k-1} \langle z_k, v_l \rangle v_l}{\|z_k - \sum_{l=1}^{k-1} \langle z_k, v_l \rangle v_l\|}, \qquad k = 2, \dots, m.$$

Note that for $k = 2, \dots, m$,

$$v_k - z_k = \left(\frac{1}{\|z_k - \sum_{l=1}^{k-1} \langle z_k, v_l \rangle v_l\|} - 1 \right) z_k - \frac{\sum_{l=1}^{k-1} \langle z_k, v_l \rangle v_l}{\|z_k - \sum_{l=1}^{k-1} \langle z_k, v_l \rangle v_l\|}$$

and

$$1 \geqslant \left\| z_k - \sum_{l=1}^{k-1} \langle z_k, v_l \rangle v_l \right\| \geqslant 1 - \sum_{l=1}^{k-1} |\langle z_k, v_l \rangle| \, .$$

Hence if

$$\sum_{l=1}^{k-1} |\langle z_k, v_l \rangle| < 1, \tag{5.14}$$

then

$$\|v_k - z_k\| \leqslant \frac{2 \sum_{l=1}^{k-1} |\langle z_k, v_l \rangle|}{1 - \sum_{l=1}^{k-1} |\langle z_k, v_l \rangle|}. \tag{5.15}$$

Also, for $s = k, \dots, m$

$$|\langle z_s, v_k \rangle| \leqslant \frac{|\langle z_s, z_k \rangle| + \sum_{l=1}^{k-1} |\langle z_k, v_l \rangle|}{1 - \sum_{l=1}^{k-1} |\langle z_k, v_l \rangle|}. \tag{5.16}$$

Step 4. We prove that for all $k = 2, \dots, m$

$$\|v_k - z_k\| \leqslant 3 \cdot 5^{k-1} \|P_U - P_V\|, \tag{5.17}$$

$$|\langle z_s, v_k \rangle| \leqslant 3 \cdot 5^{k-1} \|P_U - P_V\| \, , \quad s = k+1, \dots, m, \tag{5.18}$$

provided that

$$\|P_U - P_V\| \leqslant \frac{2}{3} 5^{-k+1} \, . \tag{5.19}$$

We prove this by induction. If $k = 2$, then, by (5.13) and (5.19),

$$|\langle z_2, v_1 \rangle| = |\langle z_2, z_1 \rangle| \leqslant 3 \|P_U - P_V\| \leqslant \tfrac{2}{5}.$$

Hence, by (5.15),

$$\|v_2 - z_2\| \leqslant \frac{6 \|P_U - P_V\|}{1 - 3\|P_U - P_V\|} \leqslant 15 \|P_U - P_V\|$$

and, by (5.16) and (5.13), for $s = 3, \dots, m$

$$|\langle z_s, v_2\rangle| \leqslant \frac{6\|P_U - P_V\|}{1 - 3\|P_U - P_V\|} \leqslant 15\|P_U - P_V\|.$$

Let $2 \leqslant k \leqslant m-1$. Suppose that the inequalities (5.17) and (5.18) under the assumption (5.19) are satisfied for all $2 \leqslant j \leqslant k$. By assuming the validity of (5.19) for $k+1$ and using (5.15), we obtain

$$\|v_{k+1} - z_{k+1}\| \leqslant \frac{6(\sum_{j=1}^k 5^{j-1})\|P_U - P_V\|}{1 - 3(\sum_{j=1}^k 5^{j-1})\|P_U - P_V\|}.$$

Since $\sum_{l=1}^k 5^{l-1} \leqslant 5^k/4$, by (5.19) with k replaced by $k+1$,

$$3\sum_{l=1}^k 5^{l-1}\|P_U - P_V\| \leqslant 1/2.$$

Hence

$$\|v_{k+1} - z_{k+1}\| \leqslant 3 \cdot 5^k\|P_U - P_V\|.$$

Similarly, by (5.16) and (5.13), for all $s = k+2, \dots m$

$$|(z_s, v_{k+1})| \leqslant \frac{3\|P_U - P_V\| + 3(\sum_{j=1}^k 5^{j-1})\|P_U - P_V\|}{1 - 3(\sum_{j=1}^k 5^{j-1})\|P_U - P_V\|} \leqslant 3 \cdot 5^k\|P_U - P_V\|.$$

Step 5. To complete the proof, we note that, by (5.12),

$$\|u_1 - v_1\| \leqslant \sqrt{2}\|P_U - P_V\|.$$

For $k \geqslant 2$ we have that (5.19) implies

$$\|u_k - v_k\| \leqslant \|u_k - z_k\| + \|z_k - v_k\| \leqslant (\sqrt{2} + 3 \cdot 5^{k-1})\|P_U - P_V\|,$$

while if (5.19) does not hold, then $\|P_U - P_V\| > 10/(3 \cdot 5^k)$, and therefore

$$\|u_k - v_k\| \leqslant 2 \leqslant 3 \cdot 5^{k-1}\|P_U - P_V\|.$$

This completes the proof of the lemma. □

Lemma 5.5. *Let* (A) *be satisfied. Let λ be a nonzero eigenvalue of H of multiplicity m, and let $n \in \mathbf{N}$ be such that $\lambda = \lambda_n[H] = \ldots = \lambda_{n+m-1}[H]$. Then the following statements hold.*

(i) *There exists $c_1 > 0$ depending only on N, τ, θ, α, c^*, $\lambda_{n-1}[H]$, λ, and $\lambda_{n+m}[H]$ such that the following is true: if*

$$\delta_\infty(\phi, \widetilde{\phi}) \leqslant c_1^{-1}$$

and $\psi_n[\widetilde{H}], \dots, \psi_{n+m-1}[\widetilde{H}]$ *are orthonormal eigenfunctions of* $\widetilde{H}$ *in the space* $L^2(\Omega, \widetilde{g}dx)$, *then there are orthonormal eigenfunctions* $\psi_n[H], \dots, \psi_{n+m-1}[H]$ *of* H *in* $L^2(\Omega, gdx)$ *such that*

$$\|\psi_k[H] - \psi_k[\widetilde{H}]\|_{L^2(\Omega)} \leqslant c_1\delta_\infty(\phi, \widetilde{\phi}), \tag{5.20}$$

for all $k = n, \dots, n+m-1$.

(ii) *Let, in addition,* (P) *be satisfied by the operators* H, $\widetilde{H}$, *and* T^*ST *for the same* q_0, γ *and* C. *Let* $s = [q_0/(q_0-2)]\max\{2, \alpha+2\gamma\}$. *Then there exists* $c_2 > 0$ *depending only on* $N, \tau, \theta, \alpha, c^*, q_0, C, \gamma, |\Omega|, \lambda_{n-1}[H], \lambda, \lambda_{n+m}[H]$ *such that the following is true: if*

$$\delta_s(\phi, \widetilde{\phi}) \leqslant c_2^{-1}$$

and $\psi_n[\widetilde{H}], \dots, \psi_{n+m-1}[\widetilde{H}]$ *are orthonormal eigenfunctions of* $\widetilde{H}$ *in the space* $L^2(\Omega, \widetilde{g}dx)$, *then there are orthonormal eigenfunctions* $\psi_n[H], \dots, \psi_{n+m-1}[H]$ *of* H *in* $L^2(\Omega, gdx)$ *such that*

$$\|\psi_k[H] - \psi_k[\widetilde{H}]\|_{L^2(\Omega)} \leqslant c_2\delta_s(\phi, \widetilde{\phi}) \tag{5.21}$$

for all $k = n, \dots, n+m-1$.

Proof. We prove only statement (ii) since the proof of (i) is similar. We first note that $f_k := w^{-1}\psi_k[\widetilde{H}]$, $k = n, \dots, n+m-1$, are orthonormal eigenfunctions in $L^2(\Omega, g\,dx)$ of $w^{-1}\widetilde{H}w$ corresponding to the eigenvalues $\lambda_n[\widetilde{H}], \dots, \lambda_{n+m-1}[\widetilde{H}]$. By Theorem 5.2 and Lemma 5.4, there exists $c > 0$ such that if $\delta_s(\phi, \widetilde{\phi}) < c^{-1}$, then there exist eigenfunctions $\psi_n[H], \dots, \psi_{n+m-1}[H]$ of H corresponding to the eigenvalue λ such that

$$\|\psi_k[H] - f_k\|_{L^2(\Omega)} \leqslant c\delta_s(\phi, \widetilde{\phi}). \tag{5.22}$$

To complete the proof, it suffices to note that

$$\begin{aligned}\|f_k - \psi_k[\widetilde{H}]\|_{L^2(\Omega)} &\leqslant \|1 - w^{-1}\|_{L^s(\Omega)}\|\psi_k[\widetilde{H}]\|_{L^{2s/(s-2)}(\Omega)} \\ &\leqslant c\|\nabla\phi - \nabla\widetilde{\phi}\|_{L^s(\Omega)}.\end{aligned}$$

The lemma is proved. □

In the following theorem, we estimate the deviation of the eigenfunctions $\psi_k[\widetilde{L}]$ of $\widetilde{L}$ from the eigenfunctions $\psi_k[L]$ of L. We adopt the convention that $\psi_k[L]$ and $\psi_k[\widetilde{L}]$ are extended by zero outside $\phi(\Omega)$ and $\widetilde{\phi}(\Omega)$ respectively.

Theorem 5.6 (stability of eigenfunctions). *Let (A) be satisfied. Let λ be a nonzero eigenvalue of L of multiplicity m, and let $n \in \mathbf{N}$ be such that $\lambda = \lambda_n[L] = \ldots = \lambda_{n+m-1}[L]$. Then the following statements hold.*

(i) *There exists $c_1 > 0$ depending only on N, τ, θ, α, c^*, $\lambda_{n-1}[L]$, λ, $\lambda_{n+m}[L]$ such that if $\delta_\infty(\phi, \widetilde{\phi}) \leqslant c_1^{-1}$ and $\psi_n[\widetilde{L}], \ldots, \psi_{n+m-1}[\widetilde{L}]$ are orthonormal eigenfunctions of $\widetilde{L}$ in $L^2(\widetilde{\phi}(\Omega))$, then there exist orthonormal eigenfunctions $\psi_n[L], \ldots, \psi_{n+m-1}[L]$ of L in $L^2(\phi(\Omega))$ such that*

$$\|\psi_k[L] - \psi_k[\widetilde{L}]\|_{L^2(\phi(\Omega)\cup\widetilde{\phi}(\Omega))} \leqslant c\big(\delta_\infty(\phi, \widetilde{\phi}) + \\ + \|\psi_k[L] \circ \phi - \psi_k[L] \circ \widetilde{\phi}\|_{L^2(\Omega)} + \|\psi_k[\widetilde{L}] \circ \phi - \psi_k[\widetilde{L}] \circ \widetilde{\phi}\|_{L^2(\Omega)}\big), \quad (5.23)$$

for all $k = n, \ldots, n + m - 1$.

(ii) *Let, in addition, (P) be satisfied by the operators L, $\widetilde{L}$, and $\widehat{L}$ for the same q_0, γ, and C. Let $s = [q_0/(q_0 - 2)] \max\{2, \alpha + 2\gamma\}$. Then there exists $c_2 > 0$ depending only on N, τ, θ, α, c^*, q_0, C, γ, $|\Omega|$, $\lambda_{n-1}[L]$, λ, and $\lambda_{n+m}[L]$ such that the following is true: if $\delta_s(\phi, \widetilde{\phi}) \leqslant c_1^{-1}$ and $\psi_n[\widetilde{L}], \ldots$, $\psi_{n+m-1}[\widetilde{L}]$ are orthonormal eigenfunctions of $\widetilde{L}$ in $L^2(\widetilde{\phi}(\Omega))$, then there exist orthonormal eigenfunctions $\psi_n[L], \ldots, \psi_{n+m-1}[L]$ of L in $L^2(\phi(\Omega))$ such that*

$$\|\psi_k[L] - \psi_k[\widetilde{L}]\|_{L^2(\phi(\Omega)\cup\widetilde{\phi}(\Omega))} \leqslant c\big(\delta_s(\phi, \widetilde{\phi}) + \|\psi_k[L] \circ \phi \\ - \psi_k[L] \circ \widetilde{\phi}\|_{L^2(\Omega)} + \|\psi_k[\widetilde{L}] \circ \phi - \psi_k[\widetilde{L}] \circ \widetilde{\phi}\|_{L^2(\Omega)}\big), \quad (5.24)$$

for all $k = n, \ldots, n + m - 1$.

Remark 5.7. We note that if, in addition, the semigroup e^{-Lt} is ultracontractive, then the eigenfunctions are bounded hence

$$\|\psi_k[L] \circ \phi - \psi_k[L] \circ \widetilde{\phi}\|_{L^2(\Omega)} + \|\psi_k[\widetilde{L}] \circ \phi - \psi_k[\widetilde{L}] \circ \widetilde{\phi}\|_{L^2(\Omega)} \leqslant c(\lambda)|\mathcal{D}|^{1/2},$$

where $\mathcal{D} = \{x \in \Omega : \phi(x) \neq \widetilde{\phi}(x)\}$.

Proof of Theorem 5.6. We set

$$\psi_k[\widetilde{H}] = \psi_k[\widetilde{L}] \circ \widetilde{\phi}$$

for all $k = n, \ldots, n + m - 1$, so that $\psi_n[\widetilde{H}], \ldots, \widetilde{\psi}_{n+m-1}[\widetilde{H}]$ are orthonormal eigenfunctions in $L^2(\Omega, \widetilde{g}dx)$ of the operator $\widetilde{H}$ corresponding to the eigenvalues $\lambda_n[\widetilde{H}], \ldots, \lambda_{n+m-1}[\widetilde{H}]$. By Lemma 5.5 (i), there exists $c_1 > 0$ such that if $\delta_\infty(\phi, \widetilde{\phi}) < c_1^{-1}$, then there exist orthonormal eigenfunctions $\psi_n[H], \ldots$, $\psi_{n+m-1}[H]$ in $L^2(\Omega, gdx)$ of H corresponding to the eigenvalue λ such that the inequality (5.20) is satisfied. We now set

$$\psi_k[L] = \psi_k[H] \circ \phi^{(-1)}$$

for all $k = n, \dots, n+m-1$, so that $\psi_n[L], \dots, \psi_{n+m-1}[L]$ are orthonormal eigenfunctions in $L^2(\phi(\Omega))$ of L corresponding to the eigenvalue λ. Changing variables in integrals, we obtain

$$\begin{aligned}\|\psi_k[L] - \psi_k[\widetilde{L}]\|_{L^2(\widetilde{\phi}(\Omega))} &\leqslant \|\psi_k[L] \circ \widetilde{\phi} - \psi_k[\widetilde{L}] \circ \widetilde{\phi}\|_{L^2(\Omega)} \\ &\leqslant c\big(\|\psi_k[L] \circ \widetilde{\phi} - \psi_k[L] \circ \phi\|_{L^2(\Omega)} + \|\psi_k[L] \circ \phi - \psi_k[\widetilde{L}] \circ \widetilde{\phi}\|_{L^2(\Omega)}\big) \\ &= c\big(\|\psi_k[L] \circ \widetilde{\phi} - \psi_k[L] \circ \phi\|_{L^2(\Omega)} + \|\psi_k[H] - \psi_k[\widetilde{H}]\|_{L^2(\Omega)}\big).\end{aligned}$$

In the same way,

$$\begin{aligned}\|\psi_k[L] - \psi_k[\widetilde{L}]\|_{L^2(\phi(\Omega))} \leqslant c(&\|\psi_k[\widetilde{L}] \circ \widetilde{\phi} - \psi_k[\widetilde{L}] \circ \phi\|_{L^2(\Omega)} \\ &+ \|\psi_k[H] - \psi_k[\widetilde{H}]\|_{L^2(\Omega)}).\end{aligned}$$

Hence (5.23) and (5.24) follow from (5.20) and (5.21) respectively. □

6 On the Regularity of Eigenfunctions

In this section, we obtain sufficient conditions for the validity of conditions (P1) and (P2). We begin by recalling the following known result based on the notion of ultracontractivity which guarantees the validity of property (P1) under rather general assumptions, namely under the assumption that a Sobolev-type embedding theorem holds for the space $\mathcal{V}$.

Lemma 6.1. *Let Ω be a domain in $\mathbf{R}^N$ of finite measure and $\mathcal{V}$ a closed subspace of $W^{1,2}(\Omega)$ containing $W_0^{1,2}(\Omega)$. Assume that there exist $p > 2$ and $D > 0$ such that*

$$\|u\|_{L^p(\Omega)} \leqslant D\|u\|_{W^{1,2}(\Omega)} \tag{6.1}$$

for all $u \in \mathcal{V}$. Then the following statements hold.

(i) *The condition* (3.2) *is satisfied for any $\alpha > \frac{p}{p-2}$.*

(ii) *The eigenfunctions of the operators H, $\widetilde{H}$ and T^*ST satisfy* (P1) *with $q_0 = \infty$, $\gamma = \frac{p}{2(p-2)}$, where C depends only on p, D, τ, θ, and c^*.*

Proof. For the proof of statement (i) we refer to [3, Theorem 7], where the case $\mathcal{V} = W^{1,2}(\Omega)$ is considered. The proof works word by word also in the slightly more general case considered here. The proof of statement (ii) is the same as in [3, Theorem 7], where it is proved that for the Neumann Laplacian property (P1) is satisfied if (6.1) holds: this proof can be easily adapted to the operators H, $\widetilde{H}$, and T^*ST. □

We now give conditions for the validity of property (P2). We consider first the case where an a priori estimate holds for the operators L and $\widetilde{L}$, which is typically the case of sufficiently smooth open sets and coefficients. Then we consider a more general situation based on an approach which goes back to Meyers [19].

The regular case. Recall that an open set in $\mathbf{R}^N$ satisfies the interior cone condition with the parameters $R > 0$ and $h > 0$ if for all $x \in \Omega$ there exists a cone $K_x \subset \Omega$ with the point x as vertex congruent to the cone

$$K(R,h) = \left\{ x \in \mathbf{R}^N : \ 0 < \left(\sum_{i=1}^{N-1} x_i^2 \right)^{1/2} < \frac{R x_N}{h} < R \right\}.$$

In this paper, the cone condition is used in order to guarantee the validity of the standard Sobolev embedding.

The next theorem is a simplified version of Theorem 5.1 in [6].

Theorem 6.2. *Let $R > 0$, $h > 0$. Let U be an open set in $\mathbf{R}^N$ satisfying the interior cone condition with the parameters R and h, and let E be an operator in $L^2(U)$ satisfying the following a priori estimate: there exists $B > 0$ such that if $2 \leqslant p < N+2$ and $u \in \mathrm{Dom}(E)$, $Eu \in L^p(U)$, then $u \in W^{2,p}(U)$ and*

$$\|u\|_{W^{2,p}(U)} \leqslant B \left(\|Eu\|_{L^p(U)} + \|u\|_{L^2(U)} \right). \tag{6.2}$$

Assume that $E\psi = \lambda\psi$ for some $\psi \in \mathrm{Dom}(E)$ and $\lambda \in \mathbf{C}$. Then there exists $c > 0$ depending only on R, h, N, and B such that for $\mu = 0, 1$,

$$\|\psi\|_{W^{\mu,\infty}(U)} \leqslant c(1+|\lambda|)^{\frac{N}{4}+\frac{\mu}{2}} \|\psi\|_{L^2(U)}. \tag{6.3}$$

Theorem 6.3. *Let* (A) *be satisfied, and let $\phi(\Omega)$ and $\widetilde{\phi}(\Omega)$ be open sets satisfying the interior cone condition with the same parameters R, h. If the operators L, $\widetilde{L}$ satisfy the a priori estimate* (6.2) *with the same B, then the operators H, $\widetilde{H}$, and T^*ST satisfy property* (P) *with $q_0 = \infty$, $\gamma = N/4$ and C depending only on τ, R, h, c^*, θ, and B.*

Proof. Recall that H, $\widetilde{H}$, and T^*ST are the operators obtained by pulling-back to Ω the operators L, $\widetilde{L}$, and $\widehat{L}$ respectively. It is clear that $\widehat{L}$ also satisfies the a priori estimate (6.2). Thus, by Theorem 6.2, the eigenfunctions of the operators L, $\widetilde{L}$, and $\widehat{L}$ satisfy the condition (6.3). Hence, by pulling such eigenfunctions back to Ω, it follows that the eigenfunctions of H, $\widetilde{H}$, and T^*ST satisfy (P1) and (P2) with $q_0 = \infty$, $\gamma = N/4$ and C as in the statement. □

The general case. Here, we assume that $\mathcal{V} = \mathrm{cl}_{W^{1,2}(\Omega)} \mathcal{V}_0$, where $\mathcal{V}_0$ is a space of functions defined in Ω such that $C_c^\infty(\Omega) \subset \mathcal{V}_0 \subset W^{1,\infty}(\Omega)$. Moreover, for all $1 < q < \infty$ we set

$$V_q = \mathrm{cl}_{W^{1,q}(\Omega)} \mathcal{V}_0.$$

Let $-\Delta_q : V_q \to (V_{q'})'$ be the operator defined by

$$(-\Delta_q u, \psi) = \int_\Omega \nabla u \cdot \nabla \psi dx$$

for all $u \in V_q$, $\psi \in V_{q'}$.

The following theorem is a variant of a result of Gröger [13] (cf. also [2]).

Theorem 6.4. *Let* (A) *be satisfied. Assume that there exists* $q_1 > 2$ *such that the operator* $I - \Delta_q : V_q \to (V_{q'})'$ *has a bounded inverse for all* $2 \leqslant q \leqslant q_1$*. Then there exist* $q_0 > 2$ *and* $c > 0$ *depending only on* $\mathcal{V}_0$, τ, *and* θ *such that if* u *is an eigenfunction of one of the operators* H, $\widetilde{H}$, T^*ST *and* λ *is the corresponding eigenvalue, then*

$$\|\nabla u\|_q \leqslant c(1+\lambda)\|u\|_q \tag{6.4}$$

for all $2 \leqslant q \leqslant q_0$.

Moreover, if Ω *is such that the interior cone condition holds, then there exists* $c > 0$ *depending only on* $\mathcal{V}_0$, τ, *and* θ *such that*

$$\|\nabla u\|_q \leqslant c(1+\lambda)\|u\|_{\frac{Nq}{N+q}} \tag{6.5}$$

for all $2 < q \leqslant q_0$.

Proof. We prove the statement for the operator T^*ST, the other cases being similar. We divide the proof into three steps.

Step 1. We define

$$Q(u,\psi) = \int_\Omega u\psi g dx + \int_\Omega \widetilde{a} \nabla u \cdot \nabla \psi \widetilde{g} dx,$$

$$Q_0(u,\psi) = \int_\Omega u\psi dx + \int_\Omega \nabla u \cdot \nabla \psi dx$$

for all $u \in V_q$, $\psi \in V_{q'}$. Since[2]

$$|Q_0(u,\psi) - \beta Q(u,\psi)| \leqslant \max\{\|1-\beta g\|_{L^\infty(\Omega)}, \|I - \beta \widetilde{a}\widetilde{g}\|_{L^\infty(\Omega)}\} \times \|u\|_{W^{1,q}(\Omega)} \|\psi\|_{W^{1,q'}(\Omega)},$$

there exist $\beta > 0$ and $0 < c < 1$ depending only on N, τ, and θ such that

$$|Q_0(u,\psi) - \beta Q(u,\psi)| \leqslant c\|u\|_{W^{1,q}(\Omega)} \|\psi\|_{W^{1,q'}(\Omega)} \tag{6.6}$$

for all $u \in W^{1,q}(\Omega)$ and $\psi \in W^{1,q'}(\Omega)$.

[2] Here, we use $\|f\|^p_{W^{1,p}(\Omega)} = \|f\|^p_{L^p(\Omega)} + \| |\nabla f| \|^p_{L^p(\Omega)}$ as the norm in $W^{1,p}(\Omega)$.

Step 2. Using the fact that $\|(I-\Delta_2)^{-1}\| = 1$ and that $q \mapsto \|(I-\Delta_q)^{-1}\|$ is continuous and taking into account that $2/(c+1) > 1$, we find that there exists $q_0 > 2$ such that

$$\|(I-\Delta_q)^{-1}\| < \frac{2}{c+1} \tag{6.7}$$

for all $2 \leqslant q \leqslant q_0$. By (6.6), for all $2 \leqslant q \leqslant q_0$

$$\begin{aligned} &\inf_{\|u\|_{W^{1,q}(\Omega)}=1} \sup_{\|\psi\|_{W^{1,q'}(\Omega)}=1} Q(u,\psi) \\ &\qquad \geqslant \frac{1}{\beta} \inf_{\|\psi\|_{W^{1,q'}(\Omega)}=1} \sup_{\|u\|_{W^{1,q}(\Omega)}=1} Q_0(u,\psi) - \frac{c}{\beta} \\ &\qquad = \frac{1}{\beta}\|(I-\Delta_q)^{-1}\|^{-1} - \frac{c}{\beta} > \frac{1-c}{2\beta} > 0. \end{aligned} \tag{6.8}$$

Step 3. By (6.8), the operator $I + (T^*ST)_q$ from V_q to $V'_{q'}$ defined by

$$(I + (T^*ST)_q u, \psi) = Q(u,\psi) \tag{6.9}$$

has a bounded inverse such that

$$\|(I + (T^*ST)_q)^{-1}\| = \Big(\inf_{\|u\|_{W^{1,q}(\Omega)}=1} \sup_{\|\psi\|_{W^{1,q'}(\Omega)}=1} Q(u,\psi) \Big)^{-1} < \frac{2\beta}{1-c}. \tag{6.10}$$

Then (6.4) follows from (6.9), (6.10), and the relation

$$Q(u,\psi) = (1+\lambda) \int_\Omega u\psi g\, dx \tag{6.11}$$

for all $\psi \in V_{q'}$.

Now, if Ω satisfies the interior cone condition, then the standard Sobolev embedding holds. Thus, if $q > 2$, then $q' < 2 \leqslant N$. Hence $V_{q'}$ is continuously embedded into $L^{\frac{Nq'}{N-q'}}(\Omega)$. By (6.11), we have

$$\begin{aligned} \|u\|_{W^{1,q}(\Omega)} &\leqslant (1+\lambda)\|(I + (T^*ST)_q)^{(-1)}\| \sup_{\|\psi\|_{W^{1,q'}(\Omega)}=1} \Big| \int_\Omega u\psi g\, dx \Big| \\ &\leqslant \frac{2\beta}{1-c}(1+\lambda)\|g\|_{L^\infty(\Omega)}\|u\|_{L^{\frac{Nq}{N+q}}(\Omega)} \sup_{\|\psi\|_{W^{1,q'}(\Omega)}=1} \|\psi\|_{L^{\frac{Nq'}{N-q'}}(\Omega)}, \end{aligned} \tag{6.12}$$

and the last supremum is finite due to the Sobolev embedding. □

Remark 6.5. If Ω satisfies the interior cone condition, then the inequality (6.1) is satisfied with $p = 2N/(N-2)$ if $N \geqslant 3$ and with any $p > 2$ if $N = 2$. Then, by Lemma 6.1, the condition (3.2) holds for any $\alpha > N/2$ and the operators H, $\widetilde{H}$, T^*ST, L, $\widetilde{L}$, and $\widehat{L}$ satisfy property (P1) with $q_0 = \infty$, $\gamma = N/4$ if $N \geqslant 3$ and any $\gamma > 1/2$ if $N = 2$. In fact, if $N = 2$, property (P1) is also satisfied for $\gamma = 1/2$. This follows from [11, Theorem 2.4.4] and [3, Lemma 10]. Thus, by the second part of Theorem 6.4, both properties (P1) and (P2) are satisfied for some $q_0 > 2$ and $\gamma = N(q_0 - 2)/(4q_0)$ for any $N \geqslant 2$.

If Ω is of class $C^{0,\nu}$ (i.e., Ω is locally a subgraph of $C^{0,\nu}$ functions) with $0 < \nu < 1$, then the inequality (6.1) is satisfied with $p = 2(N+\nu-1)/(N-\nu-1)$ for any $N \geqslant 2$ (cf. also [3]). Thus, Lemma 6.1 implies that the condition (3.2) holds for any $\alpha > (N+\nu-1)/(2\nu)$ and the operators H, $\widetilde{H}$, T^*ST, L, $\widetilde{L}$, $\widehat{L}$ satisfy property (P1) with $q_0 = \infty$ and $\gamma = (N+\nu-1)/(4\nu)$.

7 Estimates via Lebesgue Measure

In this section, we consider two examples to which we apply the results of the previous sections in order to obtain stability estimates via the Lebesgue measure.

Let $A_{ij} \in L^\infty(\mathbf{R}^N)$ be real-valued functions satisfying $A_{ij} = A_{ji}$ for all $i, j = 1, \dots, N$ and the condition (2.2). Let Ω be a bounded domain in $\mathbf{R}^N$ of class $C^{0,1}$, and let Γ be an open subset of $\partial\Omega$ with Lipschitz boundary in $\partial\Omega$ (cf. Definition 7.1 below). We consider the eigenvalue problem with the mixed Dirichlet–Neumann boundary conditions

$$\begin{cases} -\sum_{i,j=1}^N \frac{\partial}{\partial x_i}\left(A_{ij}(x)\frac{\partial u}{\partial x_j}\right) = \lambda u \text{ in } \Omega, \\ u = 0 \qquad \text{on } \Gamma, \\ \sum_{i,j=1}^N A_{ij}\frac{\partial u}{\partial x_j}\nu_i = 0 \qquad \text{on } \partial\Omega \setminus \Gamma, \end{cases} \tag{7.1}$$

where ν denotes the exterior unit normal to $\partial\Omega$. Note that our analysis comprehends the "simpler" cases $\Gamma = \partial\Omega$ (Dirichlet boundary conditions) or $\Gamma = \varnothing$ (Neumann boundary conditions), as well as all other cases where Γ is a connected component of $\partial\Omega$ (the boundary of Γ in $\partial\Omega$ is empty) (cf. [13] for details).

We denote by $\lambda_n[\Omega, \Gamma]$ the sequence of eigenvalues of the problem (7.1) and by $\psi_n[\Omega, \Gamma]$ the corresponding orthonormal system of eigenfunctions in $L^2(\Omega)$. In this section, we compare the eigenvalues and the eigenfunctions corresponding to open sets Ω and $\widetilde{\Omega}$ and the associate portions of the boundaries $\Gamma \subset \partial\Omega$ and $\widetilde{\Gamma} \subset \partial\widetilde{\Omega}$. To do so, we think of Ω as a fixed reference domain and we apply the results of the previous sections to transformations ϕ and

$\widetilde{\phi}$ defined in Ω, where $\phi = Id$ and $\widetilde{\phi}$ is a suitably constructed bi-Lipschitz homeomorphism such that $\widetilde{\Omega} = \widetilde{\phi}(\Omega)$ and $\widetilde{\Gamma} = \widetilde{\phi}(\Gamma)$.

Before doing so, we recall the weak formulation of the problem (7.1) in Ω. For $\Gamma \subset \partial\Omega$ we consider the space $W^{1,2}_{\Gamma}(\Omega)$ obtained by taking the closure of $C^{\infty}_{\Gamma}(\overline{\Omega})$ in $W^{1,2}(\Omega)$, where $C^{\infty}_{\Gamma}(\overline{\Omega})$ denotes the space of functions in $C^{\infty}(\overline{\Omega})$ vanishing in a neighborhood of Γ. Then the eigenvalues and eigenfunctions of the problem (7.1) in Ω are the eigenvalues and eigenfunctions of the operator L associated with the sesquilinear form Q_L defined on $\mathcal{W} := W^{1,2}_{\Gamma}(\Omega)$ as in (2.3).

Definition 7.1. Let Ω be a bounded open set in $\mathbf{R}^N$ of class $C^{0,1}$, and let Γ be an open subset of $\partial\Omega$. We say that Γ has *Lipschitz continuous boundary* $\partial\Gamma$ in $\partial\Omega$ if for all $x \in \partial\Gamma$ there exists an open neighborhood U of x in $\mathbf{R}^N$ and $\phi \in \Phi(U)$ such that

$$\begin{aligned}\phi(U \cap (\Omega \cup \Gamma)) = \{x \in \mathbf{R}^N : |x| < 1,\, x_N < 0\} \\ \cup \{x \in \mathbf{R}^N : |x| < 1,\, x_N \leqslant 0,\, x_1 > 0\}.\end{aligned}$$

7.1 Local perturbations

In this subsection, we consider open sets belonging to the following class.

Definition 7.2. Let V be a bounded open cylinder, i.e., there exists a rotation R such that $R(V) = W \times]a, b[$, where W is a bounded convex open set in $\mathbf{R}^{N-1}$. Let $M, \rho > 0$. We say that a bounded open set $\Omega \subset \mathbf{R}^N$ belongs to $\mathcal{C}^{m,1}_M(V, R, \rho)$ if Ω is of class $C^{m,1}$ (i.e., Ω is locally a subgraph of $C^{m,1}$ functions) and there exists a function $g \in C^{m,1}(\overline{W})$ such that $a + \rho \leqslant g \leqslant b$, $|g|_{m,1} := \sum_{0<|\alpha|\leqslant m+1} \|D^{\alpha} g\|_{L^{\infty}(W)} \leqslant M$, and

$$R(\Omega \cap V) = \{(\overline{x}, x_N) \ : \ \overline{x} \in W\,,\, a < x_N < g(\overline{x})\}. \tag{7.2}$$

Let $\Omega, \widetilde{\Omega} \in \mathcal{C}^{0,1}_M(V, R, \rho)$ be such that $\Omega \cap (V_\rho)^c = \widetilde{\Omega} \cap (V_\rho)^c$. We assume that the corresponding sets $\Gamma \subset \partial\Omega$, $\widetilde{\Gamma} \subset \partial\widetilde{\Omega}$, where the Dirichlet boundary conditions are imposed, are such that

$$\Gamma \cap V^c = \widetilde{\Gamma} \cap V^c \quad \text{and} \quad P_{R^{(-1)}W}(\Gamma \cap V) = P_{R^{(-1)}W}(\widetilde{\Gamma} \cap V), \tag{7.3}$$

where $P_{R^{(-1)}W}$ denotes the orthogonal projection onto $R^{(-1)}W$. Given Γ, the condition (7.3) uniquely determines $\widetilde{\Gamma}$.

Theorem 7.3. *Let $\Omega \in \mathcal{C}^{0,1}_M(V, R, \rho)$, and let Γ be an open subset of $\partial\Omega$ with Lipschitz continuous boundary in $\partial\Omega$. Then there exists $2 < q_0 \leqslant \infty$ such that for any $r > \max\{2, N(q_0 - 1)/q_0\}$ the following statements hold.*

(i) *There exists* $c_1 > 0$ *such that*

$$\left(\sum_{n=1}^{\infty} \left| \frac{1}{\lambda_n[\widetilde{\Omega}, \widetilde{\Gamma}] + 1} - \frac{1}{\lambda_n[\Omega, \Gamma] + 1} \right|^r \right)^{1/r} \leqslant c_1 |\Omega \vartriangle \widetilde{\Omega}|^{\frac{q_0 - 2}{r q_0}} \tag{7.4}$$

for all $\widetilde{\Omega} \in C_M^{0,1}(V, R, \rho)$ *such that* $\widetilde{\Omega} \cap (V_\rho)^c = \Omega \cap (V_\rho)^c$, $|\Omega \vartriangle \widetilde{\Omega}| \leqslant c_1^{-1}$, *where* $\widetilde{\Gamma} \subset \partial\widetilde{\Omega}$ *is determined by the condition* (7.3).

(ii) *Let* $\lambda[\Omega, \Gamma]$ *be an eigenvalue of multiplicity* m, *and let* $n \in \mathbf{N}$ *be such that* $\lambda[\Omega, \Gamma] = \lambda_n[\Omega, \Gamma] = \cdots = \lambda_{n+m-1}[\Omega, \Gamma]$. *There exists* $c_2 > 0$ *such that the following is true: if* $\widetilde{\Omega} \in C_M^{0,1}(V, R, \rho)$, $\Omega \cap (V_\rho)^c = \widetilde{\Omega} \cap (V_\rho)^c$, $|\Omega \vartriangle \widetilde{\Omega}| \leqslant c_2^{-1}$, *and* $\widetilde{\Gamma} \subset \partial\widetilde{\Omega}$ *is determined by* (7.3), *then, given orthonormal eigenfunctions* $\psi_n[\widetilde{\Omega}, \widetilde{\Gamma}], \ldots, \psi_{n+m-1}[\widetilde{\Omega}, \widetilde{\Gamma}]$, *there exist corresponding orthonormal eigenfunctions* $\psi_n[\Omega, \Gamma]$, $\ldots$, $\psi_{n+m-1}[\Omega, \Gamma]$ *such that*

$$\|\psi_n[\Omega, \Gamma] - \psi_n[\widetilde{\Omega}, \widetilde{\Gamma}]\|_{L^2(\Omega \cup \widetilde{\Omega})} \leqslant c_2 |\Omega \vartriangle \widetilde{\Omega}|^{\frac{q_0 - 2}{r q_0}}.$$

If, in addition, $A_{ij} \in C^{0,1}(\mathbf{R}^N)$, $\Omega, \widetilde{\Omega} \in C_M^{1,1}(V, R, \rho)$ *and* Γ *is a connected component of* $\partial\Omega$, *then statements* (i) *and* (ii) *hold with* $q_0 = \infty$.

For the proof we need the following variant of Lemma 4.1 in [5].

Lemma 7.4. *Let* W *be a bounded convex open set in* $\mathbf{R}^{N-1}$, *and let* $M > 0$. *Let* $0 < \rho < b - a$ *and* g_1, g_2 *be Lipschitz continuous functions from* $\overline{W}$ *to* $\mathbf{R}$ *such that*

$$a + \rho < g_1(\overline{x}), \quad g_2(\overline{x}) < b \tag{7.5}$$

for all $\overline{x} \in \overline{W}$ *and such that* $\mathrm{Lip} g_1$, $\mathrm{Lip} g_2 \leqslant M$. *Suppose that* $\delta = \frac{\rho}{2(b-a)}$, $g_3 = \min\{g_1, g_2\} - \delta |g_1 - g_2|$, *and*

$$\mathcal{O}_k := \{(\overline{x}, x_N) : \ \overline{x} \in W, \ a < x_N < g_k(\overline{x})\} \tag{7.6}$$

for $k = 1, 2, 3$. *Let* Φ *be the map from* $\overline{\mathcal{O}}_1$ *into* $\overline{\mathcal{O}}_2$ *defined as follows:*

if $g_2(\overline{x}) \leqslant g_1(\overline{x})$, *then*

$$\Phi(\overline{x}, x_N) \equiv \begin{cases} (\overline{x}, x_N) & \text{if } (\overline{x}, x_N) \in \overline{\mathcal{O}}_3, \\ (\overline{x}, g_2(\overline{x}) + \frac{\delta}{\delta+1}(x_N - g_1(\overline{x}))) & \text{if } (\overline{x}, x_N) \in \overline{\mathcal{O}}_1 \setminus \overline{\mathcal{O}}_3; \end{cases} \tag{7.7}$$

if $g_2(\overline{x}) > g_1(\overline{x})$, *then*

$$\Phi(\overline{x}, x_N) \equiv \begin{cases} (\overline{x}, x_N) & \text{if } (\overline{x}, x_N) \in \overline{\mathcal{O}}_3, \\ (\overline{x}, g_2(\overline{x}) + \frac{\delta+1}{\delta}(x_N - g_1(\overline{x}))) & \text{if } (\overline{x}, x_N) \in \overline{\mathcal{O}}_1 \setminus \overline{\mathcal{O}}_3. \end{cases} \tag{7.8}$$

Then $\varnothing \neq \mathcal{O}_3 \subset \mathcal{O}_1 \cap \mathcal{O}_2$,

$$|\{x \in \mathcal{O}_1 : \ \Phi(x) \neq x\}| = |\mathcal{O}_1 \setminus \mathcal{O}_3| \leqslant 2 |\mathcal{O}_1 \vartriangle \mathcal{O}_2|, \tag{7.9}$$

and Φ is a bi-Lipschitz homeomorphism of $\overline{\mathcal{O}}_1$ onto $\overline{\mathcal{O}}_2$. Moreover, $\Phi \in \Phi_\tau(\Omega)$, where τ depends only on N, M, δ.

Proof. The proof is the same as that of Lemma 4.1. in [5], where the case $g_2 \leqslant g_1$ was considered: here we simply replace $g_1 - g_2$ with $|g_1 - g_2|$. □

Proof of Theorem 7.3. We apply Theorems 4.11 and 5.6 with $\phi = Id$ and $\widetilde{\phi}$ given by

$$\widetilde{\phi}(x) = \begin{cases} x, & x \in \overline{\Omega} \setminus V, \\ R^{(-1)} \circ \Phi \circ R(x), & x \in \overline{\Omega} \cap V. \end{cases} \tag{7.10}$$

Here, Φ is defined as in Lemma 7.4 for $g_1 = g$ and $g_2 = \widetilde{g}$, where $g, \widetilde{g}$ are the functions describing the boundaries in V of Ω, $\widetilde{\Omega}$ respectively, as in Definition 7.2. Then clearly $\phi, \widetilde{\phi} \in \Phi_\tau(\Omega)$, where τ depends only on N, V, M, ρ. It is clear that $\phi(\Omega) = \Omega$ and $\widetilde{\phi}(\Omega) = \widetilde{\Omega}$. Moreover, $\widetilde{\phi}(\Gamma) = \widetilde{\Gamma}$. Hence

$$C_{\widetilde{\phi}}[W^{1,2}_{\widetilde{\Gamma}}(\widetilde{\Omega})] = C_\phi[W^{1,2}_{\Gamma}(\Omega)].$$

Moreover, the condition (3.2) is satisfied for any $\alpha > N/2$ (cf. Remark 3.1). Hence assumption (A) is satisfied. Note that, by (7.9) and the boundedness of the coefficients A_{ij},

$$\begin{aligned} \delta_p(\phi, \widetilde{\phi})^p &\leqslant c \int_{\{x \in \Omega : \phi(x) \neq \widetilde{\phi}(x)\}} \left(|\nabla \phi - \nabla \widetilde{\phi}|^p + |A \circ \phi - A \circ \widetilde{\phi}|^p\right) dx \\ &\leqslant c |\Omega \vartriangle \widetilde{\Omega}|. \end{aligned} \tag{7.11}$$

By [13, Theorem 3], the assumption of Theorem 6.4 is satisfied for the space $\mathcal{V}_0 = C^\infty_\Gamma(\overline{\Omega})$ for some $2 < q_1 < \infty$. Thus, by Remark 6.5, the operators L, $\widetilde{L}$, and $\widehat{L}$ satisfy properties (P1) and (P2) for some $2 < q_0 < \infty$ and $\gamma = N(q_0 - 2)/(4q_0)$. Thus, statement (i) follows from Theorem 4.11 (ii) with $p = q_0/(q_0 - 2)$. Moreover, Theorem 5.6 (ii) provides the existence of orthonormal eigenfunctions $\psi_k[\Omega, \Gamma]$ satisfying the estimate (5.24) with $s = [q_0/(q_0-2)] \max\{2, N(q_0-1)/q_0\}$. By Lemma 6.1, the functions $\psi_k[\Omega, \Gamma]$, $\psi_k[\widetilde{\Omega}, \widetilde{\Gamma}]$ are bounded. Hence, by (7.9),

$$\|\psi_k[\Omega, \Gamma] \circ \phi - \psi_k[\Omega, \Gamma] \circ \widetilde{\phi}\|^2_{L^2(\Omega)}, \|\psi_k[\widetilde{\Omega}, \widetilde{\Gamma}] \circ \phi - \psi_k[\widetilde{\Omega}, \widetilde{\Gamma}] \circ \widetilde{\phi}\|^2_{L^2(\Omega)} \leqslant c |\Omega \vartriangle \widetilde{\Omega}|. \tag{7.12}$$

Thus, statement (ii) follows from the estimates (5.24) and (7.12).

Finally, if $A_{ij} \in C^{0,1}(\mathbf{R}^N)$, $\Omega, \widetilde{\Omega} \in C^{1,1}_M(V, R, \rho)$, and Γ is a connected component of $\partial\Omega$, by Troianiello [25, Thm. 3.17 (ii)], the operators L and $\widetilde{L}$ satisfy the a priori estimate (6.2) in Ω and $\widetilde{\Omega}$ respectively. Thus, by Theorem 6.3, the operators L, $\widetilde{L}$, and $\widehat{L}$ satisfy properties (P1) and (P2) with $q_0 = \infty$ and $\gamma = N/4$, and the result follows as above. □

7.2 Global normal perturbations

Let Ω be a bounded domain with C^2 boundary. By the Tubular Neighborhood Theorem, there exists $t > 0$ such that for each $x \in (\partial\Omega)^t := \{x \in \mathbf{R}^N : \operatorname{dist}(x, \partial\Omega) < t\}$ there exists a unique couple $(\overline{x}, s) \in \partial\Omega \times]-t, t[$ such that $x = \overline{x} + s\nu(\overline{x})$; moreover, $\overline{x}$ is the (unique) nearest to x point of the boundary and $s = \operatorname{dist}(x, \partial\Omega)$. One can see that, by possibly reducing the value of t, the map $x \mapsto (\overline{x}, s)$ is a bi-Lipschitz homeomorphism of $(\partial\Omega)^t$ onto $\partial\Omega \times]-t, t[$. Accordingly, we often use the coordinates $(\overline{x}, s)$ to represent the point $x \in (\partial\Omega)^t$.

In this subsection, we consider deformations $\widetilde{\Omega}$ of Ω of the form

$$\widetilde{\Omega} = (\Omega \setminus (\partial\Omega)^t) \cup \{(\overline{x}, s) \in (\partial\Omega)^t \,:\, s < g(\overline{x})\} \tag{7.13}$$

for appropriate functions g on $\partial\Omega$.

Definition 7.5. Let Ω and t be as above. Let $0 < \rho < t$ and $M > 0$. We say that the domain $\widetilde{\Omega}$ belongs to the class $\mathcal{C}_M^{m,1}(\Omega, t, \rho)$, $m = 0$ or 1, if $\widetilde{\Omega}$ is given by (7.13) for some $C^{m,1}$ function g on $\partial\Omega$ which takes values in $]-t+\rho, t[$ and satisfies $|g|_{m,1} \leqslant M$.

For $\Gamma \subset \partial\Omega$ and $\widetilde{\Omega} \in \mathcal{C}_M^{m,1}(\Omega, t, \rho)$ the set $\widetilde{\Gamma} \subset \partial\widetilde{\Omega}$, where the homogeneous Dirichlet boundary conditions are imposed, is given by

$$\widetilde{\Gamma} = \{(\overline{x}, g(\overline{x})) \,:\, \overline{x} \in \Gamma\}. \tag{7.14}$$

Theorem 7.6. *Let Ω be an open set of class C^2, and let $t > 0$ be as above. Let Γ be an open subset of $\partial\Omega$ with Lipschitz continuous boundary in $\partial\Omega$. Then there exists $2 < q_0 \leqslant \infty$ such that for any $r > \max\{2, N(q_0 - 1)/q_0\}$ the following statements hold.*

(i) *There exists $c_1 > 0$ such that*

$$\left(\sum_{n=1}^{\infty}\left|\frac{1}{\lambda_n[\widetilde{\Omega}, \widetilde{\Gamma}] + 1} - \frac{1}{\lambda_n[\Omega, \Gamma] + 1}\right|^r\right)^{1/r} \leqslant c_1 |\Omega \vartriangle \widetilde{\Omega}|^{\frac{q_0 - 2}{r q_0}} \tag{7.15}$$

for all $\widetilde{\Omega} \in \mathcal{C}_M^{0,1}(\Omega, t, \rho)$ such that, $|\Omega \vartriangle \widetilde{\Omega}| \leqslant c_1^{-1}$, where $\widetilde{\Gamma} \subset \widetilde{\Omega}$ is given by (7.14).

(ii) *Let $\lambda[\Omega, \Gamma]$ be an eigenvalue of multiplicity m, and let $n \in \mathbf{N}$ be such that $\lambda[\Omega, \Gamma] = \lambda_n[\Omega, \Gamma] = \cdots = \lambda_{n+m-1}[\Omega, \Gamma]$. There exists $c_2 > 0$ such that the following is true: if $\widetilde{\Omega} \in \mathcal{C}_M^{0,1}(\Omega, t, \rho)$, $|\Omega \vartriangle \widetilde{\Omega}| \leqslant c_2^{-1}$, and $\widetilde{\Gamma} \subset \partial\widetilde{\Omega}$ is given by (7.14), then, given orthonormal eigenfunctions $\psi_n[\widetilde{\Omega}, \widetilde{\Gamma}], \ldots, \psi_{n+m-1}[\widetilde{\Omega}, \widetilde{\Gamma}]$, there exist orthonormal eigenfunctions $\psi_n[\Omega, \Gamma], \ldots, \psi_{n+m-1}[\Omega, \Gamma]$ such that*

$$\|\psi_n[\Omega,\Gamma]-\psi_n[\widetilde{\Omega},\widetilde{\Gamma}]\|_{L^2(\Omega\cup\widetilde{\Omega})}\leqslant c_2|\Omega\vartriangle\widetilde{\Omega}|^{\frac{q_0-2}{rq_0}}.$$

If, in addition, $A_{ij}\in C^{0,1}(\mathbf{R}^N)$, $\widetilde{\Omega}\in C^{1,1}_M(\Omega,t,\rho)$ *and* Γ *is a connected component of* $\partial\Omega$*, then statements* (i) *and* (ii) *hold with* $q_0=\infty$.

Proof. The proof is essentially a repetition of the proof of Theorem 7.3: the transformation Φ is defined as in Lemma 7.4, with $\partial\Omega$ replacing W and curvilinear coordinates $(\overline{x},s)$ replacing the local euclidean coordinates $(\overline{x},x_N)$. □

8 Appendix

In this section, we briefly discuss how Theorem 4.6 can be used to obtain stability estimates for the solutions of the Poisson problem.

Theorem 8.1. *Let* (A) *be satisfied. Let the operators* L, $\widetilde{L}$, *and* $\widehat{L}$ *satisfy* (P), *and let* Ω *satisfy the interior cone condition. Let* $f\in L^2(\mathbf{R}^N)$, *and let* $v\in\mathcal{W},\widetilde{v}\in\widetilde{\mathcal{W}}$ *be such that*

$$\begin{aligned}(L+1)v&=f\quad\text{in }\phi(\Omega),\\ (\widetilde{L}+1)\widetilde{v}&=f\quad\text{in }\widetilde{\phi}(\Omega).\end{aligned}$$

Let $s=[q_0/(q_0-2)]\max\{2,\alpha+2\gamma\}$. *If* $N\geqslant 3$, *then there exists* $c>0$ *depending only on* N, τ, α, c^*, q_0, C, γ, Ω *such that*

$$\begin{aligned}\|v-\widetilde{v}\|_{L^2(\phi(\Omega)\cup\widetilde{\phi}(\Omega))}\leqslant c\big((&|\mathcal{D}|^{1/N}+\delta_s(\phi,\widetilde{\phi}))\|f\|_{L^2(\mathbf{R}^N)}\\ &+\|f\circ\phi-f\circ\widetilde{\phi}\|_{L^2(\Omega)}\big),\end{aligned}$$

where $\mathcal{D}=\{x\in\Omega:\phi(x)\neq\widetilde{\phi}(x)\}$. *The same is true if* $N=2$ *provided that* $|\mathcal{D}|^{1/N}$ *is replaced with* $|\mathcal{D}|^{\frac{1}{2}-\epsilon}$, $\epsilon>0$.

Proof. Note that

$$(H+1)(v\circ\phi)=f\circ\phi\quad\text{in }\Omega,\qquad(\widetilde{H}+1)(\widetilde{v}\circ\widetilde{\phi})=f\circ\widetilde{\phi}\quad\text{in }\Omega.$$

Hence

$$\begin{aligned}\|v\circ\phi-\widetilde{v}\circ\widetilde{\phi}\|_{L^2(\Omega)}\leqslant&\|f\circ\phi-f\circ\widetilde{\phi}\|_{L^2(\Omega)}\\ &+\|(\widetilde{H}+1)^{-1}-(H+1)^{-1}\|\|f\circ\widetilde{\phi}\|_{L^2(\Omega)}.\end{aligned}$$

Proceeding as in the proof of Theorem 5.6, it is easy to see that

$$\begin{aligned}&\|v-\widetilde{v}\|_{L^2(\phi(\Omega)\cup\widetilde{\phi}(\Omega))}\\ &\qquad\leqslant c\big(\|v\circ\phi-v\circ\widetilde{\phi}\|_{L^2(\Omega)}+\|\widetilde{v}\circ\phi-\widetilde{v}\circ\widetilde{\phi}\|_{L^2(\Omega)}\end{aligned}$$

$$+\|f\circ\phi-f\circ\widetilde{\phi}\|_{L^2(\Omega)}+\|(\widetilde{H}+1)^{-1}-(H+1)^{-1}\|\|f\circ\widetilde{\phi}\|_{L^2(\Omega)}\big).$$

By the Sobolev embedding, it follows that if $N\geqslant 3$

$$\begin{aligned}&\|v\circ\phi-v\circ\widetilde{\phi}\|_{L^2(\Omega)},\|\widetilde{v}\circ\phi-\widetilde{v}\circ\widetilde{\phi}\|_{L^2(\Omega)}\\&\qquad\leqslant c|\mathcal{D}|^{1/N}(\|u\|_{L^2(\Omega)}+\|\nabla u\|_{L^2(\Omega)})\leqslant c|\mathcal{D}|^{1/N}\|f\|_{L^2(\mathbf{R}^N)}.\end{aligned}$$

The same is true for $N=2$ provided $|\mathcal{D}|^{1/N}$ is replaced with $|\mathcal{D}|^{\frac12-\epsilon}$, $\epsilon>0$. Moreover, by Theorem 4.6,

$$\|(\widetilde{H}+1)^{-1}-(H+1)^{-1}\|\|f\circ\widetilde{\phi}\|_{L^2(\Omega)}\leqslant c\delta_s(\phi,\widetilde{\phi})\|f\|_{L^2(\mathbf{R}^N)}.$$

Thus, the statement follows by combining the estimates above. □

We now apply the previous theorem in order to estimate $\|u-\widetilde{u}\|_{L^2(\Omega\cup\widetilde{\Omega})}$, where u and $\widetilde{u}$ are the solutions to the following mixed boundary valued problems and $\widetilde{\Omega}$ is either a local perturbation of Ω as in Section 7.1 or a global normal perturbation as in Section 7.2:

$$\begin{cases}-\sum_{i,j=1}^N\frac{\partial}{\partial x_i}\big(A_{ij}(x)\frac{\partial u}{\partial x_j}\big)=f\text{ in }\ \Omega,\\ u=0 & \text{on }\ \Gamma,\\ \sum_{i,j=1}^N A_{ij}\frac{\partial u}{\partial x_j}\nu_i=0 & \text{on }\ \partial\Omega\setminus\Gamma,\end{cases}$$

$$\begin{cases}-\sum_{i,j=1}^N\frac{\partial}{\partial x_i}\big(A_{ij}(x)\frac{\partial \widetilde{u}}{\partial x_j}\big)=f\text{ in }\ \widetilde{\Omega},\\ \widetilde{u}=0 & \text{on }\ \widetilde{\Gamma},\\ \sum_{i,j=1}^N A_{ij}\frac{\partial \widetilde{u}}{\partial x_j}\nu_i=0 & \text{on }\ \partial\widetilde{\Omega}\setminus\widetilde{\Gamma}.\end{cases}$$

For any $s>0$ we set

$$\mathcal{M}_f(s)=\sup_{\substack{A\subset\mathbf{R}^N\\|A|\leqslant s}}\Big(\int_A|f|^2dx\Big)^{1/2}.$$

The next theorem is a simple consequence of Theorem 8.1 and the inequality (7.11).

Theorem 8.2. *Let Ω, $\widetilde{\Omega}$, Γ, $\widetilde{\Gamma}$ be either as in Theorem* 7.3 *or as in* Theorem 7.6. *Then the following is true: there exists $2<q_0\leqslant\infty$ such that for any $r>\max\{2,N(q_0-1)/q_0\}$ there exists $c>0$ such that $|\Omega\triangle\widetilde{\Omega}|<c^{-1}$ implies*

$$\|u-\widetilde{u}\|_{L^2(\Omega\cup\widetilde{\Omega})}\leqslant c\big(|\Omega\vartriangle\widetilde{\Omega}|^{\frac{q_0-2}{rq_0}}\|f\|_{L^2(\mathbf{R}^N)}+\mathcal{M}_f(c|\Omega\vartriangle\widetilde{\Omega}|)\big).\qquad(8.1)$$

If, in addition, $A_{ij} \in C^{0,1}(\mathbf{R}^N)$, $\Omega, \widetilde{\Omega} \in C^{1,1}$ *and* Γ *is a connected component of* $\partial\Omega$, *then the estimate* (8.1) *holds with* $q_0 = \infty$.

Acknowledgments. This work was supported by the research project "Problemi di stabilità per operatori differenziali" of the University of Padova, Italy. The third author expresses his gratitude to the Department of Mathematics of the University of Athens for the kind hospitality during the preparation of this paper.

References

1. Barbatis, G.: Spectral stability under L^p-perturbation of the second-order coefficients. J. Differ. Equ. **124**, 302–323 (1996)
2. Barbatis, G.: Stability and regularity of higher order elliptic operators with measurable coefficients. J. London Math. Soc. **58**, 342–352 (1998)
3. Burenkov, V.I., Davies, E.B.: Spectral stability of the Neumann Laplacian. J. Differ. Equ. **186**, 485–508 (2002)
4. Burenkov, V.I., Lamberti, P.D.: Spectral stability of higher order uniformly elliptic operators. In: Maz'ya, V. (ed.), Sobolev Spaces in Mathematics. II: Applications in Analysis and Partial Differential Equations. Springer, New York; Tamara Rozhkovskaya Publisher, Novosibirsk. International Mathematical Series **9**, 69–102 (2009)
5. Burenkov, V.I., Lamberti, P.D.: Spectral stability of Dirichlet second order uniformly elliptic operators. J. Differ. Equ. **244**, 1712–1740 (2008)
6. Burenkov, V.I., Lamberti, P.D.: Spectral stability of general nonnegative selfadjoint operators with applications to Neumann-type operators. J. Differ. Equ. **233**, 345–379 (2007)
7. Burenkov, V.I., Lamberti, P.D., Lanza de Cristoforis, M.: Spectral stability of nonnegative selfadjoint operators (Russian). Sovrem. Mat. Fundam. Napravl. **15**, 76–111 (2006); English transl.: J. Math. Sci., New York **149**, 1417–1452 (2008)
8. Burenkov, V.I., Lanza de Cristoforis, M.: Spectral Stability of the Robin Laplacian. Proc. Steklov Inst. Math. **260**, 68–89 (2008)
9. Davies, E.B.: Sharp boundary estimates for elliptic operators. Math. Proc. Camb. Phil. Soc. **129**, 165–178 (2000)
10. Davies, E.B.: Eigenvalue stability bounds via weighted Sobolev spaces. Math. Z. **214**, 357–371 (1993)
11. Davies, E.B.: Heat Kernels and Spectral Theory. Cambridge Univ. Press, Cambridge (1989)
12. Deift, P.: Applications of a commutation formula. Duke Math. J. **45**, 267–310 (1978)
13. Gröger, K.: A $W^{1,p}$-estimate for solutions to mixed boundary value problems for second order elliptic differential equations. Math. Ann. **283**, 679–687 (1989)
14. Henry, D.: Perturbation of the Boundary in Boundary-Value Problems of Partial Differential Equations. Cambridge Univ. Press, Cambridge (2005)

15. Kozlov, V., Maz'ya, V., Movchan, A.: Asymptotic Analysis of Fields in Multistructures. Clarendon Press, Oxford Univ. Press, New York (1999)
16. Lamberti, P.D., Lanza de Cristoforis, M.: A global Lipschitz continuity result for a domain dependent Dirichlet eigenvalue problem for the Laplace operator. Z. Anal. Anwend. **24**, 277–304 (2005)
17. Lamberti, P.D., Lanza de Cristoforis, M.: A global Lipschitz continuity result for a domain-dependent Neumann eigenvalue problem for the Laplace operator. J. Differ. Equ. **216**, 109–133 (2005)
18. Maz'ya, V., Nazarov, S., Plamenevskii, B.: Asymptotic Theory of Elliptic Boundary Value Problems in Singularly Perturbed Domains. I. II. Birkhäuser, Basel (2000)
19. Meyers, N.G.: An L^p-estimate for the gradient of solutions of second order elliptic divergence equations. Ann. Scuola Norm Sup. Pisa **17**, 189–206 (1963)
20. Netrusov, Y., Safarov, Y.: Weyl asymptotic formula for the Laplacian on domains with rough boundaries. Commun. Math. Phys. **253**, 481–509 (2005)
21. Pang, M.M.H.: Approximation of ground state eigenvalues and eigenfunctions of Dirichlet Laplacians. Bull. London Math. Soc. **29**, 720–730 (1997)
22. Reed, M., Simon, B.: Methods of Modern Mathematical Physics II. Fourier Analysis, Self-Adjointness. Academic Press (1975)
23. Savaré, G., Schimperna, G.: Domain perturbations and estimates for the solutions of second order elliptic equations. J. Math. Pures Appl. (9) **81**, 1071–1112 (2002)
24. Simon, B.: Trace Ideals and their Applications. Cambridge Univ. Press, Cambridge etc. (1979)
25. Troianiello, G.M.: Elliptic Differential Equations and Obstacle Problems. Plenum Press, New York (1987)

Operator Pencil in a Domain with Concentrated Masses. A Scalar Analog of Linear Hydrodynamics

Gregory Chechkin

Abstract The problem describing low-frequency oscillations of a heavy viscous fluid in a vessel with a fine-meshed net on the fluid surface is studied in the case where the fluid density is inhomogeneous near the net. The obtained spectral problem for the operator pencil is treated by means of the Krein scheme [18]. To construct a homogenization for the quadratic operator pencil, the method of matching asymptotic expansions is used.

1 Introduction

The homogenization theory and asymptotic methods are successfully used in the study of problems in micro-homogeneous media (cf., for example, [28, 4, 33, 49, 3, 17, 43, 35, 11] and the references therein). In particular, problems in strips, sectors etc. were considered in the monograph [33] by Maz'ya, Nazarov, and Plamenevskii.

Based on the homogenization method and asymptotic analysis, it is also possible to treat the case of inhomogeneous media and study problems under singular perturbations of geometry [33, 35, 32, 31, 37, 38], coefficients [30, 36], or boundary conditions [12, 25, 13, 5, 6, 11]. In particular, homogenization of differential operators on a periodic curvilinear mesh was considered by Maz'ya and Slutskii [36]. Maz'ya and Hänler [30] studied the homogenization problem in a domain with porous layer. (cf. also the recent work [31] of Maz'ya and Movchan). Some problems of hydrodynamics were treated by means of homogenization, for example, in [23, 24, 19, 20, 48, 46, 42, 47, 2].

In this paper, we consider a problem describing low-frequency oscillations of a heavy viscous fluid in a vessel with a fine-meshed net on the fluid surface

Gregory Chechkin
Department of Mathematics, Moscow State University, 119992 Moscow, Russia
e-mail: chechkin@mech.math.msu.su

A. Laptev (ed.), *Around the Research of Vladimir Maz'ya II: Partial Differential Equations*, 61
International Mathematical Series 12, DOI 10.1007/978-1-4419-1343-2_3,

in the case where the fluid density is inhomogeneous near the mesh. We consider normal oscillations, i.e., the fluid velocity depends on time as $e^{-\lambda t}$. The obtained spectral problem for the operator pencil is treated by means of the Krein scheme [18]. Some results presented below were announced in [7]. We refer to [29] and [50] for other problems with such operator pencils.

2 Statement of the Problem

Let Ω be a smooth domain in $\mathbb{R}^d$, $d \geqslant 3$, with boundary $\partial\Omega$. We assume that $\partial\Omega = \Gamma_1 \cup \Gamma_2$, where Γ_2 lies in the hyperplane $x_d = 0$ and consists of two parts Γ_ε and $\gamma_\varepsilon = \bigcup_{i=1}^{N_\delta} \gamma_\varepsilon^i$. We denote by $B_\varepsilon = \bigcup_{i=1}^{N_\delta} B_\varepsilon^i$ the union of half-balls located inside Ω. Let us explain the construction. Let B_ε be a homothetic contraction, δB^ε, B_0^ε is the half-ball $\{(\xi_1, \dots, \xi_d) \mid \xi_1^2 + \dots + \xi_d^2 < \varepsilon^2,\ \xi_d < 0\}$ in the stretched space $\mathbb{R}^d$, $\xi = \frac{x}{\delta}$, $\gamma_0^\varepsilon = \{(\xi_1, \dots, \xi_d) \mid \xi_1^2 + \dots + \xi_{d-1}^2 < \varepsilon^2,\ \xi_d = 0\}$, B^ε is the domain obtained by integer shifts of B_0^ε along the hyperplane $\xi_d = 0$ with centers $\widetilde{\xi}_k = (k_1, \dots, k_{d-1}, 0)$, $k_1, \dots, k_{d-1} \in \mathbb{N}$. We also set $\gamma_0 = \{(\zeta_1, \dots, \zeta_d) \mid \zeta_1^2 + \dots + \zeta_{d-1}^2 < 1,\ \zeta_d = 0\}$ and $\zeta = \xi\varepsilon^{-1}$. The geometric configuration on the hyperplane $\{x_d = 0\}$ is similar to that described in [26, 27, 6] (cf. Fig. 1). Furthermore, $\gamma_\varepsilon = \overline{B}_\varepsilon \cap \partial\Omega$. We consider the case where the parameter $\delta(\varepsilon)$ defining the characteristic distance between γ_ε^i on the boundary tends to zero as $\varepsilon \to 0$. It is easy to see that $N_\delta = O\big(\frac{1}{\delta^{d-1}}\big)$.

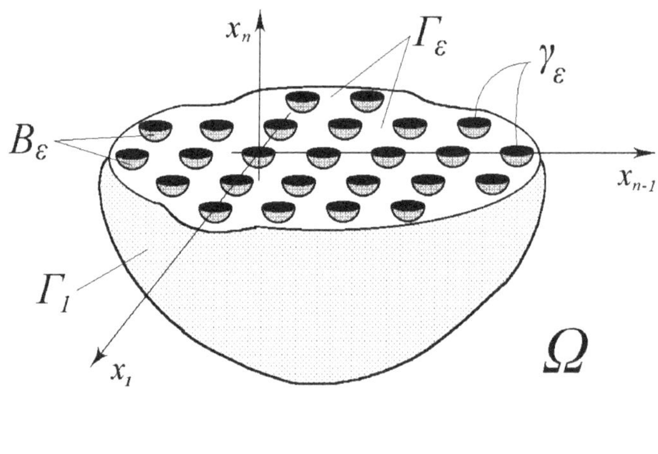

Fig. 1 A body with many concentrated masses distributed along the boundary.

We consider the following spectral problems:

$$\begin{aligned} &\Delta s_\varepsilon^k = -\lambda_\varepsilon^k \rho^\varepsilon(x) s_\varepsilon^k \quad \text{in } \Omega, \\ &s_\varepsilon^k = 0 \quad \text{on } \Gamma_1 \cup \gamma_\varepsilon, \\ &\frac{\partial s_\varepsilon^k}{\partial x_d} = 0 \quad \text{on } \Gamma_\varepsilon \end{aligned} \tag{2.1}$$

and

$$\begin{aligned} &\Delta u_\varepsilon^k = -\lambda_\varepsilon^k \rho^\varepsilon(x) u_\varepsilon^k \quad \text{in } \Omega, \\ &u_\varepsilon^k = 0 \quad \text{on } \Gamma_1 \cup \gamma_\varepsilon, \\ &\lambda_\varepsilon^k \frac{\partial u_\varepsilon^k}{\partial x_d} - q u_\varepsilon^k = 0 \quad \text{on } \Gamma_\varepsilon, \end{aligned} \tag{2.2}$$

where

$$\rho^\varepsilon(x) = \begin{cases} 1 & \text{in } \Omega \backslash B_\varepsilon, \\ 1/(\varepsilon\delta)^m & \text{in } B_\varepsilon. \end{cases}$$

Assume that $q \equiv \text{const} > 0$ and $m < 2$. We are interested in study the asymptotic behavior of eigenelements of these problems as $\varepsilon \to 0$. We will prove a homogenization theorem and estimate deviations for the egienvalues.

To study the asymptotic behavior of the eigenvalues and eigenfunctions of the problem (2.2), we show that (2.2) is equivalent to a problem about the asymptotic behavior of the eigenelements of an operator pencil. Owing to the scheme in [18], we can reduce the problem under consideration to two simpler spectral problems (one of which is (2.1)) for the Laplace operator. We also show that the problem (2.2) is a scalar analog of a classical problem in linear hydrodynamics.

3 Homogenization of Boundary Value Problems in Domains with Concentrated Masses Periodically Distributed along the Boundary

3.1 Notation and the main results

Consider the spectral problem (2.1). As is known, this problem has discrete spectrum of countably many eigenvalues of finite multiplicity. Assume that the eigenvalues $\{\lambda_\varepsilon^k\}$ are enumerated in nondescedning order, i.e., $0 < \lambda_\varepsilon^1 < \lambda_\varepsilon^2 \leqslant \ldots \leqslant \lambda_\varepsilon^k \leqslant \ldots$, with accounting their multiplicity. Moreover,

$$\int_\Omega \rho^\varepsilon(x) s_\varepsilon^k s_\varepsilon^l dx = \delta_{kl}.$$

We set

$$P := \lim_{\varepsilon \to 0} \frac{\varepsilon^{d-2}}{\delta(\varepsilon)}$$

and

$$D = \left\{ \xi \in \mathbb{R}^d \mid -\frac{1}{2} < \xi_i < \frac{1}{2}, \ i = 1, \dots, d-1, \ \xi_d < 0 \right\},$$

$$\Sigma = \left\{ \xi \in \mathbb{R}^d \mid -\frac{1}{2} < \xi_i < \frac{1}{2}, \ i = 1, \dots, d-1, \ \xi_d = 0 \right\},$$

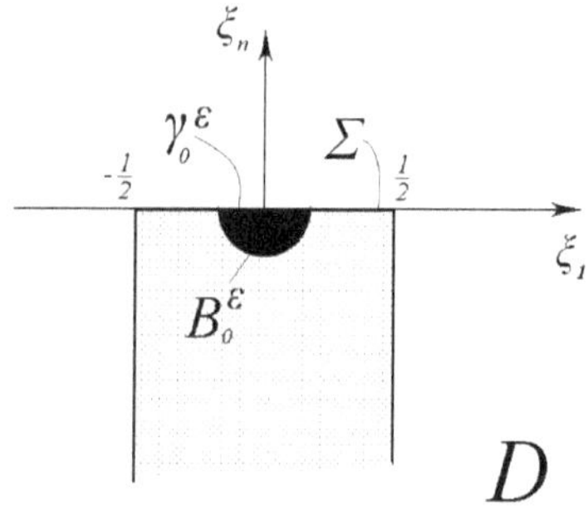

Fig. 2 The periodicity cell.

$$D_\rho = \{\xi \in D \mid \xi_d < -\rho\}, \quad \Sigma_\rho = \{\xi \in D \mid \xi_d = -\rho\}, \quad D_{\rho_1 \rho_2} = D_{\rho_1} \backslash \overline{D}_{\rho_2}.$$

Let a periodic function Θ^ε of $\xi_1, \dots, \xi_{d-1}$ be the first eigenfunction of the Steklov type problem

$$\begin{aligned} &\Delta \Theta^\varepsilon = 0 \quad \text{in } D, \\ &\Theta^\varepsilon = 0 \quad \text{on } \gamma_0^\varepsilon, \\ &\frac{\partial \Theta^\varepsilon}{\partial \xi_d} = \varsigma_\varepsilon \Theta^\varepsilon \quad \text{on } \Sigma \backslash \gamma_0^\varepsilon. \end{aligned} \tag{3.1}$$

We define the function ϑ_ε by the equality

$$\vartheta_\varepsilon(x) = 1 + \psi(x_d) \left(\Theta^\varepsilon \left(\frac{x}{\delta} \right) - 1 \right)$$

and extend it as a periodic function. Here, $\psi(t)$ is a smooth cut-off function of one variable, $0 \leqslant \psi \leqslant 1$, $\psi \equiv 1$ in some sufficiently small neighborhood of Γ_2. The properties of Θ^ε (and, consequently, of ϑ_ε) are listed in Proposition 3.1 and are discussed in [6] in detail.

To formulate the main results of this paper, we consider the boundary-value problem

$$\begin{aligned} &\Delta s^\varepsilon = -\rho^\varepsilon(x) f^\varepsilon \quad \text{in } \Omega, \\ &s^\varepsilon = 0 \quad \text{on } \Gamma_1 \cup \gamma_\varepsilon, \\ &\frac{\partial s^\varepsilon}{\partial x_d} = 0 \quad \text{on } \Gamma_\varepsilon \end{aligned} \tag{3.2}$$

corresponding to the spectral problem (2.1), and

$$
\begin{aligned}
&\Delta s^0 = -f^0(x) \quad \text{in } \Omega, \\
&s^0 = 0 \quad \text{on } \partial\Omega \quad (P = +\infty), \\
&\left[\begin{array}{c} \dfrac{\partial s^0}{\partial x_d} + P\dfrac{\sigma_d c_{\gamma_0^\varepsilon}}{2} s^0 = 0 \text{ on } \Gamma_2, \\ s^0 = 0 \text{ on } \Gamma_1 \end{array}\right] \quad P < +\infty),
\end{aligned}
\tag{3.3}
$$

where σ_d is the area of the unit d-dimensional sphere and $c_{\gamma_0^\varepsilon} := \operatorname{cap}(\gamma_0^\varepsilon)$ is the harmonic capacity of $(d-1)$-dimensional disk γ_0^ε (cf., for example, [45, 22]).

For the sake of simplicity, we assume that $f^\varepsilon \equiv f^0 \equiv f$. The case $f^\varepsilon \to f^0$ can be treated in a similar way.

Theorem 3.1. *If $P < +\infty$ s^ε, and s^0 are weak solutions to the problems* (3.2) *and* (3.3) *respectively, then there exists a constant K_1 independent of ε and δ such that for sufficiently small ε*

$$
\|s^0 \vartheta_\varepsilon - s^\varepsilon\|_{H^1(\Omega)} \leqslant K_1 \Big(\varepsilon^{\frac{d-2}{2}} + \Big| \frac{\varepsilon^{d-2}}{\delta} - P \Big| + \varepsilon^{2-m} \delta^{2-m} \Big).
$$

If $P = +\infty$, then there exists K_2 independent of ε and δ such that

$$
\|s^0 \vartheta_\varepsilon - s^\varepsilon\|_{H^1(\Omega)} \leqslant K_2 \Big(\varepsilon^{\frac{d-2}{2}} + \frac{\delta^{\frac{1}{2}}}{\varepsilon^{\frac{d-2}{2}}} + \varepsilon^{2-m} \delta^{2-m} \Big).
$$

Consider the spectral problem

$$
\begin{aligned}
&\Delta s_0^k = -\lambda_0^k s_0^k \quad \text{in } \Omega, \\
&s_0^k = 0 \quad \text{on } \partial\Omega \quad (P = +\infty), \\
&\left[\begin{array}{c} \dfrac{\partial s_0^k}{\partial x_d} + P\dfrac{\sigma_d c_{\gamma_0^\varepsilon}}{2} s_0^k = 0 \quad \text{on } \Gamma_2, \\ s_0^k = 0 \quad \text{on } \Gamma_1 \end{array}\right] \quad (P < +\infty), \\
&\int_\Omega s_0^k s_0^l \, dx = \delta_{kl}, \quad 0 < \lambda_0^1 \leqslant \lambda_0^2 \leqslant \ldots .
\end{aligned}
\tag{3.4}
$$

Theorem 3.2. *Let λ_0^k and λ_ε^k be eigenvalues of the problems* (3.4) *and* (2.1) *respectively. Then*

$$
|\lambda_0^k - \lambda_\varepsilon^k| \leqslant C_k^1 \Big(\varepsilon^{\frac{d-2}{2}} + \Big| \frac{\varepsilon^{d-2}}{\delta} - P \Big| + \varepsilon^{2-m} \delta^{2-m} \Big) \quad \text{if } P < \infty,
$$

$$
|\lambda_0^k - \lambda_\varepsilon^k| \leqslant C_k^2 \Big(\varepsilon^{\frac{d-2}{2}} + \frac{\delta^{\frac{1}{2}}}{\varepsilon^{\frac{d-2}{2}}} + \varepsilon^{2-m} \delta^{2-m} \Big) \quad \text{if } P = +\infty,
$$

where the constants C_k^1 and C_k^2 are independent of ε and δ.

If the multiplicity of the eigenvalue λ_0^l of the problem (3.4) is equal to r, i.e., $\lambda_0^l = \lambda_0^{l+1} = \ldots = \lambda_0^{l+r}$, then for any eigenfunction s_0^l of the problem (3.4) corresponding to the eigenvalue λ_0^l, $\|s_0\|_{L_2(\Omega)} = 1$, there exists a linear combination $\overline{s^\varepsilon}$ of eigenfunctions of the problem (2.1) corresponding to the eigenvalue $\lambda_\varepsilon^{l+1}, \ldots, \lambda_\varepsilon^{l+r}$ such that

$$\Big(\int_\Omega \rho^\varepsilon(x)|\overline{s^\varepsilon} - s_0^l|^2 dx\Big)^{\frac{1}{2}} \leqslant C_l^1\Big(\varepsilon^{\frac{d-2}{2}} + \Big|\frac{\varepsilon^{d-2}}{\delta} - P\Big| + \varepsilon^{2-m}\delta^{2-m}\Big) \quad \text{if } P < \infty,$$

$$\Big(\int_\Omega \rho^\varepsilon(x)|\overline{s^\varepsilon} - s_0^l|^2 dx\Big)^{\frac{1}{2}} \leqslant C_l^2\Big(\varepsilon^{\frac{d-2}{2}} + \frac{\delta^{\frac{1}{2}}}{\varepsilon^{\frac{d-2}{2}}} + \varepsilon^{2-m}\delta^{2-m}\Big) \quad \text{if } P = \infty,$$

where the constants C_l^1 and C_l^2 are independent of ε and s_0^l.

3.2 Preliminaries

We study the behavior of solutions to the spectral problem (2.1) as $\varepsilon \to 0$. The existence and uniqueness of a solution s_ε^k to the problem in $H^1(\Omega, \gamma_\varepsilon)$ can be established by the Lax–Milgram theorem (cf., for example, [53] and also [11]). The space $H^1(\Omega, \gamma_\varepsilon)$ is defined as the completion of functions in $C^\infty(\overline{\Omega})$ vanishing in a neighborhood of $\Gamma_1 \cup \gamma_\varepsilon$ in the $H^1(\Omega)$-norm. The space $H^1(D, \gamma_0^\varepsilon)$ is the completion in the norm

$$\|u\|_1 \equiv \Big(\int_D |\nabla_\xi u|^2 d\xi + \int_\Sigma u^2 d\widehat{\xi}\Big)^{\frac{1}{2}}$$

of the set of functions of class $C^\infty(\overline{D})$ that are 1-periodic in $\widehat{\xi} \equiv (\xi_1, \ldots, \xi_{d-1})$, vanish in a neighborhood of γ_0^ε and possess the finite Dirichlet integral over D.

For the minimal eigenvalue of the problem (3.1) we have

$$\varsigma_\varepsilon = \inf_{v \in H^1(D, \gamma_0^\varepsilon)\setminus\{0\}} \left(\frac{\displaystyle\int_D |\nabla_\xi v|^2 d\xi}{\displaystyle\int_\Sigma v^2 d\widehat{\xi}} \right). \tag{3.5}$$

Theorem 3.3. *Let σ_d be the area of the unit d-dimensional sphere, and let c_{γ_0} be the harmonic capacity of γ_0. Then*

$$\varsigma_\varepsilon = \varepsilon^{d-2}\frac{\sigma_d}{2}c_{\gamma_0} + o(\varepsilon^{d-2}). \tag{3.6}$$

The proof will be given below.

We also need the following assertion (cf. [6]).

Proposition 3.1. *There exists a harmonic function $\Theta^\varepsilon(\xi) \in H^1(D, \gamma_0^\varepsilon)$ in D at which the infimum in (3.5) is attained, i.e., Θ^ε is the first eigenfunction of the problem (3.1),*

$$\int_D |\nabla_\xi \Theta^\varepsilon|^2 d\xi = \varsigma_\varepsilon, \qquad \|\Theta^\varepsilon\|_{L_2(\Sigma)} = 1;$$

moreover, the boundary condition

$$\frac{\partial \Theta^\varepsilon}{\partial \xi_d} = \varsigma_\varepsilon \Theta^\varepsilon \quad on \quad \Sigma \backslash \gamma_0^\varepsilon \tag{3.7}$$

is satisfied in the following sense:

$$\int_{\Sigma_\rho} \frac{\partial \Theta^\varepsilon}{\partial \xi_d} v d\widehat{\xi} \longrightarrow \varsigma_\varepsilon \int_\Sigma \Theta^\varepsilon v d\widehat{\xi} \quad as \quad \rho \to 0 \quad for\ any \quad v \in H^1(D, \gamma_0^\varepsilon).$$

Here, $\Sigma_\rho = \{\xi \in \mathbb{R}^d \mid 0 < \xi_i < 1,\ i = 1, \ldots, d-1, \xi_d = -\rho\}$.

By (3.7), we have

$$\frac{\partial \vartheta_\varepsilon}{\partial x_d} = \frac{\varsigma_\varepsilon}{\delta} \vartheta_\varepsilon \quad \text{on} \quad \Gamma_\varepsilon.$$

Lemma 3.1. *For $v \in H^1(\Omega, \gamma_\varepsilon)$ we have*

$$\int_{B_\varepsilon^i} v^2 dx \leqslant C(\varepsilon\delta)^2 \int_{B_\varepsilon^i} |\nabla v|^2 dx, \quad i = 1, \ldots, N_\delta. \tag{3.8}$$

The proof of Lemma 3.1 is based on the standard Friedrichs inequality for a half-sphere (cf., for example, [8]).

Lemma 3.2. *For a sequence of functions $v_\varepsilon \in H^1(\Omega, \gamma_\varepsilon)$*

$$\lim_{\varepsilon \to 0} (\varepsilon\delta)^{-2} \int_{B_\varepsilon^i} |v_\varepsilon|^2 dx = 0$$

if $\|v_\varepsilon\|_{H^1(\Omega)} \leqslant K$, where K is independent of ε.

Proof. We argue as in [43]) Let $v \in H^1(\Omega, \gamma_\varepsilon$. For any $\alpha > 0$ there exists $v_\alpha \in C^\infty(\Omega, \gamma_\varepsilon)$ such that $C^{\frac{1}{2}} \|v - v_\alpha\|_{H^1(\Omega)} < \alpha$, where C is the constant in (3.8). Using the estimate (3.8) for $v - v_\alpha$, we find

$$\begin{aligned} &\left(\int_{B_\varepsilon^i} (\varepsilon\delta)^{-2} |v|^2 dx \right)^{\frac{1}{2}} \\ &\leqslant \left(\int_{B_\varepsilon^i} (\varepsilon\delta)^{-2} |v - v_\alpha|^2 dx \right)^{\frac{1}{2}} + \left(\int_{B_\varepsilon^i} (\varepsilon\delta)^{-2} |v_\alpha|^2 dx \right)^{\frac{1}{2}} \end{aligned}$$

$$\leqslant C^{\frac{1}{2}}\Big(\int_\Omega |\nabla(v-v_\alpha)|^2dx\Big)^{\frac{1}{2}} + C_\alpha(\varepsilon\delta)^{\frac{d}{2}-1} \leqslant \alpha + C_\alpha(\varepsilon\delta)^{\frac{d}{2}-1}.$$

Since $n \geqslant 3$, Lemma 3.2 is proved. □

Lemma 3.3. *Let $s^\varepsilon \in H^1(\Omega,\gamma_\varepsilon)$ be a solution to the equation*

$$\Delta s^\varepsilon = -\rho^\varepsilon(x) f^\varepsilon(x) \quad \textit{in} \ \ \Omega.$$

Then

$$\int_\Omega |\nabla s^\varepsilon|^2 dx \leqslant C_1\Big((\varepsilon\delta)^{2-m}\int_{B_\varepsilon}\rho^\varepsilon(x)|f^\varepsilon|^2dx + \int_\Omega |f^\varepsilon|^2dx\Big).$$

Proof. Taking into account the integral identity, Lemma 3.1 and a Friedrichs type inequality (cf., for example, [39]–[41]), we find

$$\|\nabla s^\varepsilon\|^2_{L_2(\Omega)}$$

$$= \int_\Omega |\nabla s^\varepsilon|^2 dx = \int_\Omega \rho^\varepsilon f^\varepsilon s^\varepsilon dx = \int_{\Omega\setminus B_\varepsilon} f^\varepsilon s^\varepsilon dx + (\varepsilon\delta)^{-m}\int_{B_\varepsilon} f^\varepsilon s^\varepsilon dx$$

$$\leqslant \Big(\int_\Omega (f^\varepsilon)^2dx\Big)^{\frac{1}{2}}\Big(\int_\Omega (s^\varepsilon)^2dx\Big)^{\frac{1}{2}} + (\varepsilon\delta)^{-m}\Big(\int_{B_\varepsilon}(f^\varepsilon)^2dx\Big)^{\frac{1}{2}}\Big(\int_{B_\varepsilon}(s^\varepsilon)^2dx\Big)^{\frac{1}{2}}$$

$$\leqslant C_2\Big(\int_\Omega (f^\varepsilon)^2dx\Big)^{\frac{1}{2}}\|\nabla s^\varepsilon\|_{L_2(\Omega)} + C_3(\varepsilon\delta)^{-m+1}\Big(\int_{B_\varepsilon}(f^\varepsilon)^2dx\Big)^{\frac{1}{2}}\|\nabla s^\varepsilon\|_{L_2(B_\varepsilon)},$$

which implies the required inequality. □

3.3 Proof of the Estimate

In this subsection, we prove Theorem 3.3.

Upper estimate for ς_ε. We study the asymptotic behavior of ς_ε as $\varepsilon \to 0$. Let B^ρ be the half-ball of radius ρ with center at 0, lying in the lower half-space. Let $B^{\varepsilon_0} \subset \overline{D}$ for some $\varepsilon_0 \in (0,1]$, $\varepsilon < \varepsilon_0$. We set

$$v_\varepsilon(\xi) = \begin{cases} \frac{|\xi|-\varepsilon}{\varepsilon_0-\varepsilon} & \text{in } B^{\varepsilon_0}\setminus B^\varepsilon, \\ 0 & \text{in } B^\varepsilon, \\ 1 & \text{in } D\setminus B^{\varepsilon_0}. \end{cases}$$

It is easy to see that $v_\varepsilon \in H^1(D,\gamma_0^\varepsilon)$ and $v_\varepsilon = 1$ on $\Sigma\setminus\overline{B^{\varepsilon_0}}$. Further, we estimate ς_ε as follows:

$$
\begin{aligned}
\varsigma_\varepsilon &\leqslant \frac{\int_D |\nabla_\xi v_\varepsilon|^2\,d\xi}{\int_\Sigma v_\varepsilon^2\,d\xi'}\\
&= \frac{\frac{1}{(\varepsilon_0-\varepsilon)^2}\int_0^\pi \dots \int_0^\pi \int_\varepsilon^{\varepsilon_0} \left|\frac{\partial}{\partial r} r\right|^2 M(r,\varphi_1,\dots,\varphi_{d-1})\,dr d\varphi_1 \dots d\varphi_{d-1}}{\frac{2}{(\varepsilon_0-\varepsilon)^2}\int_0^\pi \dots \int_0^\pi \int_\varepsilon^{\varepsilon_0} |r-\varepsilon|^2 L(r,\varphi_2,\dots,\varphi_{d-1})\,dr d\varphi_2 \dots d\varphi_{d-1} + |\Sigma| - 2\varepsilon_0}\\
&\leqslant \frac{\pi^{\frac{d}{2}}}{2\Gamma(\frac{d}{2}+1)} \frac{\varepsilon_0^d - \varepsilon^d}{(\varepsilon_0-\varepsilon)^2} = \varepsilon^{d-2}\frac{\sigma_d}{2} c_{\gamma_0} + o(\varepsilon^{d-2}),
\end{aligned}
\tag{3.9}
$$

where $M(r,\varphi_1,\dots,\varphi_{d-1}) = r^{d-1}\sin^{d-1}\varphi_1 \sin^{d-2}\varphi_2 \dots \sin\varphi_{d-2}$, $L(r,\varphi_2,\dots,\varphi_{d-1}) = r^{d-1}\sin^{d-2}\varphi_2 \dots \sin\varphi_{d-2}$, $(r,\varphi_1,\dots,\varphi_{d-1})$ are the spherical coordinates.

Asymptotics of the minimal eigenvalue. We follow [15]. For an exact estimate of ς_ε, we need the asymptotic expansion for the minimal eigenvalue of the following auxiliary problem:

$$
-\Delta\psi_\varepsilon = \lambda_\varepsilon \psi_\varepsilon \quad \text{in } D_{0h}, \qquad \frac{\partial \psi_\varepsilon}{\partial n} = 0 \quad \text{on } \Gamma_\varepsilon, \qquad \psi_\varepsilon = 0 \quad \text{on } \gamma_0^\varepsilon, \tag{3.10}
$$

where $\Gamma_\varepsilon = \partial D_{0h} \backslash \overline{\gamma}_0^\varepsilon$ and n is the outward normal to ∂D_{0h}.

As usual (cf., for example, [21]), by a *weak solution* to the boundary value problem

$$
(\Delta + \lambda)\varpi_\varepsilon = f \quad \text{in } D_{0h}, \qquad \frac{\partial \varpi_\varepsilon}{\partial n} = 0 \quad \text{on } \Gamma_\varepsilon, \qquad \varpi_\varepsilon = 0 \quad \text{on } \gamma_0^\varepsilon, \tag{3.11}
$$

where $f \in L_2(D_{0h})$, we mean $\varpi_\varepsilon \in H^1(D_{0h}, \gamma_0^\varepsilon)$ such that for any $v \in H^1(D_{0h}, \gamma_0^\varepsilon)$

$$
-(\nabla_\xi \varpi_\varepsilon, \nabla_\xi v) + \lambda(\varpi_\varepsilon, v) = (f, v), \tag{3.12}
$$

where $(\cdot,\cdot)$ denotes the inner product in $L_2(D_{0h})$. Respectively, the minimal eigenvalue of (3.10) is also understood in a weak (variational) sense. It is clear that any solution of (3.12) is a weak solution of (3.11) satisfying the homogeneous boundary conditions outside the edges of the surface ∂D_{0h}. Conversely, any solution to the problem (3.11) of class $H^1(D_{0h}, \gamma_0^\varepsilon)$ satisfies (3.12). Therefore, all the assertions below are, in fact, stated for (3.10) and (3.11).

Lemma 3.4. *The minimal eigenvalues λ_ε converge to zero as $\varepsilon \to 0$. If λ is close to zero, then the solution to the problem* (3.11) *satisfies the estimate*

$$
\|\varpi_\varepsilon\| \leqslant C|\lambda - \lambda_\varepsilon|^{-1}\|f\|,
$$

where $\|\cdot\|$ is the $L_2(D_{0h})$-norm.

The proof is similar to that of Theorem 2.3 in [9].

The main goal of this subsection is to verify the formula

$$\lambda_\varepsilon = \varepsilon^{d-2}\frac{\sigma_d}{2h}c_{\gamma_0} + o(\varepsilon^{d-2}), \tag{3.13}$$

where $c_{\gamma_0} > 0$ is the capacity of the $(d-1)$-dimensional disk γ_0 and σ_d is the area of the unit sphere in $\mathbb{R}^d$. If $d = 3$ and γ_0 is the unit disk, then $c_{\gamma_0} = 2\pi^{-1}$.

Denote by $G(\xi, \eta, \lambda)$ the Green function of the limit problem (the Neumann problem for the operator $\Delta + \lambda$ in D_{0h}). Let

$$P_j^{(t,k)}(\nabla_\eta) = \sum_{|\alpha|=j} a_{j\alpha}^{(t,k)} \frac{\partial^j}{\partial\eta_1^{\alpha_1}\partial\eta_2^{\alpha_2}\dots\partial\eta_{d-1}^{\alpha_{d-1}}},$$

$$R_j^{(k)}(\nabla_\eta) = \sum_{i=1-\delta_{j+k}^0}^{j} P_i^{(j,k)}(\nabla_\eta),$$

where $\alpha = (\alpha_1, \alpha_2, \dots, \alpha_{d-1})$ is the multiindex, $a_{j\alpha}^{(q,k)}$ are constants, and δ_i^j is the Kronecker symbol. We omit the superscripts since they play an auxiliary role in the notation above. The following assertion directly follows from the definition of $R_j(\nabla_\eta)$.

Lemma 3.5. *The function $U(\xi, \lambda) = R_j(\nabla_\eta)G(\xi, 0, \lambda)$ is infinitely differentiable in $\overline{D}_{0h}\backslash\{0\}$, satisfies the equation $(\Delta + \lambda)U = 0$ in D_{0h} and the boundary condition $\partial U/\partial n = 0$ on ∂D_{0h} outside the origin and edges.*

Lemma 3.6. *For small ξ and λ the following representations hold:*

$$G(\xi, 0, \lambda) = -\frac{1}{h\lambda} + \sum_{i=0}^{\infty} b_i\lambda^i r^{-d+2(i+1)} + g_{00}(\xi, \lambda) + h_d(r)g_{01}(\xi, \lambda),$$

$$P_j(\nabla_\eta)G(\xi, 0, \lambda) = (-1)^j \sum_{i=0}^{\infty} b_i\lambda^i P_j(\nabla_\xi) r^{-d+2(i+1)} + g_{j0}(\xi, \lambda) + h_d(r)g_{j1}(\xi, \lambda),$$

where b_i are constants, $b_0 = \frac{\sigma_d}{2}$, $h_d(r) = 0$ if d is odd and $h_d = \ln r$ if d is even, $r = |\xi|$. For sufficiently small $\rho > 0$ and any $k \geqslant 0$ the function g_{qk} belongs to $C^k(\overline{B^\rho})$ relative to ξ and is analytic with respect to λ.

For the proof we refer to [15].

Corollary 3.1.

$$\lambda G(\xi, 0, \lambda) \to -1/h \quad \textit{in} \;\; L_2(D_{0h})$$

and

$$\lambda P_j(\nabla_\eta G(\xi, 0, \lambda) \to 0 \quad \textit{in} \;\; L_2(D_{0h}\backslash\overline{B^\rho})$$

as $\lambda \to 0$ for any $\rho > 0$ and $j \geqslant 1$. Let

$$\psi_\varepsilon(\xi,\lambda) = -\lambda R_0^{(0)}(\nabla_\eta)G(\xi,0,\lambda) - \sum_{i=1}^{\infty}\sum_{j=0}^{\left[\frac{i-1}{d-2}\right]} \varepsilon^i \ln^j \varepsilon \lambda R^{(j)}_{i-j(d-2)}(\nabla_\eta)G(\xi,0,\lambda),$$

where $[\,\cdot\,]$ *is the integer part of a number. The full power asymptotic expansions for the eigenvalue* λ_ε *and the corresponding eigenfunction* ψ_ε *have the form*

$$\lambda_\varepsilon = \sum_{i=d-2}^{\infty}\sum_{j=0}^{\left[\frac{i}{d-2}\right]} \varepsilon^i \ln^j \varepsilon \lambda_{ij}, \tag{3.14}$$

$$\psi_\varepsilon(\xi) = \psi_\varepsilon(\xi,\lambda_\varepsilon), \quad \xi \in D_{0h}\backslash D_{\sqrt{\varepsilon}}, \tag{3.15}$$

$$\psi_\varepsilon(\xi) = \sum_{i=0}^{\infty}\sum_{j=0}^{\left[\frac{i}{d-2}\right]} \varepsilon^i \ln^j \varepsilon v_{ij}\Big(\frac{\xi}{\varepsilon}\Big), \quad \xi \in B^{2\sqrt{\varepsilon}}. \tag{3.16}$$

Remark 3.1. If d is odd in (3.14)–(3.16), then the coefficients λ_{ij}, v_{ij}, and $R_q^{(j)}$ vanish (cf. Remark 3.3 below).

Boundary value problems for v_i are chosen by a standard manner (cf. [13, 16, 34]). We substitute (3.14) and (3.16) into (3.10) and make the change of variable $\zeta = \xi\varepsilon^{-1}$. Then we collect the terms at the same powers of ε and pass to the formal limit as $\varepsilon \to 0$. We obtain the following recurrent system of boundary value problems in the half-space:

$$\begin{aligned} &\Delta_\zeta v_{ij} = \sum_{q=d-2}^{i-2}\ \sum_{t=j-\left[\frac{i-q-2}{d-2}\right]}^{\left[\frac{q}{d-2}\right]} \lambda_{qt} v_{i-q-2,j-t}, \quad \zeta_d < 0, \\ &\frac{\partial v_{ij}}{\partial \zeta_d} = 0, \quad \zeta \in \Gamma_0, \qquad v_{ij} = 0, \quad \zeta \in \gamma_0. \end{aligned} \tag{3.17}$$

Here, $\Gamma_0 = \mathbb{R}^{d-1}\backslash\overline{\gamma_0}$ and $\mathbb{R}^{d-1}$ is the hyperplane $\zeta_d = 0$.

Let $T_j(\zeta)$ be a homogeneous functions of degree j that are either homogeneous polynomials or the products of homogeneous polynomials in r^{-2q-d} for some integer $q \neq 0$ and satisfy the "boundary" condition $\dfrac{\partial T_j(\zeta)}{\partial \zeta_d} = 0$ for $\zeta_d = 0$, $\zeta \neq 0$.

Denote by $\widetilde{\mathcal{A}}_j$ the set of series of the form

$$T(\zeta) = \sum_{q=0}^{j} T_q(\zeta) + \ln|\zeta| \sum_{q=0}^{j} P_q(\zeta),$$

where P_q are homogeneous polynomials of degree q such that $\dfrac{\partial P_j}{\partial \zeta_d} = 0$ for $\zeta_d = 0$. On the sums $\psi_\varepsilon(\xi,\lambda_\varepsilon)$, where λ_ε is an arbitrary function admitting the asymptotics (3.14), we define the operator K_q as follows (cf. [13, 16]). We represent $\psi_\varepsilon(\xi,\lambda_\varepsilon)$ as series as $r \to 0$ and make the change of variables $\zeta = \xi\varepsilon^{-1}$. In the obtained double series, we take the sum of terms of the form $\varepsilon^j \ln^i \varepsilon \Phi(\zeta)$ for $j \leqslant q$ and equate $K_q(\psi_\varepsilon(\xi,\lambda_\varepsilon))$ to this sum. From Lemmas 3.5 and 3.6 we obtain the following assertion.

Lemma 3.7. *Let λ_ε be an arbitrary function possessing the asymptotic expansion* (3.14). *Then for any integer $N \geqslant 0$*

$$K_N(\psi_\varepsilon(\xi,\lambda_\varepsilon)) = \sum_{i=0}^{N} \sum_{j=0}^{\left[\frac{i}{d-2}\right]} \varepsilon^i \ln^j \varepsilon \, V_{ij}(\zeta).$$

The series V_{ij} belong to $\widetilde{\mathcal{A}}_{i-j(d-2)}$ and are formal asymptotic solutions to the boundary value problem (3.17) *as $|\zeta| \to \infty$, where v_{ij} are replaced with V_{ij} satisfying the representations*

$$V_{00}(\zeta) = R_0^{(0)}\frac{1}{h} - 2\sigma_d^{-1}\lambda_{d-2,0}(R_0^{(0)} + \Sigma_0^0),$$

$$V_{kq}(\zeta) = \widetilde{V}_{kq}(\zeta) - 2\sigma_d^{-1}\big(\lambda_{d-2+k,q}(R_0^{(0)} + \Sigma_0^0) + \lambda_{d-2,0}\Sigma_q^k\big),$$

$$\Sigma_k^p = \sum_{i=1} (-1)^i P_i^{(i+p,k)}(\nabla_\zeta)|\zeta|^{-d+2},$$

where the series $\widetilde{V}_{kq}$ are independent of $\lambda_{d-2+t,s}$ and $P_i^{(i+t,s)}$ for $t \neq k$.

Corollary 3.2. *The series $\widetilde{V}_{kq}$ are formal asymptotic solutions to the boundary value problem* (3.17) *as $|\zeta| \to \infty$, where v_{ij} are replaced with V_{ij} on the right-hand side.*

Remark 3.2. The formulation of the lemma is rather rough. It is easy to see (cf. Lemma 3.6) that $\widetilde{V}_{k0} \equiv 0$ for $1 \leqslant k \leqslant d-2$.

Thus, to agree the series (3.15) and (3.16), it suffices to prove the existence of solutions v_{ij} to the boundary value problems (3.17) that admit asymptotic expansions as $|\zeta| \to \infty$ coinciding with V_{ij}. For this purpose, we choose the constants λ_{kq} and differential polynomials $R_k^{(t)}(\nabla_\eta)$ in a suitable way.

Before treating the problem (3.17), we formulate some obvious assertions which will be used below.

Lemma 3.8. *Let Z_j be homogeneous harmonic polynomials of degree j such that $\dfrac{\partial(Z_j|\zeta|^{-2j-d+2})}{\partial\zeta_d} = 0$ for $\zeta_d = 0$ and $\zeta \neq 0$. Then there exists a homogeneous differential polynomial $P_j(\nabla)$ such that*

$$Z_j|\zeta|^{-2i-d+2} = P_j(\nabla)|\zeta|^{-d+2}.$$

Hereinafter, by Z_j we understand harmonic polynomials satisfying the assumption of the lemma. Of course, the "boundary" condition in the lemma is equivalent to the condition $\dfrac{\partial Z_j}{\partial \zeta_d} = 0$ for $\zeta_d = 0$, but the above formulation is more convenient for our further purposes.

We denote by $\mathbb{R}^d_-$ the half-space $\zeta_d < 0$ and by $\mathcal{A}_q$ the set of functions that are defined in $\mathbb{R}^d_-$, belong to $H^1(B^R)$ cap $C(\overline{B^R}\backslash\gamma_0)$ for any R, and admit differentiable asymptotics in $\widetilde{\mathcal{A}}_q$ as $|\zeta| \to \infty$. The following assertion was proved in [15].

Lemma 3.9. *Let $F \in \mathcal{A}_q$, and let $V \in \widetilde{\mathcal{A}}_{q+2}$ be a formal asymptotic solution to the equation $\Delta V = F$ as $|\zeta| \to \infty$. Then there exists a function $v \in \mathcal{A}_{q+2}$ that is a solution to the problem*

$$\Delta v = F,\ \zeta \in \mathbb{R}^d_-, \quad \frac{\partial v}{\partial \zeta_d} = 0, \quad \zeta \in \Gamma_0, \quad v = 0,\ \zeta \in \gamma_0$$

that admits the following asymptotic expansion as $|\zeta| \to \infty$:

$$v(\zeta) = V(\zeta) + \sum_{j=0} Z_j(\zeta)|\zeta|^{-d-2j+2}. \tag{3.18}$$

Corollary 3.3. *There exists a harmonic function $X \in \mathcal{A}_0$ in $\mathbb{R}^d_-$ that satisfies the boundary conditions $\frac{\partial X}{\partial \zeta_d} = 0$ on Γ_0, $X = 0$ on γ_0 and admits the following asymptotic expansion as $|\zeta| \to \infty$:*

$$X(\zeta) = 1 - c_{\gamma_0} r^{-d+2} + \sum_{j=1} Z_j(\zeta)|\zeta|^{-d-2j+2}.$$

We note that for this model solution the constant c_{γ_0} is the capacity of the disk γ_0 (cf. [45, 22]).

Let us formulate the main result of [15] about matching asymptotic expansions.

Theorem 3.4. *There exists a function λ_ε admitting the asymptotic expansion* (3.14), *and the series* (3.15) *and* (3.16) *such that v_{ij} belong to $\mathcal{A}_{i-j(d-2)}$, are solutions to the boundary value problems* (3.17),

$$\lambda_{d-2,0} = \frac{\sigma_d}{2h} c_{\gamma_0}, \quad R_0^{(0)} = \sqrt{h}, \quad v_{00} = \frac{1}{\sqrt{h}} \tag{3.19}$$

and for any integer $N \neq 0$

$$K_N(\psi_\varepsilon(\xi, \lambda_\varepsilon)) = \sum_{i=o}^{N} \sum_{j=o}^{\left[\frac{i}{d-2}\right]} \varepsilon^i \ln^j \varepsilon v_{ij}(\zeta), \quad |\zeta| \to \infty. \tag{3.20}$$

Remark 3.3. It is easy to see that the appearance and increase of powers of $\ln\varepsilon$ in (3.14)–(3.16) happen when the asymptotic expansions for $P_j(\nabla_\eta)G(\xi,0,\lambda)$ at zero are re-written in the "internal" variable ζ, but only if d is even (cf. Lemma 3.6). In the asymptotic expansions corresponding to an odd d, there are no the powers of $\ln\varepsilon$ (cf. Remark 3.1).

Remark 3.4. Taking into account Remark 3.2, it is easy to see that $V_{k0} = \lambda_{d-2+k,0} = \Sigma_0^k = 0$, $1 \leqslant k \leqslant d-2$. Consequently, an exact estimate in (3.14) has the form $O(\varepsilon^{2(d-2)})$ for odd d and $O(\varepsilon^{2(d-2)}|\ln\varepsilon|)$ for even d.

Let $\chi(t)$ be an infinitely differentiable cut-off function that vanishes for $t<1$ and is equal to 1 for $t>2$. Denote by $v_{\varepsilon,N}(\xi/\varepsilon)$ the partial sum of the series (3.16) and set

$$\begin{aligned}\psi_{\varepsilon,N}(\xi,\lambda) &= -\lambda_\varepsilon R_0^{(0)}(\nabla_\eta)G(\xi,0,\lambda_\varepsilon)\\ &\quad -\sum_{i=1}^{N}\sum_{j=0}^{\left[\frac{i-1}{d-2}\right]} \varepsilon^i \ln^j\varepsilon\lambda_\varepsilon R^{(j)}_{i-j(d-2)}(\nabla_\eta)G(\xi,0,\lambda_\varepsilon).\end{aligned}$$

From Theorem 3.4 we obtain the following assertion (cf., for example, [13, 14]).

Lemma 3.10. *The function*

$$\psi_{\varepsilon,N}(\xi) = \chi\Big(\frac{r}{\sqrt{\varepsilon}}\Big)\psi_{\varepsilon,N}(\xi,\lambda_\varepsilon) + \Big(1-\chi\big(\frac{r}{\sqrt{\varepsilon}}\big)\Big)v_{\varepsilon,N}(\xi/\varepsilon)$$

belongs to $H^1(D_{0h},\gamma_{0\varepsilon})$ *and is a solution to the problem* (3.11) *with* $\lambda=\lambda_\varepsilon$ *and* $f=f_{\varepsilon,N}$. *Moreover,*

$$\Big\|\psi_{\varepsilon,N}-\frac{1}{\sqrt{h}}\Big\|\to 0,\quad \|f_{\varepsilon,N}\|<C_d\varepsilon^{N_1},$$

where N_1 *unboundedly increases if* N *increases.*

By Lemmas 3.4 and 3.10 and the arbitrariness of the choice of N, the series (3.14) in Theorem 3.4 is a full asymptotic expansion of the minimal eigenvalue λ_ε of the problem (3.10) and the coefficient at the leading term of the asymptotics has the form (3.13).

Thus, we obtain the asymptotic representation for the first eigenvalue λ_ε of the problem (3.10). Let us estimate ς_ε with the help of (3.13).

Lower estimate for ς_ε. To estimate ς_ε from below, we prove the following assertion.

Lemma 3.11. *For any* $v\in H^1(B_{0\rho})$ *and* $\varkappa>0$

$$\|v\|^2_{L_2(\Sigma)} \leqslant \rho^{-1}(1+\varkappa)\|v\|^2_{L_2(B_{0\rho})} + \rho\Big(1+\frac{1}{\varkappa}\Big)\|\nabla v\|^2_{L_2(B_{0\rho})}. \tag{3.21}$$

Proof. It suffices to prove (3.21) for $v \in C^\infty(\overline{B}_{0\rho})$. We note that

$$|v(\xi', 0|^2 = \left(\int_0^s \frac{\partial v}{\partial \xi_d}\, d\xi_d - v(\xi', s)\right)^2$$
$$\leqslant \left(1 + \frac{1}{\varkappa}\right)\left|\int_0^s \frac{\partial v}{\partial \xi_d}\, d\xi_d\right|^2 + (1+\varkappa)|v(\xi', s)|^2$$

$$\leqslant \left(1 + \frac{1}{\varkappa}\right)|s|\int_0^s \left|\frac{\partial v}{\partial \xi_d}\right|^2 d\xi_d + (1+\varkappa)|v(\xi', s)|^2$$
$$\leqslant \left(1 + \frac{1}{\varkappa}\right)\rho\int_0^\rho \left|\frac{\partial v}{\partial \xi_d}\right|^2 d\xi_d + (1+\varkappa)|v(\xi', s)|^2.$$

Then, integrating over $s \in [0, \rho]$, we find

$$\rho|v(\xi', 0)|^2 \leqslant \left(1 + \frac{1}{\varkappa}\right)\rho^2\int_0^\rho \left|\frac{\partial v}{\partial \xi_d}\right|^2 d\xi_d + (1+\varkappa)\int_0^\rho v^2\, d\xi_d.$$

Dividing both sides of this inequality by ρ and integrating it with respect to ξ' along Σ, we find (3.21). Lemma 3.11 is proved. □

Consider the space $\widetilde{H}^1(D, \gamma_{0\varepsilon})$ which is the completion in the norm $\|\cdot\|_1$ of the set of functions in $C^\infty(D)$ vanishing in a neighborhood of γ_0^ε and possessing the finite Dirichlet integral. By Lemma 3.11, we have

$$\begin{aligned}\varsigma_\varepsilon &\geqslant \inf_{v \in \widetilde{H}^1(D,\gamma_0^\varepsilon)\setminus\{0\}} \left(\frac{\|\nabla_\xi v\|^2_{L_2(B_{0\rho})}}{\|v\|^2_{L_2(\Sigma)}}\right)\\ &\geqslant \inf_{v \in \widetilde{H}^1(D,\gamma_0^\varepsilon)\setminus\{0\}} \left(\frac{\|\nabla_\xi v\|^2_{L_2(B_{0\rho})}}{\frac{1}{\rho}(1+\varkappa)\|v\|^2_{L_2(B_{0\rho})} + \rho\left(1+\frac{1}{\varkappa}\right)\|\nabla_\xi v\|^2_{L_2(B_{0\rho})}}\right).\end{aligned} \tag{3.22}$$

To estimate $\|\nabla_\xi v\|^2_{L_2(B_{0\rho})}$ in terms of $\|v\|^2_{L_2(B_{0\rho})}$ for $v \in \widetilde{H}^1(D, \gamma_{0\varepsilon})$, we can use the variational definition of the minimal eigenvalue and its asymptotic behavior (3.13).

Taking into account (3.13), we have

$$\|\nabla_\xi v\|^2_{L_2(B_{0\rho})} \geqslant \left(\varepsilon^{d-2}\frac{\sigma_d}{2\rho}c_{\gamma_0} + o(\varepsilon^{d-2})\right)\|v\|^2_{L_2(B_{0\rho})}. \tag{3.23}$$

We multiply the numerator and denominator on the right-hand side (3.22) by $\left(\varepsilon^{d-2}\frac{\sigma_d}{2\rho}c_{\gamma_0} + o(\varepsilon^{d-2})\right)$ and, taking into account the estimate (3.23), replace the right-hand side of (3.22) with a less number. For fixed $\varkappa > 0$ and $\rho > 0$ we have

$$\varsigma_\varepsilon \geqslant \inf \left(\frac{\left(\varepsilon^{d-2}\frac{\sigma_d}{2\rho}c_{\gamma_0} + o(\varepsilon^{d-2})\right) \|\nabla_\xi v\|^2_{L_2(B_{0\rho})}}{\left(\frac{1}{\rho}(1+\varkappa) + (1+\frac{1}{\varkappa})\left(\varepsilon^{d-2}\frac{\sigma_d}{2}c_{\gamma_0} + o(\varepsilon^{d-2})\right)\right) \|\nabla_\xi v\|^2_{L_2(B_{0\rho})}} \right)$$

$$= \varepsilon^{d-2}\frac{\sigma_d}{2}c_{\gamma_0} + o(\varepsilon^{d-2}) \quad \text{as} \quad \varepsilon \to 0. \tag{3.24}$$

where the infimum is taken over $v \in \widetilde{H}^1(D, \gamma_0^\varepsilon)\backslash\{0\}$ Taking into account (3.9) and (3.24), we conclude that ς_ε admits the asymptotic expansion (3.6). Theorem 3.3 is proved.

3.4 Proof of the main assertions

We use the scheme suggested in [43] for studying the behavior of eigenvalues and eigenfunctions of the problem (2.1). Let us recall this scheme.

Let H_ε and H_0 be separable Hilbert spaces equipped with the inner products $(u,v)_\varepsilon$, $(u,v)_0$ and the norms $\|u\|_\varepsilon$, $\|u\|_0$ respectively. Let ε be a small positive parameter. Let $A_\varepsilon \in \mathcal{L}(H_\varepsilon)$ and $A_0 \in \mathcal{L}(H_0)$ be linear continuous operators such that $\operatorname{Im} A_0 \subset V \subset H_0$, where V is a linear subspace of H_0. Assume that the following conditions hold.

C1 There exist linear continuous operators $R_\varepsilon : H_0 \to H_\varepsilon$ such that for any $f \in V$ we have $(R_\varepsilon f, R_\varepsilon f)_\varepsilon \to \varkappa(f,f)_0$ for $\varepsilon \to 0$, where $\varkappa = \text{const} > 0$ and is independent of f.

C2 The operators A_ε and A_0 are positive, compact, and selfadjoint in H_ε and H_0 respectively; moreover, $\sup_\varepsilon \|A_\varepsilon\|_{\mathcal{L}(H_\varepsilon)} < \infty$.

C3 For any $f \in V$

$$\|A_\varepsilon R_\varepsilon f - R_\varepsilon A_0 f\|_\varepsilon \to 0 \quad \text{as} \quad \varepsilon \to 0.$$

C4 The family of operators A_ε is uniformly compact in the following sense. From any sequence $f^\varepsilon \in H_\varepsilon$, $\sup_\varepsilon \|f^\varepsilon\|_\varepsilon < \infty$, it is possible to select a subsequence $f^{\varepsilon'}$ and find a function $w \in V$ such that

$$\|A_{\varepsilon'} f^{\varepsilon'} - R_{\varepsilon'} w\|_{\varepsilon'} \to 0 \quad \text{as} \quad \varepsilon' \to 0.$$

Consider the following spectral problems for A_ε and A_0:

$$A_\varepsilon s_\varepsilon^k = \mu_\varepsilon^k s_\varepsilon^k, \quad k \in \mathbb{N} \quad (s_\varepsilon^i, s_\varepsilon^j) = \delta_{ij}, \tag{3.25}$$

$$A_0 s_0^k = \mu_0^k s_0^k, \quad k \in \mathbb{N} \quad (s_0^i, s_0^j) = \delta_{ij}, \tag{3.26}$$

where δ_{ij} is the Kronecker symbol and the eigenvalues μ_ε^k, μ_0^k are enumerated in nondescedning order (with accounting their multiplicity).

Theorem 3.5 (Oleinik–Iosif'yan–Shamaev). *Assume that conditions* C1–C4 *are satisfied. Then*

$$|\mu_\varepsilon^k - \mu_0^k| \leqslant M_\varepsilon \sup_f \|A_\varepsilon R_\varepsilon f - R_\varepsilon A_0 f\|_\varepsilon, \quad k \in \mathbb{N},$$

where $\mu_\varepsilon^k, \mu_0^k$ *are the* k*th eigenvalues of the problems* (3.25), (3.26) *respectively, the supremum is taken over* $f \in N(\mu_0^k, A_0) = \{v \in H_0 : A_0 v = \mu_0^k v\}$ *such that* $\|v\|_0 = 1$, $M_\varepsilon \leqslant \mathcal{M}$ *for* $0 < \varepsilon \leqslant 1$ *and* $M_\varepsilon \to \frac{1}{\sqrt{\varkappa}}$ *as* $\varepsilon \to 0$, *where* $\mathcal{M}$ *is independent of* ε.

Assume that $k \leqslant 0$ *and* $l \leqslant 1$ *are integers and the multiplicity of the eigenvalue* μ_0^{k+1} *of the problem* (12) *is equal to* l, *i.e.*, $\mu_0^{k+1} = \dots = \mu_0^{k+l}$. *Then for any* $s_0 \in N(\mu_0^{k+1}, A_0)$ *there exists a linear combination* $\overline{s}_\varepsilon$ *of the eigenfunctions* $s_\varepsilon^{k+1}, \dots, s_\varepsilon^{k+l}$ *of the problem* (11) *such that*

$$\|\overline{s}_\varepsilon - R_\varepsilon s_0\|_\varepsilon \leqslant M_k \|A_\varepsilon R_\varepsilon s_0 - R_\varepsilon A_0 s_0\|_\varepsilon,$$

where the constant M_k *is independent of* ε.

To use the scheme, one should choose appropriate spaces H_0, H_ε, V and operators A_0, A_ε, R_ε and then check conditions C1–C4.

We recall that

$$p := P\frac{\sigma_d C_{\gamma_0^\varepsilon}}{2}.$$

Denote by H_ε and H_0 the space $L_2(\Omega)$ equipped with the inner products

$$(f^\varepsilon, g^\varepsilon)_{H_\varepsilon} \equiv \int_\Omega \rho^\varepsilon(x) f^\varepsilon g^\varepsilon dx, \quad (f^0, g^0)_{H_0} \equiv \int_\Omega f^0 g^0 dx$$

respectively. For V we take the space $H^{1\circ}(\Omega)$ if $p = +\infty$ and the space $H^1(\Omega, \Gamma_1)$ if $p < +\infty$. We set $R_\varepsilon f^0 = f^0$ for any $f^0 \in H_0$.

If $f^0 \in V$, then

$$(R_\varepsilon f^0, R_\varepsilon f^0)_{H_\varepsilon} = \int_\Omega \rho^\varepsilon(x) (f^0)^2 dx = \int_{\Omega \setminus B_\varepsilon} (f^0)^2 dx + (\varepsilon\delta)^{-m} \int_{B_\varepsilon} (f^0)^2 dx$$
$$\to \int_\Omega (f^0)^2 dx = (f^0, f^0)_{H_0} \quad \text{as } \varepsilon \to 0$$

in view of Lemma 3.2. Indeed,

$$\lim_{\varepsilon \to 0} (\varepsilon\delta)^{-m} \int_{B_\varepsilon} (f^0)^2 dx = \lim_{\varepsilon \to 0} (\varepsilon\delta)^{2-m} \int_{B_\varepsilon} (f^0)^2 (\varepsilon\delta)^{-2} dx,$$

but

$$\lim_{\varepsilon \to 0} (\varepsilon\delta)^{-2} \int_{B_\varepsilon} (f^0)^2 dx = 0,$$

$m < 2$. Hence condition C1 holds for $\varkappa = 1$.

Denote by $A_\varepsilon : H_\varepsilon \to H_\varepsilon$ the operator associating with a function $f^\varepsilon \in H_\varepsilon$ the solution $s^\varepsilon \in H^1(\Omega, \gamma_\varepsilon)$ to the problem (3.2). Denote by $A_0 : H_0 \to H_0$ the operator sending $f^0 \in H_0$ to the solution $s^0 \in H^1(\Omega, \Gamma_1)$ to the problem (3.3). It is easy to verify that the operators A_ε and A_0 are positive, compact, and selfadjoint in H_ε and H_0 respectively (cf. [6]). The uniform boundedness of the family of operators in the corresponding operator norm

$$\sup_\varepsilon \|A_\varepsilon\|_{\mathcal{L}(H_\varepsilon)} < M$$

follows from Lemma 3.3 since, in the case $m < 2$, Lemmas 3.1 and 3.3 and the Friedrichs inequality imply

$$\|A_\varepsilon f^\varepsilon\|^2_{H_\varepsilon} = \int_\Omega \rho^\varepsilon(x)(s^\varepsilon)^2 dx$$

$$\leqslant \Big(\int_{\Omega \setminus B_\varepsilon} (s^\varepsilon)^2 dx + (\varepsilon\delta)^{-2} \int_{B_\varepsilon} (s^\varepsilon)^2 dx \Big) \leqslant C_4 \int_\Omega |\nabla s^\varepsilon|^2 dx$$

$$\leqslant C_5 \Big((\varepsilon\delta)^{2-m} \int_{B_\varepsilon} \rho^\varepsilon(x) \big(f^\varepsilon\big)^2 dx + \int_{\Omega \setminus B_\varepsilon} \rho^\varepsilon(x) \big(f^\varepsilon\big)^2 dx + \int_{B_\varepsilon} \big(f^\varepsilon\big)^2 dx \Big)$$

$$\leqslant C_6 \Big(\int_{\Omega \setminus B_\varepsilon} \rho^\varepsilon(x) \big(f^\varepsilon\big)^2 dx + \int_{B_\varepsilon} \rho^\varepsilon(x) \big(f^\varepsilon\big)^2 dx \Big) = C_6 \|f^\varepsilon\|^2_{H_\varepsilon}.$$

Thus, condition C2 is satisfied.

To show that condition C3 is satisfied, we set $f^0 \in H_0$. Then

$$A_\varepsilon R_\varepsilon f^0 = s^\varepsilon, \quad R_\varepsilon A_0 f^0 = s^0,$$

where

$$\Delta s^\varepsilon = -\rho^\varepsilon(x) f^0 \quad \text{in } \Omega, \quad s^\varepsilon \in H^1(\Omega, \gamma_\varepsilon),$$
$$\Delta s^0 = -f^0 \quad \text{in } \Omega, \quad s^0 \in H^1(\Omega, \Gamma_1).$$

Further, we assume that $f \in L_\infty(\Omega)$. in (3.2) and (3.3). If $s^0 \in H^1(\Omega)$ is a weak solution to the problem (3.3), then, by the regularity of solutions to elliptic equations [1], we have $s^0 \in W^2_l(\Omega)$ for any $l > 1$. By the embedding theorems [21], [52], [51], $s^0, \nabla s^0 \in L_\infty(\Omega)$. The first equation in (3.3) implies $\Delta s^0 \in L_\infty(\Omega)$; moreover,

$$\|s^0\|_{L_\infty(\Omega)} + \|\nabla s^0\|_{L_\infty(\Omega)} + \|\Delta s^0\|_{L_\infty(\Omega)} \leqslant \text{const } \|f\|_{L_\infty(\Omega)}.$$

Similar estimates for $f \in L_2(\Omega)$ hold (cf., for example, [44, 10]). The weak convergence of s^ε to s^0 in $H^1(\Omega)$ as $\varepsilon \to 0$ is proved in [6, Theorem 1] (cf. also [26]).

We recall that

$$p := \lim_{\varepsilon\to 0} \frac{\varepsilon^{d-2}\frac{\sigma_d c_{\gamma_0^\varepsilon}}{2}}{\delta(\varepsilon)}.$$

Theorem 3.6. *If $p < +\infty$, s^ε, and s^0 are weak solutions to the problems (3.2) and (3.3) respectively, then there exists a constant K_3 independent of ε and δ such that for sufficiently small ε*

$$\|s^0\vartheta_\varepsilon - s^\varepsilon\|_{H^1(\Omega)} \leqslant K_3\Big((\varsigma_\varepsilon)^{\frac{1}{2}} + \Big|\frac{\varsigma_\varepsilon}{\delta} - p\Big| + (\varepsilon\delta)^{2-m}\Big).$$

Proof. By the properties of s^0, ϑ_ε, and s^ε, for any $v \in H^1(\Omega, \gamma_\varepsilon)$ we have

$$\int_\Omega (\nabla(s^0\vartheta_\varepsilon - s^\varepsilon), \nabla v)dx = \int_\Omega (\vartheta_\varepsilon - 1)\big(2\Delta s^0 v + 2(\nabla v, \nabla s^0) + \rho^\varepsilon f v\big)dx$$
$$+ \Big|\frac{\varsigma_\varepsilon}{\delta} - p\Big| \int_{\partial\Omega} s^0\vartheta_\varepsilon v ds + 2\int_{\partial\Omega} (\vartheta_\varepsilon - 1)ps^0 v ds - \int_\Omega s^0 \Delta\vartheta_\varepsilon v dx.$$

Now, we estimate the integrals on the right-hand side of the last equality. We have

$$\Big|\int_\Omega (\vartheta_\varepsilon - 1)\big(2\Delta s^0 v + 2(\nabla v, \nabla s^0) + \rho^\varepsilon f v\big)dx\Big|$$
$$\leqslant \Big|\int_\Omega 2(\vartheta_\varepsilon - 1)\Delta s^0 v dx\Big| + \Big|\int_\Omega 2(\vartheta_\varepsilon - 1)(\nabla v, \nabla s^0)dx\Big|$$
$$+ \Big|\int_{\Omega\setminus B_\varepsilon} (\vartheta_\varepsilon - 1) f v dx\Big| + \Big|\int_{B_\varepsilon} (\vartheta_\varepsilon - 1)(\varepsilon\delta)^{-m} f v dx\Big|$$
$$\leqslant \|\vartheta_\varepsilon - 1\|_{L_2(\Omega)}\bigg(\Big(\int_\Omega 4(\Delta s^0 v)^2 dx\Big)^{\frac{1}{2}} + \Big(\int_\Omega 4(\nabla s^0, \nabla v)^2 dx\Big)^{\frac{1}{2}}$$
$$+ \Big(\int_{\Omega\setminus B_\varepsilon} (fv)^2 dx\Big)^{\frac{1}{2}}\bigg) + (\varepsilon\delta)^{2-m}\|v\|_{H^1(\Omega)}\|f\|_{L_\infty(\Omega)}$$
$$\leqslant C_7(\varsigma_\varepsilon\delta)^{\frac{1}{2}}\|f\|_{L_\infty(\Omega)}\|v\|_{H^1(\Omega)} + (\varepsilon\delta)^{2-m}\|f\|_{L_\infty(\Omega)}\|v\|_{H^1(\Omega)},$$

where we used Lemma 3.1 and the properties of ϑ_ε. From the properties of s^0 and ϑ_ε we find

$$\Big|\int_\Omega s^0 \Delta\vartheta_\varepsilon v dx\Big| \leqslant C_8(\varsigma_\varepsilon\delta)^{\frac{1}{2}}\|v\|_{H^1(\Omega)}\|f\|_{L_\infty(\Omega)},$$

$$\Big|\frac{\varsigma_\varepsilon}{\delta} - p\Big| \ \Big|\int_{\partial\Omega} s^0\vartheta_\varepsilon v ds\Big| \leqslant C_9\Big|\frac{\varsigma_\varepsilon}{\delta} - p\Big| \ \|v\|_{H^1(\Omega)}\|f\|_{L_\infty(\Omega)}.$$

The remaining integral is estimated as follows:

$$\Big|\int_{\partial\Omega}(\vartheta_\varepsilon-1)ps^0vds\Big|\leqslant C_{10}(\varsigma_\varepsilon)^{\frac{1}{2}}\|v\|_{H^1(\Omega)}\|f\|_{L_\infty(\Omega)}.$$

Substituting $v=s^0\vartheta_\varepsilon-s^\varepsilon$ and using Lemma 3.2, we find

$$\|s^0\vartheta_\varepsilon-s^\varepsilon\|_{H^1(\Omega)}\leqslant K_3\Big((\varsigma_\varepsilon)^{\frac{1}{2}}+\Big|\frac{\varsigma_\varepsilon}{\delta}-p\Big|+(\varepsilon\delta)^{2-m}\Big).$$

The theorem is proved. □

Theorem 3.7. *If $p=+\infty$, then there exists a constant K_4 independent of ε and δ such that*

$$\|s^0\vartheta_\varepsilon-s^\varepsilon\|_{H^1(\Omega)}\leqslant K_4\Big((\varsigma_\varepsilon)^{\frac{1}{2}}+\Big(\frac{\delta}{\varsigma_\varepsilon}\Big)^{\frac{1}{2}}+(\varepsilon\delta)^{2-m}\Big).$$

This theorem is proved as above, but with the help of the results of [6].

We continue the verification of condition C3. By Lemma 3.1 and the Friedrichs inequality, we obtain

$$\begin{aligned}
&\|A_\varepsilon R_\varepsilon f^0-R_\varepsilon A_0f^0\|^2_{H_\varepsilon}=\int_\Omega\rho^\varepsilon(x)|s^\varepsilon-s^0|^2dx\\
&\leqslant\int_{\Omega\setminus B_\varepsilon}|s^\varepsilon-s^0|^2dx+\int_{B_\varepsilon}(\varepsilon\delta)^{-m}|s^\varepsilon-s^0|^2dx\\
&\leqslant C_{11}\int_\Omega|\nabla(s^\varepsilon-s^0)|^2dx+C_{12}(\varepsilon\delta)^{2-m}\int_{B_\varepsilon}|\nabla(s^\varepsilon-s^0)|^2dx\\
&\leqslant C_{13}\int_\Omega|\nabla(s^\varepsilon-s^0)|^2dx\leqslant\int_\Omega|\nabla(s^\varepsilon-s^0\vartheta_\varepsilon)|^2dx\\
&+\int_\Omega\Big(|\nabla s^0|^2(1-\vartheta_\varepsilon)^2+(s^0)^2|\nabla(1-\vartheta_\varepsilon)|^2\Big)dx.
\end{aligned}\tag{3.27}$$

Taking into account the properties of s^0 and ϑ_ε, Theorems 3.6, 3.7, Proposition 3.1, and Lemma 3.2 with $m<2$, we conclude that condition C3 is satisfied.

Let us show that condition C4 is satisfied. If $\sup_\varepsilon\|f^\varepsilon\|_{H\varepsilon}<\infty$, then Lemma 3.3 and the Friedrichs inequality imply

$$\sup_\varepsilon\|s^\varepsilon\|_{H^1(\Omega,\gamma_\varepsilon)}<\infty,$$

where s^ε is a solution to the problem (3.2). Consequently, in view of the Rellich theorem, there exists $S^*\in V$ and a subsequence $\varepsilon'\to0$ such that

$$s^{\varepsilon'}\to S^*\quad\text{weakly in }\ H^1(\Omega)\quad\text{and strongly in }\ L_2(\Omega).\tag{3.28}$$

By Lemma 3.1, we have

$$\|A_\varepsilon f^\varepsilon - R_\varepsilon S^*\|^2_{H_\varepsilon} = \int_\Omega \rho^\varepsilon(x)|s^\varepsilon - S^*|^2 dx$$

$$\leqslant \int_{\Omega\setminus B_\varepsilon} |s^\varepsilon - S^*|^2 dx + (\varepsilon\delta)^{-m}\int_{B_\varepsilon} |s^\varepsilon - S^*|^2 dx$$

$$\leqslant \int_\Omega |s^\varepsilon - S^*|^2 dx + C_{14}(\varepsilon\delta)^{2-m}\int_\Omega |\nabla(s^\varepsilon - S^*)|^2 dx,$$

where $s^\varepsilon = A_\varepsilon f^\varepsilon$ and C_{14} is independent of ε. Hence condition C4 holds because $m < 2$ and we have the convergence (3.28).

Thus, conditions C1–C4 hold and we can apply the theorem about the convergence of the spectra of a sequence of operators defined in different Hilbert spaces (Oleinik–Iosif'yan–Shamaev).

The problem about eigenvalues of the operator A_0 takes the form (3.26), where $\mu_0^k = \frac{1}{\lambda_0^k}$ and λ_0^k are eigenvalues of the problem (3.4). Thus, the following assertion holds.

Theorem 3.8. *Let λ_0^k and λ_ε^k be eigenvalues of the problems* (3.4) *and* (2.1) *respectively. Then*

$$|\lambda_0^k - \lambda_\varepsilon^k| \leqslant C_k^3\Big((\varsigma_\varepsilon)^{\frac{1}{2}} + (\varepsilon\delta)^{2-m} + \Big|\frac{\varsigma_\varepsilon}{\delta} - p\Big|\Big) \quad \textit{if } p < \infty,$$

$$|\lambda_0^k - \lambda_\varepsilon^k| \leqslant C_k^4\Big((\varsigma_\varepsilon)^{\frac{1}{2}} + \Big(\frac{\delta}{\varsigma_\varepsilon}\Big)^{\frac{1}{2}} + (\varepsilon\delta)^{2-m}\Big) \quad \textit{if } p = +\infty,$$

where the constants C_k^3 and C_k^4 are independent of ε.

If the multiplicity of the eigenvalue λ_0^l of the problem (3.4) *is equal to r, i.e., $\lambda_0^l = \lambda_0^{l+1} = \ldots = \lambda_0^{l+r}$, then for any eigenfunction s_0^l of the problem* (3.4) *corresponding to the eigenvalue λ_0^l, $\|s_0\|_{L_2(\Omega)} = 1$, there exists a linear combination $\overline{s^\varepsilon}$ of eigenfunctions of the problem* (2.1) *corresponding to the eigenvalue $\lambda_\varepsilon^{l+1}, \ldots, \lambda_\varepsilon^{l+r}$ such that*

$$\Big(\int_\Omega \rho^\varepsilon(x)|\overline{s^\varepsilon} - s_0^l|^2 dx\Big)^{\frac{1}{2}} \leqslant C_l^3\Big((\varsigma_\varepsilon)^{\frac{1}{2}} + (\varepsilon\delta)^{2-m} + \Big|\frac{\varsigma_\varepsilon}{\delta} - p\Big|\Big) \quad \textit{if } p < \infty,$$

$$\Big(\int_\Omega \rho^\varepsilon(x)|\overline{s^\varepsilon} - s_0^l|^2 dx\Big)^{\frac{1}{2}} \leqslant C_l^4\Big((\varsigma_\varepsilon)^{\frac{1}{2}} + \Big(\frac{\delta}{\varsigma_\varepsilon}\Big)^{\frac{1}{2}} + (\varepsilon\delta)^{2-m}\Big) \quad \textit{if } p = +\infty,$$

where the constants C_l^3 and C_l^4 are independent of ε and s_0^l.

Substituting (3.6) into the inequalities in Theorems 3.6, 3.7, and 3.8, we obtain the estimate in Theorems 3.1 and 3.2. Thus, Theorems 3.1 and 3.2 are proved.

4 Asymptotic Behavior of Eigenvalues and Eigenfunctions of the Operator Pencil

4.1 Reduction of the problem to the case of an operator pencil

We return to the study of the problem (2.2). A function $u_\varepsilon \in H^1(\Omega, \Gamma_1 \cup \gamma_\varepsilon)$ is a solution to the problem if it satisfies the integral identity

$$\int_\Omega \nabla u_\varepsilon(x) \nabla v(x) dx - \int_{\Gamma_\varepsilon} \frac{q}{\lambda_\varepsilon} u_\varepsilon(x) v(x) dl = \int_\Omega \lambda_\varepsilon \rho^\varepsilon(x) u_\varepsilon(x) v(x) dx$$

for any $v \in H^1(\Omega, \Gamma_1 \cup \gamma_\varepsilon)$. We equip the space $H^1(\Omega, \Gamma_1 \cup \gamma_\varepsilon)$ with the inner product

$$(u, v)_{1,0} = \int_\Omega \nabla u \nabla v \, dx.$$

Denote by $L_{2,\rho}(\Omega)$ and $H^1_\rho(\Omega)$ weight Hilbert spaces equipped with the inner products

$$(u, \overline{v})_{0,\rho} \equiv \int_\Omega \rho_\varepsilon \, u \, \overline{v} \, dx$$

and

$$(u, \overline{v})_{1,\rho} \equiv \int_\Omega ((\nabla u, \nabla \overline{v}) + \rho_\varepsilon \, u \, \overline{v}) \, dx$$

respectively. We write the integral identity in the form

$$(u_\varepsilon, v)_{1,0} - \frac{q}{\lambda_\varepsilon}(u_\varepsilon, v)_{L_2(\Gamma_\varepsilon)} = \lambda_\varepsilon (u_\varepsilon, v)_{0,\rho}. \tag{4.1}$$

By the Riesz theorem, there exist linear selfadjoint positive compact operators $\mathbf{A}_\varepsilon$ and $\mathbf{B}_\varepsilon$ such that

$$\int_\Omega \nabla \mathbf{A}_\varepsilon[u_\varepsilon] \nabla v \, dx = \int_\Omega \rho^\varepsilon u_\varepsilon v dx \tag{4.2}$$

and

$$\int_\Omega \nabla \mathbf{B}_\varepsilon[u_\varepsilon] \nabla v \, dx = \int_{\Gamma_\varepsilon} u_\varepsilon v ds. \tag{4.3}$$

Moreover, the integral identity (4.1) takes the form

$$(u_\varepsilon, v)_{1,0} - \frac{q}{\lambda_\varepsilon}(\mathbf{B}_\varepsilon[u_\varepsilon], v)_{1,0} = \lambda_\varepsilon (\mathbf{A}_\varepsilon[u_\varepsilon], v)_{1,0}.$$

Introduce the notation

$$L_\varepsilon(\lambda_\varepsilon) = I - \lambda_\varepsilon \mathbf{A}_\varepsilon - \frac{q}{\lambda_\varepsilon} \mathbf{B}_\varepsilon, \tag{4.4}$$

where I is the identity operator.

Let us find boundary value problems for the operators $\mathbf{A}_\varepsilon$ and $\mathbf{B}_\varepsilon$. We write the integral identity corresponding to boundary value problem (3.2):

$$\int_\Omega \nabla s_\varepsilon \nabla v \, dx = \int_\Omega \rho^\varepsilon f \, v dx \tag{4.5}$$

for all $v \in H^1(\Omega, \Gamma_1 \cup \gamma_\varepsilon)$. Taking into account the integral identity (4.5) and the equality (4.2), we conclude that the operator $\mathbf{A}_\varepsilon$ provides us with a weak solution to the boundary value problem (3.2), i.e., $\mathbf{A}_\varepsilon[f] = s_\varepsilon$, where s_ε is a solution to the problem (3.2). Note that the inverse $\mathbf{A}_\varepsilon^{-1}$ has discrete spectrum consisting of countably many positive eigenvalues $\{\lambda_\varepsilon^n(\mathbf{A}_\varepsilon^{-1})\}_{n=1}^\infty$, $\lim_{n\to\infty} \lambda_\varepsilon^n(\mathbf{A}_\varepsilon^{-1}) = +\infty$ which are eigenevalues of the problem (2.1).

Consider the boundary value problem

$$\begin{aligned} &\Delta w_\varepsilon(x) = 0 \quad \text{in } \Omega, \\ &w_\varepsilon(x) = 0 \quad \text{on } \Gamma_1 \cup \gamma_\varepsilon, \\ &\frac{\partial w_\varepsilon(x)}{\partial x_d} = \varphi \quad \text{on } \Gamma_\varepsilon \end{aligned} \tag{4.6}$$

and write the corresponding integral identity

$$\int_\Omega \nabla w_\varepsilon \nabla v \, dx = \int_{\Gamma_\varepsilon} \varphi v ds \tag{4.7}$$

for $v \in H^1(\Omega, \Gamma_1 \cup \gamma_\varepsilon)$. Taking into account the integral identity (4.7) and the equality (4.3), we conclude that the operator $\mathbf{B}_\varepsilon$ provides us with a solution to the boundary value problem (4.6), i.e., $\mathbf{B}_\varepsilon[\varphi] = w_\varepsilon$, where w_ε is a solution to the problem (4.6). Note that the operator $\mathbf{B}_\varepsilon^{-1}$ has discrete spectrum consisting of countably many positive eigenvalues $\{\lambda_\varepsilon^n(\mathbf{B}_\varepsilon^{-1})\}_{n=1}^\infty$, $\lim_{n\to\infty} \lambda_\varepsilon^n(\mathbf{B}_\varepsilon^{-1}) = +\infty$, called the *eigenvalues* of the spectral problem

$$\begin{aligned} &\Delta w_\varepsilon^k = 0 \quad \text{in } \Omega, \\ &w_\varepsilon^k = 0 \quad \text{on } \Gamma_1 \cup \gamma_\varepsilon, \\ &\frac{\partial w_\varepsilon^k}{\partial x_d} = -\lambda_\varepsilon^k w_\varepsilon^k \quad \text{on } \Gamma_\varepsilon. \end{aligned} \tag{4.8}$$

4.2 Properties of the operator pencils

In this subsection, we collect definitions and assertions (cf. [18]) which will be required to study the original homogenization problem in a domain with masses concentrated along the boundary.

The operator pencil. Let E be a separable Hilbert space. Consider the problem

$$\varphi = \lambda \mathbf{A}[\varphi] + \lambda^{-1}\mathbf{B}[\varphi] \quad (\lambda \neq 0), \tag{4.9}$$

where $\mathbf{A} : E \to E$ is a positive compact operator and $\mathbf{B} : E \to E$ is a nonnegative compact operator. A function φ is called an *eigenfunction* of the problem (4.9) if it satisfies (4.9) with some λ, called an *eigenvalue*.

The *operator pencil* of the problem (4.9) is defined as

$$L(\lambda) = I - \lambda \mathbf{A} - \lambda^{-1}\mathbf{B}. \tag{4.10}$$

Theorem 4.1. *If*

$$4\|\mathbf{A}\|\,\|\mathbf{B}\| < 1, \tag{4.11}$$

then the spectrum of the operator pencil $L(\lambda)$ in $\mathbb{R}$ consists of countably many eigenvalues of finite multiplicity. The eigenvalues belong to the intervals $(0, r_-)$, $(r_+, +\infty)$ of the real axis, where

$$r_\pm = \frac{1 \pm \sqrt{1 - 4\|\mathbf{A}\|\,\|\mathbf{B}\|}}{2\|\mathbf{A}\|}, \quad 0 < r_- < r_+,$$

and are divided into two families $\{\lambda_k^+\}_{k=1}^\infty$ and $\{\lambda_k^-\}_{k=1}^\infty$ with the accumulation points $+\infty$ and 0 respectively, i.e., $\lambda_n^- \to 0$ as $n \to \infty$ and $\lambda_n^+ \to +\infty$ as $n \to \infty$; moreover, $\lambda_k^- \neq \lambda_j^+$, $k, j \in \mathbb{N}$.

Then the following relations hold:

$$\lambda_n(\mathbf{B}) \leqslant \lambda_n^- \leqslant \frac{\lambda_n(\mathbf{B})}{1 - 2\lambda_n(\mathbf{B})\|\mathbf{A}\|} \quad (n = 1, 2, \ldots),$$

i.e., $\lambda_n^- = \lambda_n(\mathbf{B})(1 + o(1))$ as $n \to +\infty$, and

$$\frac{1}{\lambda_n(\mathbf{A})} - 2\|\mathbf{B}\| \leqslant \lambda_n^+ \leqslant \frac{1}{\lambda_n(\mathbf{A})} \quad (n = 1, 2, \ldots),$$

i.e., $\lambda_n^+ = \dfrac{1}{\lambda_n(\mathbf{A})} + O(1)$ as $n \to +\infty$.

4.3 Remarks

Thus, the original homogenization problem for the operator pencil (4.4) is divided into two homogenization problems for two classical boundary value problems (2.1) and (4.8), which corresponds to the scheme in [18].

By the properties of auxiliary problems, the operator $\mathbf{A}_\varepsilon$ is positive and compact and the operator $\mathbf{B}_\varepsilon$ is nonnegative and compact. Consequently, (4.4) is a special case of the operator pencil (4.10) with $\mathbf{A} = \mathbf{A}_\varepsilon$, $\mathbf{B} = q\mathbf{B}_\varepsilon$.

We choose q in such a way to satisfy the condition (4.11). From Theorem 4.1 we obtain the following assertion.

Theorem 4.2. *The problem* (4.4) *possesses the following properties.*

- *The spectrum of the problem* (4.4) *is discrete and consists of countably many real eigenvalues of finite multiplicity which belong to the intervals* $(0, r_-)$ *and* $(r_+, +\infty)$ *of the real axis, where*

$$r_\pm = \frac{1 \pm \sqrt{1 - 4q\|\mathbf{A}_\varepsilon\|\,\|\mathbf{B}_\varepsilon\|}}{2\|\mathbf{A}_\varepsilon\|}, \qquad 0 < r_- < r_+,$$

and is divided into two families $\{(\lambda_\varepsilon^k)^+\}_{k=1}^\infty$ *and* $\{(\lambda_\varepsilon^k)^-\}_{k=1}^\infty$ *with accumulation points* $+\infty$ *and* 0 *respectively, i.e.,* $(\lambda_\varepsilon^k)^- \to 0$ *as* $k \to \infty$ *and* $(\lambda_\varepsilon^k)^+ \to +\infty$ *as* $k \to \infty$; *moreover,* $\lambda_k^- \neq \lambda_j^+$, $k, j \in \mathbb{N}$.

- *The following estimates and asymptotic expansions hold:*

$$\frac{1}{\lambda_\varepsilon^k(\mathbf{A}_\varepsilon)} - 2q\|\mathbf{B}_\varepsilon\| \leqslant (\lambda_\varepsilon^k)^+ \leqslant \frac{1}{\lambda_\varepsilon^k(\mathbf{A}_\varepsilon)}, \qquad k = 1, 2, \ldots,$$

$$(\lambda_\varepsilon^k)^+ = \frac{1}{\lambda_\varepsilon^k(\mathbf{A}_\varepsilon)}[1 + o(1)] \quad \text{as } k \to +\infty,$$

$$q\lambda_\varepsilon^k(\mathbf{B}_\varepsilon) \leqslant (\lambda_\varepsilon^k)^- \leqslant \frac{q\lambda_\varepsilon^k(\mathbf{B}_\varepsilon)}{1 - 2q\lambda_\varepsilon^k(\mathbf{B}_\varepsilon)\|\mathbf{A}_\varepsilon^{-1}\|}, \qquad k = 1, 2, \ldots,$$

$$(\lambda_\varepsilon^k)^- = q\lambda_\varepsilon^k(\mathbf{B}_\varepsilon)[1 + o(1)] \quad \text{as } k \to +\infty.$$

4.4 Homogenization theorem

We assume that $\lim_{\varepsilon \to 0} \dfrac{\varepsilon^{d-2}\sigma_d c_{\gamma_0}}{2\delta(\varepsilon)} = p \in [0, +\infty]$, where σ_d is the area of the unit d-dimensional sphere and c_{γ_0} is the harmonic capacity of the $(d-1)$-dimensional disk γ_0. We recall that $P := p\frac{2}{\sigma_d c_{\gamma_0}}$.

Based on the previous section and the results of [6], we can obtain the following homogenized problems for (3.2) and (4.6) respectively, as $\varepsilon \to 0$:

$$\begin{aligned} &\Delta s^0(x) = -f(x) \quad \text{in } \Omega, \qquad s^0(x) = 0 \quad \text{on } \Gamma_1, \\ &\frac{\partial s^0(x)}{\partial x_d} + ps^0 = 0 \quad \text{on } \Gamma_2 \quad (p < \infty), \\ &s^0 = 0 \quad \text{on } \partial\Omega \quad (p = \infty) \end{aligned} \tag{4.12}$$

and

$$\begin{aligned} &\Delta w^0(x) = 0 \quad \text{in } \Omega, \qquad w^0(x) = 0 \quad \text{on } \Gamma_1, \\ &\frac{\partial w^0(x)}{\partial x_d} + pw^0 = \varphi \quad \text{on } \Gamma_2 \quad (p < \infty), \\ &w^0 = 0 \quad \text{on } \partial\Omega \quad (p = \infty). \end{aligned} \tag{4.13}$$

We write the integral identity corresponding to the boundary value problem (4.12). We have

$$\int_{\Omega} \nabla s^0 \nabla v \, dx + p \int_{\Gamma_2} s^0 v \, dl = \int_{\Omega} f \, v \, dx \tag{4.14}$$

for all $v \in H^1(\Omega, \Gamma_1)$ in the case $p < \infty$ and

$$\int_{\Omega} \nabla s^0 \nabla v \, dx = \int_{\Omega} f \, v \, dx \tag{4.15}$$

for all $v \in H^{1\circ}(\Omega)$ in the case $p = +\infty$. We equip the space $H^1(\Omega, \Gamma_1)$ with the equivalent inner product

$$(u, v)_{1,p} = \int_{\Omega} \nabla u \nabla v \, dx + p \int_{\Gamma_2} uv \, dl, \quad p < \infty.$$

Taking into account the above notation, we write the integral identity (4.14) and (4.15) as follows:

$$\begin{aligned} (s^0, v)_{1,p} &= (f, v)_{L_2(\Omega)}, \quad p < \infty, \\ (s^0, v)_{1,0} &= (f, v)_{L_2(\Omega)}, \quad p = \infty. \end{aligned}$$

By the Riesz theorem, for a linear continuous functional $(f, v)_{L_2(\Omega)}$ in $H^1(\Omega, \Gamma_1)$ (if $p < \infty$), then in the space $H^{1\circ}(\Omega)$ (in the case $p = +\infty$) there exists an operator $\widehat{\mathbf{A}}$ such that

$$\begin{aligned} (\widehat{\mathbf{A}}[f], v)_{1,p} &= (f, v)_{L_2(\Omega)}, \quad p < \infty, \\ (\widehat{\mathbf{A}}[f], v)_{1,0} &= (f, v)_{L_2(\Omega)}, \quad p = \infty. \end{aligned} \tag{4.16}$$

Taking into account the integral identity (4.14) (respectively (4.15)) and the equality (4.16), we conclude that the operator $\widehat{\mathbf{A}}$ provides us with a weak solution to the boundary value problem (4.12), i.e., $\widehat{\mathbf{A}}[f] = s^0$, where s^0 is a solution to the boundary value problem (4.12). We note that the inverse $\widehat{\mathbf{A}}^{-1}$ has discrete spectrum consisting of countably many positive eigenvalues $\{\lambda_0^n(\widehat{\mathbf{A}}^{-1})\}_{n=1}^{\infty}$, $\lim_{n\to\infty} \lambda_0^n(\widehat{\mathbf{A}}^{-1}) = +\infty$, called the *eigenvalues* of the problem

$$\begin{aligned} &\Delta s^0(x) = -\lambda s^0 \quad \text{in} \ \Omega, \\ &s^0(x) = 0 \quad \text{on} \ \Gamma_1, \\ &\frac{\partial s^0(x)}{\partial x_d} + p s^0 = 0 \quad \text{on} \quad \Gamma_2 \quad (p < \infty), \\ &s^0 = 0 \quad \text{on} \ \partial\Omega \quad (p = \infty). \end{aligned}$$

We write the integral identity for the boundary value problem (4.13) in the case $p < \infty$ (note that the problem (4.13) has only the trivial solution if

$p = +\infty$). We have

$$\int_{\Omega} \nabla w^0 \nabla v \, dx + p \int_{\Gamma_2} w^0 v \, dl = \int_{\Gamma_2} \varphi \, v \, dl \tag{4.17}$$

for all $v \in H^1(\Omega, \Gamma_1)$. Taking into account the above notation, we write (4.17) as follows:

$$(w^0, v)_{1,p} = (\varphi, v)_{L_2(\Gamma_2)}.$$

By the Riesz theorem, for a linear continuous functional $(\varphi, v)_{L_2(\Gamma_2)}$ in $H^1(\Omega, \Gamma_1)$ there exists an operator $\widehat{\mathbf{B}}$ such that

$$(\widehat{\mathbf{B}}[\varphi], v)_{1,p} = (\varphi, v)_{L_2(\Gamma_2)}. \tag{4.18}$$

Taking into account the integral identity (4.17) and the equality (4.18), we conclude that the operator $\widehat{\mathbf{B}}$ provides us with a weak solution to the boundary value problem (4.13), i.e., $\widehat{\mathbf{B}}[\varphi] = w^0$ if $p < \infty$, where w^0 is a solution to the boundary value problem (4.13). If $p < \infty$, the operator $\widehat{\mathbf{B}}^{-1}$ has discrete spectrum consisting of countably many positive eigenvalues $\{\lambda_0^n(\widehat{\mathbf{B}}^{-1})\}_{n=1}^{\infty}$, $\lim_{n\to\infty} \lambda_0^n(\widehat{\mathbf{B}}^{-1}) = +\infty$, called the *eigenvalues* of the spectral problem

$$\begin{aligned} &\Delta w^0(x) = 0 \quad \text{in } \Omega, \\ &w^0(x) = 0 \quad \text{on } \Gamma_1, \\ &\frac{\partial w^0(x)}{\partial x_d} + p w^0 = \lambda w^0 \quad \text{on } \Gamma_2. \end{aligned}$$

In the case $p = +\infty$, the operator $\widehat{\mathbf{B}}$ vanishes identically.

The following assertions hold (cf. also [6] and [43]).

Theorem 4.3. *Assume that $\lambda_\varepsilon^k(\mathbf{A}_\varepsilon)$ are eigenvalues of the operator $\mathbf{A}_\varepsilon$ such that $\lambda_\varepsilon^1(\mathbf{A}_\varepsilon) > \lambda_\varepsilon^2(\mathbf{A}_\varepsilon) \geqslant \ldots > 0$, $\lambda_\varepsilon^k(\mathbf{B}_\varepsilon)$ are eigenvalues of the operator $\mathbf{B}_\varepsilon$ such that $\lambda_\varepsilon^1(\mathbf{B}_\varepsilon) > \lambda_\varepsilon^2(\mathbf{B}_\varepsilon) \geqslant \ldots > 0$, $\lambda_0^k(\widehat{\mathbf{A}})$ are eigenvalues of the operator $\widehat{\mathbf{A}}$ such that $\lambda_0^1(\widehat{\mathbf{A}}) > \lambda_0^2(\widehat{\mathbf{A}}) \geqslant \ldots > 0$, and $\lambda_0^k(\widehat{\mathbf{B}})$ are eigenvalues of the operator $\widehat{\mathbf{B}}$ in the case $p < \infty$ such that $\lambda_0^1(\widehat{\mathbf{B}}) > \lambda_0^2(\widehat{\mathbf{B}}) \geqslant \ldots > 0$. We enumerate the eigenvalues taking into account their multiplicity.*

Then there exist constants C_1, C_2, C_3, C_4 depending only on k such that

$$\Big|\frac{1}{\lambda_\varepsilon^k(\mathbf{A}_\varepsilon)} - \frac{1}{\lambda_0^k(\widehat{\mathbf{A}})}\Big| \leqslant C_1\Big(\varepsilon^{\frac{d-2}{2}} + (\varepsilon\delta)^{2-m} + \Big|\frac{\varepsilon^{d-2}}{\delta} - p\frac{2}{\sigma_d c_{\gamma_0^\varepsilon}}\Big|\Big) \quad \text{if } p < \infty,$$

$$\Big|\frac{1}{\lambda_\varepsilon^k(\mathbf{A}_\varepsilon)} - \frac{1}{\lambda_0^k(\widehat{\mathbf{A}})}\Big| \leqslant C_2\Big(\varepsilon^{\frac{d-2}{2}} + (\varepsilon\delta)^{2-m} + \Big(\frac{\sqrt{\delta}}{\varepsilon^{\frac{d-2}{2}}}\Big)\Big) \quad \text{if } p = \infty,$$

$$\Big|\lambda_\varepsilon^k(\mathbf{B}_\varepsilon) - \lambda_0^k(\widehat{\mathbf{B}})\Big| \leqslant C_3\Big(\varepsilon^{\frac{d-2}{2}} + (\varepsilon\delta)^{2-m} + \Big|\frac{\varepsilon^{d-2}}{\delta} - p\frac{2}{\sigma_d c_{\gamma_0^\varepsilon}}\Big|\Big) \quad \text{if } p < \infty,$$

$$\left|\lambda_\varepsilon^k(\mathbf{B}_\varepsilon)\right| \leqslant C_4\Big(\varepsilon^{\frac{d-2}{2}} + (\varepsilon\delta)^{2-m} + \Big(\frac{\sqrt{\delta}}{\varepsilon^{\frac{d-2}{2}}}\Big)\Big) \quad \textit{if } p = \infty.$$

Theorem 4.4. *Let u_ε be a solution to the problem (2.2). Then $u_\varepsilon \rightharpoonup u^0$ weakly in $H^1(\Omega)$ as $\varepsilon \to 0$, where*

$$\begin{aligned}
&\Delta u^0 = -\lambda u^0 \quad \textit{in } \Omega,\\
&u^0 = 0 \quad \textit{on } \Gamma_1,\\
&\lambda\Big(\frac{\partial u^0}{\partial x_d} + pu^0\Big) - qu^0 = 0 \quad \textit{on } \Gamma_2 \quad p < \infty,\\
&u^0 = 0 \quad \textit{on } \partial\Omega \quad (p = \infty).
\end{aligned} \tag{4.19}$$

Taking into account the above notation, we write the integral identity for the problem (4.19):

$$\begin{aligned}
&(u^0, v)_{1,p} - \frac{q}{\lambda}(u^0, v)_{L_2(\Gamma_2)} = \lambda(u^0, v)_{L_2(\Omega)}, \quad p < \infty,\\
&(u^0, v)_{1,0} = \lambda(u^0, v)_{L_2(\Omega)}, \quad p = \infty.
\end{aligned} \tag{4.20}$$

Taking into account (4.16) and (4.18), we write (4.20) as

$$\begin{aligned}
&(u^0, v)_{1,p} - \frac{q}{\lambda}(\widehat{\mathbf{B}}[u^0], v)_{1,p} = \lambda(\widehat{\mathbf{A}}[u^0], v)_{1,p}, \quad p < \infty,\\
&(u^0, v)_{1,0} = \lambda(\widehat{\mathbf{A}}[u^0], v)_{1,0}, \quad p = \infty.
\end{aligned}$$

The corresponding operator pencil has the form

$$\begin{aligned}
&\widehat{L}(\lambda) = I - \lambda\widehat{\mathbf{A}} - \frac{q}{\lambda}\widehat{\mathbf{B}} \quad (p < \infty),\\
&\widehat{L}(\lambda) = I - \lambda\widehat{\mathbf{A}} \quad (p = \infty).
\end{aligned}$$

Definition 4.1. An operator pencil $\widehat{L}(\lambda)$ is called the *homogenized* operator pencil $L_\varepsilon(\lambda_\varepsilon)$ defined in (4.4).

4.5 Small oscillations of a viscous inhomogeneous fluid in a stationary open vessel with a rigid net on the surface. Normal oscillations

In this subsection, we use [18, Chapter 7, Section 1]. We assume that there is a fine-meshed rigid net on the open fluid surface, We assume that Assume that a heavy viscous fluid of density ρ occupies a stationary open vessel and, at the rest state, occupies the domain Ω bounded by the solid wall S, rigid net γ_ε on the surface, and free surface Γ_ε. The classical setting of a problem about small motions close to the rest state involves the linearized Navier–Stokes equations

$$\frac{\partial \mathbf{u}}{\partial t} = -\frac{1}{\rho}\nabla p + \nu \Delta \mathbf{u} + f, \quad \operatorname{div} \mathbf{u} = 0 \quad \text{in } \Omega, \tag{4.21}$$

where $\mathbf{u} = \mathbf{u}(t,x) = (u_1(t,x), u_2(t,x), u_3(t,x))$ is the velocity field, $p = p(t,x)$ is the dynamical pressure, $f(t,x)$ is the exterior force field, and ν is the fluid viscosity.

On S and γ_ε, the following adherence condition is imposed:

$$\mathbf{u} = 0 \quad \text{on } S \cup \gamma_\varepsilon. \tag{4.22}$$

Dynamic and kinematic conditions are also imposed on the free surface Γ_ε. If Γ_ε is described by the unknown function $x_d = \Xi(t, x_1, x_2)$, then these conditions take the form

$$\begin{aligned} &\left(\frac{\partial u_i}{\partial x_3} + \frac{\partial u_3}{\partial x_i}\right) = 0, \quad i = 1,2, \\ &-p + 2\rho\nu \frac{\partial u_3}{\partial x_3} = \rho g \Xi, \quad \frac{\partial \Xi}{\partial t} = -\gamma_n \mathbf{u} \quad \text{on } \Gamma_\varepsilon, \end{aligned} \tag{4.23}$$

where $\gamma_n \mathbf{u} := u_1 n_1 + u_2 n_2 + u_3 n_3$ and $n = (n_1, n_2, n_3)$ is the unit outward normal to the boundary of the domain.

Consider normal oscillations, i.e., solutions to the homogeneous problem (4.21)–(4.23) such that

$$\begin{aligned} &\mathbf{u}(t,x) = e^{-\lambda t}\, \mathbf{u}(x), \\ &p(t,x) = e^{-\lambda t} p(x), \\ &\Xi(t, x_1, x_2) = e^{-\lambda t}\, \Xi(x_1, x_2). \end{aligned} \tag{4.24}$$

Substituting (4.24) into Equation (4.21) and the boundary conditions (4.22), (4.23), we obtain the problem

$$\begin{aligned} &\frac{1}{\rho}\nabla p - \nu \Delta \mathbf{u} = \lambda \mathbf{u} \quad \operatorname{div} \mathbf{u} = 0 \quad \text{in } \Omega, \\ &\mathbf{u} = 0 \quad \text{on } S \cup \gamma_\varepsilon, \\ &\left(\frac{\partial u_i}{\partial x_3} + \frac{\partial u_3}{\partial x_i}\right) = 0, \quad i = 1,2, \\ &-p + 2\rho\nu \frac{\partial u_3}{\partial x_3} = \rho g \frac{1}{\lambda} \gamma_n \mathbf{u} \quad \text{on } \Gamma_\varepsilon. \end{aligned} \tag{4.25}$$

Further, we consider two auxiliary problems.

The first problem includes the homogeneous condition on Γ_ε :

$$\begin{aligned}
&\nabla p_1 - \nu \Delta \mathbf{s} = \Psi, \quad \operatorname{div} \mathbf{s} = 0 \quad \text{in } \Omega, \\
&\mathbf{s} = 0 \quad \text{on } S \cup \gamma_\varepsilon, \\
&\left(\frac{\partial s_i}{\partial x_3} + \frac{\partial s_3}{\partial x_i} \right) = 0, \quad i = 1, 2, \\
&- p_1 + 2\nu \frac{\partial s_3}{\partial x_3} = 0 \quad \text{on } \Gamma_\varepsilon.
\end{aligned} \tag{4.26}$$

The second problem has a homogeneous right-hand side of the equation and inhomogeneous condition on Γ_ε :

$$\begin{aligned}
&\nabla p_2 - \nu \Delta \mathbf{w} = 0, \quad \operatorname{div} \mathbf{w} = 0 \quad \text{in } \Omega, \\
&\mathbf{w} = 0 \quad \text{on } S \cup \gamma_\varepsilon, \\
&\left(\frac{\partial w_i}{\partial x_3} + \frac{\partial w_3}{\partial x_i} \right) = 0, \quad i = 1, 2, \\
&- p_2 + 2\nu \frac{\partial w_3}{\partial x_3} = \psi \quad \text{on } \Gamma_\varepsilon, \quad \int_{\Gamma_\varepsilon} \psi d\Gamma = 0.
\end{aligned} \tag{4.27}$$

Let $\mathbf{u} = \mathbf{u}(x)$ and $p = p(x)$ be classical solutions to the problem (4.25). Then we can write

$$\mathbf{u}(x) = \mathbf{s}(x) + \mathbf{w}(x), \quad \frac{1}{\rho} p(x) = p_1(x) + p_2(x), \tag{4.28}$$

where $\mathbf{s}(x)$, $p_1(x)$ is a solution to the boundary value problem (4.26) with right-hand side $\Psi = \lambda \mathbf{u}$, whereas $\mathbf{w}(x)$, $p_2(x)$ is a solution to the boundary value problem (4.27) with inhomogeneity $\psi = g \frac{1}{\lambda} \gamma_n \mathbf{u}$.

We write the problem (4.25) in the operator form. Let $\mathbf{A}$ be an operator such that $\nu^{-1}\mathbf{A}[\Psi] = \mathbf{s}$, where $\mathbf{s}$ is a solution to the problem (4.26) and $\mathbf{B}$ is an operator such that $\mathbf{B}[\psi] = \mathbf{w}$, where $\mathbf{w}$ is a solution to the problem (4.27). We set $\Psi = \lambda \mathbf{u}$ and $\psi = g \frac{1}{\lambda} \gamma_n \mathbf{u}$. Then the problems (4.26) and (4.27) can be written in the operator form as follows:

$$\begin{aligned}
&\nu \mathbf{A}^{-1}[\mathbf{s}] = \lambda \mathbf{u}, \\
&\nu \mathbf{w} = g \frac{1}{\lambda} \mathbf{B}[\mathbf{u}], \\
&\mathbf{u} = \mathbf{s} + \mathbf{w}.
\end{aligned} \tag{4.29}$$

Removing $\mathbf{u}(x)$ from (4.29), we find

$$\begin{aligned}
&- \lambda \mathbf{s} = -\nu \mathbf{A}^{-1}[\mathbf{s}] + g\nu^{-1} \mathbf{B}[\mathbf{s} + \mathbf{w}], \\
&- \lambda \mathbf{w} = -g\nu^{-1} \mathbf{B}[\mathbf{s} + \mathbf{w}].
\end{aligned} \tag{4.30}$$

Adding both equations, we find

$$\nu \mathbf{s} = \lambda \mathbf{A}[\mathbf{s} + \mathbf{w}], \quad \nu \mathbf{w} = \tfrac{g}{\lambda} \mathbf{B}[\mathbf{s} + \mathbf{w}].$$

Adding again the equations and taking into account (4.28), we obtain the main spectral problem

$$\nu\,\mathbf{u} = \lambda\mathbf{A}[\,\mathbf{u}\,] + \frac{g}{\lambda}\mathbf{B}[\,\mathbf{u}\,]. \tag{4.31}$$

Here, the operator **A** is positive and compact, whereas **B** is an infinite-dimensional nonnegative compact operator with infinite-dimensional kernel.

The operator pencil corresponding to the problem (4.31) has the form

$$L(\lambda) = \nu I - \lambda\mathbf{A} - \frac{g}{\lambda}\mathbf{B}. \tag{4.32}$$

Remark 4.1. For this pencil Theorem 4.1 remains valid. The eigenvalues λ_k^+ correspond to the normal oscillations of a viscous fluid, called *internal dissipative waves*. The eigenvalues λ_k^- correspond to the normal oscillations of a viscous fluid, called textitsurface gravity waves.

The operator pencil (4.4) has the same structure and the same properties of coefficients as the operator pencil (4.32). Therefore, we can regard the problem (2.2) as a scalar analog of equations of linear hydrodynamics.

Acknowledgement. It was an idea to Professor S.A. Stepin to combine the Krein scheme with homogenization methods to study a scalar analog of linear hydrodynamics in the case of micro-inhomogeneity on the boundary.

The work is partially supported by Russian Foundation for Basic Research (grant no. 09-01-00530) and Program of Support of Leading Scientific Schools (grant no. NSh -1698.2008.1).

References

1. Agmon, S., Douglis, A., Nirenberg, L.: Estimates near the boundary for solutions of elliptic partial differential equations satisfying general boundary conditions. I. Commun. Pure Appl. Math. **12**, 623–727 (1959)
2. Amirat, Y., Chechkin, G.A., Romanov, M.S.: On Multiscale Homogenization Problems in Boundary Layer Theory. Preprint (2009)
3. Bakhvalov, N.S., Panasenko, G.P.: Homogenization: Averaging Processes in Periodic Media. Kluwer Academic Publisher, Dordrecht (1990)
4. Bensoussan, A., Lions, J.-L., Papanicolau, G.: Asymptotic Analysis for Periodic Structures. North-Holland, Amsterdam (1978)
5. Chechkin, G.A.: Boundary value problems for an elliptic second-order equation with oscillating boundary conditions (Russian). in: Non-classical Partial Differential Equations, pp. 95–104. Novosibirsk (1988)
6. Chechkin, G.A.: Averaging of boundary value problems with a singular perturbation of the boundary conditions (Russian). Mat. Sb. **184**, no. 6, 99–150 (1993); English transl.: Sb. Math. **79**, no. 1, 191–222 (1994)

7. Chechkin, G.A.: Vibration of Fluids in a Vessel with a Net on the Surface. In: Cioranescu, D. et al. (eds.), Homogenization and Applications to Material Sciences. GAKUTO Intern. Series **9**, pp. 95–112. Gakkotosho, Tokyo (1997)
8. Chechkin, G.A.: On the estimation of solutions of boundary value problems in domains with concentrated masses periodically distributed along the boundary. The case of "light" masses (Russian). Mat. Zametki **76**, no. 6, 928–944 (2004); English transl.: Math. Notes **76**, no. 5–6, 865–879 (2004)
9. Chechkin, G.A.: Asymptotic expansions of the eigenvalues and eigenfunctions of an elliptic operator in a domain with many "light" concentrated masses near the boundary. The two-dimensional case (Russian). Izv. Ross. Akad. Nauk, Ser. Mat. **69**, no. 4, 161–204 (2005); English transl.: Izv. Math. **69**, no. 4, 805–846 (2005)
10. Chechkin, G.A., Pérez, M. E., Yablokova, E.I.: Non–Periodic Boundary Homogenization and "Light" Concentrated Masses. Indiana Univ. Math. J. **54**, no. 2, 321–348 (2005)
11. Chechkin, G.A., Piatnitski, A.L., Shamaev, A.S.: Homogenization: Methods and Applications (Russian). Tamara Rozhkovskaya Publisher, Novosibirsk (2007); English transl.: Am. Math. Soc., Providence, RI (2007)
12. Damlamian, A., Li, Ta-Tsien (Li, Daqian): Boundary homogenization for elliptic problems. J. Math. Pure Appl. **66**, 351–361 (1987)
13. Gadyl'shin, R.R.: Asymptotic properties of an eigenvalue of a problem for a singularly perturbed self-adjoint elliptic equation with a small parameter in the boundary conditions (Russian). Differ. Uravn. **22**, no. 4, 640–652 (1986); English transl.: Differ. Equations **22**, 474–483 (1986)
14. Gadyl'shin, R.R.: Surface potentials and the method of matching asymptotic expansions in the Helmholtz resonator problem (Russian). Algebra Anal. **4**, no. 2, 88–115 (1992); English transl.: St. Petersburg Math. J. **4**, no. 2, 273–296 (1993)
15. Gadyl'shin, R.R., Chechkin, G.A.: A boundary value problem for the Laplacian with rapidly changing type of boundary conditions in a multi-dimensional domain (Russian). Sib. Mat. Zh. **40**, no. 2, 271–287 (1999); English transl.: Sib. Math. J. **40**, no. 2, 229–244 (1999)
16. Il'in, A.M.: Matching of Asymptotic Expansions of Solutions of Boundary Value Problems (Russian). Nauka, Moscow (1989); English transl.: Am. Math. Soc., Providence, RI (1992)
17. Jikov, V.V., Kozlov, S.M., Oleinik, O.A.: Homogenization of Differential Operators and Integral Functionals. Springer, Berlin etc. (1994)
18. Kopachevskii, N.D., Krein, S G., Ngo Huy Can.: Operator Methods in Linear Hydrodynamics. Evolution and Spectral Problems (Russian). Nauka, Moscow (1989)
19. Kulonen, G.A.: A method of finding the laminar boundary layer on a porous surface (Russian). Vestnik Leningrad. Univ. **16**, no. 7, 123–135 (1961)
20. Kulonen, G.A., Kulonen, L.A.: Response of a laminar boundary layer to suction through a set of slots of finite width (Russian). Prikl. Mekh. **14**, no. 9, 83–88 (1978); English transl.: Sov. Appl. Mech. **14**, 968–973 (1979)
21. Ladyzhenskaya, O.A.: Boundary Value Problems of Mathematical Physics (Russian). Nauka, Moscow (1973)
22. Landkof, N.S.: Foundations of Modern Potential Theory. Springer, New York etc. (1972)
23. Lin, C.C.: On periodically oscillating wakes in the Oseen approximation. In: Studies in Mathematics and Mechanics Presented to Richard von Mises, pp. 170–176. Academic Press, New York (1954)

24. Lin, C.C.: Motion in the boundary layer with a rapidly oscillating external flow. In: Proc. 9-th Intern. Congress of Appl. Mech. **4**, pp. 155–167 (1957)
25. Lobo, M., Pérez, M.E.: Asymptotic behavior of an elastic body with a surface having small stuck regions. Math Model. Numerical Anal. **22**, no. 4, 609–624 (1988)
26. Lobo, M., Pérez, M.E.: On vibrations of a body with many concentrated masses near the boundary. Math. Model. Method. Appl. Sci. **3**, no. 2, 249–273 (1993)
27. Lobo, M., Pérez, M.E.: Vibrations of a body with many concentrated masses near the boundary: high frequency vibrations. In: Spectral Analysis of Complex Structures, pp. 85–101. Hermann, Paris (1995)
28. Marchenko, V.A., Khruslov, E.Ya.: Boundary Value Problems in Domains with Finegrained Boundary (Russian). Naukova Dumka, Kiev (1974)
29. Marletta, M., Shkalikov, A.A., Tretter, C.: Pencils of differential operators containing the eigenvalue parameter in the boundary conditions. Proc. Roy. Soc. Edinburgh Sect. A **133**, no. 4, 893–917 (2003)
30. Maz'ya, V.G., Hänler, M.: Approximation of solutions to the Neumann problem in disintegrating domains. Math. Nachr. **162**, 261–278 (1993)
31. Maz'ya, V.G., Movchan, A.S.: Asymptotic treatment of perforated domains without homogenization. Math. Nachr. [To appear]
32. Maz'ya, V.G., Nazarov, S.A.: Paradoxes of limit passage in solutions of boundary value problems involving the approximation of smooth domains by polygonal domains (Russian). Izv. Akad. Nauk SSSR Ser. Mat. **50**, no. 6, 1156–1177, 1343 (1986); English transl.: Math. USSR, Izv. **29**, 511–533 (1987)
33. Maz'ya, V.G., Nazarov, S.A., Plamenevskii, B.A.: Asymptotics of Solutions to Elliptic Boundary Value Problems under Singular Perturbation of Domains (Russian). Tbilisi Univ. Press, Tbilisi (1981)
34. Maz'ya, V.G., Nazarov, S.A., Plamenevskii, B.A.: Asymptotic expansions of the eigenvalues of boundary value problems for the Laplace operator in domains with small holes (Russian). Izv. Akad. Nauk SSSR, Ser. Mat. **48**, no. 2, 347–371 (1984); English transl.: Math. USSR, Izv. **24**, 321–345 (1985)
35. Maz'ya, V., Nazarov, S., Plamenevskii, B.: Asymptotic Theory of Elliptic Boundary Value Problems in Singularly Perturbed Domains. I. II. Birkhäuser, Basel (2000)
36. Maz'ya, V.G., Slutskii, A.S.: Homogenization of a differential operator on a fine periodic curvilinear mesh. Math. Nachr. **133**, 107–133 (1986)
37. Mel'nik, T.A., Nazarov, S.A.: The asymptotics of the solution to the Neumann spectral problem in a domain of the "dense-comb" type (Russian). Tr. Semin. I.G. Petrovskogo **19**, 138–173 (1996); English transl.: J. Math. Sci., New York **85**, no. 6, 2326–2346 (1997)
38. Nazarov, S.A.: Junctions of singularly degenerating domains with different limit dimensions I. II (Russian). Tr. Semin. I.G. Petrovskogo **18**, 3–78 (1995); **20**, 155–195 (1997); English transl.: J. Math. Sci., New York **80**, no. 5, 1989–2034 (1996); **97**, no. 3, 4085–4108 (1999)
39. Oleinik, O.A., Chechkin, G.A.: Boundary value problems for elliptic equations with rapidly changing type of boundary conditions (Russian). Usp. Mat. Nauk **48**, no. 6, 163–164 (1993); English transl.: Russ. Math. Surv. **48**, no. 6, 173–175 (1993)
40. Oleinik, O.A., Chechkin, G.A.: On a problem of boundary averaging for a system of elasticity theory (Russian). Usp. Mat. Nauk. **49**, no. 1, 114 (1994); English transl.: Russ. Math. Surv. **49**, no. 4, 113–114 (1994)

41. Oleinik, O.A., Chechkin, G.A.: On asymptotics of solutions and eigenvalues of the boundary value problems with rapidly alternating boundary conditions for the system of elasticity. Atti Accad. Naz. Lincei, Cl. Sci. Fis. Mat. Nat., IX. Ser., Rend. Lincei, Mat. Appl. **7**, no. 1, 5–15 (1996)
42. Oleinik, O.A., Samokhin, V.N.: Mathematical Models in Boundary Layer Theory. Chapman and Hall/CRC, London etc. (1999)
43. Oleinik, O.A., Shamaev, A.S., Yosifian, G.A.: Mathematical Problems in Elasticity and Homogenization. North-Holland, Amsterdam (1992)
44. Pérez, M.E., Chechkin, G.A., Yablokova, E.I.: On eigenvibrations of a body with "light" concentrated masses on a surface (Russian). Usp. Mat. Nauk **57**, no. 6, 195–196 (2002); English transl.: Russ. Math. Surv. **57**, no. 6, 1240–1242 (2002)
45. Pólya, G., Szegö, G.: Isoperimetric Inequalities in Mathematical Physics. Princeton Univ. Press, Princeton, NJ (1951)
46. Romanov, M.S.: Homogenization of a Problem with Many Scales in Magnetic Hydrodynamics (Russian). Probl. Mat. Anal. **35**, 133–138 (2007); English transl.: J. Math. Sci., New York **144**, no. 6, 4665–4670 (2007)
47. Romanov, M.S., Samokhin, V.N., Chechkin, G.A.: On the rate of convergence of solutions to the Prandtl equations in rapidly oscillating magnetic field (Russian). Dokl. Math. [To appear]
48. Samokhin, V.N.: Averaging of a system of Prandtl equations (Russian). Differents. Uravn. **26**, no. 3, 495–501 (1990); English transl.: Differ. Equ. **26**, no. 3, 369–374 (1990)
49. Sánchez-Palencia, E.: Homogenization Techniques for Composite Media. Springer, Berlin etc. (1987)
50. Shkalikov, A.A.: Operator pencils arising in elasticity and hydrodynamics: the instability index formula. Oper. Theory Adv. Appl. **87**, 358–385 (1996)
51. Sobolev, S.L.: Selected Problems in the Theory of Function Spaces and Generalized Functions (Russian). Nauka, Moscow (1989)
52. Sobolev, S.L.: Some Applications of Functional Analysis in Mathematical Physics. Am. Math. Soc., Providence, RI (1991)
53. Yosida, K.: Functional Analysis. Springer, Berlin (1995)

Selfsimilar Perturbation near a Corner: Matching Versus Multiscale Expansions for a Model Problem

Monique Dauge, Sébastien Tordeux, and Grégory Vial

Abstract We consider the Laplace–Dirichlet equation in a polygonal domain perturbed at the small scale ε near a vertex. We assume that this perturbation is selfsimilar, i.e., derives from the same pattern for all relevant values of ε. We construct and validate asymptotic expansions of the solution in powers of ε via two different techniques, namely the method of multiscale expansions and the method of matched asymptotic expansions. Then we show how the terms of each expansion can be split into a finite number of sub-terms in order to reconstruct the other expansion. Compared with the fairly general approach of Maz'ya, Nazarov, and Plamenevskii relying on multiscale expansions, the novelty of our paper is the rigorous validation of the method of matched asymptotic expansions, and its comparison with the multiscale method. The consideration of a model problem allows us to simplify the exposition of these rather complicated two techniques.

1 Introduction

The perturbations under consideration concern the space domain, they have the same small scale ε in every direction and they are *selfsimilar*, which means that there exists a reference point x_0 and a pattern Ω such that the ε-

Monique Dauge
IRMAR, Université de Rennes 1, Campus de Beaulieu, 35042 Rennes, France
e-mail: monique.dauge@univ-rennes1.fr

Sébastien Tordeux
Institut de Mathématiques de Toulouse, INSA-Toulouse, 31077 Toulouse, France
e-mail: sebastien.tordeux@insa-toulouse.fr

Grégory Vial
IRMAR, ENS Cachan Bretagne, CNRS, UEB, 35170 Bruz, France
e-mail: gvial@bretagne.ens-cachan.fr

A. Laptev (ed.), *Around the Research of Vladimir Maz'ya II: Partial Differential Equations*, International Mathematical Series 12, DOI 10.1007/978-1-4419-1343-2_4,

perturbation is given by the set of points x such that $(x - x_0)\varepsilon^{-1}$ belongs to Ω. Although a local perturbation of the metric of a Riemannian manifold could be of interest as well, we only investigate in this paper the case where the perturbation involves the boundary of the domain. We are more particularly interested in the influence of corners, both in the unperturbed domain ω and in the perturbation pattern Ω.

An example of such a perturbation is given by *rounded corners*. Here, the unperturbed domain ω is a domain with conical points, the perturbation pattern Ω is a smooth domain, and the limiting point x_0 of the perturbation is a conical point (cf. Fig. 1). The *fillets* in material engineering precisely enter this framework.

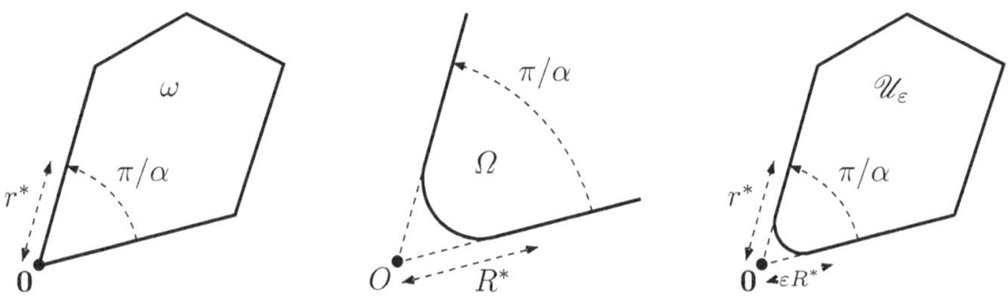

Fig. 1 Rounded corner: Domains ω, Ω and $\mathscr{U}_\varepsilon$.

The interesting and, at first glance puzzling, feature of such domains is the following: If one considers the solutions u_ε of a common elliptic problem posed on such a family of domains $\mathscr{U}_\varepsilon$ with rounded corners, each solution u_ε is smooth, but the sequence u_ε converges as $\varepsilon \to 0$ to the limiting solution in the corner domain ω which should contain singularities (we refer to the fundamental papers [12, 20] and the books [9, 6, 22, 13] for the vast topic of singularities).

Conversely, one can consider smooth limiting domains ω and associated patterns Ω with corners (cf. Fig. 2). In this case, the limit solution is smooth and each solution u_ε has corner singularities.

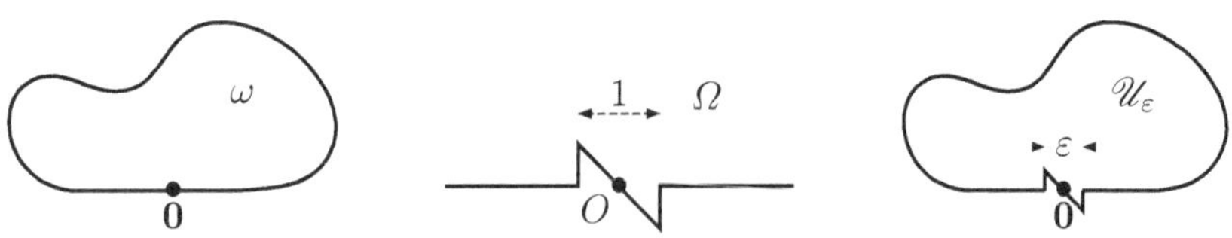

Fig. 2 Corner perturbation originating from a smooth boundary point ($\alpha = 1$).

More generally, selfsimilar perturbations may include numerous different situations. Let us mention, for example, small cracks originating from a boundary point of the limiting domain (cf. Fig. 5) and also small junctions between several connected components of ω (cf. Fig. 4).

For such singular perturbation problems the method of *matched asymptotic expansions* is widely used. This method, spread by [25], consists in constructing two distinct complete expansions of the solution in different regions with different scalings, and to match them in an intermediate region. It was used in [15] for the situation of Fig. 1 (cf. also [10] for a general framework). Although intuitive, this method is difficult to justify rigorously (cf. [24, 11] for such a more recent justification in the case of thin slots).

An alternative is given by the *multiscale expansion* technique consisting of a superposition of terms via cut-off functions which involve different scales. An optimal rigorous error analysis can be performed for such a method. This analysis was performed V.G. Maz'ya and coauthors in [17, 18] and written in a very general framework in the monograph [19].

In this paper, we mainly investigate, as a model case, the solutions u_ε of the Dirichlet problem for the Laplace operator set on a family of plane selfsimilar domains $\mathscr{U}_\varepsilon$. For each fixed ε the regularity properties of u_ε can be very different from those of their limit u_0 (more or less regular, depending on different configurations; cf. Figs. 1 and 2 respectively). An asymptotic expansion of u_ε as ε tends to 0 is the right way of understanding the mechanism of this transformation.

Our aim in this work is twofold:

(i) Provide the complete constructions and validations of the two different expansions provided by the two methods of *multiscale expansion* and *matched asymptotic expansions* for the same simple example, so that everything is made explicit and as clear as possible,

(ii) Compare the two expansions with each other, i.e., split each term of each expansion into sub-terms, and re-assemble them to reconstruct the terms of the other expansion.

Our paper is organized as follows. In Section 2, we define the families of selfsimilar domains and the problems under consideration, and next we provide an outline of our results, giving the structure of the first terms of both expansions. In Section 3, we state some preliminary results on limit problems in spaces with asymptotics, which we call *super-variational problems*. Section 4 is devoted to the method of *multiscale expansion*, like in [4, 26, 3], where optimal remainder estimates are proved. In Section 5, we present the method of *matched asymptotic expansions*, with the construction of the terms and matching conditions, and, by the technique of [24, 11], the validation of the expansion by remainder estimates. Sections 4 and 5 may be read independently. We compare the expansions obtained by these two techniques in Section 6, providing formulas for the translation of the terms of each expansion into the terms of the other one. In Section 7, we mention how expansions can be generalized to other situations (more general domains, data, operators etc.). We conclude in Section 8 with the definition of a "compound expansion" with the application to the study of the first singularity coefficient as $\varepsilon \to 0$.

2 Notation. Outline of Results

2.1 Selfsimilar perturbations

The families $(\mathscr{U}_\varepsilon)_{\varepsilon>0}$ under consideration are defined with the help of two domains, ω the *limit* (or unperturbed) domain, and Ω the *pattern* (or profile) of the perturbation. We denote by x and X the Cartesian coordinates in ω and Ω respectively and by $\mathbf{0}$ and O the corresponding origin of coordinates.

To simplify the exposition, we assume without restriction of the analysis that there is *one* perturbation and the corresponding reference point x_0 coincides with the origin $\mathbf{0}$. Indeed, ω and Ω do not "live" in the same world. The x coordinates are the *slow variables* and $X = \frac{x}{\varepsilon}$ are the *fast variables.*

The junction set. The connection between ω and Ω is realized by a plane sector K with vertex at the origin. Let $\frac{\pi}{\alpha}$ be the opening of K, including the situations of a half-plane ($\alpha = 1$) or a crack ($\alpha = \frac{1}{2}$). Thus, K is a dilation invariant set and makes sense in both systems of coordinates x and X.

We denote by $\mathscr{B}_\rho$ and B_ρ the ball centered at the origin with radius ρ in the x and X coordinates respectively. Let (r, θ) and (R, θ) be polar coordinates corresponding to variables x and X respectively and such that

$$K = \big\{x \in \mathbb{R}^2;\ r > 0,\ \theta \in (0, \tfrac{\pi}{\alpha})\big\} = \big\{X \in \mathbb{R}^2;\ R > 0,\ \theta \in (0, \tfrac{\pi}{\alpha})\big\}.$$

The limit domain. Let ω be a bounded domain of $\mathbb{R}^2$ containing the origin $\mathbf{0}$ in its boundary $\partial\omega$. Assume that there exists $r^* > 0$ such that

$$\omega \cap \mathscr{B}_{r^*} = K \cap \mathscr{B}_{r^*}.$$

The perturbing pattern. Let Ω be an unbounded domain of $\mathbb{R}^2$ such that there exists $R^* > 0$ for which

$$\Omega \cap \complement_{\mathbb{R}^2} B_{R^*} = K \cap \complement_{\mathbb{R}^2} B_{R^*}.$$

The perturbed domains. Let ε_0 be such that $\varepsilon_0 R^* = r^*$. For any $\varepsilon < \varepsilon_0$ we denote by $\mathscr{U}_\varepsilon$ the bounded domain

$$\mathscr{U}_\varepsilon \ = \ \{x \in \omega\ ;\ |x| > \varepsilon R^*\} \cup \{x \in \varepsilon\Omega\ ;\ |x| < r^*\}. \tag{2.1}$$

The domain $\mathscr{U}_\varepsilon$ coincides with the limit domain ω except in an ε-neighborhood of the origin, where its shape is given by the ε-dilation of the domain Ω (cf. Figs. 1 and 2). In the *intermediate region* $\varepsilon R^* \leqslant |x| \leqslant r^*$, $\mathscr{U}_\varepsilon$ coincides with K

$$\mathscr{U}_\varepsilon \cap \{\varepsilon R^* \leqslant |x| \leqslant r^*\} = K \cap \{\varepsilon R^* \leqslant |x| \leqslant r^*\}. \tag{2.2}$$

Note that Ω is the limit as $\varepsilon \to 0$ of the domain $\mathscr{U}_\varepsilon/\varepsilon$, whereas ω is the limit of $\mathscr{U}_\varepsilon$.

For the most part of this work, we do not assume any particular regularity for ω and Ω, except the coincidence with the sector K in the matching regions.

2.2 The Dirichlet problem and its singularities

As the simplest, and nevertheless typical, example of an elliptic boundary value problem on a family $(\mathscr{U}_\varepsilon)$ of selfsimilarly perturbed domains, we consider the *Laplace–Dirichlet problem.* We are interested in asymptotic expansions with respect to ε of the solution u_ε of the problem

$$\textit{Find } u_\varepsilon \in \mathrm{H}^1_0(\mathscr{U}_\varepsilon) \textit{ such that } -\Delta u_\varepsilon = f|_{\mathscr{U}_\varepsilon} \textit{ in } \mathscr{U}_\varepsilon. \tag{2.3}$$

Here, f is a fixed function belonging to $\mathrm{L}^2(\mathbb{R}^2)$. We assume for simplicity[1] that

$$f \equiv 0 \quad \text{in} \quad \mathscr{B}_{r^*}. \tag{2.4}$$

Thus, the support of f is contained in $\mathscr{U}_\varepsilon \setminus \mathscr{B}_{r^*}$, which coincides with $\omega \setminus \mathscr{B}_{r^*}$, hence independent of ε. Without risk of misunderstanding, we denote simply by f the right-hand side of (2.3).

When ε tends to 0, we expect the solution u_ε of (2.3) to converge to the solution u_0 of the limit problem

$$\textit{Find } u_0 \in \mathrm{H}^1_0(\omega) \textit{ such that } -\Delta u_0 = f \textit{ in } \omega. \tag{2.5}$$

In the following, we will derive the full asymptotic expansion of u_ε into powers of ε. The nature of the terms in this expansion depends on the asymptotics as $r \to 0$ and $R \to \infty$ of solutions to the Dirichlet problem on the limit domain ω and the pattern domain Ω respectively.

Both asymptotics involve the *singular functions* of the Laplace–Dirichlet problem in the sector K, which solve the homogeneous problem

$$\mathfrak{s} = 0 \ \text{ on } \ \partial K \quad \text{and} \quad -\Delta\mathfrak{s} = 0 \ \text{ in } \ K. \tag{2.6}$$

---

[1] This assumption may be removed (cf. Subsection 7.1.

For the sector opening $\frac{\pi}{\alpha}$ a generating set for all solutions of (2.6) on the sector K is given in the polar coordinates (ρ, θ) by (cf., for example, [12, 9])

$$\mathfrak{s}^{p\alpha}(\rho,\theta) = \rho^{p\alpha}\sin(p\alpha\theta) \quad \forall p \in \mathbb{Z}^*. \tag{2.7}$$

2.3 Outline of results

As a result of our two methods of analysis, this expansion is described by two different formulas, the first terms of which we present now.

- *The powers of* ε appearing in both formulas are the exponents $p\alpha$ of the singularities (2.7).
- *The remainders* in the following formulas are of the form $\mathcal{O}_{\mathrm{H}^1}(\varepsilon^\alpha)$, which means that their norms in $\mathrm{H}^1(\mathscr{U}_\varepsilon)$ are uniformly bounded by $C\varepsilon^\alpha$ as $\varepsilon \to 0$.

Multiscale Expansion. The method of Multiscale Expansion consists in looking for an expansion of u_ε in powers of ε with "coefficients" $v^\alpha(x)$ and $V^\alpha(\frac{x}{\varepsilon})$ in slow and rapid variables respectively, and so that these terms are combined with each other by cut-off functions $\chi(\frac{x}{\varepsilon})$ and $\psi(x)$ in rapid and slow variables respectively.

We choose a smooth function $X \mapsto \chi(X)$ which equals 1 except in a neighborhood of O and another smooth function $x \mapsto \psi(x)$ with compact support and equal to 1 in a neighborhood of $\mathbf{0}$. The first step of the multiscale expansion yields that

$$u_\varepsilon = \chi(\tfrac{x}{\varepsilon})\, v^0(x) + \mathcal{O}_{\mathrm{H}^1}(\varepsilon^\alpha) \tag{2.8}$$

with $v^0 = u_0$, which makes precise in what sense u_0 is the limit of u_ε. By the cut-off function $\chi(\frac{x}{\varepsilon})$, the term $\chi(\frac{x}{\varepsilon})\, u^0(x)$ is well defined on $\mathscr{U}_\varepsilon$ and is zero on the boundary $\mathscr{U}_\varepsilon$ in any configuration (cf. Figs. 1 and 2 for example). The next step of this method yields the two-term asymptotics

$$u_\varepsilon = \chi(\tfrac{x}{\varepsilon})\, v^0(x) + \psi(x)\, \varepsilon^\alpha V^\alpha(\tfrac{x}{\varepsilon}) + \mathcal{O}_{\mathrm{H}^1}(\varepsilon^{2\alpha}), \tag{2.9}$$

which proves, in particular, that the remainder in (2.8) is optimal. The general terms in the multiscale expansion are $\chi(\frac{x}{\varepsilon})\, v^{p\alpha}(x)$ and $\psi(x)\, V^{p\alpha}(\frac{x}{\varepsilon})$, for integers $p = 2, \dots$ (cf. Theorem 4.1 for an optimal estimate of remainders). The slow terms $v^\lambda(x)$, $\lambda = 2\alpha, 3\alpha, \dots$, are also solution of *variational problems* in the limiting domain ω, while the profiles $V^\lambda(X)$, $\lambda = \alpha, 2\alpha, \dots$ solve variational problems in the pattern Ω.

The cut-off functions are used with a scale opposite to the associated terms of the asymptotic expansion. As a consequence, the transition region where both terms $v^0(x)$ and $V^\lambda(\frac{x}{\varepsilon})$ contribute together to the asymptotics is the full domain (2.2), where $\mathscr{U}_\varepsilon$ coincides with the sector K. A wide range of problems can be treated like this (cf. [19, Chapter 4]). The slow-rapid product Ansatz can also be compared with the homogenization and asymptotic expansions in periodic structures (cf. [23]).

Matched Asymptotic Expansions. The method of Matched Asymptotic Expansions consists in constructing two different expansions (the inner and outer expansions) of u_ε in rapid variables (near the perturbation) and slow variables (outside the perturbation). A priori, none of these expansions is unique or valid everywhere. They have to be *matched* inside an intermediate zone contained in the region (2.2). The method consists in matching the asymptotics as $X = \frac{x}{\varepsilon} \to \infty$ of the inner expansion with the asymptotics as $x \to \mathbf{0}$ of the outer expansion.

Following the approach of [8] or [24], it is possible to construct an asymptotics of u_ε *valid everywhere* with the help of a smooth cut-off function φ at an intermediate scale ε^δ, with a fixed $\delta \in (0,1)$. Let ϕ be such that $\phi(\rho) = 0$ for $\rho \leqslant 1$ and $\phi(\rho) = 1$ for $\rho \geqslant 2$. By the method of Matched Asymptotic Expansions, we find the following first terms (cf. Theorem 5.2):

$$u_\varepsilon = \phi\big(r/\varepsilon^\delta\big)\, u^0(x) + \big(1 - \phi\big(r/\varepsilon^\delta\big)\big)\, \varepsilon^\alpha U^\alpha\big(\tfrac{x}{\varepsilon}\big) + \mathcal{O}_{\mathrm{H}^1}(\varepsilon^{2\alpha\beta}) \tag{2.10}$$

with $\beta = \min\{\delta, (1-\delta)\}$. The remainders are optimized if we choose $\delta = \frac{1}{2}$. Here, again, the first term u^0 coincides with the limit u_0. The general asymptotics involve outer terms $u^{p\alpha}$ defined in ω and inner terms $U^{p\alpha}$ defined in the pattern Ω. All of them are solution of what we call *super-variational problems*, i.e., problems set in spaces larger than the variational spaces (cf. Equations (3.9) and (3.17)) and where standard formulations would have nonunique solutions.

Comparison. The terms $v^{2\alpha}$, V^α, $u^{2\alpha}$, and U^α exchange with each other via two singular terms colinear to the singular functions $\mathfrak{s}^\alpha$ and $\mathfrak{s}^{-\alpha}$ (cf. (2.7)). Then (cf. Theorem 6.1)

$$\begin{aligned} U^\alpha(X) &= \quad V^\alpha(X) + \chi(X)\, b_1^0\, \mathfrak{s}^\alpha(X), \quad X \in \Omega, \\ u^{2\alpha}(x) &= \quad v^{2\alpha}(x) + \psi(x)\, B_1^1\, \mathfrak{s}^{-\alpha}(x), \quad x \in \omega. \end{aligned} \tag{2.11}$$

Here, b_1^0 and B_1^1 are the first coefficients of singularities for v^0 and V^α respectively. More generally, all terms of the matched inner and outer expansions can be reconstructed from the terms of the multiscale expansion, and vice versa. The pros and contras of each method are:

- The multiscale technique gives by construction a global approximation of the solution, with optimal estimates of the remainder, whereas twice as much terms are needed in the case of matched asymptotic expansions if one wants the same order for the remainder.
- The matched asymptotic expansions method builds outer and inner terms which are canonical, i.e., they do depend only on the domains ω and Ω, and not on cut-off functions, as it is the case for the multiscale technique.

3 Super-Variational Problems

In this section, we define the precise functional framework in which we will build the asymptotic expansions. The objects we define here are needed to derive rigorously both expansions.

All the terms in (2.8)–(2.10) appear as solutions of Dirichlet problems on ω or Ω. We first recall their variational framework before considering their solutions in larger spaces.

3.1 Variational problems

The variational space $V(\omega)$ for the Dirichlet problem on the bounded domain ω is $\mathrm{H}_0^1(\omega)$ and for f in its dual space, the variational formulation is

$$\begin{cases} \textit{Find } u \in V(\omega) \textit{ such that} \\ \displaystyle\int_\omega \nabla u(x)\, \nabla v(x)\, dx = \int_\omega f(x)\, v(x)\, dx \quad \forall v \in V(\omega). \end{cases} \tag{3.1}$$

The problem (3.1) has a unique solution. As a classical consequence of the angular Poincaré inequality, we find that the variational space is embedded into a weighted Sobolev space

$$V(\omega) = \mathrm{H}_0^1(\omega) \subset \mathrm{W}_0^1(\omega) := \{u \in \mathrm{H}^1(\omega)\,;\ r^{-1}u \in \mathrm{L}^2(\omega)\}. \tag{3.2}$$

The variational space $V(\Omega)$ for the Dirichlet problem on the unbounded domain Ω is the weighted space

$$V(\Omega) = \{U \in \mathrm{L}^2_{\mathrm{loc}}(\Omega)\,;\ \langle R\rangle^{-1}U \in \mathrm{L}^2(\Omega),\ \nabla U \in \mathrm{L}^2(\Omega),\ U|_{\partial\Omega} = 0\}, \tag{3.3}$$

where $\langle R\rangle = \sqrt{R^2+1}$. Then, for f in the dual of $V(\Omega)$, the variational problem below has a unique solution:

$$\begin{cases} \textit{Find } U \in V(\Omega) \textit{ such that} \\ \displaystyle\int_\Omega \nabla U(X)\, \nabla V(X)\, dX = \int_\Omega f(X)\, V(X)\, dX \quad \forall V \in V(\Omega). \end{cases} \tag{3.4}$$

One can refer, for example, to [3] for more details.

3.2 Super-variational problems in ω. Behavior at the origin

First, we introduce some function spaces to specify the behavior near the origin.

Definition 3.1. (i) Let $V_{\mathrm{loc},0}(\omega)$ be the space of distributions

$$V_{\mathrm{loc},0}(\omega) = \left\{u \in \mathscr{D}'(\omega)\ ;\ \ \varphi u \in \mathrm{H}^1_0(\omega)\ \ \forall\varphi \in \mathscr{D}(\mathbb{R}^2 \setminus \{\mathbf{0}\})\right\}.$$

(ii) For $m \in \mathbb{N}$ and $s \in \mathbb{R}$ let $\mathrm{W}^m_s(\omega)$ be the weighted Sobolev space

$$\mathrm{W}^m_s(\omega) = \left\{u \in \mathscr{D}'(\omega)\ ;\ \ r^{|\beta|-s-1}\,\partial^\beta_x u \in \mathrm{L}^2(\omega)\ \ \forall\beta,\ |\beta| \leqslant m\right\}.$$

Then we particularize the meaning of $\mathcal{O}(r^s)$ as follows:

Notation 3.2. For $s \in \mathbb{R}$ a function $u : \omega \to \mathbb{R}$ is said to be $\mathcal{O}_{r\to 0}(r^s)$ and we write $u = \mathcal{O}_{r\to 0}(r^s)$ if there exists a neighborhood $\mathscr{V}$ of $\mathbf{0}$ in $\mathbb{R}^2$ such that

$$\forall m, n \in \mathbb{N}, \quad \exists C > 0, \quad |r^m \partial^m_r \partial^n_\theta u| \leqslant C r^s \ \text{ in } \ \omega \cap \mathscr{V}.$$

Combining the change of variables $x \mapsto (t = \log r, \theta)$ with Sobolev embeddings, we prove:

$$u \in \mathrm{W}^m_s(\omega \cap \mathscr{V})\ \ \forall m \in \mathbb{N} \quad \Longrightarrow \quad u = \mathcal{O}_{r\to 0}(r^s). \tag{3.5}$$

Note that the converse implication is not true: the function $x \mapsto r^s$ is $\mathcal{O}_{r\to 0}(r^s)$, but does not even belong to $\mathrm{W}^0_s(\omega \cap \mathscr{V})$.

For functions harmonic in a neighborhood of the corner $\mathbf{0}$ the following assertion holds.

Lemma 3.3. *Let $u \in V_{\mathrm{loc},0}(\omega)$ be such that $\Delta u = 0$ in $\omega \cap \mathscr{V}$ for a neighborhood $\mathscr{V}$ of $\mathbf{0}$. Then for any real number s we have the implication*

$$u \in \mathrm{W}^1_s(\omega) \quad \Longrightarrow \quad u = \mathcal{O}_{r\to 0}(r^s). \tag{3.6}$$

Proof. Let $u \in \mathrm{W}^1_s(\omega)$ satisfy the assumptions of the lemma. Let $\rho' \in (0, r^*]$ be such that the finite sector $K_{\rho'} := \omega \cap \mathscr{B}_{\rho'}$ is contained in $\mathscr{V}$. Let $\rho \in (0, \rho')$, and let $m \in \mathbb{N}$ be fixed. Let us prove that u belongs to $\mathrm{W}^{m+2}_s(K_\rho)$, where $K_\rho = \omega \cap \mathscr{B}_\rho$.

For this purpose, we consider two sectorial annuli, $\mathscr{A}$ and $\mathscr{A}'$, defined as

$$\mathscr{A} = \{x \in \omega\ \ \rho_0 < |x| < \rho\} \quad \text{and} \quad \mathscr{A}' = \{x \in \omega\ \ \rho_0' < |x| < \rho'\}$$

with $\rho_0' < \rho_0 < \rho/2$, whence $\mathscr{A} \subset \mathscr{A}'$. A standard local elliptic estimate reads, for u satisfying $u \in \mathrm{W}^1_s(K_{\rho'})$, $\Delta u \in \mathrm{W}^m_{s+2}(K_{\rho'})$, and $u = 0$ on $\partial\omega \cap \mathscr{B}_{\rho'}$ (cf. [1]),

$$\|u\|_{\mathrm{H}^{m+2}(\mathscr{A})} \leqslant C\big(\|\Delta u\|_{\mathrm{H}^m(\mathscr{A}')} + \|u\|_{\mathrm{H}^1(\mathscr{A}')}\big). \tag{3.7}$$

Applying this estimate to the functions $u_k(x) = u(2^{-k}x)$ and summing up over k the obtained inequalities (multiplied by 2^{-sk}), we get the following estimate from dyadic partition equivalence:

$$\|u\|_{\mathrm{W}_s^{m+2}(K_\rho)} \leqslant C\big(\|\Delta u\|_{\mathrm{W}_{s+2}^m(K_{\rho'})} + \|u\|_{\mathrm{W}_s^1(K_{\rho'})}\big). \tag{3.8}$$

The conclusion then follows from (3.5). □

We can now state about the solvability of super-variational problems on ω, i.e., in spaces containing some of the *dual singular functions* $\mathfrak{s}^{-p\alpha}$ for $p \geqslant 1$. If we know the dual singular part of a function $u \in V_{\mathrm{loc},0}(\omega)$ and its Laplacian Δu, then this function is uniquely defined.

Proposition 3.4. *For any data $f \in \mathrm{H}^{-1}(\omega)$, $f \equiv 0$ in a neighborhood of $\mathbf{0}$, and any finite sequence $(a_p)_{1\leqslant p\leqslant P}$ of real numbers there exists a unique solution u to the "super-variational problem"*

$$\begin{cases} \text{Find } u \in V_{\mathrm{loc},0}(\omega) \text{ such that} \\ -\Delta u = f \text{ in } \omega \quad \text{and} \quad u - \displaystyle\sum_{p=1}^{P} a_p\, \mathfrak{s}^{-p\alpha} = \mathcal{O}_{r\to 0}(1). \end{cases} \tag{3.9}$$

Remark 3.5. If the sequence of coefficients $(a_p)_p$ is empty, the problem (3.9) is nothing but the variational problem (3.1).

Proof. Let a smooth cut-off function ψ satisfy $\psi(x) = 1$ for $|x| < r^*/2$ and $\psi(x) = 0$ for $|x| > r^*$. We set $v = \psi \sum_p a_p\, \mathfrak{s}^{-p\alpha}$, which obviously satisfies

$$v \in V_{\mathrm{loc},0}(\omega), \quad \text{and} \quad \Delta v = 0 \omega \cap \mathscr{B}_{r^*/2}. \tag{3.10}$$

Hence the problem to find w such that $-\Delta w = f + \Delta v$ in ω admits a unique variational solution $w \in V(\omega) = \mathrm{H}_0^1(\omega)$. Moreover, (3.2) gives that w belongs to $\mathrm{W}_0^1(\omega)$, and by localization near point $\mathbf{0}$, w is $\mathcal{O}_{r\to 0}(1)$ thanks to (3.6); the function $u = w + v$ meets then the requirements. □

On the other hand, every solution of the Laplace–Dirichlet equation can be expanded near the corner point $\mathbf{0}$ in terms of the singular functions, compare with the results, for example, in [12, 20, 21, 9].

Proposition 3.6. *Let $s \geqslant 0$ be a real number. We define P as the integer part of s/α. For any $u \in V_{\mathrm{loc},0}(\omega)$ for which there is a neighborhood $\mathscr{V}$ of $\mathbf{0}$ such that*

$$\Delta u = 0 \ \text{ in } \ \omega \cap \mathscr{V} \quad \text{and} \quad u = \mathcal{O}_{r\to 0}(r^{-s}), \tag{3.11}$$

there exists a unique finite sequence $(a_p)_{1\leqslant p\leqslant P}$ *and a unique sequence* $(b_p)_{p\in\mathbb{N}^*}$ *(generically infinite) such that for all* $N \in \mathbb{N}^*$

$$u(x) = \sum_{p=1}^{P} a_p\, \mathfrak{s}^{-p\alpha}(r,\theta) + \sum_{p=1}^{N} b_p\, \mathfrak{s}^{p\alpha}(r,\theta) + \mathcal{O}_{r\to 0}(r^{(N+1)\alpha}). \tag{3.12}$$

Notation 3.7. In the situation of Proposition 3.6, we write

$$u(x) \underset{r\to 0}{\simeq} \sum_{p=1}^{P} a_p\, \mathfrak{s}^{-p\alpha}(r,\theta) + \sum_{p=1}^{\infty} b_p\, \mathfrak{s}^{p\alpha}(r,\theta). \tag{3.13}$$

Proof. One can prove this lemma using the Mellin transform (cf. [12]). In the particular case we are interested in, an argument based on separation of variables via angular Fourier series also leads to the result. □

In accordance with the literature on corner asymptotics [21, 7, 5], we can call the sum $\sum a_p\, \mathfrak{s}^{-p\alpha}$ the *dual singular part* of u, whereas $\sum b_p\, \mathfrak{s}^{p\alpha}$ represents the asymptotics of the variational part of u and can be called the *primal singular part* of u.

In the particular case of an opening angle equal to π, i.e., $\alpha = 1$, the asymptotics of the variational part contains polynomials only – it is a Taylor expansion, but the dual singular part is actually singular. More generally, if the opening angle has the form $\frac{\pi}{n}$ with a positive integer n, i.e., $\alpha = n$, the asymptotics of the variational part is polynomial and can be regarded as *regular*.

3.3 Super-variational problems in Ω. Behavior at infinity

We give for the pattern domain Ω similar definitions and results as in the previous section, $r \to 0$ being replaced with $R \to +\infty$.

Definition 3.8. (i) Let $V_{\mathrm{loc},\infty}(\Omega)$ be the space of distributions

$$V_{\mathrm{loc},\infty}(\Omega) = \left\{U \in \mathscr{D}'(\Omega)\ ;\ \ \varphi U \in \mathrm{H}^1_0(\Omega)\ \ \forall \varphi \in \mathscr{D}(\mathbb{R}^2)\right\}.$$

(ii) For $m \in \mathbb{N}$ and $s \in \mathbb{R}$ let $\mathrm{W}^m_s(\Omega)$ be the weighted Sobolev space

$$\mathrm{W}^m_s(\Omega) = \left\{U \in \mathscr{D}'(\Omega)\ ;\ \ \langle R\rangle^{|\beta|-s-1}\, \partial^\beta_X U \in \mathrm{L}^2(\Omega)\ \ \forall \beta,\ |\beta| \leqslant m\right\},$$

where $\langle R\rangle = \sqrt{R^2+1}$.

In the following, we say that W is a neighborhood of infinity if there exists a ball B_R of radius R such that

$$\complement_{\mathbb{R}^2} B_R \subset W. \tag{3.14}$$

We introduce, similarly to Notation 3.2

Notation 3.9. For $s \in \mathbb{R}$ a function $U : \Omega \to \mathbb{R}$ is said to be $\mathcal{O}_{R\to\infty}(R^s)$ and we write $U = \mathcal{O}_{R\to\infty}(R^s)$ if there exists a neighborhood W of infinity such that

$$\forall m, n \in \mathbb{N}, \quad \exists C > 0, \quad |R^m \partial_R^m \partial_\theta^n U(R,\theta)| \leqslant C R^s \ \text{ in } \ \Omega \cap W.$$

We have the implication

$$u \in \mathrm{W}_s^m(\Omega \cap W) \ \ \forall m \in \mathbb{N} \quad \Longrightarrow \quad u = \mathcal{O}_{R\to\infty}(R^s). \tag{3.15}$$

By a similar shift result as for Lemma 3.3, we get the following assertion.

Lemma 3.10. *Let $U \in V_{\mathrm{loc},\infty}(\Omega)$ be such that $\Delta U = 0$ in $\Omega \cap W$ for a neighborhood W of infinity. Then for any real number s we have the implication*

$$U \in \mathrm{W}_s^1(\Omega) \quad \Longrightarrow \quad U = \mathcal{O}_{R\to\infty}(R^s). \tag{3.16}$$

The following two propositions are counterparts of Propositions 3.4 and 3.6. The dual singular functions at infinity in Ω are now $\mathfrak{s}^{p\alpha}$ for positive integers p.

Proposition 3.11. *For any $F \in \mathrm{H}^{-1}(\Omega)$ with compact support in $\overline{\Omega}$ and any finite sequence $(A_p)_{1\leqslant p\leqslant P}$ of real numbers there exists a unique solution U to the "super-variational problem"*

$$\begin{cases} \text{Find } U \in V_{\mathrm{loc},\infty}(\Omega) \text{ such that} \\ -\Delta U = F \ \text{ in } \Omega \quad \text{and} \quad U - \displaystyle\sum_{p=1}^{P} A_p\, \mathfrak{s}^{p\alpha} = \mathcal{O}_{R\to\infty}(1). \end{cases} \tag{3.17}$$

Proof. It is very similar to Proposition 3.4, the suitable variational space being here $V(\Omega) = \mathrm{W}_0^1(\Omega)$. □

Remark 3.12. If the sequence of coefficients (A_p) is empty, the problem (3.17) is nothing but the variational problem (3.4).

Proposition 3.13. *Let $s \geqslant 0$ be a real number. We define P as the integer part of s/α. For any $U \in V_{\mathrm{loc},\infty}(\Omega)$ for which there is a neighborhood W of infinity such that*

$$\Delta U = 0 \quad in \quad \Omega \cap W \quad and \quad U = \mathcal{O}_{R\to\infty}(R^s), \tag{3.18}$$

there exists a unique finite sequence $(A_p)_{1\leqslant p\leqslant P}$ *and a unique sequence* $(B_p)_{p\in\mathbb{N}^*}$ *(generically infinite) such that for all* $N \in \mathbb{N}^*$

$$U(X) = \sum_{p=1}^{P} A_p\, \mathfrak{s}^{p\alpha}(R,\theta) + \sum_{p=1}^{N} B_p\, \mathfrak{s}^{-p\alpha}(R,\theta) + \mathcal{O}_{R\to\infty}(R^{-(N+1)\alpha}). \tag{3.19}$$

Notation 3.14. In the situation of Proposition 3.13, we write

$$U(X) \underset{R\to\infty}{\simeq} \sum_{p=1}^{P} A_p\, \mathfrak{s}^{p\alpha}(R,\theta) + \sum_{p=1}^{\infty} B_p\, \mathfrak{s}^{-p\alpha}(R,\theta). \tag{3.20}$$

4 Multiscale Expansion

The multiscale expansion in the domain $\mathscr{U}_\varepsilon$ is composed of two different types of terms: the slow terms involving the original variable x and the *profiles* appearing in the rapid scaled variable $\frac{x}{\varepsilon}$. They are superposed via cut-off functions according to the Ansatz

$$u_\varepsilon(x) = \chi(\tfrac{x}{\varepsilon}) \sum_{\ell=0}^{n} \varepsilon^{\ell\alpha} v^{\ell\alpha}(x) + \psi(x) \sum_{\ell=0}^{n} \varepsilon^{\ell\alpha} V^{\ell\alpha}(\tfrac{x}{\varepsilon}) + o(\varepsilon^{n\alpha}), \tag{4.1}$$

where the functions χ and ψ are smooth and satisfy

$$\begin{aligned} &\chi(X) = 1 \ \text{ for } \ |X| > 2R^* \quad \text{and}\chi(X) = 0 \text{ for } \ |X| < \frac{3R^*}{2}, \\ &\psi(x) = 1 \text{ for } \ |x| < \frac{r^*}{2} \qquad \text{and}\psi(x) = 0 \text{ for } \ |x| > r^*. \end{aligned} \tag{4.2}$$

The first sum in (4.1) has its support away from an ε-neighborhood of the limit point $\mathbf{0}$ and, conversely, the second brings a contribution in a neighborhood of $\mathbf{0}$ (independent of ε). The *transition region* is the common support of the two sums which, thanks to (2.2), for any $\varepsilon \leqslant \varepsilon_0/2$ satisfies

$$\begin{aligned} \mathscr{U}_\varepsilon \cap \Big(\text{supp}\chi(\frac{\cdot}{\varepsilon}) \cap \text{supp}\psi\Big) &\subset \{x \in \mathscr{U}_\varepsilon,\ \varepsilon R^* \leqslant |x| \leqslant r^*\} \\ &= \{x \in K,\ \varepsilon R^* \leqslant |x| \leqslant r^*\}. \end{aligned}$$

The construction principles of the terms is as follows: $v^{\ell\alpha}$ and $V^{\ell\alpha}$ are solutions of variational problems in slow variables $x \in \omega$ and fast variables $X \in \Omega$. The cut-off by $\chi(\frac{x}{\varepsilon}) = \chi(X)$ and $\psi(x)$ introduces an error in fast and slow variables. These errors can be corrected with the help of the expansions as $r \to 0$ of the terms $v^{\ell\alpha}$ and as $R \to \infty$ of the terms $V^{\ell\alpha}$. Both expansions in homogeneous terms do make sense in fast *and* slow variables simultaneously, which allows us to bridge the terms in the two sums in (4.1).

4.1 The construction of the first terms

Step 0. Let $v^0 = u_0$ be the solution of the limit variational problem (2.5). Since v^0 is defined in the domain ω, and not on $\mathscr{U}_\varepsilon$, we choose to consider the truncated function $\widetilde{v}^0 = \chi(\frac{x}{\varepsilon})v^0$ instead. We note that $\widetilde{v}^0$ satisfies the Dirichlet boundary condition $\widetilde{v}^0 = 0$ on $\partial\mathscr{U}_\varepsilon$ and belongs to $\mathrm{H}^1_0(\mathscr{U}_\varepsilon)$. We consider the first remainder r^0_ε defined as

$$u_\varepsilon(x) = \chi(\tfrac{x}{\varepsilon})v^0(x) + r^0_\varepsilon(x).$$

Thus, the support of Δr^0_ε is contained in the support of $\nabla\chi(\frac{x}{\varepsilon})$. Using the commutator $[\Delta, \phi]$ defined by $[\Delta, \phi]f := \Delta(\phi f) - \phi\Delta f$, we find

$$\begin{aligned} -\Delta r^0_\varepsilon(x) &= \big([\Delta, \chi(\tfrac{\cdot}{\varepsilon})]v^0\big)(x) \\ &= 2\nabla_x v^0(x) \cdot \nabla_x\big(\chi(\tfrac{x}{\varepsilon})\big) + v^0(x)\Delta_x\big(\chi(\tfrac{x}{\varepsilon})\big). \end{aligned} \tag{4.3}$$

Since $f \equiv 0$ in a neighborhood of $\mathbf{0}$, according to Proposition 3.6 (and using Notation 3.7) there exists a sequence $\big(\mathbf{b}^0_p\big)_{p\geqslant 1}$ such that v^0 expands as $r \to 0$ as

$$v^0(x) \underset{r\to 0}{\simeq} \sum_{p=1}^{\infty} \mathbf{b}^0_p\, \mathfrak{s}^{p\alpha}(x). \tag{4.4}$$

We insert the expansion (4.4) into (4.3). For each of its terms we use the fundamental relation which allows us to convert the commutator in fast variables

$$\big[\Delta, \chi(\tfrac{x}{\varepsilon})\big]\, \mathfrak{s}^{p\alpha}(x) = \varepsilon^{-2}\, \varepsilon^{p\alpha}\, \big(\big[\Delta_X, \chi\big]\, \mathfrak{s}^{p\alpha}\big)\, (\tfrac{x}{\varepsilon}). \tag{4.5}$$

Thus, the remainder (4.3) can be written as

$$\Delta r^0_\varepsilon(x) \underset{r\to 0}{\simeq} -\varepsilon^{-2} \sum_{p=1}^{\infty} \varepsilon^{p\alpha}\, \mathbf{b}^0_p\, \big(\big[\Delta_X, \chi\big]\, \mathfrak{s}^{p\alpha}\big)(\tfrac{x}{\varepsilon}). \tag{4.6}$$

To complete step 0, we set $V^0 = 0$ and we are going to consider the further terms for $p = 1, \ldots,$ as *right-hand sides* of a problem on Ω in the fast variable $X = \frac{x}{\varepsilon}$.

Step 1. The first term in the remainder asymptotics (4.6) is

$$\varepsilon^{-2}\, \varepsilon^{\alpha}\, \mathbf{b}^0_1\, \big(\big[\Delta_X, \chi\big]\, \mathfrak{s}^{\alpha}\big)(X). \tag{4.7}$$

This function is smooth with compact support. Let V^α be the solution of the variational problem in Ω (cf. (3.4)):

$$\textit{Find } V^\alpha \in V(\Omega) \textit{ such that } -\Delta_X V^\alpha = \mathbf{b}_1^0 \big[\Delta_X, \chi\big]\, \mathfrak{s}^\alpha \textit{ in } \Omega. \tag{4.8}$$

Then it is clear that $\Delta_x\big(\varepsilon^\alpha V^\alpha(\frac{x}{\varepsilon})\big)$ coincides with the function (4.7). Therefore, a better start for the asymptotic expansion of u_ε reads

$$\chi(\tfrac{x}{\varepsilon})v^0(x) + \psi(x)\varepsilon^\alpha V^\alpha(\tfrac{x}{\varepsilon}),$$

which satisfies the Dirichlet boundary conditions on $\partial\mathscr{U}_\varepsilon$, and the associated remainder r_ε^α is defined as

$$u_\varepsilon(x) = \chi(\tfrac{x}{\varepsilon})v^0(x) + \psi(x)\varepsilon^\alpha V^\alpha(\tfrac{x}{\varepsilon}) + r_\varepsilon^\alpha(x).$$

Since $\psi \equiv 1$ on the support of the right-hand side (4.7), we find

$$\Delta r_\varepsilon^\alpha(x) = -\big[\Delta, \chi(\tfrac{x}{\varepsilon})\big]\big(v^0(x) - \mathbf{b}_1^0\mathfrak{s}^\alpha(x)\big) - \big[\Delta, \psi\big]\varepsilon^\alpha V^\alpha(\tfrac{x}{\varepsilon}). \tag{4.9}$$

Again, the commutator $\big[\Delta, \chi(\frac{x}{\varepsilon})\big]\big(v^0(x) - \mathbf{b}_1^0\mathfrak{s}^\alpha(x)\big)$ will be converted in rapid variables, and since

$$v^0(x) - \mathbf{b}_1^0\mathfrak{s}^\alpha(x) \underset{r\to 0}{\simeq} \sum_{p=2}^{\infty} \mathbf{b}_p^0\mathfrak{s}^{p\alpha}(\tfrac{x}{\varepsilon}), \tag{4.10}$$

we have gained one power of ε^α.

Next, we express the other part of the remainder (4.9) in slow variables. By Lemma 3.10, we have $V^\alpha(X) = \mathcal{O}_{R\to\infty}(1)$. Thus, Proposition 3.13 yields that V^α expands at infinity as

$$V^\alpha(X) \underset{R\to\infty}{\simeq} \sum_{p=1}^{\infty} \mathbf{B}_p^1\,\mathfrak{s}^{-p\alpha}(X). \tag{4.11}$$

Since $\Delta\mathfrak{s}^{-p\alpha} = 0$, we find

$$\big[\Delta, \psi\big]\varepsilon^\alpha V^\alpha(\tfrac{x}{\varepsilon}) \underset{\varepsilon\to 0}{\simeq} \sum_{p=1}^{\infty} \varepsilon^{(1+p)\alpha}\mathbf{B}_p^1\,\big[\Delta, \psi\big]\mathfrak{s}^{-p\alpha}(x). \tag{4.12}$$

The terms in (4.12) start with $\varepsilon^{2\alpha}$. They can be compensated by the solution of problems in ω. We set $v^\alpha = 0$.

Step 2. Next we define $v^{2\alpha}$ as the solution of the problem in slow variables in ω

$$\textit{Find } v^{2\alpha} \in \mathrm{H}_0^1(\omega) \textit{ such that } -\Delta_x v^{2\alpha} = \mathbf{B}_1^1\,\big[\Delta, \psi\big]\,\mathfrak{s}^{-\alpha}, \tag{4.13}$$

and $V^{2\alpha}$ as the solution of the problem in fast variables in Ω (cf. (4.8))

$$\textit{Find } V^{2\alpha} \in V(\Omega) \textit{ such that } -\Delta_X V^{2\alpha} = \mathbf{b}_2^0 \left[\Delta, \chi\right] \mathfrak{s}^{2\alpha}. \tag{4.14}$$

4.2 The general construction

The construction is done by induction. Let us assume the asymptotic expansion built up to order $n-1$, i.e.,

$$u_\varepsilon(x) = \chi(\tfrac{x}{\varepsilon}) \sum_{\ell=0}^{n-1} \varepsilon^{\ell\alpha} v^{\ell\alpha}(x) + \psi(x) \sum_{\ell=1}^{n-1} \varepsilon^{\ell\alpha} V^{\ell\alpha}(\tfrac{x}{\varepsilon}) + r_\varepsilon^{(n-1)\alpha}(x) \tag{4.15}$$

with $v^{\ell\alpha} \in \mathrm{H}_0^1(\omega)$ and $V^{\ell\alpha} \in V(\Omega)$ whose Laplacians vanish in a neighborhood of zero and ∞ respectively. For $\ell = 0, \dots, n-1$ we expand the term $v^{\ell\alpha}$ into singular functions at the corner point (cf. Proposition 3.6)

$$v^{\ell\alpha}(x) \underset{r\to 0}{\simeq} \sum_{p=1}^{+\infty} \mathbf{b}_p^\ell \, \mathfrak{s}^{p\alpha}(x), \tag{4.16}$$

and we also expand the profiles $V^{\ell\alpha}$ into dual singular functions at infinity (cf. Proposition 3.13)

$$V^{\ell\alpha}(X) \underset{R\to +\infty}{\simeq} \sum_{p=1}^{+\infty} \mathbf{B}_p^\ell \, \mathfrak{s}^{-p\alpha}(X). \tag{4.17}$$

The definitions for the next terms $v^{n\alpha}$ and $V^{n\alpha}$ generalize (4.13) and (4.14). The function $v^{n\alpha} \in \mathrm{H}_0^1(\omega)$ solves

$$\Delta v^{n\alpha}(x) = -\Delta \left[\psi(x) \sum_{\ell=1}^{n-1} \mathbf{B}_{n-\ell}^\ell \mathfrak{s}^{-(n-\ell)\alpha}(x) \right], \tag{4.18}$$

and $V^{n\alpha} \in V(\Omega)$ satisfies

$$\Delta V^{n\alpha}(X) = -\Delta \left[\chi(X) \sum_{\ell=0}^{n-1} \mathbf{b}_{n-\ell}^\ell \mathfrak{s}^{(n-\ell)\alpha}(X) \right]. \tag{4.19}$$

Let us calculate the residual. The Laplacian of the remainder is given by

$$\Delta r_\varepsilon^{(n-1)\alpha}(x) = \Delta u^\varepsilon - \sum_{\ell=0}^{n-1} \varepsilon^{\ell\alpha} \Big[\Delta\big(\chi(\tfrac{x}{\varepsilon}) v^{\ell\alpha}(x)\big) + \Delta\big(\psi(x) V^{\ell\alpha}(\tfrac{x}{\varepsilon})\big) \Big]. \tag{4.20}$$

Next, we expand this relation using (4.16), (4.17), and the relations (4.18), (4.19) with n replaced by $1, 2, \dots, n-1$. We obtain

$$\Delta r_\varepsilon^{(n-1)\alpha} = -\sum_{\ell=0}^{n-1} \varepsilon^{\ell\alpha} \Big[\big[\Delta, \chi(\tfrac{\cdot}{\varepsilon})\big] v_{n-1-\ell}^{\ell\alpha} + \big[\Delta, \psi\big] V_{n-1-\ell}^{\ell\alpha}(\tfrac{\cdot}{\varepsilon}) \Big], \tag{4.21}$$

with

$$\begin{aligned} v_k^{\ell\alpha}(x) &:= v^{\ell\alpha}(x) - \sum_{p=1}^{k} \mathbf{b}_p^\ell\, \mathfrak{s}^{p\alpha}(x) \underset{r\to 0}{\simeq} \sum_{p=k+1}^{+\infty} \mathbf{b}_p^\ell\, \mathfrak{s}^{p\alpha}(x) \\ V_k^{\ell\alpha}(X) &:= V^{\ell\alpha}(X) - \sum_{p=1}^{k} \mathbf{B}_p^\ell\, \mathfrak{s}^{-p\alpha}(X) \underset{R\to\infty}{\simeq} \sum_{p=k+1}^{+\infty} \mathbf{B}_p^\ell\, \mathfrak{s}^{-p\alpha}(X). \end{aligned} \tag{4.22}$$

The leading term of the remainder $\Delta r_\varepsilon^{(n-1)\alpha}$ corresponds to the lowest terms in the sums on the right-hand sides of the identities (4.22) and is therefore

$$\Delta \left[\sum_{\ell=0}^{n-1} \varepsilon^{\ell\alpha} \mathbf{b}_{n-\ell}^\ell\, \mathfrak{s}^{(n-\ell)\alpha}(x) \chi(\tfrac{x}{\varepsilon}) \right] + \Delta \left[\sum_{\ell=1}^{n-1} \varepsilon^{\ell\alpha} \mathbf{B}_{n-\ell}^\ell\, \mathfrak{s}^{-(n-\ell)\alpha}(\tfrac{x}{\varepsilon}) \psi(x) \right],$$

which leads after scaling to (cf. (4.18) and (4.19))

$$\varepsilon^{n\alpha} \left(\Delta \left[\sum_{\ell=0}^{n-1} \mathbf{b}_{n-\ell}^\ell\, \mathfrak{s}^{(n-\ell)\alpha}(\tfrac{x}{\varepsilon}) \chi(\tfrac{x}{\varepsilon}) \right] + \Delta \left[\sum_{\ell=1}^{n-1} \mathbf{B}_{n-\ell}^\ell\, \mathfrak{s}^{-(n-\ell)\alpha}(x) \psi(x) \right] \right).$$

4.3 Optimal error estimate

Theorem 4.1. *The solution u_ε of the problem (2.3) admits the following multiscale expansion into powers of ε (recall that π/α is the opening angle of ω at $\mathbf{0}$) :*

$$u_\varepsilon(x) = \chi(\tfrac{x}{\varepsilon}) \sum_{\ell=0}^{n} \varepsilon^{\ell\alpha} v^{\ell\alpha}(x) + \psi(x) \sum_{\ell=0}^{n} \varepsilon^{\ell\alpha} V^{\ell\alpha}(\tfrac{x}{\varepsilon}) + r_\varepsilon^{n\alpha}(x), \tag{4.23}$$

where the terms $v^{\ell\alpha}$ and $V^{\ell\alpha}$ do not depend on ε and are defined in ω and Ω by Equations (4.18) and (4.19) respectively. Moreover, the remainder $r_\varepsilon^{n\alpha}$ satisfies the estimate

$$\|r_\varepsilon^{n\alpha}\|_{\mathrm{H}^1(\mathscr{U}_\varepsilon)} \leqslant C\varepsilon^{(n+1)\alpha}. \tag{4.24}$$

Proof. A basic technique to estimate the remainder consists in investigating the Laplace–Dirichlet problem it solves. By construction, $r_\varepsilon^{n\alpha}$ satisfies the homogeneous Dirichlet condition and belongs to $\mathrm{H}_0^1(\mathscr{U}_\varepsilon)$. By uniform coercivity, there exists $C_0 > 0$ such that

$$\|r_\varepsilon^{n\alpha}\|_{\mathrm{H}^1(\mathscr{U}_\varepsilon)} \leqslant C_0 \|\Delta r_\varepsilon^{n\alpha}\|_{\mathrm{H}^{-1}(\mathscr{U}_\varepsilon)} \quad \forall \varepsilon \leqslant \varepsilon_0. \tag{4.25}$$

Since $\Delta r_\varepsilon^{n\alpha}$ has the expression (4.21) (with $n-1$ replaced by n), we have to estimate each of its terms in the $\mathrm{H}^{-1}(\mathscr{U}_\varepsilon)$-norm.

- For all v the commutator of Δ and $\chi(\frac{\cdot}{\varepsilon})$ is given by

$$\big([\Delta, \chi(\tfrac{\cdot}{\varepsilon})]v\big)(x) = 2\varepsilon^{-1}\nabla v(x) \cdot (\nabla\chi)(\tfrac{x}{\varepsilon}) + \varepsilon^{-2} v(x)(\Delta\chi)(\tfrac{x}{\varepsilon}). \tag{4.26}$$

Hence the support of $[\Delta, \chi(\frac{\cdot}{\varepsilon})]v$ is included in the annulus $3R^*\varepsilon/2 \leqslant r \leqslant 2R^*\varepsilon$. For $v_k^{\ell\alpha}$, which is $\mathcal{O}_{r\to 0}(r^{(k+1)\alpha})$, one obtains the L^∞-bound

$$\big\|[\Delta, \chi(\tfrac{\cdot}{\varepsilon})]\, v_k^{\ell\alpha}\big\|_{\mathrm{L}^\infty(\mathscr{U}_\varepsilon)} \leqslant C\,\varepsilon^{(k+1)\alpha-2}. \tag{4.27}$$

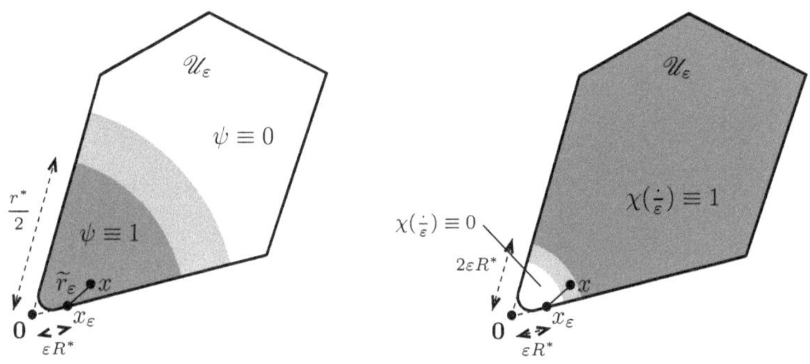

Fig. 3 Point x_ε, distance $\widetilde{r}_\varepsilon$, and supports of cut-off functions ψ and $\chi(\frac{\cdot}{\varepsilon})$.

Let us choose X_0 such that $X_0 \in \partial\Omega$ and $|X_0| = R^*$ (such a point does exist since Ω coincides with K in the region $R > R^*$). Then the point $x_\varepsilon = X_0/\varepsilon$ belongs to $\partial\mathscr{U}_\varepsilon$ (cf. Fig. 3). Moreover, if we set

$$\widetilde{r}_\varepsilon(x) = |x - x_\varepsilon|,$$

we find that $\widetilde{r}_\varepsilon$ is equivalent to r in the support of $[\Delta, \chi(\frac{\cdot}{\varepsilon})]$ uniformly in ε. Since $x_\varepsilon \in \partial\mathscr{U}_\varepsilon$, we have

$$\Big\|\frac{w}{\widetilde{r}_\varepsilon}\Big\|_{\mathrm{L}^2(\mathscr{U}_\varepsilon)} \leqslant C_1 \|w\|_{\mathrm{H}^1(\mathscr{U}_\varepsilon)} \quad \forall w \in \mathrm{H}_0^1(\mathscr{U}_\varepsilon)$$

with a constant C_1 independent of $\varepsilon < \varepsilon_0/2$ and w. Let $w \in \mathrm{H}^1_0(\mathscr{U}_\varepsilon)$. We deduce via Hölder inequality

$$\begin{aligned}\Big\langle \big[\Delta, \chi(\tfrac{\cdot}{\varepsilon})\big] v_k^{\ell\alpha}, w \Big\rangle &= \Big\langle \widetilde{r}_\varepsilon \big[\Delta, \chi(\tfrac{\cdot}{\varepsilon})\big] v_k^{\ell\alpha}, \frac{w}{\widetilde{r}_\varepsilon} \Big\rangle \\ &\leqslant C \big\| \big[\Delta, \chi(\tfrac{\cdot}{\varepsilon})\big] v_k^{\ell\alpha} \big\|_{\mathrm{L}^\infty(\mathscr{U}_\varepsilon)} \big\| \widetilde{r}_\varepsilon \big\|_{\mathrm{L}^2(\mathscr{U}_\varepsilon)} \big\| w \big\|_{\mathrm{H}^1(\mathscr{U}_\varepsilon)} \\ &\leqslant C\, \varepsilon^{(k+1)\alpha - 2} \varepsilon^2 \big\| w \big\|_{\mathrm{H}^1(\mathscr{U}_\varepsilon)}.\end{aligned}$$

Hence

$$\big\| \big[\Delta, \chi(\tfrac{\cdot}{\varepsilon})\big] v_k^{\ell\alpha} \big\|_{\mathrm{H}^{-1}(\mathscr{U}_\varepsilon)} \leqslant C\, \varepsilon^{(k+1)\alpha}. \tag{4.28}$$

• Using that the function $V_k^{\ell\alpha}$ is $\mathscr{O}_{R\to\infty}(R^{(k+1)\alpha})$, we easily deduce the estimate

$$\big\| [\Delta, \psi] V_k^{\ell\alpha}(\tfrac{\cdot}{\varepsilon}) \big\|_{\mathrm{L}^2(\mathscr{U}_\varepsilon)} \leqslant C\, \varepsilon^{(k+1)\alpha}.$$

Hence

$$\big\| [\Delta, \psi] V_k^{\ell\alpha}(\tfrac{\cdot}{\varepsilon}) \big\|_{\mathrm{H}^{-1}(\mathscr{U}_\varepsilon)} \leqslant C\, \varepsilon^{(k+1)\alpha}. \tag{4.29}$$

One deduces immediately from (4.21), (4.28), and (4.29)

$$\big\| \Delta r_\varepsilon^{n\alpha} \big\|_{\mathrm{H}^{-1}(\mathscr{U}_\varepsilon)} \leqslant C\, \varepsilon^{(k+1)\alpha}, \tag{4.30}$$

and using the *a priori* estimate (4.25), we obtain the bound (4.24), which completes the proof. □

5 Matching of Asymptotic Expansions

5.1 Formal derivation of the asymptotic expansions

We represent the solution u_ε as a *formal series* in each zone of interest, i.e., the *corner expansion* (or inner expansion) near the origin **0** and the *outer expansion* away from **0**. We write these two formal series in the form

$$u_\varepsilon(x) \simeq \sum_{\ell=-\infty}^{+\infty} \varepsilon^{\ell\alpha}\, U^{\ell\alpha}(\tfrac{x}{\varepsilon}) \quad \text{and} \quad u_\varepsilon(x) \simeq \sum_{\ell=-\infty}^{+\infty} \varepsilon^{\ell\alpha}\, u^{\ell\alpha}(x). \tag{5.1}$$

This *Ansatz* is suggested by the homogeneity of the singular functions (cf. (2.7)). We will give a sense to the infinite sums in terms of asymptotic expansions later on.

Since the H^1-norm of u_ε is uniformly bounded with respect to ε, we know that all the $u^{\ell\alpha}$ and $U^{\ell\alpha}$ for $\ell < 0$ are just zero. Moreover, it is clear that the terms of the asymptotic expansions must satisfy

$$\begin{cases} -\Delta u^0 = f \text{ in } \omega \text{ and } u^0 = 0 \text{ on } \partial\omega, \\ \forall \ell > 0, \quad \Delta u^{\ell\alpha} = 0 \text{ in } \omega \text{ and } u^{\ell\alpha} = 0 \text{ on } \partial\omega \setminus \{\mathbf{0}\}, \\ \forall \ell \geqslant 0, \quad \Delta U^{\ell\alpha} = 0 \text{ in } \Omega \text{ and } U^{\ell\alpha} = 0 \text{ on } \partial\Omega. \end{cases} \tag{5.2}$$

Now, we need to ensure the matching of the two formal series in the transition zone

$$\varepsilon \ll r \ll 1. \tag{5.3}$$

To do so, we expand the terms $u^{\ell\alpha}$ and $U^{\ell\alpha}$. By Propositions 3.6 and 3.13 – note that $r \ll 1$ and $\frac{r}{\varepsilon} \gg 1$ – these expansions read [2]

$$\begin{aligned} u^{\ell\alpha}(x) &= \sum_{p=1}^{+\infty} \Big(a_p^\ell \, \mathfrak{s}^{-p\alpha}(r,\theta) \; + \; b_p^\ell \, \mathfrak{s}^{p\alpha}(r,\theta) \Big), \\ U^{\ell\alpha}(X) &= \sum_{p=1}^{+\infty} \Big(A_p^\ell \, \mathfrak{s}^{p\alpha}(R,\theta) \; + \; B_p^\ell \, \mathfrak{s}^{-p\alpha}(R,\theta) \Big). \end{aligned} \tag{5.4}$$

We use the homogeneity of the functions $\mathfrak{s}^{p\alpha}$ and transform the rapid variable $\frac{r}{\varepsilon}$ into the slow one r. Ensuring the equality of the two formal series (5.1), we get

$$\begin{aligned} &\sum_{\ell=-\infty}^{+\infty} \Big(\varepsilon^{\ell\alpha} \sum_{p=1}^{+\infty} \Big(a_p^\ell \mathfrak{s}^{-p\alpha}(r,\theta) \; + \; b_p^\ell \mathfrak{s}^{p\alpha}(r,\theta) \Big) \Big) \\ &= \sum_{\ell=-\infty}^{+\infty} \Big(\varepsilon^{\ell\alpha} \sum_{p=1}^{+\infty} \Big(A_p^\ell \, \mathfrak{s}^{p\alpha}(\frac{r}{\varepsilon},\theta) \; + \; B_p^\ell \, \mathfrak{s}^{-p\alpha}(\frac{r}{\varepsilon},\theta) \Big) \Big) \\ &= \sum_{\ell=-\infty}^{+\infty} \Big(\varepsilon^{\ell\alpha} \sum_{p=1}^{+\infty} \Big(\varepsilon^{-p\alpha} \, A_p^\ell \, \mathfrak{s}^{p\alpha}(r,\theta) \; + \; \varepsilon^{p\alpha} \, B_p^\ell \, \mathfrak{s}^{-p\alpha}(r,\theta) \Big) \Big) \\ &= \sum_{\ell=-\infty}^{+\infty} \Big(\varepsilon^{\ell\alpha} \sum_{p=1}^{+\infty} \Big(A_p^{\ell+p} \, \mathfrak{s}^{p\alpha}(r,\theta) \; + \; B_p^{\ell-p} \, \mathfrak{s}^{-p\alpha}(r,\theta) \Big) \Big). \end{aligned} \tag{5.5}$$

Identifying the terms of the two series leads to

$$b_p^\ell = A_p^{\ell+p} \quad \text{and} \quad a_p^\ell = B_p^{\ell-p}, \tag{5.6}$$

i. e.

[2] Note that we do not use the boldface notation for the coefficients b_p^0 because we do not yet know whether they coincide with the coefficients $\mathbf{b}_p^0$ already defined in Section 4. The coincidence will be shown in Section 6.

$$
\begin{aligned}
a_p^\ell &= B_p^{\ell-p} \text{ if } p \leqslant \ell \quad \text{and} \quad a_p^\ell = 0 \text{ if } p > \ell, \\
A_p^\ell &= b_p^{\ell-p} \text{ if } p \leqslant \ell \quad \text{and} \quad A_p^\ell = 0 \text{ if } p > \ell,
\end{aligned} \tag{5.7}
$$

knowing that $b_p^{\ell-p} = B_p^{\ell-p} = 0$ for $p > \ell$ since the terms $u^{n\alpha}$ and $U^{n\alpha}$ are 0 for $n < 0$.

Remark 5.1. Here, we have chosen to derive the matching relations without any knowledge of the matched asymptotic technique. However, one can derive the relations (5.7) using the Van Dyke principle (cf. [25]).

5.2 Definition of the asymptotic terms

For $\ell \in \mathbb{N}$ the functions $u^{\ell\alpha}$ and $U^{\ell\alpha}$ are defined inductively. The following algorithm defines step by step $u^{\ell\alpha} : \omega \to \mathbb{R}$, $U^{\ell\alpha} : \Omega \to \mathbb{R}$, $b^\ell = \left(b_p^\ell\right)_{p\in\mathbb{N}^*}$, and $B^\ell = \left(B_p^\ell\right)_{p\in\mathbb{N}^*}$ for $\ell \in \mathbb{N}$.

Step 0. $u^0 \in V_{\mathrm{loc},0}(\omega)$ is defined via Proposition 3.4 (in the particular case of Remark 3.5) as a unique function satisfying

$$
\Delta u^0 = -f \ \text{ in } \omega \quad \text{and} \quad u^0 = \mathcal{O}_{r\to 0}(1). \tag{5.8}
$$

Moreover, U^0 is chosen to be 0. Let b^0 be the sequence of numbers defined by Proposition 3.6, and let B^0 be zero:

$$
b^0 = \left(b_p^0\right)_{p\in\mathbb{N}^*} \quad \text{and} \quad B^0 = \left(B_p^0\right)_{p\in\mathbb{N}^*} = 0. \tag{5.9}
$$

Step ℓ. We denote by $a^\ell = \left(a_p^\ell\right)_{p\in\mathbb{N}^*}$ and $A^\ell = \left(A_p^\ell\right)_{p\in\mathbb{N}^*}$ two finite sequences of real numbers such that

$$
\begin{aligned}
a_p^\ell &= B_p^{\ell-p} \text{ if } 1 \leqslant p \leqslant \ell-1 \quad \text{and} \quad a_p^\ell = 0 \text{ if } p \geqslant \ell, \\
A_p^\ell &= b_p^{\ell-p} \text{ if } 1 \leqslant p \leqslant \ell \quad \text{and} \quad A_p^\ell = 0 \text{ if } p \geqslant \ell+1.
\end{aligned} \tag{5.10}
$$

The functions $u^{\ell\alpha}$ and $U^{\ell\alpha}$ are defined via Propositions 3.4 and 3.11 as the unique solutions of the problems

$$
\left\{
\begin{array}{l}
\textit{Find } u^{\ell\alpha} \in V_{\mathrm{loc},0}(\omega) \textit{ such that} \\[2ex]
\Delta u^{\ell\alpha} = 0 \ \textit{ in } \omega \quad \textit{and} \quad u^{\ell\alpha} - \displaystyle\sum_{p=1}^{\ell-1} a_p^\ell \, \mathfrak{s}^{-p\alpha} = \mathcal{O}_{r\to 0}(1)
\end{array}
\right. \tag{5.11}
$$

and

$$\begin{cases} \textit{Find } U^{\ell\alpha} \in V_{\mathrm{loc},\infty}(\Omega) \textit{ such that} \\ \Delta U^{\ell\alpha} = 0 \textit{ in } \Omega \textit{ and } U^{\ell\alpha} - \sum_{p=1}^{\ell} A_p^{\ell}\, \mathfrak{s}^{p\alpha} = \mathcal{O}_{R\to\infty}(1). \end{cases} \tag{5.12}$$

Finally, we define the sequences b^ℓ and B^ℓ associated with $u^{\ell\alpha}$ and $U^{\ell\alpha}$ in Propositions 3.6 and 3.13

$$b^\ell = \left(b_p^\ell\right)_{p\in\mathbb{N}^*} \quad \text{and} \quad B^\ell = \left(B_p^\ell\right)_{p\in\mathbb{N}^*}. \tag{5.13}$$

5.3 Global error estimates

The main idea to prove error estimates is to define a global approximation $\widehat{u}^\varepsilon_{n\alpha} \in \mathrm{H}^1_0(\mathscr{U}_\varepsilon)$ of u_ε by the formula

$$\widehat{u}^\varepsilon_{n\alpha}(x) = \varphi\left(\tfrac{r}{\eta(\varepsilon)}\right) \sum_{\ell=0}^{n} \varepsilon^{\ell\alpha}\, u^{\ell\alpha}(x) + \left(1 - \varphi\left(\tfrac{r}{\eta(\varepsilon)}\right)\right) \sum_{\ell=1}^{n} \varepsilon^{\ell\alpha}\, U^{\ell\alpha}\left(\tfrac{x}{\varepsilon}\right), \tag{5.14}$$

where φ is a smooth cut-off function with $\varphi(\rho) = 0$ for $\rho < 1$ and $\varphi(\rho) = 1$ for $\rho > 2$ and η is a smooth function of ε such that

$$\lim_{\varepsilon\to 0} \eta(\varepsilon) = 0 \quad \text{and} \quad \lim_{\varepsilon\to 0} \frac{\eta(\varepsilon)}{\varepsilon} = +\infty. \tag{5.15}$$

Theorem 5.2. *There exists a constant C such that*

$$\|u_\varepsilon - \widehat{u}^\varepsilon_{n\alpha}\|_{\mathrm{H}^1(\mathscr{U}_\varepsilon)} \leqslant C \left[\left(\eta(\varepsilon)\right)^{(n+1)\alpha} + \left(\frac{\varepsilon}{\eta(\varepsilon)}\right)^{(n+1)\alpha} \right]. \tag{5.16}$$

Remark 5.3. One can optimize the estimate (5.16) by choosing the best η: For $\eta(\varepsilon) = \varepsilon^{1/2}$, there exists a constant C such that

$$\|u_\varepsilon - \widehat{u}^\varepsilon_{n\alpha}\|_{H^1(\mathscr{U}_\varepsilon)} \leqslant C\, \varepsilon^{(n+1)\alpha/2}. \tag{5.17}$$

Proof. First, we denote by $\widehat{e}^\varepsilon_{n\alpha}$ the approximation error at step n

$$\widehat{e}^\varepsilon_{n\alpha}(x) = \widehat{u}^\varepsilon_{n\alpha}(x) - u_\varepsilon(x)$$

and by $\mathcal{E}^\varepsilon_{n\alpha}$ the corresponding matching error

$$\mathcal{E}^\varepsilon_{\alpha n}(x) = \sum_{\ell=0}^{n} \varepsilon^{\ell\alpha} \left[u^{\ell\alpha}(x) - U^{\ell\alpha}\left(\tfrac{x}{\varepsilon}\right)\right].$$

Of course, the matching error makes sense and is small only in the intermediate region; we shall express the H^1-norm of $\widehat{e}^{\varepsilon}_{n\alpha}$ over $\mathscr{U}_\varepsilon$ in terms of $\mathcal{E}^{\varepsilon}_{n\alpha}$ in this region. By harmonicity of u_ε, $u^{\ell\alpha}$, and $U^{\ell\alpha}$, we obtain

$$\Delta\widehat{e}^{\varepsilon}_{n\alpha}(x) = \tfrac{2}{\eta(\varepsilon)}\,[\nabla\varphi](\frac{r}{\eta(\varepsilon)})\,\nabla\mathcal{E}^{\varepsilon}_{n\alpha}(x) + \frac{1}{[\eta(\varepsilon)]^2}[\Delta\varphi](\frac{r}{\eta(\varepsilon)})\,\mathcal{E}^{\varepsilon}_{n\alpha}(x).$$

Since $\widehat{e}^{\varepsilon}_{n\alpha}$ belongs to $\mathrm{H}^1_0(\mathscr{U}_\varepsilon)$, the Green formula leads to

$$\begin{aligned}\int_{\mathscr{U}_\varepsilon}\left(\nabla\widehat{e}^{\varepsilon}_{n\alpha}\right)^2 dx &= \tfrac{2}{\eta(\varepsilon)}\int_{\mathscr{U}_\varepsilon}[\nabla\varphi](\tfrac{r}{\eta(\varepsilon)})\,\nabla\mathcal{E}^{\varepsilon}_{n\alpha}\,\widehat{e}^{\varepsilon}_{n\alpha}\\ &\quad+\tfrac{1}{[\eta(\varepsilon)]^2}\int_{\mathscr{U}_\varepsilon}[\Delta\varphi](\tfrac{r}{\eta(\varepsilon)})\,\mathcal{E}^{\varepsilon}_{n\alpha}\,\widehat{e}^{\varepsilon}_{n\alpha}\,dx\\ &\leqslant \frac{C}{[\eta(\varepsilon)]^2}\left[\|\mathcal{E}^{\varepsilon}_{n\alpha}\|_{\infty,\eta(\varepsilon)}+\eta(\varepsilon)\|\nabla\mathcal{E}^{\varepsilon}_{n\alpha}\|_{\infty,\eta(\varepsilon)}\right]\|\widehat{e}^{\varepsilon}_{n\alpha}\|_{1,\eta(\varepsilon)}\end{aligned}$$

with the notation, for $p\in[1,\infty]$

$$\|u\|_{p,\eta(\varepsilon)} = \|u\|_{\mathrm{L}^p(\{x\in\omega\,;\,\eta(\varepsilon)\leqslant r\leqslant 2\eta(\varepsilon)\})}. \tag{5.18}$$

Using the Poincaré inequality on $\mathscr{U}_\varepsilon$ (uniform with respect to ε), we get

$$\|\widehat{e}^{\varepsilon}_{n\alpha}\|^2_{\mathrm{H}^1(\mathscr{U}_\varepsilon)} \leqslant \frac{C}{(\eta(\varepsilon))^2}\left[\|\mathcal{E}^{\varepsilon}_{n\alpha}\|_{\infty,\eta(\varepsilon)}+\eta(\varepsilon)\|\nabla\mathcal{E}^{\varepsilon}_{n\alpha}\|_{\infty,\eta(\varepsilon)}\right]\times\|\widehat{e}^{\varepsilon}_{n\alpha}\|_{1,\eta(\varepsilon)}.$$

The conclusion is obtained from the following two lemmas (proved below). □

Lemma 5.4. *There exists a constant C such that for all $u\in\mathrm{H}^1_0(\mathscr{U}_\varepsilon)$ the norm $\|u\|_{1,\eta(\varepsilon)}$, defined in* (5.18), *can be estimated as follows:*

$$\|u\|_{1,\eta(\varepsilon)} \leqslant C\,[\eta(\varepsilon)]^2\,\|u\|_{\mathrm{H}^1(\mathscr{U}_\varepsilon)}. \tag{5.19}$$

Lemma 5.5. *There exists a constant C such that (for the definition of the norms cf.* (5.18))

$$\|\mathcal{E}^{\varepsilon}_{n\alpha}\|_{\infty,\eta(\varepsilon)} \leqslant C\left[\Big(\eta(\varepsilon)\Big)^{(n+1)\alpha}+\Big(\frac{\varepsilon}{\eta(\varepsilon)}\Big)^{(n+1)\alpha}\right] \tag{5.20}$$

and

$$\|\nabla\mathcal{E}^{\varepsilon}_{n\alpha}\|_{\infty,\eta(\varepsilon)} \leqslant C\,\frac{1}{\eta(\varepsilon)}\left[\Big(\eta(\varepsilon)\Big)^{(n+1)\alpha}+\Big(\frac{\varepsilon}{\eta(\varepsilon)}\Big)^{(n+1)\alpha}\right]. \tag{5.21}$$

Proof of Lemma 5.4. For all $u\in\mathrm{H}^1_0(\mathscr{U}_\varepsilon)$ and $r\in[\eta(\varepsilon),2\eta(\varepsilon)]$

$$\int_0^{\frac{\pi}{\alpha}} |u(r,\theta)|\, d\theta \leqslant \int_0^{\frac{\pi}{\alpha}} \Big[\int_0^{\theta} \Big|\frac{\partial u}{\partial \theta}(r,\theta')\Big|\, d\theta' \Big]\, d\theta \leqslant \frac{\pi}{\alpha} \int_0^{\frac{\pi}{\alpha}} \Big|\frac{\partial u}{\partial \theta}(r,\theta)\Big|\, d\theta.$$

Hence

$$\int_{\eta(\varepsilon)}^{2\eta(\varepsilon)} \int_0^{\frac{\pi}{\alpha}} |u(r,\theta)|\, rdr\, d\theta \leqslant \frac{\pi}{\alpha} \int_{\eta(\varepsilon)}^{2\eta(\varepsilon)} \int_0^{\frac{\pi}{\alpha}} \Big|\frac{\partial u}{\partial \theta}(r,\theta)\Big|\, rdr\, d\theta$$

$$\leqslant \frac{\pi}{\alpha} \int_{\eta(\varepsilon)}^{2\eta(\varepsilon)} \int_0^{\frac{\pi}{\alpha}} \frac{2\eta(\varepsilon)}{r} \Big|\frac{\partial u}{\partial \theta}(r,\theta)\Big|\, rdr\, d\theta \leqslant C\, \eta(\varepsilon)\, \|\nabla u\|_{1,\eta(\varepsilon)}.$$

We conclude, using the Cauchy–Schwarz inequality, that

$$\|u\|_{1,\eta(\varepsilon)} \leqslant C\, \eta(\varepsilon)\, \|\nabla u\|_{1,\eta(\varepsilon)} \leqslant C\, [\eta(\varepsilon)]^2\, \|\nabla u\|_{2,\eta(\varepsilon)}. \qquad \square$$

Proof of Lemma 5.5. We prove (5.20). The inequality (5.21) can be obtained by the same technique. The first step is to expand the $u^{\ell\alpha}$ and $U^{\ell\alpha}$ by using (3.12) and (3.19). By the definition of $u^{\ell\alpha}$ and $U^{\ell\alpha}$ (cf. (5.11) and (5.12)) and taking (5.10) into account, one finds

$$u^{\ell\alpha}(x) = \sum_{p=1}^{\ell} B_p^{\ell-p}\, \mathfrak{s}^{-p\alpha}(r,\theta) + \sum_{p=1}^{n-\ell} b_p^{\ell}\, \mathfrak{s}^{p\alpha}(r,\theta) + \mathcal{O}_{r\to 0}(r^{(n+1-\ell)\alpha})$$

and

$$U^{\ell\alpha}(X) = \sum_{p=1}^{\ell} b_p^{\ell-p}\, \mathfrak{s}^{p\alpha}(R,\theta) + \sum_{p=1}^{n-\ell} B_p^{\ell}\, \mathfrak{s}^{-p\alpha}(R,\theta) + \mathcal{O}_{R\to\infty}(R^{(\ell-n-1)\alpha}).$$

Since $\eta(\varepsilon)$ tends to 0 and $\eta(\varepsilon)/\varepsilon$ tends to $+\infty$ when ε tends to 0, for $\eta(\varepsilon) \leqslant r \leqslant 2\eta(\varepsilon)$

$$\begin{aligned} &\Big| u^{\ell\alpha}(x) - \sum_{p=1}^{\ell} B_p^{\ell-p}\, \mathfrak{s}^{-p\alpha}(r,\theta) - \sum_{p=1}^{n-\ell} b_p^{\ell}\, \mathfrak{s}^{p\alpha}(r,\theta) \Big| \leqslant C\, [\eta(\varepsilon)]^{(n+1-\ell)\alpha}, \\ &\Big| U^{\ell\alpha}(\tfrac{x}{\varepsilon}) - \sum_{p=1}^{\ell} b_p^{\ell-p}\, \mathfrak{s}^{p\alpha}(\tfrac{r}{\varepsilon},\theta) - \sum_{p=1}^{n-\ell} B_p^{\ell}\, \mathfrak{s}^{-p\alpha}(\tfrac{r}{\varepsilon},\theta) \Big| \leqslant C[\tfrac{\varepsilon}{\eta(\varepsilon)}]^{(n+1-\ell)\alpha}. \end{aligned} \tag{5.22}$$

Let S be given by

$$S = \sum_{\ell=0}^{n} \varepsilon^{\ell\alpha} \Big(\sum_{p=1}^{\ell} B_p^{\ell-p}\, \mathfrak{s}^{-p\alpha}(r,\theta) + \sum_{p=1}^{n-\ell} b_p^{\ell}\, \mathfrak{s}^{p\alpha}(r,\theta) \Big)$$
$$- \sum_{\ell=0}^{n} \varepsilon^{\ell\alpha} \Big(\sum_{p=1}^{\ell} b_p^{\ell-p}\, \mathfrak{s}^{p\alpha}(\tfrac{r}{\varepsilon},\theta) + \sum_{p=1}^{n-\ell} B_p^{\ell}\, \mathfrak{s}^{-p\alpha}(\tfrac{r}{\varepsilon},\theta) \Big). \tag{5.23}$$

From (5.22) and triangle inequalities, we obtain

$$\|\mathcal{E}_{n\alpha}^{\varepsilon}(r,\theta) - S\|_{\infty,\eta(\varepsilon)}$$
$$\leqslant C \Big\{ \sum_{\ell=0}^{n} \varepsilon^{\ell\alpha}\, [\eta(\varepsilon)]^{(n+1-\ell)\alpha} + \sum_{\ell=0}^{n} \varepsilon^{\ell\alpha}\, [\tfrac{\varepsilon}{\eta(\varepsilon)}]^{(n+1-\ell)\alpha} \Big\}$$
$$\leqslant C \Big\{ \sum_{\ell=0}^{n} [\tfrac{\varepsilon}{\eta(\varepsilon)}]^{\ell\alpha} [\eta(\varepsilon)]^{(n+1)\alpha} + \sum_{\ell=0}^{n} \eta(\varepsilon)^{\ell\alpha} [\tfrac{\varepsilon}{\eta(\varepsilon)}]^{(n+1)\alpha} \Big\}$$
$$\leqslant C \Big\{ [\eta(\varepsilon)]^{(n+1)\alpha} + [\tfrac{\varepsilon}{\eta(\varepsilon)}]^{(n+1)\alpha} \Big\}. \tag{5.24}$$

Now it remains to show that $S = 0$. By definition (cf. (2.7)), the singular functions $\mathfrak{s}^{\pm p\alpha}$ satisfy the homogeneity property

$$\mathfrak{s}^{-p\alpha}(\tfrac{r}{\varepsilon},\theta) = \varepsilon^{p\alpha}\, \mathfrak{s}^{-p\alpha}(r,\theta) \quad \text{and} \quad \mathfrak{s}^{p\alpha}(r,\theta) = \varepsilon^{p\alpha}\, \mathfrak{s}^{p\alpha}(\tfrac{r}{\varepsilon},\theta).$$

Therefore, S is given by

$$S = \sum_{\ell=0}^{n}\sum_{p=1}^{\ell} \varepsilon^{(\ell-p)\alpha} B_p^{\ell-p}\, \mathfrak{s}^{-p\alpha}(\tfrac{r}{\varepsilon},\theta) + \sum_{\ell=0}^{n}\sum_{p=1}^{n} \varepsilon^{\ell\alpha} b_p^{\ell}\, \mathfrak{s}^{p\alpha}(r,\theta)$$
$$- \sum_{\ell=0}^{n}\sum_{p=1}^{\ell} \varepsilon^{(\ell-p)\alpha} b_p^{\ell-p}\, \mathfrak{s}^{p\alpha}(r,\theta) - \sum_{\ell=0}^{n}\sum_{p=1}^{n} \varepsilon^{\ell\alpha} B_p^{\ell}\, \mathfrak{s}^{-p\alpha}(\tfrac{r}{\varepsilon},\theta).$$

The change of variables $\ell - p \mapsto \ell$ in the first and third terms leads to $S = 0$. □

5.4 Local error estimates

In this subsection, $\mathscr{B}_r$ denotes the ball of radius r and of center O. Starting from the global error estimates obtained in (5.17), it is easy to get estimates far from the corner and near the corner.

Theorem 5.6. *For any $r_0 > 0$ there exists $C > 0$ such that*

$$\left\| u_\varepsilon(r,\theta) - \sum_{\ell=0}^{n} \varepsilon^{\ell\alpha}\, u^{\ell\alpha}(r,\theta) \right\|_{\mathrm{H}^1(\omega\setminus\mathscr{B}_{r_0})} = \mathcal{O}(\varepsilon^{(n+1)\alpha}). \tag{5.25}$$

For any $R_0 > 0$ there exists $C > 0$ such that

$$\left\| u_\varepsilon(\varepsilon R,\theta) - \sum_{\ell=0}^{n} \varepsilon^{\ell\alpha}\, U^{\ell\alpha}(R,\theta) \right\|_{\mathrm{H}^1(\Omega\cap\mathscr{B}_{R_0})} = \mathcal{O}(\varepsilon^{(n+1)\alpha}). \tag{5.26}$$

Proof. To prove (5.25), we remark that, for ε small enough, the only contribution comes from the terms $u^{\ell\alpha}$

$$\widehat{u}^{\varepsilon}_{n\alpha} = \sum_{\ell=1}^{n} \varepsilon^{\ell\alpha}\, u^{\ell\alpha} \quad \text{in} \quad \mathscr{U}_\varepsilon \setminus \mathscr{B}_{r_0} = \omega \setminus \mathscr{B}_{r_0}\,. \tag{5.27}$$

Consequently,

$$\begin{aligned}
&\|u_\varepsilon - \widehat{u}^{\varepsilon}_{n\alpha}\|_{\mathrm{H}^1(\omega\setminus\mathscr{B}_{r_0})} \\
&\quad \leqslant \|u_\varepsilon - \widehat{u}^{\varepsilon}_{(2n+2)\alpha}\|_{\mathrm{H}^1(\omega\setminus\mathscr{B}_{r_0})} + \|\widehat{u}^{\varepsilon}_{(2n+2)\alpha} - \widehat{u}^{\varepsilon}_{n\alpha}\|_{\mathrm{H}^1(\omega\setminus\mathscr{B}_{r_0})} \\
&\quad \leqslant \|u_\varepsilon - \widehat{u}^{\varepsilon}_{(2n+2)\alpha}\|_{\mathrm{H}^1(\mathscr{U}_\varepsilon)} + \|\widehat{u}^{\varepsilon}_{(2n+2)\alpha} - \widehat{u}^{\varepsilon}_{n\alpha}\|_{\mathrm{H}^1(\omega\setminus\mathscr{B}_{r_0})}.
\end{aligned} \tag{5.28}$$

On the other hand, from (5.27) it follows that

$$\widehat{u}^{\varepsilon}_{(2n+2)\alpha} - \widehat{u}^{\varepsilon}_{n\alpha} = \sum_{\ell=n+1}^{2n+2} \varepsilon^{\ell\alpha}\, u^{\ell\alpha} \quad \text{in} \quad \omega \setminus \mathscr{B}_{r_0}\,, \tag{5.29}$$

and, since the $u^{\ell\alpha}$'s do not depend on ε,

$$\|\widehat{u}^{\varepsilon}_{(2n+2)\alpha} - \widehat{u}^{\varepsilon}_{n\alpha}\|_{\mathrm{H}^1(\omega\setminus\mathscr{B}_{r_0})} \leqslant C\, \varepsilon^{(n+1)\alpha}. \tag{5.30}$$

Due to (5.17), one finally has

$$\|u_\varepsilon - \widehat{u}^{\varepsilon}_{(2n+2)\alpha}\|_{\mathrm{H}^1(\omega\setminus\mathscr{B}_{r_0})} \leqslant C\, \varepsilon^{(n+1)\alpha}. \tag{5.31}$$

The estimate (5.25) follows from (5.27), (5.28), (5.30), and (5.31). The same technique leads to (5.26) as well. A scaling is needed ($R = r/\varepsilon$) to recover a domain independent of ε. □

Remark 5.7. Due to estimates (5.25) and (5.26), the outer and corner expansions are unique. Moreover, as the remainders are of the same orders as the first neglected term in the outer and corner expansions, these estimates are optimal. The outer and corner expansions can be seen as Taylor expansions of the exact solution expressed in the (r,θ) or $(r/\varepsilon,\theta)$ coordinates.

6 Comparison of the Two Expansions

In Section 5, starting from the outer and corner (matched) expansions, we were able to build a global asymptotic expansion for the solution u_ε of problem (2.3) (cf. the expression (5.14)). Using the multiscale technique, we proved in Section 4 another asymptotic expansion, which is also valid in the whole domain $\mathscr{U}_\varepsilon$. The global error estimates given in Theorems 5.2 and 4.1 allow us to compare these expansions.

Theorem 6.1. *The expansions* (5.14) *and* (4.23) *compare in the following way:*

- *The terms $u^{n\alpha}$ and $v^{n\alpha}$ coincide away from the corner point, i.e., for $r \geqslant r^*$.*

- *The profiles $U^{n\alpha}$ and $V^{n\alpha}$ coincide in the corner region, i.e., for $R \leqslant R^*/2$.*

More precisely, we have the identities

$$\begin{aligned} v^{n\alpha}(x) = & \quad u^{n\alpha}(x) - \psi(x) \sum_{p=1}^{n-1} a_p^n \, \mathfrak{s}^{-p\alpha}(x), \\ V^{n\alpha}(X) = & \quad U^{n\alpha}(X) - \chi(X) \sum_{p=1}^{n} A_p^n \, \mathfrak{s}^{p\alpha}(X). \end{aligned} \tag{6.1}$$

where the coefficients a_p^n and A_p^n are those defined in Section 5.2.

Proof. The first two statements follow directly from the optimal estimates (5.25), (5.26), (4.23), and (4.24) via localization. To get formula (6.1), we start from the problem (4.19) which defines $V^{n\alpha}$. We set

$$\begin{aligned} \widetilde{U}^{n\alpha}(X) &= V^{n\alpha}(X) + \chi(X) \sum_{\ell=0}^{n-1} \mathbf{b}_{n-\ell}^{\ell} \, \mathfrak{s}^{(n-\ell)\alpha}(X) \\ &= V^{n\alpha}(X) + \chi(X) \sum_{p=1}^{n} \mathbf{b}_p^{n-p} \, \mathfrak{s}^{p\alpha}(X). \end{aligned} \tag{6.2}$$

By the definition of $V^{n\alpha}$ (cf. (4.19)), $\widetilde{U}^{n\alpha}$ satisfies $\Delta \widetilde{U}^{n\alpha} = 0$ in Ω. Hence

$$\begin{cases} \widetilde{U}^{n\alpha} - U^{n\alpha} \in C^\infty(\Omega), \\ \Delta[\widetilde{U}^{n\alpha} - U^{n\alpha}] = 0 \ \text{ in } \ \Omega, \\ \widetilde{U}^{n\alpha}(X) - U^{n\alpha}(X) = 0 \ \text{ for } \ R \leqslant R^*/2. \end{cases} \tag{6.3}$$

Since $\widetilde{U}^{n\alpha} - U^{n\alpha}$ is harmonic, it is analytic in Ω. Hence, by unique continuation Theorem, $U^{n\alpha} = \widetilde{U}^{n\alpha}$. Moreover, as $V^{n\alpha}$ is a $\mathcal{O}_{R\to\infty}(1)$, one has $A_p^n = \mathbf{b}_p^{n-p}$

$$U^{n\alpha}(X) = V^{n\alpha}(X) + \chi(X) \sum_{p=1}^{n} A_p^n \, \mathfrak{s}^{p\alpha}(X). \tag{6.4}$$

The same argumentation can be done for $u^{n\alpha}$. □

Remark 6.2. As can be seen in (6.2), another formula linking the two expansions is

$$\begin{aligned} u^{n\alpha}(x) = &\quad v^{n\alpha}(x) + \psi(x) \sum_{p=1}^{n-1} \mathbf{B}_p^{n-p} \, \mathfrak{s}^{-p\alpha}(x), \\ U^{n\alpha}(X) = &\quad V^{n\alpha}(X) + \chi(X) \sum_{p=1}^{n} \mathbf{b}_p^{n-p} \, \mathfrak{s}^{p\alpha}(X). \end{aligned} \tag{6.5}$$

Moreover, as $A_p^n = \mathbf{b}_p^{n-p}$ and due to the matching condition (5.10), one has

$$\mathbf{B}_p^\ell = B_p^\ell \quad \text{and} \quad \mathbf{b}_p^\ell = b_p^\ell \quad \forall \ell \in \mathbb{N} \;\; \forall p \in \mathbb{N}^*. \tag{6.6}$$

Remark 6.3. The mechanism to switch from expansion (4.23) to expansion (5.14) consists in using the homogeneity of the singular functions $\mathfrak{s}^{p\alpha}$ to pass them from fast variables into slow variables:

$$\begin{aligned} \sum_{\ell=0}^{n} \varepsilon^{\ell\alpha} V^{\ell\alpha}(\tfrac{x}{\varepsilon}) &= \sum_{\ell=0}^{n} \varepsilon^{\ell\alpha} \Big[U^{\ell\alpha}(\tfrac{x}{\varepsilon}) - \chi(\tfrac{x}{\varepsilon}) \sum_{p=1}^{\ell} A_p^\ell \, \mathfrak{s}^{p\alpha}(\tfrac{x}{\varepsilon}) \Big] \\ &= \sum_{\ell=0}^{n} \varepsilon^{\ell\alpha} U^{\ell\alpha}(\tfrac{x}{\varepsilon}) - \chi(\tfrac{x}{\varepsilon}) \sum_{\ell=0}^{n} \sum_{p=1}^{\ell} \varepsilon^{(\ell-p)\alpha} A_p^\ell \, \mathfrak{s}^{p\alpha}(x) \\ &= \sum_{\ell=0}^{n} \varepsilon^{\ell\alpha} U^{\ell\alpha}(\tfrac{x}{\varepsilon}) - \chi(\tfrac{x}{\varepsilon}) \sum_{j=0}^{n} \varepsilon^{j\alpha} \sum_{p=0}^{n-j} \underbrace{A_p^{p+\ell}}_{=b_p^\ell} \mathfrak{s}^{p\alpha}(x). \end{aligned}$$

The second term involves the slow variable and will contribute to the terms $(u^{\ell\alpha})$ in the intermediate region.

7 Extensions and Generalizations

The above results can be more or less easily generalized to other situations of interest. For the convenience of the reader, we briefly present a few of them:

1. Smooth right-hand sides f without condition of support
2. Neumann boundary conditions
3. Small holes and multiple junctions
4. Helmholtz operator

7.1 Smooth data without condition of support

Until now, we assumed that the right-hand side f of the problem (2.3) is zero in a neighborhood of the limit point $\mathbf{0}$ of the ε-perturbation. This assumption can be relaxed by considering functions f which are restrictions to $\mathscr{U}_\varepsilon$ of a $\mathcal{C}^\infty$ function $\overline{f}$ defined in a neighborhood of $\cup_{\varepsilon\leqslant\varepsilon_0}\mathscr{U}_\varepsilon$. In this case, we can write

$$f(x) \underset{r\to 0}{\simeq} \sum_{q=0}^{+\infty} f^q(r,\theta) \quad \text{with} \quad f^q(x) = \varepsilon^q\, f^q(\tfrac{x}{\varepsilon}). \tag{7.1}$$

The asymptotic expansion (7.1) of the right-hand side introduces new terms with integer exponents in the asymptotics of $u_0 = v^0$ as $r \to 0$: Instead of the infinite expansion (4.4), we have now

$$v^0(r,\theta) \underset{r\to 0}{\simeq} \sum_{p=1}^{\infty} \mathbf{b}_p^0\, \mathfrak{s}^{p\alpha}(r,\theta) + \sum_{q=1}^{\infty} \mathfrak{T}^q(r,\theta), \tag{7.2}$$

where $\mathfrak{T}^q(r,\theta)$ is the sum of terms of the form $r^q\varphi_0(\theta)$ and $r^q \log r\varphi_1(\theta)$ (with $\varphi_1 = 0$ if α is not rational). In turn, these new terms induce new factors with integer powers of ε in rapid and slow expansions. If α is not a rational number, the expansion of u_ε takes the form (cf. (4.23))

$$u_\varepsilon = \sum_{\substack{p,\, q\in\mathbb{N} \\ p\alpha+q<s}} \varepsilon^{p\alpha+q}\Big(\chi(\tfrac{x}{\varepsilon})\, v^{p\alpha+q}(x) + \psi(x)\, V^{p\alpha+q}(\tfrac{x}{\varepsilon})\Big) + \mathcal{O}_{\mathrm{H}^1}(\varepsilon^s). \tag{7.3}$$

The asymptotics obtained by matched asymptotics expansion contains the same powers of ε as (7.3). In the case where α is rational, logarithms may appear via the scale $\varepsilon^{p\alpha+q}\log\varepsilon$. For more details, we refer to [19, 3].

7.2 Neumann boundary conditions

Instead of (2.3), let us consider the problem

Find $u_\varepsilon \in \mathrm{H}^1(\mathscr{U}_\varepsilon)$ *such that* $\forall v \in \mathrm{H}^1(\mathscr{U}_\varepsilon)$,

$$\int_{\mathscr{U}_\varepsilon} \nabla u_\varepsilon \cdot \nabla v \mathrm{d}x = \int_{\mathscr{U}_\varepsilon} f\, v \mathrm{d}x. \tag{7.4}$$

The solvability of this problem needs the compatibility condition

$$\int_{\mathscr{U}_\varepsilon} f \mathrm{d}x = 0 \quad \forall \varepsilon < \varepsilon_0. \tag{7.5}$$

Let us assume that $f \equiv 0$ in a neighborhood of $\mathbf{0}$ and

$$\int_\omega f \mathrm{d}x = 0.$$

This implies the condition (7.5) for ε small enough. To ensure uniqueness, we require

$$\int_{\mathscr{U}_\varepsilon} u_\varepsilon \mathrm{d}x = 0 \quad \forall \varepsilon < \varepsilon_0. \tag{7.6}$$

The construction of the multiscale expansion for u_ε relies on the solution of variational Neumann problems in ω and Ω. In the unbounded domain Ω, the variational space $V(\Omega)$ is defined as

$$\{U \in \mathscr{D}'(\Omega)\ ;\ \ \nabla U \in \mathrm{L}^2(\Omega),\ \ (1+R)^{-1}(\log(2+R))^{-1}\, U \in \mathrm{L}^2(\Omega)\}. \tag{7.7}$$

The bilinear form

$$(U, V) \longmapsto \int_\Omega \nabla U \cdot \nabla V \mathrm{d}x$$

is continuous and coercive on the quotient space $V(\Omega)/\mathbb{R}$ (cf. [2] or [3, Proposition 3.22]). Therefore, like in ω, the solution of the Neumann problem in Ω with right-hand side F requires the compatibility condition

$$\int_\Omega F \mathrm{d}X = 0.$$

Thus, new features have to be taken into account:

1. *Compatibility conditions.* The right-hand sides which occur during the construction have the form $[\Delta, \psi]\, \mathfrak{s}^{-p\alpha}$ in ω and $[\Delta_X, \chi]\, \mathfrak{s}^{p\alpha}$ in Ω, with the Neumann singularities $\mathfrak{s}^{p\alpha} = \rho^{p\alpha} \cos p\alpha\theta$. Since $\mathfrak{s}^{p\alpha}$ is harmonic, these right-hand sides are equal to $\Delta(\psi\, \mathfrak{s}^{-p\alpha})$ and $\Delta_X(\chi\, \mathfrak{s}^{p\alpha})$ respectively. For $\psi\, \mathfrak{s}^{-p\alpha}$ and $\chi\, \mathfrak{s}^{p\alpha}$ satisfying the Neumann boundary condition on $\partial\omega$ and $\partial\Omega$ respectively, we can show that the compatibility conditions are fulfilled for all integer $p \geqslant 1$.

2. *The role of constants.* (i) The asymptotic expansion of v^0 at O starts with $\mathbf{b}_0^0 \mathfrak{s}^0$, which is a constant. The associated problem in fast variables is (cf. (4.8))
$$-\Delta V^0 = \mathbf{b}_0^0 \Delta_X \chi \ \text{ in } \ \Omega \quad \text{and} \quad \partial_n V^0 = 0 \ \text{ on } \ \partial\Omega. \tag{7.8}$$

We choose the solution $V^0 = \mathbf{b}_0^0(1-\chi)$. Thus, in particular, $\psi(x)V^0(\frac{x}{\varepsilon}) = V^0(\frac{x}{\varepsilon})$: the cut-off by ψ does not introduce any error. Let us note that the function $\chi(\frac{x}{\varepsilon}) + \psi(x)(1-\chi)(\frac{x}{\varepsilon})$ represents the extension by 1 from ω to $\mathscr{U}_\varepsilon$.
(ii) For problems in Ω giving $V^{p\alpha}$, $p \geqslant 1$, we choose the variational solution which tends to 0 as $R \to \infty$.

3. *The condition for uniqueness.* By construction, the slow variable terms $v^{p\alpha}$ have the zero integral on ω. Using their asymptotics as $r \to 0$, we find

$$\int_{\mathscr{U}_\varepsilon} \chi(\tfrac{x}{\varepsilon})\, v^{p\alpha}(x)\mathrm{d}x = \beta_p\, \varepsilon^2, \quad \beta_p \in \mathbb{R}.$$

For fast variable terms we find

$$\int_{\mathscr{U}_\varepsilon} \psi(x)\, V^{p\alpha}(\tfrac{x}{\varepsilon})\mathrm{d}x = \beta'_p\, \varepsilon^2, \quad \beta'_p \in \mathbb{R}.$$

We compensate the possible nonzero integral of the multiscale expansion by a series of constant functions – with values $\gamma_{p,n} \in \mathbb{R}$, $p \in \mathbb{N}$, $n \in \mathbb{N}^*$ – associated with the gauge functions $\varepsilon^{p\alpha+2n}$. By the equality

$$\int_{\mathscr{U}_\varepsilon} \mathrm{d}x = \operatorname{meas}\,\omega + \gamma\varepsilon^2,$$

the $\gamma_{p,n}$'s are defined by forcing the formal equality

$$\sum_{p=0}^{+\infty} \varepsilon^{p\alpha+2}\left(\beta_p + \beta'_p\right) + \left(\operatorname{meas}\omega + \gamma\varepsilon^2\right) \sum_{p=0}^{+\infty}\sum_{n=1}^{+\infty} \varepsilon^{2n+p\alpha}\gamma_{p,n} = 0. \tag{7.9}$$

At the end, we obtain a multiscale expansion of the form

$$u_\varepsilon = \sum_{\substack{p\in\mathbb{N}\\ p\alpha<s}} \varepsilon^{p\alpha}\Big(\chi(\tfrac{x}{\varepsilon})\, v^{p\alpha}(x) + \psi(x)\, V^{p\alpha}(\tfrac{x}{\varepsilon}) + \sum_{\substack{n\in\mathbb{N}^*\\ p\alpha+2n<s}} \gamma_{p,n}\varepsilon^{2n}\Big) + \mathcal{O}_{\mathrm{H}^1}(\varepsilon^s). \tag{7.10}$$

7.3 Small holes

The set $K = \mathbb{R}^2$ may also be convenient as a junction set. It allows us to consider the case of small holes (or small cracks) of size ε inside $\mathscr{U}_\varepsilon$. This is indeed the first case considered in the book [19, Section 2.4.1]. Let us consider the Dirichlet case. Then we are in a situation which shares common features with the Dirichlet case investigated in the most part of this paper and the Neumann case considered above.

Indeed, the limit problem in ω is uniquely solvable. But the limit problem in Ω is not coercive in the subspace of $\mathrm{W}_0^1(\Omega)$ with zero trace on $\partial\Omega$. The correct variational space is the subspace of the space (7.7) with zero trace on $\partial\Omega$. Nevertheless, arguments are slightly different from the Neumann case (like in [3], the asymptotic behavior $\log R$ as $R \to \infty$ has to be considered). The outcome of the analysis is the appearance of terms $(\log \varepsilon)^{-1}$. Finally, the cut-off functions $\chi(\frac{x}{\varepsilon})$ and $\psi(x)$ can be simply omitted since $\mathscr{U}_\varepsilon$ is a subset of ω and $\varepsilon\Omega$.

7.4 Domains with multiply connected junction sectors

This is a situation where the family of domains $(\mathscr{U}_\varepsilon)$ is defined like in Section 2.1, where we relax the assumption on that the set K is a plane sector. Our results extend to the case where K is the finite disjoint union of plane sectors $K_1, \dots, K_m$ with common vertex. Accordingly, we relax the assumption on ω which is still open and bounded, but can be multiply connected. The unbounded open set Ω can also be multiply connected. The open sets $\mathscr{U}_\varepsilon$ have still to be connected. If $m = 2$, this requires that either ω or Ω should be connected. Of course, the interesting case occurs when Ω is connected (cf. Fig. 4).

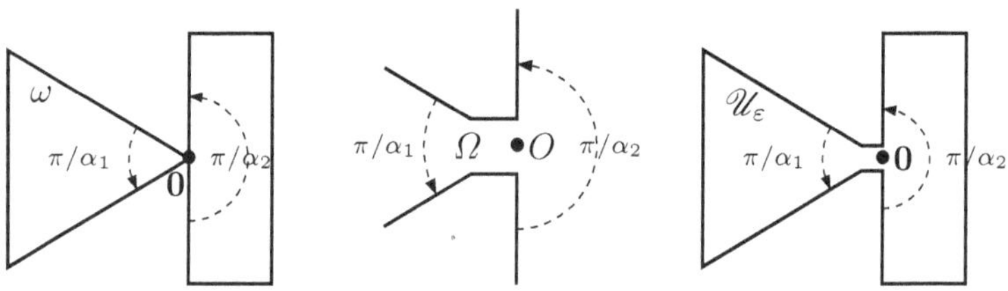

Fig. 4 Example of domains ω, Ω and $\mathscr{U}_\varepsilon$ in the multiply connected case ($\alpha_1 = 3$, $\alpha_2 = 1$).

The generalization of our expansions to this situation is straightforward. We denote by $\pi/\alpha_1, \dots, \pi/\alpha_m$ the openings of the sectors $K_1, \dots, K_m$. The multiscale expansion of u_ε solution of the Dirichlet problem (2.3) with right-hand side $f \equiv 0$ in a neighborhood of O is as follows. For all real number $s > 0$

$$u_\varepsilon = \sum_{\substack{p_1,\dots,p_m \in \mathbb{N} \\ p_1\alpha_1+\dots+p_m\alpha_m < s}} \varepsilon^{p_1\alpha_1+\dots+p_m\alpha_m} \Big(\chi(\tfrac{x}{\varepsilon})\, v^{p_1\alpha_1+\dots+p_m\alpha_m}(x) + \psi(x)\, V^{p_1\alpha_1+\dots+p_m\alpha_m}(\tfrac{x}{\varepsilon}) \Big) + \mathcal{O}_{\mathrm{H}^1}(\varepsilon^s). \quad (7.11)$$

Here, $V^0 = 0$ and $v^{\alpha_j} = 0$ for $j = 1, \ldots, m$. The matched asymptotics expansion is similar.

7.5 Nonhomogeneous operators. Helmholtz equation

The investigation of the Helmholtz operator in a singularly perturbed domain is of major importance for applications (cf. [11, 24] for an example of wave propagation in a domain with thin slots). We want to give here the key arguments to derive and justify the matched asymptotic of the solution of the following model problem, posed in the domain $\mathscr{U}_\varepsilon$ defined in (2.1):

$$\textit{Find } u_\varepsilon \in \mathrm{H}_0^1(\mathscr{U}_\varepsilon) \textit{ such that } -\Delta u_\varepsilon - k^2 u_\varepsilon = f|_{\mathscr{U}_\varepsilon} \textit{ in } \mathscr{U}_\varepsilon, \tag{7.12}$$

where for the sake of simplicity we suppose that

(i) α is not a rational,

(ii) $-k^2$ is not an eigenvalue of the Dirichlet Laplacian in the limit domain ω.

The first assumption avoids the occurrence of a logarithmic gauge function in the asymptotic expansions. The second leads to a well-posed limit problem.

This situation is rather more technical since this operator is not selfsimilar:

$$\Delta_x + k^2 = \frac{1}{\varepsilon^2}\Big(\Delta_X + \varepsilon^2 k^2\Big). \tag{7.13}$$

A second difficulty comes from the loss of coercivity. The proofs of existence and convergence need then to be modified (cf., for example, [19, Chapter 4] and [11, 24]).

Some preliminaries on super-variational problems.

According to a common usage, we denote by $J_{p\alpha}$ and $Y_{p\alpha}$ the Bessel function of first and second kind of order $p\alpha$ respectively (cf., for example, [14]).

Proposition 7.1. *Under condition* (ii), *for any finite sequence* $(a_p)_{1\leqslant p\leqslant P}$ *of real numbers there exists a unique solution* u *to the "super-variational problem"*

$$\begin{cases} \textit{Find } u \in V_{\mathrm{loc},0}(\omega) \textit{ such that} \\ \Delta u + k^2 u = 0 \quad \textit{in} \quad \omega \quad \textit{and} \quad u(x) = \displaystyle\sum_{p=1}^{P} a_p Y_{p\alpha}(kr)\sin(p\alpha\theta) + \mathcal{O}_{r\to 0}(1). \end{cases}$$

In a neighborhood of $\mathbf{0}$, *this solution can be expanded as follows:*

$$u(x) \underset{r\to 0}{\simeq} \sum_{p=1}^{P} a_p\, Y_{p\alpha}(kr)\, \sin(p\alpha\theta) + \sum_{p=1}^{+\infty} b_p\, J_{p\alpha}(kr)\, \sin(p\alpha\theta).$$

Let $J_{p\alpha,\ell}$ and $Y_{p\alpha,\ell}$ be the coefficients of the generalized Taylor series of the Bessel functions $J_{p\alpha}$ and $Y_{p\alpha}$ (the coefficients for odd ℓ are all zero):

$$J_{p\alpha}(z) = z^{p\alpha} \sum_{\ell\in\mathbb{N}} J_{p\alpha,\ell}\, z^{\ell} \quad \text{and} \quad Y_{p\alpha}(z) = \frac{1}{z^{p\alpha}} \sum_{\ell\in\mathbb{N}} Y_{p\alpha,\ell}\, z^{\ell}$$

Proposition 7.2. *Under condition* (i), *for any finite sequence* (A_p^m), $1 \leqslant p \leqslant P$, $0 \leqslant m \leqslant M$, *of real numbers there exists a unique solution* $(U^m)_{0\leqslant m\leqslant M}$ *of the "super-variational system"*

$$\begin{cases} \textit{For } m = 0,\dots,M \textit{ find } U^m \in V_{\mathrm{loc},\infty}(\Omega) \textit{ such that} \\ \Delta U^m + k^2 U^{m-2} = 0 \ \textit{ in } \ \Omega, \ \ \textit{with convention } \ U^m = 0 \ \textit{ if } \ m < 0, \\ U^m(X) = \displaystyle\sum_{p=1}^{P}\sum_{\ell=0}^{m} A_p^{m-\ell}\, J_{p\alpha,\ell}\,(kR)^{p\alpha+\ell}\, \sin(p\alpha\theta) \\ \qquad + \displaystyle\sum_{p=1}^{+\infty}\sum_{\ell=1}^{m} B_p^{m-\ell}\, Y_{p\alpha,\ell}\,(kR)^{-p\alpha+\ell}\, \sin(p\alpha\theta) + \mathcal{O}_{R\to\infty}(1), \end{cases}$$

where $(B_p^m)_{0\leqslant p\leqslant+\infty,\ 0\leqslant m\leqslant M}$ *are the coefficients such that, in a neighborhood of infinity,*

$$U^m(X) \underset{R\to+\infty}{\simeq} \sum_{p=1}^{P}\sum_{\ell=0}^{m} A_p^{m-\ell}\, J_{p\alpha,\ell}\,(kR)^{p\alpha+\ell}\, \sin(p\alpha\theta)$$
$$+ \sum_{p=1}^{+\infty}\sum_{\ell=0}^{m} B_p^{m-\ell}\, Y_{p\alpha,\ell}\,(kR)^{-p\alpha+\ell}\, \sin(p\alpha\theta).$$

Remark 7.3. If condition (i) is not satisfied, the expansions of $J_{p\alpha}$ and $Y_{p\alpha}$ do not only include terms of the form r^{μ}, but also terms of the form $r^{\mu}(\ln r)^{\nu}$.

Definition of the matched asymptotic expansions.

The gauge functions appearing in the asymptotic expansions of u_ε are of the form $\varepsilon^{m+n\alpha}$, i.e., we look for two asymptotic expansions of the form

$$u_\varepsilon(x) \simeq \sum_{(m,n)\in\mathbb{N}^2} \varepsilon^{m+n\alpha} U^{m,n}\left(\tfrac{x}{\varepsilon}\right) \quad \text{and} \quad u_\varepsilon(x) \simeq \sum_{(m,n)\in\mathbb{N}^2} \varepsilon^{m+n\alpha} u^{m,n}(x).$$

The coefficients of these expansions can be defined hierarchically as the unique solutions of the coupled problem

$$\begin{cases} \textit{Find } u^{m,n} \in V_{\mathrm{loc},0}(\omega) \textit{ such that} \\ \Delta u^{m,n} + k^2 u^{m,n} = 0 \ \textit{ in } \omega \quad (\textit{or } -f \textit{ if } m=n=0) \\ u^{m,n} - \displaystyle\sum_{p=1}^{n} a_p^{m,n}\, Y_{p\alpha}(kr) \sin(p\alpha\theta) = \mathcal{O}_{r\to 0}(1), \end{cases}$$

$$\begin{cases} \textit{Find } U^{m,n} \in V_{\mathrm{loc},\infty}(\Omega) \textit{ such that} \\ \Delta U^{m,n} + k^2 U^{m-2,n} = 0 \ \textit{ in } \Omega, \quad (\textit{with } U^{m,n} = 0 \textit{ if } m<0) \\ U^{m,n}(X) - \displaystyle\sum_{p=1}^{n}\sum_{\ell=0}^{m} A_p^{m-\ell,n}\, J_{p\alpha,\ell}\,(kR)^{p\alpha+\ell} \sin(p\alpha\theta) \\ \displaystyle\sum_{p=1}^{+\infty}\sum_{\ell=1}^{m} B_p^{m-\ell,n}\, Y_{p\alpha,\ell}\,(kR)^{-p\alpha+\ell} \sin(p\alpha\theta) = \mathcal{O}_{R\to\infty}(1), \end{cases}$$

together with the matching conditions

$$a_p^{m,n} = B_p^{m,n-p} \quad \text{and} \quad A_p^{m,n} = b_p^{m,n-p} \ \text{ if } \ 1 \leqslant p \leqslant n,$$

where the coefficients $b_p^{m,n}$ are defined through the sub-variational expansion of $u^{m,n}$:

$$u^{m,n}(x) \underset{r\to 0}{\simeq} \sum_{p=1}^{n} a_p^{m,n}\, Y_{p\alpha}(kr) \sin(p\alpha\theta) + \sum_{p=1}^{+\infty} b_p^{m,n}\, J_{p\alpha}(kr) \sin(p\alpha\theta).$$

Error Estimates.

Theorem 7.4. *Let I_N be the set of indices corresponding to gauge functions of order lower than N*

$$I_N = \left\{(m,n) \in \mathbb{N}^2 : m + n\alpha \leqslant N\right\}. \tag{7.14}$$

The global approximation is defined by

$$\widehat{u}_N^\varepsilon(x) = \varphi\big(\tfrac{r}{\eta(\varepsilon)}\big) \sum_{(m,n)\in I_N} \varepsilon^{m+n\alpha}\, u^{m,n}(x)$$

$$+ \Big(1 - \varphi\big(\tfrac{r}{\eta(\varepsilon)}\big)\Big) \sum_{(m,n)\in I_N} \varepsilon^{m+n\alpha}\, U^{m,n}\big(\tfrac{x}{\varepsilon}\big), \tag{7.15}$$

where φ is a smooth cut-off function with $\varphi(\rho) = 0$ for $\rho < 1$ and $\varphi(\rho) = 1$ for $\rho > 2$ and η is smooth and satisfies

$$\lim_{\varepsilon\to 0} \eta(\varepsilon) = 0 \quad \textit{and} \quad \lim_{\varepsilon\to 0} \frac{\eta(\varepsilon)}{\varepsilon} = +\infty. \tag{7.16}$$

There exists a constant C such that

$$\|u_\varepsilon - \widehat{u}_N^\varepsilon\|_{\mathrm{H}^1(\mathscr{U}_\varepsilon)} \leqslant C\left[\Big(\eta(\varepsilon)\Big)^N + \Big(\frac{\varepsilon}{\eta(\varepsilon)}\Big)^N\right]. \tag{7.17}$$

8 Conclusion: A Practical Two-Term Expansion

In order to investigate the influence of singular perturbations of the domain on a local functional ϕ_ε acting over the solution u_ε, it is valuable to use a compound version of the asymptotic expansion, in between multiscale and matched asymptotic expansions.

8.1 Compound expansion

Indeed, using (2.9) and the relation (2.11) between the profiles V^α and U^α, we get

$$u_\varepsilon = \chi(\tfrac{x}{\varepsilon})\, u_0(x) + \psi(x)\, \varepsilon^\alpha \big[U^\alpha(\tfrac{x}{\varepsilon}) - \chi(\tfrac{x}{\varepsilon})\, A\, \mathfrak{s}^\alpha(\tfrac{x}{\varepsilon})\big] + \mathcal{O}_{\mathrm{H}^1}(\varepsilon^{2\alpha}),$$

which can be written, thanks to the homogeneity of the singular function $\mathfrak{s}^\alpha$

$$u_\varepsilon = \chi(\tfrac{x}{\varepsilon})\, \big[u_0(x) - A\, \psi(x)\mathfrak{s}^\alpha(x)\big] + \psi(x)\, \varepsilon^\alpha U^\alpha(\tfrac{x}{\varepsilon}) + \mathcal{O}_{\mathrm{H}^1}(\varepsilon^{2\alpha}).$$

Let us introduce the first "canonical" profile U^α_Ω as the solution of the super-variational Dirichlet problem on Ω

$$\begin{cases} \textit{Find } U^\alpha_\Omega \in V_{\mathrm{loc},\infty}(\Omega) \textit{ such that} \\ \Delta U^\alpha_\Omega = 0 \ \textit{ in } \ \Omega \quad \textit{and} \quad U^\alpha_\Omega - \mathfrak{s}^\alpha = \mathcal{O}_{R\to\infty}(1). \end{cases} \tag{8.1}$$

We have $U^\alpha = A\, U^\alpha_\Omega$ and hence

$$u_\varepsilon = \chi(\tfrac{x}{\varepsilon})\,[u_0(x) - A\,\psi(x)\mathfrak{s}^\alpha(x)] + \psi(x)\varepsilon^\alpha A\,U^\alpha_\Omega(\tfrac{x}{\varepsilon}) + \mathcal{O}_{\mathrm{H}^1}(\varepsilon^{2\alpha}). \qquad (8.2)$$

In (8.2), only canonical objects are involved: the limit term u_0, its first singularity coefficient A, and the first profile U^α_Ω. The contribution near the corner is fully contained in the profile AU^α_Ω, whereas the "far-field" information is carried out by $u_0 - A\psi\mathfrak{s}^\alpha$ corresponding to the limit term without its first singularity. In a sense, the strongest singularity of u_0 is "chopped off" for $\varepsilon > 0$ via the cut-off function $\chi(\frac{x}{\varepsilon})$, and is replaced with the profile AU^α_Ω, which connects the local geometry around O with the plane sector of opening α at infinity.

8.2 Application: asymptotics of coefficients of singularities

An interesting application of the expansion (8.2) is the determination of Stress Intensity Factors at the tip of a short crack emanating from a sharp of a rounded V-notch (cf. [16]). More generally, the question is the determination of the asymptotic behavior of the coefficients of singularities of u_ε associated with the corners (or cracks) of the domain $\mathscr{U}_\varepsilon$ *inside its perturbed region.* The functional $\phi_\varepsilon(u_\varepsilon)$ is then defined as the value of this coefficient of singularity corresponding to a corner whose position depends on ε.

Indeed, to each corner point (or crack tip) d of the perturbation pattern Ω corresponds a corner point (or crack tip) d_ε of the perturbed domain $\mathscr{U}_\varepsilon$. In Fig. 5, two such points are involved, both associated with angle 2π.

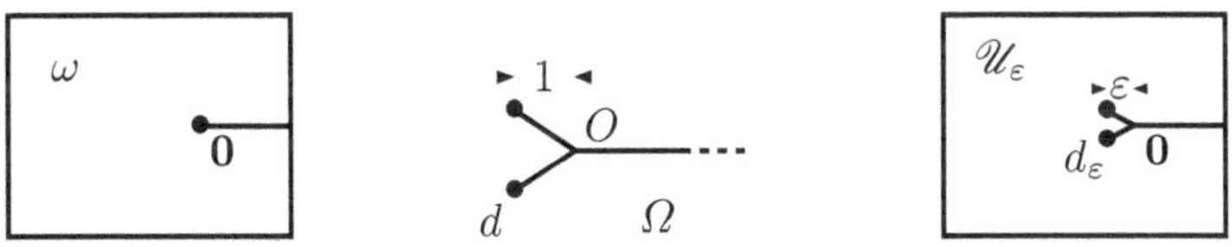

Fig. 5 Crack tips: Domains ω, Ω and $\mathscr{U}_\varepsilon$.

The solution u_ε of the Laplace–Dirichlet problem (2.3) is singular at the point d_ε, with the following first order approximation:

$$u_\varepsilon(x) = \gamma_\varepsilon r_\varepsilon^\mu \sin(\mu\theta_\varepsilon) + o_\varepsilon(r_\varepsilon^\mu) \quad \text{as} \quad r_\varepsilon \to 0, \qquad (8.3)$$

where $(r_\varepsilon, \theta_\varepsilon)$ denote the polar coordinates around d_ε. The exponent μ is the singular exponent corresponding to d_ε ($\mu = \pi/\vartheta$ for a corner of opening ϑ and $\mu = 1/2$ for a crack). The functional ϕ_ε is defined as

$$\phi_\varepsilon(u_\varepsilon) = \gamma_\varepsilon.$$

Our results allow us to give an asymptotic expansion of the singular coefficient γ_ε as $\varepsilon \to 0$: we still denote by α the singular exponent associated with the limit problem in ω. Using (8.2), we get

$$u_\varepsilon(x) = \varepsilon^\alpha A U_\Omega^\alpha(\tfrac{x}{\varepsilon}) + \text{ higher order profiles if } |x| \leqslant \varepsilon r_*. \tag{8.4}$$

But the first canonical profile U_Ω^α has a singularity at point d associated with exponent μ:

$$U_\Omega^\alpha(X) = \varGamma R_d^\mu \sin(\mu\Theta_d) + o(R_d^\mu) \text{ as } R_d \to 0, \tag{8.5}$$

where (R_d, Θ_d) are the polar coordinates around the point d. We have

$$r_\varepsilon = \varepsilon R_d \quad \text{and} \quad \theta_\varepsilon = \Theta_d. \tag{8.6}$$

Back to the variable x, the relations (8.4) to (8.6) lead to

$$u_\varepsilon(x) = \varepsilon^{\alpha-\mu} A\varGamma r_\varepsilon^\mu \sin(\mu\theta_\varepsilon) + o(\varepsilon^{\alpha-\mu} r_\varepsilon^\mu) \text{ if } |x| \leqslant \varepsilon r_*. \tag{8.7}$$

Putting (8.3) and (8.7) together, we obtain the expression of the singular coefficient γ_ε:

$$\gamma_\varepsilon = \varepsilon^{\alpha-\mu} A\varGamma + o(\varepsilon^{\alpha-\mu}). \tag{8.8}$$

It is worth noticing that the coefficient associated with a stronger singularity than the limit singularity ($\mu < \alpha$) will go to 0, whereas it will blow up to infinity for weaker singularities. It has to be linked to the appearance of singularities discussed above. In the case of Fig. 5, we have $\alpha = 1/2$ and $\mu = 1/2$. The coefficient associated with the v-notch cracks is $\mathcal{O}(1)$.

The above analysis also applies in the framework of elasticity, and is the foundation of the investigation in [16]. We stress that a rigorous derivation of (8.8) with an optimal estimate for the remainder requires more efforts in studying the singular-regular expansion of u_ε.

The expansion (8.2) could also be used to investigate the asymptotic behavior of other local functionals $\phi_\varepsilon(u_\varepsilon)$ relating, for example, to the maximal values of the stress tensor in elasticity.

References

1. Agmon, S.: Lectures on Elliptic Boundary Value Problems. Van Nostrand, Princeton, NJ (1965)
2. Amrouche, C., Girault, V., Giroire, J.: Dirichlet and Neumann exterior problems for the n-dimensional Laplace operator. An approach in weighted Sobolev spaces. J. Math. Pures Appl., IX Ser. **76**, no. 1, 55–81 (1997)
3. Caloz, G., Costabel, M., Dauge, M., Vial, G.: Asymptotic expansion of the solution of an interface problem in a polygonal domain with thin layer. Asymptotic Anal. [To appear]

4. Costabel, M., Dauge, M.: A singularly perturbed mixed boundary value problem. Commun. Partial Differ. Equ. **21**, 1919–1949 (1996)
5. Costabel, M., Dauge, M., Yosibash, Z.: A quasi-dual function method for extracting edge stress intensity functions. SIAM J. Math. Anal. **35**, no. 5, 1177–1202 (electronic) (2004)
6. Dauge, M.: Elliptic Boundary Value Problems in Corner Domains – Smoothness and Asymptotics of Solutions. Lect. Notes Math. **1341**, Springer, Berlin (1988)
7. Dauge, M., Nicaise, S., Bourlard, M., Lubuma, J.M.S.: Coefficients des singularités pour des problèmes aux limites elliptiques sur un domaine à points coniques. I. Résultats généraux pour le problème de Dirichlet. RAIRO Modél. Math. Anal. Numér. **24**, no. 1, 27–52 (1990)
8. Gadyl'shin, R.R.: Surface potentials and the method of matching asymptotic expansions in the problem of the Helmholtz resonator (Russian). Algebra Anal. **4**, no. 2, 88–115 (1992); English transl.: St. Petersburg Math. J. **4**, no. 1, 273–296 (1993)
9. Grisvard, P.: Boundary Value Problems in Non-Smooth Domains. Pitman, London (1985)
10. Il'in, A.M.: Matching of Asymptotic Expansions of Solutions of Boundary Value Problems. Am. Math. Soc., Providence, RI (1992)
11. Joly, P., Tordeux, S.: Matching of asymptotic expansions for wave propagation in media with thin slots I: The asymptotic expansion. Multiscale Model. Simul. **5**, no. 1, 304–336 (2006)
12. Kondrat'ev, V.A.: Boundary value problems for elliptic equations in domains with conical or angular points. Trans. Moscow Math. Soc. **16**, 227–313 (1967)
13. Kozlov, V.A., Maz'ya, V.G., Rossmann, J.: Elliptic Boundary Value Problems in Domains with Point Singularities. Am. Math. Soc., Providence, RI (1997)
14. Lebedev, N.N.: Special Functions and Their Applications. Dover Publications Inc., New York (1972)
15. Leguillon, D., Sanchez-Palencia, E.: Computation of Singular Solutions in Elliptic Problems and Elasticity. Masson, Paris (1987)
16. Leguillon, D., Yosibash, Z.: Crack onset at a V-Notch. Influence of the notch tip radius. Int. J. Fracture **122**, 1–21 (2003)
17. Maz'ya, V.G., Nazarov, S.A., Plamenevskii, B.A.: Asymptotic solutions of elliptic boundary value problems when the domain is varied close to conical points (Russian). Dokl. Akad. Nauk SSSR **249**, no. 1, 94–96 (1979); English transl.: Sov. Phys. Dokl. **24**, 904–905 (1979)
18. Maz'ya, V.G., Nazarov, S.A., Plamenevskii, B.A.: On the asymptotic behavior of solutions of elliptic boundary value problems with irregular perturbations of the domain (Russian). Probl. Mat. Anal. **8**, 72–153 (1981)
19. Maz'ya, V., Nazarov, S., Plamenevskii, B.: Asymptotic Theory of Elliptic Boundary Value Problems in Singularly Perturbed Domains. I. II. Birkhäuser, Basel (2000)
20. Maz'ya, V.G., Plamenevskii, B.A.: Estimates in L^p and in Hölder classes and the Miranda–Agmon maximum principle for solutions of elliptic boundary value problems in domains with singular points on the boundary. Transl., Ser. 2, Am. Math. Soc. **123**, 1–56 (1984)
21. Maz'ya, V.G., Plamenevskii, B.A.: On the coefficients in the asymptotic of solutions of the elliptic boundary problem in domains with conical points. Transl., Ser. 2, Am. Math. Soc. **123**, 57–88 (1984)
22. Nazarov, S.A., Plamenevskii, B.A.: Elliptic Problems in Domains with Piecewise Smooth Boundaries. Walter de Gruyter, Berlin (1994)

23. Oleinik, O.A., Shamaev, A.S., Yosifian, G.A.: Mathematical Problems in Elasticity and Homogenization. North-Holland, Amsterdam (1992)
24. Tordeux, S.: Méthodes asymptotiques pour la propagation des ondes dans les milieux comportant des fentes. Thès. doctorat. (2004)
25. Van Dyke, M.: Perturbation Methods in Fluid Mechanics. The Parabolic Press, Stanford, California (1975)
26. Vial, G.: Analyse multi-échelle et conditions aux limites approchées pour un problème de couche mince dans un domaine à coin. Thès. doctorat. (2003)

Stationary Navier–Stokes Equation on Lipschitz Domains in Riemannian Manifolds with Nonvanishing Boundary Conditions

Martin Dindoš

Abstract In the previous work, the author and M. Mitrea presented a method of solving the stationary Navier–Stokes equation on Lipschitz domains in Riemannian manifolds via the boundary integral technique, where only the vanishing Dirichlet boundary condition was considered. In this paper, more sophisticated estimates are developed, which allows us to consider arbitrary large (dim $M \leqslant 4$) Dirichlet boundary data for this equation.

1 Introduction

The Navier–Stokes equations are one of the most studied nonlinear partial differential equations. They model the flow of an incompressible viscous fluid. The long–time existence of smooth solutions (in 3d) is one of the most challenging open problems in mathematics.

In this paper, we take on the task to develop a theory on the existence of a solution to the stationary Navier–Stokes equation on domains with boundary. More precisely, we solve the equation

$$\mathcal{L}u + \nabla_u u + dp = f \tag{1.1}$$

for a divergence free 1-form u in $L^2_1(\Omega)$ such that Tr $u = g$ on $\partial\Omega$. As is customary, we identify 1-forms with vector fields on Ω, i.e., we do not distinguish between them. More on this can be found in the next section.

There is a connection between our work and earlier results [17]–[18] of Maz'ya and Rossmann who studied the Stokes system and related stationary

Martin Dindoš
School of Mathematics, Maxwell Institute of Mathematics Sciences, University of Edinburgh, JCMB King's buildings, Mayfield Rd, Edinbugh EH9 3JZ, UK
e-mail: `m.dindos@ed.ac.uk`

A. Laptev (ed.), *Around the Research of Vladimir Maz'ya II: Partial Differential Equations*, International Mathematical Series 12, DOI 10.1007/978-1-4419-1343-2_5,

Navier–Stokes equation in domains of polyhedral type. In particular, L^∞ estimates for solutions of the stationary Navier–Stokes equation with arbitrary large boundary data in 3d are established in [18].

The approach we present here also works for arbitrary large data in dimensions $n = \dim M \leqslant 4$ and small data for higher dimensions. Our considerations are based on the results of the paper [8], which dealt with the Stationary Navier–Stokes equation on Lipschitz domains in Riemannian manifolds, and also the paper [20] on the Stokes operator. We maintain the notation similar to these papers.

Let us now momentarily digress and explain why we find it both natural and important to study these (and related) problems on *Riemannian manifolds*. For starters, note that our setting applies to the case of a bounded Euclidean domain in $\mathbb{R}^n$ since such a domain can be embedded in a (sufficiently large) flat torus (equipped with the canonical metric). However, the main advantage of working with a *general* Riemannian metric tensor is that this gives rise to a context in which *variable-coefficient* operators arise naturally. For example, any (scalar) divergence-form operator

$$Lu = \sum_{1\leqslant i,j\leqslant n} \partial_i(a_{ij}(x)\partial_j u), \tag{1.2}$$

induced by a positive definite matrix $A(x) = (a_{ij}(x))_{ij}$, can be viewed (modulo a multiple of suitable power of det $[A(x)]$) as the Laplace–Beltrami operator Δ_g associated with the Riemannian metric tensor

$$g(x) := (\det[A(x)])^{1/(n-2)} \sum_{1\leqslant i,j\leqslant n} a^{ij}(x)\, dx_i \otimes dx_j, \tag{1.3}$$

where a^{ij} are the entries of A^{-1}, for $n \geqslant 3$.

Formulating and studying the Navier–Stokes equations on Riemannian manifolds has a fairly rich history, going back to the influential paper [9] by Ebin and Marsden, where the correct form of these equations was first identified. One of the important observations made in [9] is as follows. While in the flat, Euclidean setting, the operator $\mathcal{L}$ in (1.1) is the ordinary, constant coefficient Laplacian, the correct replacement – in the context of a Riemannian manifold – is not the Hodge–Laplacian on forms, but rather the deformation-Laplacian 2Def*Def (the two do not coincide unless the manifold is Ricci-flat). Other papers dealing with fluid dynamics problems on Riemannian manifolds are [6, 19, 21, 22, 25, 26, 27]. In addition, a number of authors dealt with special geometric settings, such as that of a sphere (cf. [29, 30]).

There is a rather extensive literature devoted to the study of the Stokes system in domains exhibiting a limited amount of smoothness.

When $\partial\Omega \in C^{1,1}$, the classical approach based on the Agmon–Douglis–Nirenberg theory [1] applies. This point of view was exploited in [2], which also made essential use of the resolvent estimates from [15] (cf. also [28] and the approach via pseudodifferential operators from [27]). Certain lower dimensional cases (dim $\leqslant 3$) for C^2 domains were treated earlier in [7] and [13]. For indices related by $s + 1/p = 1$, Euclidean domains with a small Lipschitz constant were dealt with in [14] based on flattening the boundary and *a priori* estimates. L^p estimates in conical domains were derived in [5, 4] (cf. also [3] for regularity issues related to the Stokes system in Lipschitz domains).

An important remark is that in dimensions 3 and higher the domains of polyhedral type studied in [18] are not a subset of Lipschitz domains. In fact, there are domains of polyhedral type that are not Lipschitz and vice versa. It would be very interesting to find an approach unifying both cases so that they can be treated in the same way.

In the case of smooth domains in $\mathbb{R}^n$ with $n \leqslant 6$, solutions to the stationary Navier–Stokes equations were constructed for arbitrary smooth f (and $h = 0$, $g = 0$) by Frehse, Ružička and Struwe in a series of papers [11]–[10] and [23, 24]. In the case of periodic boundary conditions, the same result is true up to dimension 15.

The solutions these authors produce are smooth in the interior even for large (smooth) data. Their approach avoids using perturbation techniques and, instead, relies on a suitable maximum principle. This is particularly relevant in dimensions $\geqslant 5$, where the standard methods are no longer applicable (for large data). Let us also mention that the issue of establishing regularity *up to the boundary* is open even for smooth large data.

2 Statements of the Main Results

In all what follows, we assume that M is an n-dimensional Riemannian manifold with a smooth metric and $\Omega \subset M$ an open connected set with Lipschitz boundary.

We denote by $\mathcal{L}$ the second order partial differential operator $\mathcal{L} = 2\,\mathrm{Def}^*\mathrm{Def}$ acting on 1-forms. The symbols L^p_s and $B^{p,p}_s$ denote the usual Sobolev and Besov spaces respectively.

The first result we have works in any dimension and is for the linearized version of the stationary Navier–Stokes equation.

Theorem 2.1. *Let Ω be a connected Lipschitz domain in M, and let $\omega \in L^n(\Omega)$ be a divergence free vector field such that ω has an L^{n-1} trace on $\partial\Omega$ and $\omega_{\mathrm{nor}} = \langle \omega, \nu \rangle = 0$. Consider the following boundary value problem:*

$$\begin{aligned} &u \in L^2_1(\Omega), \quad \mathcal{L}u + \nabla_\omega u + dp = f \in L^2_{-1,0}(\Omega), \\ &\delta u = 0 \ \ in \ \ \Omega, \\ &\mathrm{Tr} \in B^{2,2}_{1/2}(\partial\Omega), \quad g_{\mathrm{nor}} = \langle g, \nu \rangle = 0. \end{aligned} \tag{2.1}$$

Then there is $\alpha < 1$ *such that the problem* (2.1) *has a unique solution* $u \in L^2_1(\Omega)$ *satisfying the estimate*

$$\|u\|_{L^2_1(\Omega)} \leqslant C\|\omega\|^{\alpha}_{L^n(\Omega)}(\|g\|_{B^{2,2}_{1/2}(\partial\Omega)} + \|f\|_{L^2_{-1}(\Omega)}) \tag{2.2}$$

for some $C = C(\Omega) > 0$.

As a corollary, using fixed point technique (cf. below), we get the following theorem on the existence of solutions to the stationary Navier–Stokes equation for arbitrary large data in dimension $n \leqslant 4$.

Theorem 2.2. *Assume that* M *is a compact Riemannian manifold of dimension* $n \leqslant 4$ *with smooth metric tensor. Let* $\Omega \subset M$ *be a connected Lipschitz domain in* M *with nonempty boundary. Then the boundary value problem*

$$\begin{aligned} &u \in L^2_1(\Omega), \quad \mathcal{L}u + \nabla_u u + dp = f \in L^2_{-1,0}(\Omega),\\ &\delta u = 0 \quad in \quad \Omega,\\ &\mathrm{Tr}\ u = g \in B^{2,2}_{1/2}(\partial\Omega), \quad g_{\mathrm{nor}} = \langle g, \nu\rangle = 0. \end{aligned} \tag{2.3}$$

has at least one solution u.

3 Linear Theory Revisited

As was mentioned above, we are going to use certain results from the paper [8], where the authors used fixed point technique to solve the stationary Navier–Stokes equation on Lipschitz domains in Riemannian manifolds. The idea is to define a map $T : \omega \to u$ by solving the linear problem

$$\begin{aligned} &u \in L^p_{s+1/p}(\Omega), \quad \mathcal{L}u + \nabla_\omega u + dp = f \in L^p_{s+1/p-2}(\Omega),\\ &\delta u = 0 \ \text{ in } \ \Omega,\\ &\mathrm{Tr}\ u = g \in B^{p,p}_s(\partial\Omega). \end{aligned} \tag{3.1}$$

As was shown in the paper, if ω is an $L^n(\Omega)$ divergence free vector field, then Equation (3.1) is well defined and has a unique solution for $0 < s < 1$ and p close to 2, provided that the domain Ω is Lipschitz. Assuming the C^1 regularity of the boundary, the statement remains true for all $1 < p < \infty$.

Given this, a fixed point of the map T solves the nonlinear stationary Navier–Stokes equation

$$\begin{aligned} &u \in L^p_{s+1/p}(\Omega), \quad \mathcal{L}u + \nabla_u u + dp = f \in L^p_{s+1/p-2}(\Omega),\\ &\delta u = 0 \ \text{ in } \ \Omega,\\ &\mathrm{Tr}\ u = g \in B^{p,p}_s(\partial\Omega). \end{aligned} \tag{3.2}$$

Hence it suffices to prove that the map T has a fixed point. Here, one of the key ingredients is to show that for some $R > 0$ the ball

$$B_R = \{u \in L^p_{s+1/p}(\Omega, \Lambda^1 TM); \delta u = 0; \|u\|_{L^p_{s+1/p}(\Omega)} \leqslant R\}$$

is mapped by T into itself. As was shown in [8], there exists $C > 0$ independent of ω such that

$$\|u\|_{L^2_1(\Omega)} \leqslant C\|f\|_{L^2_{-1}(\Omega)} \tag{3.3}$$

if $g \equiv 0$. If $g \neq 0$, an estimate corresponding to (3.3) is missing in [8]. The goal of this section is to rectify this situation. Equation (3.2) represents the stationary Navier–Stokes equation with viscosity 1.

We start with a lemma proof of which can be found in [8].

Lemma 3.1. *If $1 < p < \infty$, then there exist $C_1(p) > 0$ and $C_2(p) > 0$ such that*

$$C_1\|u\|_{L^p_{1,0}(\Omega)} \leqslant \|\operatorname{Def}\ u\|_{L^p(\Omega)} \leqslant C_2\|u\|_{L^p_{1,0}(\Omega)}, \tag{3.4}$$

for all $u \in L^p_1(\Omega)$ with $\operatorname{Tr} u = 0$.

Equipped with this lemma, we consider (for fixed $g \in B^{2,2}_{1/2}(\partial\Omega)$) the following two solutions:

$$\begin{aligned} &u, v \in L^2_1(\Omega),\\ &\mathcal{L}u + \nabla_\omega u + dp^1 = 0 \quad \text{in } \Omega, \qquad \mathcal{L}v + dp^2 = 0 \quad \text{in } \Omega,\\ &\delta u = \delta v = 0 \quad \text{in} \quad \Omega,\\ &\operatorname{Tr} u = \operatorname{Tr} v = g. \end{aligned} \tag{3.5}$$

Here, ω is a divergence free vector field. The solutions u and v exist (cf. [8, Theorem 5.1]). It follows that $w = u - v$ is a solution of the elliptic system

$$w \in L^2_{1,0}(\Omega), \quad \mathcal{L}w + dp = -\nabla_\omega u = f, \quad \delta w = 0. \tag{3.6}$$

Note that if $\omega \in L^n(\Omega)$, then the 1-form f on the right-hand side belongs to $L^q(\Omega, \Lambda^1 TM)$, where

$$\frac{1}{q} = \frac{1}{2} + \frac{1}{n}, \quad \text{i.e.,} \quad q = \frac{2n}{n+2} > 1. \tag{3.7}$$

This implies (if the boundary $\partial\Omega$ is C^1 or better) that $w \in L^q_2(\Omega, \Lambda^1 TM)$ and $p \in L^q_1(\Omega)$. In particular, it follows that we can multiply both sides of Equation (3.6) by u and integrate over Ω since $\mathcal{L}w.u \in L^1$ and $dp.u \in L^1$. If the domain Ω is just Lipschitz, we only have that $w, p \in L^q_{2,\mathrm{loc}}(\Omega)$.

We now want to integrate by parts. We need to be careful however, as in the case of a Lipschitz domain there might not be enough regularity of the solution w at the boundary. To avoid this difficulty, we approximate the domain Ω by an increasing sequence of smooth connected domains $\Omega_1 \subset$

$\Omega_2 \subset \cdots \subset \Omega$. On each subdomain, $w, p \in L^q_2(\Omega_n)$. Hence we can multiply (3.6) and integrate over the domain. This gives

$$\int_{\Omega_n} \langle \mathcal{L}w, u\rangle \, d\,\mathrm{Vol} + \int_{\Omega_n} \langle \nabla_\omega u, u\rangle \, d\,\mathrm{Vol} + \int_{\Omega_n} \langle dp, u\rangle \, d\,\mathrm{Vol} = 0.$$

Now we integrate by parts. Note that the second integral yields

$$-\int_{\Omega_n} \langle \nabla_\omega u, u\rangle d\,\mathrm{Vol} + \text{boundary terms}.$$

Hence it can be expressed by using only boundary terms. A similar statement is true for the third term since $\delta u = 0$. This gives

$$\begin{aligned}\int_{\Omega_n} \langle \mathrm{Def}\,(w), \mathrm{Def}\,(u)\rangle \, d\,\mathrm{Vol} &= -2\int_{\partial\Omega_n} \langle \mathrm{Def}\,(w)\nu_n, u\rangle \, d\sigma \\ -\frac{1}{2}\int_{\partial\Omega_n} \langle \omega, \nu_n\rangle |u|^2 \, d\sigma &- \int_{\partial\Omega_n} p\langle u, \nu_n\rangle \, d\sigma. \end{aligned} \tag{3.8}$$

Here, ν_n is the outward pointing conormal to the boundary of Ω_n. We can assume that $\nu_n \in L^\infty(\Omega_n)$ and $\nu_n \to \nu$ almost everywhere as $n \to \infty$. Taking the limit $n \to \infty$ on both sides of (3.8) gives us formally:

$$\begin{aligned}&2\int_{\Omega} \langle \mathrm{Def}\,(w), \mathrm{Def}\,(u)\rangle \, d\,\mathrm{Vol} \\ &= -2\int_{\partial\Omega} \langle \mathrm{Def}\,(w)\nu, g\rangle d\sigma - \int_{\partial\Omega} p\langle g, \nu\rangle \, d\sigma - \frac{1}{2}\int_{\partial\Omega} \langle \omega, \nu\rangle |g|^2 \, d\sigma. \end{aligned} \tag{3.9}$$

Here, we assume that ω has a trace on $\partial\Omega$ such that $\omega_{\mathrm{nor}} = \langle \omega, \nu\rangle \in L^{n-1}(\partial\Omega)$, which guarantees that the last boundary integral in (3.9) is well defined.

At this point, we digress for a moment to make sense of the first two boundary integrals given in the assumptions of Theorem 2.1. Consider any $g \in B^{2,2}_{1/2}(\partial\Omega)$ such that $\langle g, \nu\rangle = 0$ and let also $\omega \in L^n(\Omega)$ with $\omega_{\mathrm{nor}} = \langle \omega, \nu\rangle = 0$.

Let us for the moment give more regularity to the function g, namely, let $g \in B^{2,2}_1(\partial\Omega)$. Then, as was shown in [20] and also [8], the maximal operator p^* defined by the formula

$$p^*(x) = \sup_{y \in \gamma(x)} |p(y)|, \qquad x \in \partial\Omega$$

is well defined and $p^* \in L^2(\partial\Omega)$. Here, $(\gamma(.))$ is a collection of nontangential cones inside Ω with vertices at $\partial\Omega$. Then the second integral in (3.9) is well defined and vanishes as $\langle g, \nu\rangle = 0$.

It remains to make sense of the first boundary integral. In general, having only $(w, p) \in L^2_1(\Omega) \times L^2(\Omega)$ the trace of $\mathrm{Def}\,(w)\nu$ on $\partial\Omega$ is utterly ill defined.

In our case, however, we have one extra ingredient. Namely, w is not an arbitrary divergence free 1-form, but a solution of the equation

$$\mathcal{L}w + dq = f$$

for some f on the right-hand side in $L^q(\Omega)$ with q given by (3.7). This means that the 1-form f has a unique "natural" extension to the space $L^2_{-1,0}(\Omega) = (L^2_1(\Omega))^*$ defined via the pairing

$$\langle f, \varphi \rangle = \int_\Omega f.\varphi \, d\,\mathrm{Vol}\,.$$

The function inside this integral is in $L^1(\Omega)$, and therefore well defined.

Consider now any $\phi \in B^{2,2}_{1/2}(\partial\Omega)$, $\langle \phi, \nu \rangle = 0$. Let $\widetilde{\phi} \in L^2_1(\Omega)$ be a unique extension of ϕ into Ω such that $\mathcal{L}\widetilde{\phi} + d\widetilde{q} = 0$ in Ω. Such an extension exists by [8], is unique, and

$$\|\widetilde{\phi}\|_{L^2_1(\Omega)} \leqslant C \|\phi\|_{B^{2,2}_{1/2}(\partial\Omega)}.$$

Then we put

$$2\int_{\partial\Omega} \langle \mathrm{Def}\,(w)\nu, \phi \rangle d\sigma = \langle f, \widetilde{\phi} \rangle - 2\int_\Omega \langle \mathrm{Def}\,(w), \mathrm{Def}\,(\widetilde{\phi}) \rangle \, d\,\mathrm{Vol}\,. \tag{3.10}$$

Let us observe that, provided that $f \in L^2_{-1,0}(\Omega)$, these integrals on the right-hand side are well defined since $\mathrm{Def}\,(w), \mathrm{Def}\,(\widetilde{\phi}) \in L^2(\Omega)$. It also follows that

$$\begin{aligned} \| \mathrm{Def}\,(w)\nu\big|_{\partial\Omega}\|_X &= \sup_{\|\phi\|_{B^{2,2}_{1/2}(\partial\Omega)}=1} \left| \int_{\partial\Omega} \langle \mathrm{Def}\,(w)\nu, \phi \rangle d\sigma \right| \\ &\leqslant C(\|f\|_{L^2_{-1,0}(\Omega)} + \|w\|_{L^2_1(\Omega)}) \end{aligned} \tag{3.11}$$

by (3.10). Here, X is the dual of the space $\{\phi \in B^{2,2}_{1/2}(\partial\Omega); \langle \phi, \nu \rangle = 0\}$.

Now, assuming that g has zero normal component, i.e., $g_{\mathrm{nor}} = \langle g, \nu \rangle = 0$ and also $\omega_{\mathrm{nor}} = 0$, we see that (3.9) simplifies to

$$\int_\Omega \langle \mathrm{Def}\,(w), \mathrm{Def}\,(u) \rangle \, d\,\mathrm{Vol} = -\int_{\partial\Omega} \langle \mathrm{Def}\,(w)\nu, g \rangle \, d\sigma.$$

Finally, using Lemma 3.1 and the fact that

$$\| \mathrm{Def}\,(w)\nu\|_X \leqslant C(\|f\|_{L^2_{-1,0}(\Omega)} + \|w\|_{L^2_1(\Omega)})$$

(for $f = -\nabla_\omega u$), we get

$$\begin{aligned}C_1\|w\|^2_{L^2_1(\Omega)} &\leqslant \|\operatorname{Def}(w)\|^2_{L^2(\Omega)}\\ &= -\int_\Omega \langle \operatorname{Def}(w), \operatorname{Def}(v)\rangle\, d\operatorname{Vol} - \int_{\partial\Omega} \langle \operatorname{Def}(w)\nu, g\rangle\, d\sigma\\ &\leqslant C(\|w\|_{L^2_1(\Omega)}\|v\|_{L^2_1(\Omega)} + (\|f\|_{L^2_{-1,0}(\Omega)} + \|w\|_{L^2_1(\Omega)})\|g\|_{B^{2,2}_{1/2}(\partial\Omega)}).\end{aligned}$$

By (3.5), we have $\|v\|_{L^2_1(\Omega)} \leqslant C\|g\|_{B^{2,2}_{1/2}(\partial\Omega)}$. Hence for some $C > 0$ independent of ω

$$\|w\|^2_{L^2_1(\Omega)} \leqslant C(\|f\|_{L^2_{-1,0}(\Omega)} + \|w\|_{L^2_1(\Omega)})\|g\|_{B^{2,2}_{1/2}(\partial\Omega)}, \tag{3.12}$$

where $f = -\nabla_\omega u$. It remains to estimate the $L^2_{-1,0}(\Omega)$ norm of f. We produce two different estimates. We first estimate $L^2_{-1}(\Omega)$ norm (this space is larger than $L^2_{-1,0}(\Omega)$ since the first one is the dual of $L^2_{1,0}(\Omega)$ and the later is dual of $L^2_1(\Omega)$). To do this, we just realize that

$$\|f\|_{L^2_{-1}(\Omega)} = \|\mathcal{L}w + dp\|_{L^2_{-1}(\Omega)} \approx \|w\|_{L^2_1(\Omega)}.$$

On the other hand, for q as in (3.7)

$$\begin{aligned}\|f\|_{L^q(\Omega)} &\leqslant C\|\omega\|_{L^n(\Omega)}\|u\|_{L^2_1(\omega)}\\ &\leqslant C\|\omega\|_{L^n(\Omega)}(\|w\|_{L^2_1(\omega)} + \|g\|_{B^{2,2}_{1/2}(\partial\Omega)}).\end{aligned} \tag{3.13}$$

Interpolating these two estimates, we obtain

$$\|f\|_{L^{2n/(n+2-2s)}_{-s}(\Omega)} \leqslant C\|\omega\|^{1-s}_{L^n(\Omega)}(\|w\|_{L^2_1(\omega)} + \|g\|_{B^{2,2}_{1/2}(\partial\Omega)}).$$

Now, the point is that for small $s \geqslant 0$ the spaces $L^{2n/(n+2-2s)}_{-s}(\Omega)$ and $L^{2n/(n+2-2s)}_{-s,0}(\Omega)$ coincide as both are dual of

$$L^{2n/(n-2+2s)}_{s}(\Omega) = L^{2n/(n-2+2s)}_{s,0}(\Omega)$$

which are the same spaces for s close to 0.

But, by the embedding theorem, $\|f\|_{L^{2n/(n+2-2s)}_{-s,0}(\Omega)} \subset L^2_{-1,0}(\Omega)$. Hence for some small $s > 0$

$$\|f\|_{L^2_{-1,0}(\Omega)} \leqslant C\|\omega\|^{1-s}_{L^n(\Omega)}(\|w\|_{L^2_1(\omega)} + \|g\|_{B^{2,2}_{1/2}(\partial\Omega)}).$$

By (3.12),

$$\|w\|^2_{L^2_1(\Omega)} \leqslant C\|\omega\|^{1-s}_{L^n(\Omega)}(\|w\|_{L^2_1(\Omega)} + \|g\|_{B^{2,2}_{1/2}(\partial\Omega)})\|g\|_{B^{2,2}_{1/2}(\partial\Omega)}. \tag{3.14}$$

This implies

$$\|w\|_{L^2_1(\Omega)} \leqslant C\|\omega\|^{1-s}_{L^n(\Omega)}\|g\|_{B^{2,2}_{1/2}(\partial\Omega)} \tag{3.15}$$

for some $s > 0$ and C independent of ω. We claim that this establishes (2.2) and Theorem 2.1 because $\|u\|_{L^2_1(\Omega)} \leqslant \|v\|_{L^2_1(\Omega)} + \|u\|_{L^2_1(\Omega)}$ and $\|v\|_{L^2_1(\Omega)} \leqslant C\|g\|_{B^{2,2}_{1/2}(\partial\Omega)}$. Note that we have only the considered equation (3.2) with zero right-hand side since the nonzero right-hand side is handled by the estimate (3.3) and linearity of the equation.

Proof of Theorem 2.2. As was outlined above, the map $T : L^2_1(\Omega, \Lambda^1 TM) \to L^2_1(\Omega, \Lambda^1 TM)$ (defined by $T : \omega \mapsto u$) is well defined if $\dim M = n \leqslant 4$ since $L^2_1(\Omega) \subset L^n(\Omega)$. By (2.2), for sufficiently large $R > 0$ the map T maps the ball of radius R depending on the norms of f and g into itself. It can also be checked that for $n \leqslant 3$ the map T is compact. Hence, by the Schauder fixed point theorem, T has a fixed point. If $n = 4$, the map T is not compact in the strong topology. It is true however that T is continuous in the weak topology and, in this topology, the ball of radius R is compact. Hence T again has a fixed point. For details cf. [8]. □

Acknowledgement. The author was supported in part by EU and EPSRC.

References

1. Agmon, S., Douglis, A., Nirenberg, L.: Estimates near the boundary for solutions of elliptic partial differential equations satisfying general boundary conditions. I. Commun. Pure Appl. Math. **12**, 623–727 (1959)
2. Amrouche, C., Girault, V.: On the existence and regularity of the solution of Stokes problem in arbitrary dimension. Proc. Japan Acad., Ser. A Math. Sci. **67**, 171–175 (1991)
3. Brown, R.M., Shen, Z.: Estimates for the Stokes operator in Lipschitz domains. Indiana Univ. Math. J. **44**, 1183–1206 (1995)
4. Deuring, P.: The Stokes System in an Infinite Cone. Akadem.-Verlag, Berlin (1994)
5. Deuring, P.: The Stokes resolvent in 3D domains with conical boundary points: nonregularity in L_p-spaces. Adv. Differ. Equ. **6**, 175–228 (2001)
6. Carverhill, A., Pedit, F.J.: Global solutions of the Navier–Stokes equation with strong viscosity. Ann. Global Anal. Geom. **10**, 255–261 (1992)
7. Cattabriga, L.: Su un problema al contorno relativo al sistema di equazioni di Stokes. Rend. Sem. Mat. Univ. Padova **31**, 308–340 (1961)
8. Dindoš, M., Mitrea, M.: Stationary Navier–Stokes equation on Lipschitz domain in Riemannian manifolds. Arch. Ration. Mech. Anal. **174**, no. 1, 1–47 (2004)
9. Ebin, D.G., Marsden, J.: Groups of diffeomorphisms and the notion of an incompressible fluid. Ann. Math. **92**, 102–163 (1970)
10. Frehse, J., Ružička, M.: Existence of regular solutions to the steady Navier–Stokes equations in bounded six-dimensional domains. Ann. Scuola Norm. Sup. Pisa Cl. Sci. **23**, 701–719 (1996)
11. Frehse, J., Ružička, M.: A new regularity criterion for steady Navier–Stokes equations. Differ. Integral Equations **11**, 361–368 (1998)

12. Frehse, J., Ružička, M.: Regularity for steady solutions of the Navier-Stokes equations. In: Heywood, J.G. et al. (eds.), Theory of the Navier-Stokes Equations, pp. 159–178. World Scientific, Singapore (1998)
13. Galdi, G.P., Simader, C.G.: Existence, uniqueness and L_q-estimates for the Stokes problem in an exterior domain. Arch. Ration. Mech. Anal. **112**, 291–318 (1990)
14. Galdi, G.P., Simader, C.G., Sohr, H.: On the Stokes problem in Lipschitz domains. Ann. Mat. Pura Appl. **167**, 147–163 (1994)
15. Giga, Y.: Analyticity of the semigroup generated by the Stokes operator in L_r spaces. Math. Z. **178**, 297–329 (1981)
16. Maz'ya, V., Rossmann, J.: Pointwise estimates for Green's kernel of mixed boundary value problem to the Stokes system in polyhedral cone. Math. Nachr. **278**, no. 15, 1766–1810 (2005)
17. Maz'ya, V., Rossmann, J.: Schauder estimates for solution to a mixed boundary value problem for the Stokes system in polyhedral domains. Math. Meth. Appl. Sci. **29**, 965–1017 (2006)
18. Maz'ya, V., Rossmann, J.: A Maximum Modules Estimate for Solutions of the Navier–Stokes System in Domains of Polyhedral Type. Preprint.
19. Mazzucato, A.: Besov–Morrey spaces: function space theory and applications to non-linear PDE. Trans. Am. Math. Soc. **355**, 1297–1364 (2003)
20. Mitrea. M., Taylor, M.: Navier–Stokes Equations on Lipschitz Domains in Riemannian manifolds. Math. Ann. **321**, no. 4, 955–987 (2001)
21. Nagasawa, T.: Navier–Stokes flow on Riemannian manifolds. Nonlinear Anal. **30**, 825–832 (1997)
22. Priebe, V.: Solvability of the Navier–Stokes equations on manifolds with boundary. Manuscripta Math. **83**, 145–159 (1994)
23. Struwe, M.: Regular solutions of the stationary Navier–Stokes equations on $\mathbb{R}^5$. Math. Ann. **302**, 719–741 (1995)
24. Struwe, M.: Regularity results for the stationary Navier–Stokes equations. In: Differential Equations and Mathematical Physics, pp. 213–216. Int. Press, Boston (1995)
25. Shkoller, S.: Analysis on groups of diffeomorphisms of manifolds with boundary and the averaged motion of a fluid. J. Differ. Geom. **55**, 145–191 (2000)
26. Taylor, M.: Analysis on Morrey spaces and applications to Navier–Stokes and other evolution equations. Commun. Partial Differ. Equ. **17**, 1407–1456 (1992)
27. Taylor, M.: Partial Differential Equations. I–III. Springer, New York (1996)
28. Temam, R.: Navier–Stokes Equations, Theory and Numerical Analysis. Elsevier, Amsterdam (1984)
29. Temam, R., Wang, S.: Inertial forms of Navier–Stokes equations on the sphere. J. Funct. Anal. **117**, 215–242 (1993)
30. Temam, R., Ziane, M.: Navier–Stokes equations in thin spherical domains. Contemp. Math. **209**, 281–314 (1997)

On the Regularity of Nonlinear Subelliptic Equations

András Domokos and Juan J. Manfredi

Abstract We prove C^∞ regularity results for Lipschitz solutions of nondegenerate quasilinear subelliptic equations of p-Laplacian type for a class of Hörmander vector fields that include certain nonnilpotent structures.

It is a great privilege to present this contribution in honor of Professor Vladimir Maz'ya. As one of the developers of the modern theory of Sobolev spaces and their applications to partial differential equations, we are all indebted to his dedication and vision. Professor Maz'ya's contributions permeate all areas of mathematical analysis. We have chosen to present some recent developments on the regularity of solutions to the p-Laplace equation in subelliptic structures. Professor Maz'ya has written a number of influential articles on the p-Laplacian, including a seminal article in nonlinear potential theory [15] and the first boundary regularity estimate in terms of p-capacity [13]. In addition, Professor Maz'ya has written several books from which the now classic *Sobolev Spaces* [14] and (with T.O Shaposhnikova) *Theory of multipliers in spaces of differentiable functions* [16] are obligated references that have influenced several generation of analysts.

We have benefited from discussions with Professor Maz'ya in several occasions at West Lafayette, Linköping, Pittsburgh, and Washington. We would like to express our deeper appreciation for his kindness and generosity.

András Domokos
Department of Mathematics and Statistics, California State University Sacramento, Sacramento, 95819, USA e-mail: domokos@csus.edu

Juan J. Manfredi
Department of Mathematics, University of Pittsburgh, Pittsburgh, PA 15260, USA e-mail: manfredi@pitt.edu

A. Laptev (ed.), *Around the Research of Vladimir Maz'ya II: Partial Differential Equations*, International Mathematical Series 12, DOI 10.1007/978-1-4419-1343-2_6,

1 Introduction

Consider a domain $\Omega \subset \mathbb{R}^n$, $N \leqslant n$, and a Hörmander (or bracket-generating) system of smooth vector fields $\mathfrak{X} = \{X_1, \ldots, X_N\}$ defined on Ω. We denote by $\mathfrak{X}u = (X_1 u, \ldots, X_N u)$ the $\mathfrak{X}$-gradient, or horizontal gradient, of a function u. In this paper, we study second and higher order interior regularity for weak solutions of the following quasilinear subelliptic equation:

$$\sum_{i=1}^{N} X_i^* \left(a_i(x, \mathfrak{X}u)\right) = 0 \ \text{ in } \ \Omega\,, \tag{1.1}$$

where $a_i(x,\xi)$ are differentiable functions on $\Omega \times \mathbb{R}^N$ which for some positive constants c, l and a.e. $(x,\xi) \in \Omega \times \mathbb{R}^N$ and every $\eta \in \mathbb{R}^N$ satisfy the following properties:

$$l^{-1}|\eta|^2 \leqslant \sum_{i,j=1}^{N} \frac{\partial a_i}{\partial \xi_j}(x,\xi)\, \eta_i \eta_j \leqslant l|\eta|^2\,, \tag{1.2}$$

$$\sum_{i,j=1}^{N} \left| \frac{\partial a_i}{\partial x_j}(x,\xi) \right| \leqslant c\,(1+|\xi|)\,, \tag{1.3}$$

$$\sum_{i=1}^{N} |a_i(x,\xi)| \leqslant c\,(1+|\xi|)\,. \tag{1.4}$$

The best known representative of Equation (1.1) is the subelliptic p-Laplacian equation with

$$a_i(x,\xi) = |\xi|^{p-2}\xi_i\,.$$

The assumptions (1.2)–(1.4) are satisfied if we suppose $p \geqslant 2$ and $0 < M^{-1} \leqslant |\xi| \leqslant M$ which for weak solutions u corresponds to the assumption

$$0 < M^{-1} \leqslant |\mathfrak{X}u(x)| \leqslant M \quad \text{a.e. } x \in \Omega\,. \tag{1.5}$$

In the forthcoming paper [10], we will study the quasilinear subelliptic regularity problem without assumption (1.5) and for $p \neq 2$.

As a short overview of previously proved subelliptic interior regularity results we mention Hörmander's C^∞ regularity result in the $a_i(x,\xi) = \xi_i$ linear case [11]. The Hölder continuity of weak solutions for quasilinear equations of the form

$$\sum_{i=1}^{N} X_i^* \left(a_i(x, u, \mathfrak{X}u)\right) = 0 \ \text{ in } \ \Omega \tag{1.6}$$

was proved in [4, 12], and higher regularity for

$$\sum_{i=1}^{N} X_i^* \left(a_i(x,u)X_i u\right) = 0 \quad \text{in} \quad \Omega \tag{1.7}$$

was studied in [21]. These results are valid for any system of Hörmander vector fields. However, the second order differentiability results from [5, 7, 8, 9, 18] are valid just in Heisenberg and Carnot groups and the Moser iterations leading to $C^{1,\alpha}$ regularity developed in [17] just in Heisenberg groups.

In this paper, we introduce the notion of ν-closed systems of Hörmander vector fields, which allows us to study higher order interior regularity in certain nonnilpotent structures. More precisely, this new notion includes all the previously studied nilpotent systems of vector fields including the generators of the Lie Algebra of Carnot groups, the Grušin vector fields, and extends the regularity results to the case of nonnilpotent systems of vector fields, as those generating the Lie Algebra of the rotation group SO (n) or other noncompact semisimple or solvable Lie groups. As it follows from Definition 1.1, we do not even suppose $\mathfrak{X}$ to generate the Lie algebra of a Lie group.

We denote by $\mathfrak{B}$ the set of all commutators of length up to a fixed $\nu \in \mathbb{N}$. Elements of $\mathfrak{X}$ will be considered as commutators of length 1, so we have $\mathfrak{X} \subset \mathfrak{B}$.

Definition 1.1. We say that $\mathfrak{X}$ is a *ν-closed Hörmander system* of vector fields on Ω if

(1) the set $\mathfrak{B}$ of commutators of order at most ν spans the tangent space at every $x \in \Omega$,

(2) there exists a $T_1 \in \mathfrak{B} \setminus \mathfrak{X}$ such that

$$[T_1, X_i] \in \mathfrak{X} \cup \{0\} \quad \text{for all} \quad 1 \leqslant i \leqslant N,$$

(3) if after selecting $\mathfrak{T}_k = \{T_1, \dots, T_k\}$ we still have $\mathfrak{B} \setminus \{\mathfrak{X} \cup \mathfrak{T}_k\} \neq \varnothing$, then there exists $T_{k+1} \in \mathfrak{B} \setminus \{\mathfrak{X} \cup \mathfrak{T}_k\}$ such that

$$[T_{k+1}, Y] \in \mathfrak{X} \cup \mathfrak{T}_k \cup \{0\} \quad \text{for all} \quad Y \in \mathfrak{X} \cup \mathfrak{T}_k,$$

(4) continuing the process started in (1)-(3), we cover all $\mathfrak{B} \setminus \mathfrak{X}$.

Our starting point in finding this definition was Taylor's description [20] of a subelliptic Laplacian operator in the special unitary group SU (2). For a detailed description of the following examples we refer to [10]. In $\mathbb{R}^3$, we consider linearly independent vector fields X_1, X_2, T satisfying the commutation relations

$$[X_1, X_2] = T, \quad [X_2, T] = X_1, \quad [T, X_1] = X_2. \tag{1.8}$$

In this case, the system $\mathfrak{X} = \{X_1, X_2\}$ generates a three dimensional Lie Algebra which is isomorphic to so(3), the Lie Algebra of SO(3).

This can be immediately generalized to SO(n), which has its Lie algebra so (n) spanned by a basis $\mathfrak{B} = \{X_{jk},\ 1 \leqslant j < k \leqslant n\}$ with nonzero commutation relations

$$[X_{mk}, X_{jk}] = X_{jm}, \quad [X_{jk}, X_{jm}] = X_{mk}, \quad [X_{jm}, X_{mk}] = X_{jk}$$

for all $1 \leqslant j < m < k \leqslant n$. This allows us to consider $\mathfrak{X} = \{X_{mn},\ 1 \leqslant m \leqslant n-1\}$. Then any other vector field X_{jk}, $1 \leqslant j < k \leqslant n-1$, is a commutator of length two of elements of $\mathfrak{X}$ and satisfies the following commutation relations:

$$[X_{jm}, X_{qn}] = \left\{ \begin{array}{cl} 0 & \text{if } j \neq q \text{ and } q \neq m \\ -X_{mn} & \text{if } q = j \\ X_{mn} & \text{if } q = m \end{array} \right\}. \tag{1.9}$$

Hence we can start the selection process of Definition 1.1 with any vector field not belonging to $\mathfrak{X}$ and continue it in an arbitrary order.

As other examples for Definition 1.1 we can mention the 3-dimensional Lie algebras listed in [19]. Among them we find a nilpotent (the Heisenberg Lie Algebra), a compact semisimple (so (3)), a noncompact semisimple (sl $(2, \mathbb{R})$) and a solvable (the Lie algebra of the motion group of the Lorentzian plane).

For $k \in \mathbb{N}$ we define the following subelliptic Sobolev space:

$$\mathfrak{X}W^{k,2}(\Omega) = \Big\{ u \in L^p(\Omega) : X_{i_1} \ldots X_{i_k} u \in L^2(\Omega) \ \text{ for all } \ 1 \leqslant i_j \leqslant N \Big\}.$$

Let $\mathfrak{X}W_0^{k,2}(\Omega)$ be the closure of $C_0^\infty(\Omega)$ in $\mathfrak{X}W^{k,2}(\Omega)$ with respect to its usual norm.

A function $u \in \mathfrak{X}W_{\text{loc}}^{1,2}(\Omega)$ is a weak solution of Equation (1.1) if

$$\sum_{i=1}^{N} \int_\Omega a_i(x, \mathfrak{X}u(x))\, X_i\varphi(x) dx = 0 \ \text{ for all } \ \varphi \in C_0^\infty(\Omega)\,. \tag{1.10}$$

Our main task is to prove the following theorem.

Theorem 1.1. *Assume that $\mathfrak{X}$ is a ν-closed system of Hörmander vector fields, the functions $a_i(x, \xi)$ are C^∞ in $\Omega \times \mathbb{R}^n$, and a weak solution u of Equation* (1.1) *satisfies $0 < M^{-1} \leqslant |\mathfrak{X}u| \leqslant M$ a.e. in Ω. Then $u \in C^\infty(\Omega)$.*

By Theorem 1.1, we generalize similar results obtained in Heisenberg and Carnot groups by Capogna [2, 3] and also show alternate proofs for the second order differentiability results in Theorems 2.1 and 2.2 below.

2 Second Order Horizontal Differentiability of Weak Solutions

We denote by $e^{tT}x$ the flow associated to a vector field T. We follow now a point of view that can be derived from the methods presented in [1, 11]. We think about a vector field X as a linear mapping $X : C_0^\infty(\Omega) \to C_0^\infty(\Omega)$ defined by

$$X\phi(x) = \frac{d}{dt}\bigg|_{t=0} \phi(e^{tX}x)\,.$$

For $\phi \in C_0^\infty(\Omega)$ and $t > 0$ sufficiently small let us define

$$e^{tX}\phi\,(x) = \phi(e^{tX}x)\,.$$

The Taylor series expansion at $t = 0$ gives the following formal power series representation:

$$e^{tX}\phi = \sum_{k=0}^{\infty} \frac{t^k}{k!}\, X^k\phi\,. \tag{2.1}$$

If we use the operator $\mathrm{ad}Z\,(X) = [Z, X]$, then (2.1) leads to the following three lemmas. For their proofs we refer to [10].

Lemma 2.1. *Consider an arbitrary $\phi \in C_0^\infty(\Omega)$ and $x \in \Omega$. Then for sufficiently small $s > 0$ we have*

$$X\big(e^{sZ}\phi(x)\big) = \sum_{k=0}^{\infty}(-1)^k\,\frac{s^k}{k!}\,\Big((\mathrm{ad}Z)^k(X)\,\phi\Big)\,\big(e^{sZ}x\big)\,, \tag{2.2}$$

$$X\big(e^{-sZ}\phi(x)\big) = \sum_{k=0}^{\infty} \frac{s^k}{k!}\,\Big((\mathrm{ad}Z)^k(X)\,\phi\Big)\,\big(e^{-sZ}x\big)\,. \tag{2.3}$$

Lemma 2.2. *Consider a ν-closed Hörmander system $\mathfrak{X}$ of vector fields and an arbitrary $\phi \in C_0^\infty(\Omega)$ and $x \in \Omega$.*

(i) *If $T \in \mathfrak{B} \setminus \mathfrak{X}$ such that $[T, X_i] \subset \mathfrak{X} \cup \{0\}$ for all $1 \leqslant i \leqslant N$, then there exists $\Psi_i(s) = \langle \psi_{i,1}(s), \ldots, \psi_{i,N}(s)\rangle$, where $\psi_{i,k}$ are analytic functions in s, such that*

$$X_i\big(e^{\pm sT}\phi(x)\big) = X_i\phi\big(e^{\pm sT}x\big) \pm s\Psi_i(\pm s)\cdot \mathfrak{X}\phi\big(e^{\pm sT}x\big)\,. \tag{2.4}$$

(ii) *Suppose that we already selected a set of vector fields $\mathfrak{T} = \{T_1, \ldots, T_m\}$ and continue the process by selecting $Z \in \mathfrak{B}$ such that $[Z, Y] \in \mathfrak{X} \cup \mathfrak{T} \cup \{0\}$ for all $Y \in \mathfrak{X} \cup \mathfrak{T}$. Then for every $1 \leqslant i \leqslant N$ there exist two vectors Ψ_i and Φ_i of analytic functions in s such that*

$$X_i\big(e^{\pm sZ}\phi(x)\big) = X_i\phi\big(e^{\pm sZ}x\big) \pm s\Psi_i(\pm s)\cdot\mathfrak{X}\phi\big(e^{\pm sZ}x\big) \pm s\Phi_i(\pm s)\cdot\mathfrak{T}\phi\big(e^{\pm sZ}x\big)\,. \tag{2.5}$$

Example 2.1. If we return to the vector fields X_1, X_2, T with commutation relations presented in (1.8), we find that

$$\begin{aligned} X_1\big(e^{sT}\phi(x)\big) &= X_1\phi\big(e^{sT}x\big) \\ &\quad + s\frac{-\sin s}{s}X_2\phi\big(e^{sT}x\big) + s\frac{\cos s - 1}{s}X_1\phi\big(e^{sT}x\big)\,. \end{aligned} \tag{2.6}$$

For $s > 0$ we define the following difference quotients:

$$D_{Z,s,\gamma}u(x) = \frac{u(e^{sZ}x) - u(x)}{s^\gamma},$$
$$D_{Z,-s,\gamma}u(x) = \frac{u(x) - u(e^{-sZ}x)}{s^\gamma}.$$

Lemma 2.3. (i) *Consider the vector field T from* (i) *of Lemma 2.2. If $u \in L^2(\Omega)$ has compact support and $X_i u \in L^2(\Omega)$ for all $1 \leqslant i \leqslant N$, then we have the following identity in the weak sense:*

$$X_i \Big(D_{T,\pm s,\gamma}u(x) \Big) = D_{T,\pm s,\gamma}\Big(X_i u(x) \Big) \pm s^{1-\gamma} \Psi_i(\pm s) \cdot \mathfrak{X} u\big(e^{\pm sT}x\big). \quad (2.7)$$

(ii) *Consider the vector field Z from* (ii) *of Lemma 2.2. If, in addition to* (i)*, we suppose that $T_j u \in L^2(\Omega)$ for all $1 \leqslant j \leqslant m$, then we have the following identity in the weak sense:*

$$X_i \Big(D_{Z,\pm s,\gamma}u(x) \Big) = D_{Z,\pm s,\gamma}\Big(X_i u(x) \Big) \pm s^{1-\gamma} \Psi_i(\pm s) \cdot \mathfrak{X} u\big(e^{\pm sT}x\big) \pm s^{1-\gamma} \Phi_i(\pm s) \cdot \mathfrak{T} u\big(e^{\pm sZ}x\big). \quad (2.8)$$

We are able now to prove our first regularity result.

Theorem 2.1. *Consider a weak solution $u \in XW^{1,2}_{\text{loc}}(\Omega)$ of* (1.1)*. Let $T \in \mathfrak{B} \setminus \mathfrak{X}$ be such that $[T, X_i] \subset \mathfrak{X} \cup \{0\}$ for all $1 \leqslant i \leqslant N$. Suppose that T is a commutator of length m of the horizontal vector fields. If $x_0 \in \Omega$ and $r > 0$ are such that $B\big(x_0, 3r\big) \subset \Omega$, then there exist numbers $k \in \mathbb{N}$ and $c > 0$ depending only on m and* dist $(x_0, \partial\Omega)$ *such that*

$$\int_{B\left(x_0, r/k\right)} |Tu(x)|^2 dx \leqslant c \int_{B\left(x_0, 2r\right)} \big(1 + |\mathfrak{X} u(x)|^2 + |u(x)|^2\big)\, dx \qquad (2.9)$$

and

$$\int_{B(x_0, r/k)} \left|T\mathfrak{X} u(x)\right|^2 \, dx \leqslant c \int_{B(x_0, 2r)} \big(1 + |\mathfrak{X} u(x)|^2 + |u(x)|^2\big)\, dx\,, \quad (2.10)$$

which implies $Tu \in \mathfrak{X}W^{1,2}_{\text{loc}}(\Omega)$

Proof. Denote $\gamma = 1/m$. Let η be a cut-off function between the Carnot–Carathéodory metric balls $B(x_0, \frac{r}{2})$ and $B(x_0, r)$. For the sake of simplicity, let us just use the notation B_r. Consider the test function

$$\varphi = D_{T,-s,\gamma}\left(\eta^2 D_{T,s,\gamma}u\right)$$

to get

$$\sum_{i=1}^{N} \int_{\Omega} a_i(x, \mathfrak{X}u(x)) \;\; X_i\Big(D_{T,-s,\gamma}\left(\eta^2 D_{T,s,\gamma}u\right)\Big)(x)dx = 0\,.$$

By Lemma 2.3,

$$\sum_{i=1}^{N} \int_{\Omega} a_i(x, \mathfrak{X}u(x))\, D_{T,-s,\gamma}X_i\left(\eta^2 D_{T,s,\gamma}u\right)(x)\,dx$$

$$= \sum_{i=1}^{N} \int_{\Omega} a_i(x, \mathfrak{X}u(x))\, s^{1-\gamma}\Psi_i(-s)\cdot \mathfrak{X}\left(\eta^2 D_{T,s,\gamma}u\right)\left(e^{-sT}x\right)\,dx\,. \quad (2.11)$$

We denote by $J_T^+(s,x)$ the Jacobian determinant of the transformation $x \mapsto e^{sT}x$ and use the facts that for small $s>0$ it can be bounded in the following way:

$$0 < a \leqslant \left|J_T^+(s,x)\right| \leqslant b.$$

Also, the formal adjoint of $D_{T,s,\gamma}$ is given by the formula

$$\left(D_{T,-s,\gamma}\right)^* u(x) = -D_{T,s,\gamma}u(x) + \frac{1-|J_T^+(s,x)|}{s^\gamma}u(e^{sT}x). \quad (2.12)$$

For a vector field $T=\sum_{i=1}^n \beta_i(x)\frac{\partial}{\partial x_i}$ we have

$$\lim_{s\to 0}\frac{J_T^+(s,x)-1}{s} = \sum_{i=1}^n \frac{\partial \beta_i}{\partial x_i}(x).$$

Therefore, it follows that for $0<\gamma\leqslant 1$ the function $\dfrac{1-|J_T^+(s,x)|}{s^\gamma}$ is bounded for small s on compact subsets of Ω.

Similarly, we denote by $J_T^-(s,x)$ the Jacobian determinant of the transformation $x\mapsto e^{-sT}x$ and obtain the following formula for the formal adjoint:

$$\left(D_{T,s,\gamma}\right)^* u(x) = -D_{T,-s,\gamma}u(x) + \frac{|J_T^-(s,x)|-1}{s^\gamma}u(e^{-sT}x)\,. \quad (2.13)$$

Let us return to (2.11) and continue as

$$\sum_{i=1}^{N} \int_{\Omega} D_{T,s,\gamma}a_i(x, \mathfrak{X}u(x))\; X_i\left(\eta^2 D_{T,s,\gamma}u\right)(x)\,dx$$

$$= \sum_{i=1}^{N} \int_{\Omega} \frac{1-|J_T^+(s,x)|}{s^\gamma}\, a_i(e^{sT}x, \mathfrak{X}u(e^{sT}x))\; X_i\left(\eta^2 D_{T,s,\gamma}u\right)(x)\,dx$$

$$- \sum_{i=1}^{N} \int_{\Omega} a_i(x, \mathfrak{X}u(x))\, s^{1-\gamma}\Psi_i(-s)\cdot \mathfrak{X}\left(\eta^2 D_{T,s,\gamma}u\right)\left(e^{-sT}x\right)\,dx\,.$$

Using Lemma 2.3 several times, we get the equation

$$
\begin{aligned}
&\sum_{i=1}^{N}\int_{\Omega} D_{T,s,\gamma}a_i(x,\mathfrak{X}u(x))\;\eta^2(x)\;D_{T,s,\gamma}X_iu(x)\;dx\\
&=-\sum_{i=1}^{N}\int_{\Omega} D_{T,s,\gamma}a_i(x,\mathfrak{X}u(x))\;\eta^2(x)\;s^{1-\gamma}\Psi(s)\cdot\mathfrak{X}u(e^{sT}x)\;dx\\
&-\sum_{i=1}^{N}\int_{\Omega} D_{T,s,\gamma}a_i(x,\mathfrak{X}u(x))\;2\eta(x)\;X_i\eta(x)\;D_{T,s,\gamma}u(x)\;dx\\
&+\sum_{i=1}^{N}\int_{\Omega}\frac{1-|J_T^+(s,x)|}{s^\gamma}\;a_i(e^{sT}x,\mathfrak{X}u(e^{sT}x))\;X_i(\eta^2)(x)D_{T,s,\gamma}u(x)\,dx\\
&+\sum_{i=1}^{N}\int_{\Omega}\frac{1-|J_T^+(s,x)|}{s^\gamma}\;a_i(e^{sT}x,\mathfrak{X}u(e^{sT}x))\;\eta^2(x)D_{T,s,\gamma}X_iu(x)\,dx\\
&+\sum_{i=1}^{N}\int_{\Omega}\frac{1-|J_T^+(s,x)|}{s^\gamma}\;a_i(e^{sT}x,\mathfrak{X}u(e^{sT}x))\;\eta^2(x)s^{1-\gamma}\Psi_i(s)\cdot\mathfrak{X}u(e^{sT}x)\,dx\\
&-\sum_{i=1}^{N}\int_{\Omega}|J_T^-(s,x)|\;a_i(e^{sT}x,\mathfrak{X}u(e^{sT}x))\;s^{1-\gamma}\Psi_i(-s)\cdot\mathfrak{X}(\eta^2)(x)D_{T,s,\gamma}u(x)\;dx\\
&-\sum_{i=1}^{N}\int_{\Omega}|J_T^-(s,x)|\;a_i(e^{sT}x,\mathfrak{X}u(e^{sT}x))\;s^{1-\gamma}\eta^2\Psi_i(-s)\cdot D_{T,s,\gamma}\mathfrak{X}u(x)\;dx\\
&-\sum_{i,j,k=1}^{N}\int_{\Omega}|J_T^-(s,x)|\;a_i(e^{sT}x,\mathfrak{X}u(e^{sT}x))\;s^{2-2\gamma}\eta^2\psi_{ij}(-s)\psi_{jk}(s)X_ku(x)\;dx\,.
\end{aligned}
$$

For the following estimates we use the properties (1.2)–(1.5), the relation

$$
\begin{aligned}
D_{T,s,\gamma}a_i(x,\mathfrak{X}u(x))=&\;\frac{a_i(e^{sT}x,\mathfrak{X}u(e^{sT}x)-a_i(e^{sT}x,\mathfrak{X}u(x))}{s^\gamma}\\
&\qquad+\frac{a_i(e^{sT}x,\mathfrak{X}u(x)-a_i(x,\mathfrak{X}u(x))}{s^\gamma}\,,
\end{aligned}
$$

and the fact that $\left|\dfrac{e^{sT}x-x}{s^\gamma}\right|$ is bounded on compact subsets of Ω for small s when $\gamma\leqslant 1$.

$$
\begin{aligned}
\text{Left Side }\;&\geqslant l\int_{B_r}\eta^2(x)\;|D_{T,s,\gamma}\mathfrak{X}u|^2\;dx\\
&\quad-c\int_{B_r}\eta^2(x)\;\left(1+|\mathfrak{X}u(x)|^2\right)\;dx.
\end{aligned}
$$

Right Side line 1

$$\leqslant \delta \int_{B_r} \eta^2(x)\, |D_{T,s,\gamma}\mathfrak{X}u|^2 \, dx$$

$$+ c \int_{B_r} \eta^2(x)\, \left(1 + |\mathfrak{X}u(x)|^2 + |\mathfrak{X}u(e^{sT}x)|^2\right); dx.$$

Right Side line 2

$$\leqslant \delta \int_{B_r} \eta^2 \, |D_{T,s,\gamma}\mathfrak{X}u|^2 \, dx$$

$$+ c \int_{B_r} \left(1 + |\mathfrak{X}u(x)|^2 + |D_{T,s,\gamma}u|^2\right) \, dx.$$

Right Side lines 3 and 6

$$\leqslant c \int_{B_r} \left(1 + |\mathfrak{X}u(x)|^2 + |\mathfrak{X}u(e^{sT}x)|^2 + |D_{T,s,\gamma}u|^2\right) \, dx.$$

Right Side lines 4 and 7

$$\leqslant \delta \int_{B_r} \eta^2 \, |D_{T,s,\gamma}\mathfrak{X}u|^2 \, dx$$

$$+ c \int_{B_r} \left(1 + |\mathfrak{X}u(x)|^2 + |\mathfrak{X}u(e^{sT}x)|^2\right) \, dx.$$

Right Side lines 5 and 8

$$\leqslant c \int_{B_r} \left(1 + |\mathfrak{X}u(x)|^2 + |\mathfrak{X}u(e^{sT}x)|^2\right) \, dx\,.$$

Using the fact that by Hörmander's result [11, Theorem 4.3] for $\gamma = 1/m$ we have

$$\int_{B_r} |D_{T,s,\gamma}u(x)|^2 \, dx \leqslant c \int_{B_{2r}} \left(1 + |\mathfrak{X}u(x)|^2 + |u(x)|^2\right) dx\,,$$

we find that

$$\int_{B_r} \eta^2(x) \, \left|D_{T,s,\gamma}\mathfrak{X}u(x)\right|^2 \, dx \leqslant c \int_{B_{2r}} \left(1 + |\mathfrak{X}u(x)|^2 + |u(x)|^2\right) dx\,. \tag{2.14}$$

Therefore, we obtain

$$\int_{B_r} \left|\mathfrak{X}D_{T,s,\gamma}(\eta^2 u)(x)\right|^2 \, dx \, \leqslant \, c \int_{B_{2r}} \left(1 + |\mathfrak{X}u(x)|^2 + |u(x)|^2\right) dx\,. \tag{2.15}$$

Using again Hörmander's result [11, Theorem 4.3], we find that there exists $\sigma > 0$ such that

$$\sup_{0<|s|\leqslant\sigma} \int_{B_r} \left| D_{T,-s,\frac{1}{m}} D_{T,s,\frac{1}{m}} (\eta^2 u)(x) \right|^2 dx$$
$$\leqslant c \int_{B_{2r}} \left(1 + |\mathfrak{X}u(x)|^2 + |u(x)|^2\right) dx \,.$$

If $2/m < 1$, then, by [6], we also have

$$\sup_{0\neq|s|\leqslant\sigma} \int_{B(x_0,r)} \left| D_{T,s,\frac{2}{m}} (\eta^2 u) \right|^2 dx \leqslant c \int_{B_{2r}} \left(1 + |\mathfrak{X}u(x)|^2 + |u(x)|^2\right) dx \,. \tag{2.16}$$

Hence, if we begin the proof with $\gamma_0 = 1/m$, then, by (2.16), we can restart it with $\gamma_1 = c2/m$ and, by an iteration process, we get $\gamma_j = \frac{j}{m}$. We can push γ_j over 1 after m iteration steps, and this finishes the proof. □

For the next theorem let us suppose that for the vector fields $\mathfrak{T} = \{T_1, \ldots, T_m\}$ given by Definition 1.1 the maximum length of the commutators used in $\mathfrak{T}$ is $m_{\mathfrak{T}}$.

Theorem 2.2. *Consider a weak solution* $u \in XW^{1,2}_{\text{loc}}(\Omega)$ *of* (1.1) *and suppose that the conclusions of Theorem* 2.1 *hold for the vector fields* $T_1, \ldots, T_m$. *Following Definition* 1.1, *let* $Z \in \mathfrak{B} \setminus \{\mathfrak{X} \cup \mathfrak{T}\}$ *be the next element in the selection process. Suppose that* Z *is a commutator of length* m_Z *of the horizontal vector fields. If* $x_0 \in \Omega$ *and* $r > 0$ *are such that* $B(x_0, 3r) \subset \Omega$, *then there exist numbers* $l \in \mathbb{N}$ *and* $c > 0$ *depending only on* $\operatorname{dist}(x_0, \partial\Omega)$ *and* $\max\{m_{\mathfrak{T}}, m_Z\}$ *such that*

$$\int_{B(x_0,r/l)} |Zu|^2 \, dx \leqslant c \int_{B(x_0,2r)} \left(1 + |\mathfrak{X}u|^2 + |u|^2\right) dx \tag{2.17}$$

and

$$\int_{B(x_0,r/l)} |Z\mathfrak{X}u(x)|^2 \, dx \leqslant c \int_{B(x_0,2r)} \left(1 + |\mathfrak{X}u(x)|^2 + |u(x)|^2\right) dx \,, \tag{2.18}$$

which implies $Zu \in L^2_{\text{loc}}(\Omega)$ *and* $|Z\mathfrak{X}u|$, $|\mathfrak{X}Zu| \in L^2_{\text{loc}}(\Omega)$.

Proof. We can follow the proof of Theorem 2.1 with the test function

$$\varphi = D_{Z,-s,\gamma}\left(\eta^2 D_{Z,s,\gamma} u\right) , \tag{2.19}$$

where $\gamma = \min\left\{\frac{1}{m_Z}, \frac{1}{m_{\mathfrak{T}}}\right\}$. The additional derivatives which appear because of the use of (2.8) have the required integrability properties and therefore we get

$$\int_{B(x_0,r)} \eta^2(x) \, |D_{Z,s,\gamma}\mathfrak{X}u(x)|^2 \, dx$$

$$\leqslant c \int_{B(x_0,2r)} \left(1 + |\mathfrak{X}u(x)|^2 + |u(x)|^2\right) dx\,. \tag{2.20}$$

We can finish the proof now by iteration on the order of the fractional difference quotients in the same manner as for Theorem 2.1. □

With the observation that the proof of Theorem 2.2 can be now repeated for $\gamma = 1$ and $Z = X_i$, $1 \leqslant i \leqslant N$, and the highest order of commutators used is ν we obtain the following assertion.

Theorem 2.3. *Consider a weak solution* $u \in \mathfrak{X}W^{1,2}_{\text{loc}}(\Omega)$ *of* (1.1). *If* $x_0 \in \Omega$ *and* $r > 0$ *are such that* $B\big(x_0, 3r\big) \subset \Omega$, *then there exist numbers* $l \in \mathbb{N}$ *and* $c > 0$ *depending only on* $\operatorname{dist}(x_0, \partial\Omega)$ *and* ν *such that*

$$\int_{B(x_0,r/l)} \left|\mathfrak{X}^2 u(x)\right|^2 dx \leqslant c \int_{B(x_0,2r)} \left(1 + |\mathfrak{X}u(x)|^2 + |u(x)|^2\right) dx\,. \tag{2.21}$$

3 C^∞ Regularity

For the last part of this paper we use the following function spaces.

— The space of Hölder continuous functions with respect to the Carnot–Carathéodory distance

$$\Gamma^\alpha(\Omega) = \left\{ u : \Omega \to \mathbb{R} \mid \sup_{x \neq y} \frac{|u(x) - u(y)|}{d(x,y)^\alpha} < \infty \right\}.$$

— The Morrey space $M^{p,\lambda}(\Omega)$ which contains the functions $u \in L^p_{\text{loc}}(\Omega)$ such that

$$\fint_{B(x,r)\cap\Omega} |u(y)|^p \, dy \leqslant c r^{p(\lambda-1)}$$

for all $x \in \Omega$ and $0 < r < \min\{R, \operatorname{diam}(\Omega)\}$.

The important fact given by these function spaces is that $\mathfrak{X}u \in M^{2,\lambda}(\Omega)$ implies $u \in \Gamma^\lambda_{\text{loc}}(\Omega)$.

We are ready to state the following theorem.

Theorem 3.1. *Suppose that the functions* $a_i(x,\xi)$ *are* C^∞ *in* $\Omega \times \mathbb{R}^N$. *Let* u *be a weak solution of* (1.1) *satisfying* (1.2)–(1.5). *Then*

$$u \in \mathfrak{X}W^{2,2}_{\text{loc}}(\Omega) \bigcap C^{1,\alpha}_{\text{loc}}(\Omega). \tag{3.1}$$

Proof. We follow the ideas from [3, Sec. 4]. First, selecting a test function $\varphi(x) = \eta^2(x)\big(u(x) - u(x_0)\big)$, we get $\mathfrak{X}u \in M^{2,\lambda}(\Omega)$.

In the following, we select the vector fields in the order given by Definition 1.1. Differentiating Equation (1.1) with respect to T_1, we get that $w = T_1 u$

is a weak solution of the equation

$$\sum_{i,j=1}^{N} X_i^* \Big(\frac{\partial a_i}{\partial \xi_j}(x, \mathfrak{X}u) X_j w + \beta_i \Big) = \beta \,, \tag{3.2}$$

where

$$\beta_i = \sum_{k=1}^{n} \frac{\partial a_i}{\partial x_k}(x, \mathfrak{X}u) T_i x_k + \sum_{j=1}^{N} \frac{\partial a_i}{\partial \xi_j}(x, \mathfrak{X}u)[T_1, X_j]u \tag{3.3}$$

and

$$\beta = [T_1, X_i] a_i(x, \mathfrak{X}u) + (T_1 g_i) a_i(x, \mathfrak{X}u). \tag{3.4}$$

We used the notation $X_i^* = -X_i - g_i$. By Definition 1.1, $[T_1, X_i] = 0$ or $[T_1, X_i] \subset \mathfrak{X}$, in which case the first term of β can be embedded into the left side. Therefore, the conditions for Theorem 6.32 in [3] are satisfied and we can conclude that $T_1 u \in \Gamma^{\lambda}_{\mathrm{loc}}(\Omega)$ and $\mathfrak{X} T_1 u \in M^{2,\lambda}(\Omega)$.

We can now differentiate (1.1) with respect to T_2 and, using the fact that the commutators of T_2 with the horizontal vector fields are 0 or T_1 or an element of $\mathfrak{X}$, we can find that $T_2 u \in \Gamma^{\lambda}_{\mathrm{loc}}(\Omega)$ and $\mathfrak{X} T_2 u \in M^{2,\lambda}(\Omega)$.

We continue in this way until we cover all the vector fields from $\mathfrak{B}$ and by this finish the proof. □

Once we have the $C^{1,\alpha}$ interior regularity of weak solutions, we can start using the method elaborated in [21] to iteratively obtain higher and higher differentiability properties of solutions to equations in forms similar to (1.7) and ultimately finish the proof of Theorem 1.1.

References

1. Agrachev, A.A., Sachkov, Y.L.: Control Theory from the Geometric Viewpoint. Springer, Berlin etc. (2004)
2. Capogna, L.: Regularity of quasilinear equations in the Heisenberg group. Commun. Pure Appl. Math. **50**, 867–889 (1997)
3. Capogna, L.: Regularity for quasilinear equation and 1-quasiconformal maps in Carnot groups. Math. Ann. **313**, 263–295 (1999)
4. Capogna, L., Danielli, D., Garofalo, N.: An embedding theorem and the Harnack inequality for nonlinear subelliptic equations. Commun. Partial Differ. Equ. **18**, 1765–1794 (1993)
5. Domokos, A.: Differentiability of solutions for the nondegenerate p-Laplacian in the Heisenberg group. J. Differ. Equ. **204**, 439–470 (2004)
6. Domokos, A.: Weighted function spaces of fractional derivatives for vector fields. Electron. J. Differ. Equ. **17**, 1–8 (2007)
7. Domokos, A.: On the regularity of subelliptic p-harmonic functions in Carnot groups. Nonlinear Anal. **69**, 1744–1756 (2008)
8. Domokos, A., Manfredi, J.J.: Subelliptic Cordes estimates. Proc. Am. Math. Soc. **133**, 1047–1056 (2005)

9. Domokos, A., Manfredi, J.J.: $C^{1,\alpha}$-regularity for p-harmonic functions in the Heisenberg group for p near 2. Contemp. Math. **370**, 17–23 (2005)
10. Domokos, A., Manfredi, J.J.: Subelliptic nonlinear equations. [Submitted for publication]
11. Hörmander, L.: Hypoelliptic second order differential equations. Acta Math. **119**, 147–171 (1967)
12. Lu, G.: Embedding theorems into Lipschitz and BMO spaces and applications to quasilinear subelliptic differential equations. Publ. Mat. **40**, 310–329 (1996)
13. Maz'ya, V.G.: The continuity at a boundary point of the solutions of quasilinear elliptic equations (Russian). Vestnik Leningrad. Univ. **25**, no. 13, 42–55 (1970); English transl.: Vestn. Leningrad Univ. Math. **3**, 225–242 (1976)
14. Maz'ya, V.G.: Sobolev Spaces. Springer, Berlin etc. (1985)
15. Maz'ya, V.G., Khavin, V.P.: Nonlinear potential theory (Russian). Usp. Mat. Nauk **27**, no. 6, 67–138 (1972); English transl.: Russ. Math. Surv. **27**, no. 6, 71–148 (1973)
16. Maz'ya, V.G., Shaposhnikova, T.O.: Theory of Multipliers in Spaces of Differentiable Functions. Pitman, Boston etc. (1985) Russian edition: Leningrad. Univ. Press, Leningrad (1986)
17. Manfredi, J.J., Mingione, G.: Regularity results for quasilinear elliptic equations in the Heisenberg group. Math. Ann. **339**, 485–544 (2007)
18. Marchi, S.: $C^{1,\alpha}$ local regularity for the solutions of the p-Laplacian on the Heisenberg group for $2 \leqslant p \leqslant 1+\sqrt{5}$. Z. Anal. Anwend. **20**, 617–636 (2001); Erratum: ibid **22**, 471–472 (2003)
19. Strichartz, R.S.: Sub-Riemannian Geometry, J. Differ. Geom. **24**, 221–263 (1986)
20. Taylor, M.E.: Noncommutative Harmonic Analysis. Am. Math. Soc., Providence, RI (1986)
21. Xu, C.J., Zuily, C.: Higher regularity for quasilinear subelliptic systems. Cal. Var. PDE **5**, 323–344 (1997)

Rigorous and Heuristic Treatment of Sensitive Singular Perturbations Arising in Elliptic Shells

Yuri V. Egorov, Nicolas Meunier, and Evariste Sanchez-Palencia

Abstract We consider singular perturbations of elliptic systems depending on a parameter ε such that, for $\varepsilon = 0$ the boundary conditions are not adapted to the equation (they do not satisfy the Shapiro–Lopatinskii condition). The limit holds only in very abstract spaces out of distribution theory involving complexification and nonlocal phenomena. This system appears in thin shell theory when the middle surface is elliptic and the shell is fixed on a part of the boundary and free on the rest. We use a heuristic reasoning applying some simplifications which allow us to reduce the original problem in a domain to another problem on its boundary. The novelty of this work is that we consider systems of partial differential equations while in our previous work we were dealing with single equations.

1 Introduction

This paper is devoted to a very singular kind of perturbation problems arising in thin shell theory. Up to our knowledge, it is disjoint of relevant and well known contributions of V. Maz'ya on perturbation of domains and multistruc-

Yuri V. Egorov
Laboratoire MIP, Université Paul Sabatier, 118 route de Narbonne, 31062 Toulouse Cedex 9, France
e-mail: egorov@mip.ups-tlse.fr

Nicolas Meunier
Laboratoire MAP5, Université Paris Descartes (Paris V), 45 Rue des Saints Pères, 75006 Paris, France
e-mail: nicolas.meunier@math-info.univ-paris5.fr

Evariste Sanchez-Palencia
Institut Jean Le Rond d'Alembert, case 162. CNRS and Université Pierre et Marie Curie, 4, place Jussieu, 75252 Paris, France
sanchez@lmm.jussieu.fr

A. Laptev (ed.), *Around the Research of Vladimir Maz'ya II: Partial Differential Equations*, International Mathematical Series 12, DOI 10.1007/978-1-4419-1343-2_7,

tures for elliptic problems including the Navier–Stokes system ([12, 11, 13]), as the pathological feature of our problem is concerned with ill-posedness of the limit problem, generating singularities out of the distribution space. So, it may be considered as a contribution to enlarge perturbation theory of Maz'ya. More precisely, the main purpose of this paper is to generalize the previous work done on equations (cf. [7, 15]) to systems of partial differential equations. The motivation for studying that kind of problems comes from shell theory. It appears that when the middle surface is elliptic (both principal curvatures have same sign) and is fixed on a part Γ_0 of the boundary and free on the rest Γ_1, the "limit problem" as the thickness ε tends to zero is elliptic, with boundary conditions satisfying Shapiro–Lopatinskii (SL hereafter) on Γ_0, but not satisfying it on Γ_1. In other words, the "limit problem" for $\varepsilon = 0$ is highly ill-posed. This pathological behavior arises only as $\varepsilon = 0$. In fact, for $\varepsilon > 0$ the problem is "classical."

In such kind of situations, the limit problem has no solution within classical theory of partial differential equations, which uses distribution theory. It is sometimes possible to prove the convergence of the solutions u^ε towards some limit u^0, but this "limit solution" and the topology of the convergence are concerned with abstract spaces not included in the distribution space.

The variational problem we are interested in is

$$\begin{aligned} &\text{Find } u^\varepsilon \in V \text{ such that } \forall\, v \in V \\ &a(u^\varepsilon, v) + \varepsilon^2 b(u^\varepsilon, v) = \langle f, v \rangle \end{aligned} \tag{1.1}$$

or, equivalently, the minimization in V of the functional

$$a(u, u) + \varepsilon^2 b(u, u) - 2\langle f, u \rangle,$$

where $f \in V'$ is given and the brackets denote the duality between V' and V.

This is the Koiter model of shells, ε denoting the relative thickness. The corresponding energy space V is a classical Sobolev space.

The limit boundary partial differential system associated with (1.1) when $\varepsilon = 0$ is elliptic and ill-posed.

Let us consider formally the variational problem of the membrane problem (i.e., $\varepsilon = 0$):

$$\begin{aligned} &\text{Find } u \in V_a \text{ such that } \forall\, v \in V_a \\ &a(u, v) = \langle f, v \rangle, \end{aligned} \tag{1.2}$$

where V_a is the abstract completion of the "Koiter space" V with the norm $\|v\|_a = a(v, v)^{1/2}$, it is to be noted that the elements of V_a are not necessary distributions. The term "sensitive" originates from the fact that this latter problem is unstable. Very small and smooth variations of f (even in $\mathcal{D}(\Omega)$) induce modifications of the solution which are large and singular (out of the distribution space).

The plan of the article is as follows. After recalling the Koiter shell model (Section 2), we recall the definitions of the ellipticity and Shapiro–Lopatinskii condition for systems elliptic in the Douglis–Nirenberg sense (Section 3). In Section 4, we study four systems of partial differential equations which are involved in our study of shell theory. These systems are the rigidity system, the membrane tension system, the membrane system and the Koiter shell system.

In Section 5, we study a sensitive perturbation problem arising in Koiter linear shell theory and we briefly recall some abstract convergence results. In Section 6, we report the heuristic procedure of [7]. In this latter article, we addressed a model problem including a variational structure, somewhat analogous to the shell problem studied here, but simpler, as concerning an equation instead of a system. It is shown that the limit problem involves, in particular, an elliptic Cauchy problem. This problem was handled in both a rigorous (very abstract) framework and using a heuristic procedure for exhibiting the structure of the solutions with very small ε. The reasons why the solution goes out of the distribution space as ε goes to 0 are then evident. The heuristic procedure is very much analogous to the method of construction of a parametrix in elliptic problems [20, 8]:

— Only principal (with higher differentiation order) terms are taken into account.

— Locally, the coefficients are considered to be constant, their values being frozen at the corresponding points.

— After Fourier transform ($x \to \xi$), terms with small ξ are neglected with respect to those with larger ξ (which amounts to taking into account singular parts of the solutions while neglecting smoother ones). We note that this approximation, aside with the two previous ones, lead to some kind of "local Fourier transform" which we shall use freely in the sequel.

Another important feature of the heuristics is a previous drastic restriction of the space where the variational problem is handled. In order to search for the minimum of energy, we only take into account functions such that the energy of the limit problem is very small. This is done using a boundary layer method within the previous approximations, i.e., for large $|\xi|$. This leads to an approximate simpler formulation of the problem for small ε, where it is apparent that the limit problem involves a smoothing operator and cannot have a solution within distribution theory.

The notation is standard. We denote

$$\partial_k = \frac{\partial}{\partial x_k}, \quad k = 1, 2, \tag{1.3}$$

and

$$D_k = -i\frac{\partial}{\partial x_k}, \quad k = 1, 2,$$
$$D^\alpha = D_1^{\alpha_1} D_2^{\alpha_2}, \quad \alpha = (\alpha_1, \alpha_2) \in \mathbb{Z}_+^2. \tag{1.4}$$

Moreover, the definition of the Sobolev space $H^s(\Gamma)$, $s \in \mathbb{R}$, where Γ is a one-dimensional compact manifold is classical using a partition of unity and local mappings.

The inner product and the duality products associated with a space V and its dual V' will be denoted by $(\cdot\,,\cdot)$ and $\langle\cdot\,,\cdot\rangle$ respectively.

The usual convention of summation of repeated indices is used. Greek and latin indices belong to the sets $\{1, 2\}$ and $\{1, 2, 3\}$ respectively.

2 Generalities on the Koiter Shell Model

Let Ω be a bounded open set of $\mathbb{R}^2$ with smooth boundary Γ. Let $\mathcal{E}^3$ be the Euclidean space referred to the orthonormal frame (O, e_1, e_2, e_3). We consider shell theory within the framework of the Koiter theory and, more precisely, the mathematical framework of this linear theory. The middle surface S of the shell is the image in $\mathcal{E}^3$ of Ω for the map

$$\varphi\ :\ (y^1, y^2) \in \overline{\Omega} \to \varphi(y) \in \mathcal{E}^3.$$

The two tangent vectors of S at any point y are given by

$$a_\alpha = \partial_\alpha \varphi,\ \alpha \in \{1, 2\},$$

where ∂_α denotes the differentiation with respect to y^α, while the unit normal vector is

$$a_3 = \frac{a_1 \wedge a_2}{\|a_1 \wedge a_2\|}.$$

For the sake of simplicity, we omitted y in the previous notation $(a_\alpha(y))$.

The middle surface S is assumed to be smooth $(\mathcal{C}^\infty)$, and we may consider in a neighborhood of it a system of "normal coordinates" y^1, y^2, y^3, when y^3 is the normal distance to S. More precisely, we consider a shell of constant thickness ε, i.e., it is the set

$$C = \Big\{M \in \mathcal{E}^3,\ M = \varphi(y^1, y^2) + y^3 a_3,\ (y^1, y^2) \in \Omega, -\frac{1}{2}\varepsilon < y^3 < \frac{1}{2}\varepsilon\Big\}.$$

Under these conditions, let $u = u(y^1, y^2)$ be the displacement vector of the middle surface of the shell. In the linear theory of shells, which is our framework here, the displacement vector is assumed to describe the first order term of the mathematical expression as the thickness ε is small (cf. [4, 18]).

Remark 2.1. In the sequel, "smooth" should be understood in the sense of $\mathcal{C}^\infty$.

Remark 2.2. We consider here the case where the surface is defined by only one chart, but this could be easily generalized to the case of several charts (atlas).

More precisely, since we consider the case where u is supposed to be small, the Koiter theory is described in terms of the *deformation tensor* (or strain tensor) $\gamma_{\alpha\beta}$ of the middle surface:

$$\gamma_{\alpha\beta} = \frac{1}{2}(\widetilde{a}_{\alpha\beta} - a_{\alpha\beta})$$

and the *change of curvature tensor* $\rho_{\alpha\beta}$:

$$\rho_{\alpha\beta} = \widetilde{b}_{\alpha\beta} - b_{\alpha\beta}.$$

In the previous definitions, the expressions $a_{\alpha\beta}$ (respectively, $\widetilde{a}_{\alpha\beta}$) denote the coefficients of the first fundamental form of the middle surface before (respectively, after) deformation:

$$a_{\alpha\beta} = a_\alpha \cdot a_\beta = \partial_\alpha\varphi \cdot \partial_\beta\varphi,$$

and $b_{\alpha\beta}$ (respectively, $\widetilde{b}_{\alpha\beta}$) the coefficients of the second fundamental form accounting for the curvatures before (respectively, after) deformation:

$$b_{\alpha\beta} = -a_\alpha \cdot \partial_\beta a_3 = a_3 \cdot \partial_\beta a_\alpha = a_3 \cdot \partial_\alpha a_\beta = b_{\beta\alpha},$$

due to the fact that $a_\alpha \cdot a_3 = 0$.

The dual basis a^i is defined by

$$a_i \cdot a^j = \delta_i^j,$$

where δ denotes the Kronecker symbol. The contravariant components a^{ij} of the metric tensor are

$$a^{ij} = a^i \cdot a^j,$$

and a_{ij} are used to write covariant components of vectors and tensors in the usual way. Finally, the tensors γ and ρ take the form

$$\gamma_{\beta\alpha}(u) = \gamma_{\alpha\beta}(u) = \frac{1}{2}(u_{\alpha|\beta} + u_{\beta|\alpha}) - b_{\alpha\beta}u_3, \tag{2.1}$$

$$\rho_{\alpha\beta}(u) = u_{3|\alpha\beta} + b^\lambda_{\beta|\alpha}u_\lambda + b^\lambda_\beta u_{\lambda|\alpha} + b^\lambda_\alpha u_{\lambda|\beta} - b^\lambda_\alpha b_{\lambda\beta}u_3, \tag{2.2}$$

where $\partial_\alpha a_3 = b^\gamma_\alpha a_\gamma$, $b^\beta_\alpha = a^{\beta\sigma}b_{\alpha\sigma}$, $\cdot_{|\alpha}$ denotes the *covariant differentiation* which is defined by

$$\begin{aligned} u_{\alpha|\beta} &= \partial_\beta u_\alpha - \Gamma^\lambda_{\alpha\beta} u_\lambda, \\ u_{3|\beta} &= \partial_\beta u_3 \end{aligned} \tag{2.3}$$

and

$$\begin{aligned} b^\lambda_{\alpha|\beta} &= \partial_\alpha b^\lambda_\beta + \Gamma^\lambda_{\alpha\nu} b^\nu_\beta - \Gamma^\nu_{\beta\alpha} b^\lambda_\nu = b^\lambda_{\beta|\alpha}, \\ u_{3|\alpha\beta} &= \partial_{\alpha\beta} u_3 - \Gamma^\lambda_{\alpha\beta} \partial_\lambda u_3, \end{aligned} \tag{2.4}$$

where $\Gamma^\alpha_{\beta\gamma}$ are the Christoffel symbols of the surface

$$\Gamma^\alpha_{\beta\gamma} = \Gamma^\alpha_{\gamma\beta} = a^\alpha \cdot \partial_\beta a_\gamma = a^\alpha \cdot \partial_\gamma a_\beta.$$

Let us now define the energy of the shell within the Koiter framework. It consists of two bilinear forms a and b: a corresponds to a *membrane strain energy* and b is a *bending energy* (which acts as a perturbation term). More precisely, a is defined by

$$a(u, v) = \int_S A^{\alpha\beta\lambda\mu} \gamma_{\lambda\mu}(u) \gamma_{\alpha\beta}(\overline{v}) \mathrm{d}s, \tag{2.5}$$

where $A^{\alpha\beta\lambda\mu}$ are the *membrane rigidity* coefficients which we assume to be smooth in Ω. Moreover, we assume that some symmetry holds:

$$A^{\alpha\beta\lambda\mu} = A^{\lambda\mu\alpha\beta} = A^{\mu\lambda\alpha\beta}. \tag{2.6}$$

Defining the *membrane stress tensors* by

$$T^{\alpha\beta}(u) = A^{\alpha\beta\lambda\mu} \gamma_{\lambda\mu}(u) \tag{2.7}$$

and using the symmetry of γ, we immediately see that

$$T^{\alpha\beta}(u) = T^{\beta\alpha}(u) \tag{2.8}$$

and

$$a(u, v) = \int_S T^{\alpha\beta}(u) \gamma_{\alpha\beta}(\overline{v}) \mathrm{d}s = \int_S \gamma_{\alpha\beta}(u) T^{\alpha\beta}(\overline{v}) \mathrm{d}s. \tag{2.9}$$

Furthermore, we assume that the coercivity condition holds uniformly on the surface:

$$A^{\alpha\beta\lambda\mu} \xi_{\alpha\beta} \xi_{\lambda\mu} \geqslant C \|\xi\|^2, \ C > 0. \tag{2.10}$$

Remark 2.3. Note that there are two different symmetries on A: the first one $A^{\alpha\beta\lambda\mu} = A^{\lambda\mu\alpha\beta}$ is necessary to exchange u and v in (2.9), while the second $A^{\lambda\mu\alpha\beta} = A^{\mu\lambda\alpha\beta}$ is used to obtain (2.8), but is not necessary in order to obtain (2.9) since we could use the symmetry of γ.

Analogously, we define the bilinear form b which corresponds to the bending energy of the shell and which will act as a perturbation term:

$$b(u,v) = \int_S B^{\alpha\beta\lambda\mu}\rho_{\lambda\mu}(u)\rho_{\alpha\beta}(\overline{v})\mathrm{d}s, \tag{2.11}$$

where $B^{\alpha\beta\lambda\mu}$ are the *bending rigidity* coefficients which we assume to be smooth in Ω and to have the same properties (2.6) and (2.10) as A, namely,

$$B^{\alpha\beta\lambda\mu} = B^{\lambda\mu\alpha\beta} = B^{\mu\lambda\alpha\beta} \tag{2.12}$$

and

$$B^{\alpha\beta\lambda\mu}\xi_{\alpha\beta}\xi_{\lambda\mu} \geqslant C\|\xi\|^2 \tag{2.13}$$

uniformly on the surface.

Similarly to a we can write

$$b(u,v) = \int_S M^{\alpha\beta}(u)\rho_{\alpha\beta}(\overline{v})\mathrm{d}s, \tag{2.14}$$

where the *bending stress tensors* are

$$M^{\alpha\beta}(u) = B^{\alpha\beta\lambda\mu}\rho_{\lambda\mu}(u). \tag{2.15}$$

In this paper, we restrict ourselves to the case of an elliptic surface, i.e., we always assume that the coefficients $b_{\alpha\beta}$ are such that

$$b_{11}b_{22} - b_{12}^2 > 0 \quad \text{uniformly on } S \text{ and } b_{11} > 0. \tag{2.16}$$

Let us finish this introduction by topoligical considerations, the boundary $\partial\Omega = \Gamma_0 \cup \Gamma_1$ is assumed to be smooth (i.e., of class $\mathcal{C}^\infty$) in the variable $y = (y^1, y^2)$, where Γ_0 and Γ_1 are disjoint; they are one-dimensional compact smooth manifolds without boundary, then diffeomorphic to the unit circle.

We consider the following variational problem (which has possibly only a formal sense):

$$\begin{gathered} \text{Find } u^\varepsilon \in V \text{ such that } \forall\, v \in V \\ a(u^\varepsilon, v) + \varepsilon^2 b(u^\varepsilon, v) = \langle f, v\rangle \end{gathered} \tag{2.17}$$

with a and b defined by (2.5) and (2.14), where the space V is the "energy space" with the essential boundary conditions on Γ_0

$$V = \{v;\ v_\alpha \in H^1(\Omega),\ v_3 \in H^2(\Omega);\ v_{|\Gamma_0} = 0 \text{ in the sense of trace}\}. \tag{2.18}$$

Remark 2.4. The essential boundary conditions on Γ_0 (2.18) corresponds to the case of the fixed boundary of the shell. Other boundary conditions could have been considered such as

$$\begin{aligned} V = \{v;\ v_\alpha \in H^1(\Omega),\ v_3 \in H^2(\Omega);\ v_{|\Gamma_0} = 0,& \\ \partial_\nu v_{3|\Gamma_0} = 0 \text{ in the sense of trace}\},& \end{aligned} \tag{2.19}$$

where ν is the normal to Γ_0 (i.e., the normal to the boundary which lies in the tangent plane), which corresponds to the clamped case.

The following lemma was obtained by Bernardou and Ciarlet (cf. [4]).

Lemma 2.1. *The bilinear form $a + b$ is coercive on V.*

We denote by V' the dual space of V. Here, "dual" is obviously understood in the abstract sense of the space of continuous linear functionals on V. In order to make explicit computations in terms of equation and boundary conditions, we often take f as a "function" defined in Ω in the space

$$\{f \in H^{-1}(\Omega;\mathbb{R}) \times H^{-1}(\Omega;\mathbb{R}) \times H^{-2}(\Omega;\mathbb{R}); \\ f \text{ "smooth" in a neighborhood of } \Gamma_1\} \subset V', \tag{2.20}$$

where "smooth" means allowing classical integration by parts. It is obvious that other choices for f are possible.

Moreover, we immediately obtain the following result.

Proposition 2.1. *For $\varepsilon > 0$ and f in V' the variational problem* (2.17) *is of Lax–Milgram type and is a selfadjoint problem which has a coerciveness constant larger than $c\varepsilon^2$, with $c > 0$.*

Remark 2.5. Note that the coerciveness of the previous problem disappears when $\varepsilon = 0$.

3 The Ellipticity of Systems and the Shapiro–Lopatinskii Condition

In this section, we recall some classical results on the linear boundary value problems for elliptic systems in the sense of Douglis and Nirenberg [6]. We begin with the definition of ellipticity for systems, then we recall the Shapiro–Lopatinskii condition. This latter condition states which boundary conditions are well suited in order to have well posed problems for elliptic systems. We then recall in what sense an elliptic system with Shapiro–Lopatinskii condition is "well behaved."

For brevity, from now on we abbreviate SL for the Shapiro–Lopatinskii condition.

3.1 Elliptic systems in the sense of Douglis and Nirenberg

In this paper, we deal with systems of l ($l = 3$ or $l = 6$) equations with 3 unknowns (noted here u_1, u_2, u_3) defined on an open set $\Omega \subset \mathbb{R}^2$ with smooth boundary, which has the form

$$l_{kj} u_j = f_k, \quad k = 1, \ldots, l, \tag{3.1}$$

or, equivalently,

$$L\mathbf{u} = f.$$

The coefficients $l_{kj}(x, D)$ with $D = (D_1, D_2)$ and $D_l = -i\frac{\partial}{\partial x_l}$, $l \in \{1, 2\}$, are linear differential operators with real smooth coefficients. In our systems (3.1), the highest order of differentiation is different for the three unknowns and depends on the equation. A way to take into account such differences between the various equations and unknowns is to define integer indices (s_1, s_2, s_3) attached to the equations and integer indices (t_1, t_2, t_3) attached to the unknowns (cf. [6]) so that the "higher order terms" (which will be called "principal terms") are in equation j the terms where each unknown "k" appears by its derivative of order $s_k + t_j$. More precisely, the integers (s_k, t_j) are such that

if $s_k + t_j \geqslant 0$, then the order of l_{kj} is less than or equal to $s_k + t_j$,

if $s_k + t_j < 0$, then l_{kj} is equal to zero.

The *principal part* l'_{kj} of l_{kj} is obtained by keeping the terms of order $s_k + t_j$ if $s_k + t_j \geqslant 0$ and by taking $l'_{kj} = 0$ if $s_k + t_j < 0$. The matrix $L'(x, \xi)$, $\xi = (\xi_1, \xi_2) \in \mathbb{R}^2$, obtained by substituting ξ_α for D_α in l'_{kj}, is called the *principal symbol of the system.* Since l'_{kj} are homogeneous of order $s_k + t_j$ with respect to ξ_α, the determinant of the matrix $L'(x, \xi)$, denoted by $\mathrm{D}(x, \xi)$, is homogeneous of degree $\Sigma_k s_k + \Sigma_j t_j$.

Definition 3.1. The system (3.1) is *elliptic in the sense of Douglis and Nirenberg* at a point $x \in \Omega$ if

$$\mathrm{D}(x, \xi) \neq 0 \quad \forall\, \xi \in \mathbb{R}^2 \setminus \{0\}. \tag{3.2}$$

Remark 3.1. Since the coefficients are assumed to be real, the function $\mathrm{D}(x, \xi)$ for an elliptic system is even in ξ of order $2m$ with

$$\Sigma_k s_k + \Sigma_j t_j = 2m.$$

Remark 3.2. The definition of the indices s_j and t_k for a system is slightly ambiguous. Indeed, the result is exactly the same after adding an integer n to the indices s_j and subtracting n from t_k.

Remark 3.3. Let $x_0 \in \Omega$ be such that the system (3.1) is not elliptic. Then there exists $\xi \in \mathbb{R}^2 \setminus \{0\}$ such that $\mathrm{D}(x_0, \xi) = 0$. In such a case, the system $L'(x_0, D)u = 0$, with frozen coefficients at x_0 admits a solution of the form $u(x) = ve^{i\xi x}$, with $v \in \mathbb{R}^3 \setminus \{0\}$.

Remark 3.4. Moreover, throughout this paper, *ellipticity* will be understood in the sequel as uniform, i.e., there exists a positive constant A such that

$$A^{-1}\Sigma_\alpha|\xi_\alpha|^2 \leqslant |\det L'(x,\xi)| \leqslant A\Sigma_\alpha|\xi_\alpha|^2$$

for all $x \in \Omega$ and $\xi = (\xi_1, \xi_2) \in \mathbb{R}^2$.

3.2 Shapiro–Lopatinskii conditions for elliptic systems in the sense of Douglis and Nirenberg

From now on, for the sake of simplicity, we say that a system is *elliptic* when it is elliptic in the sense of Douglis and Nirenberg [6].

Let l_{kj} (L) be an elliptic system of order $2m$ with principal part l'_{kj} (L'), and let m be boundary conditions given by

$$b_{kj}u_j = g_k, \quad k \in \{1,\dots,m\},$$

where $b_{kj}(x,D)$ are differential operators with smooth coefficients. Let us define the integers r_k (indices of the boundary conditions, $k = 1,\dots,m$) such that

if $r_k + t_j \geqslant 0$, then the order of b_{kj} is less than or equal to $r_k + t_j$,

if $r_k + t_j < 0$, b_{kj} is equal to zero.

The principal part b'_{kj} is b_{kj} if $r_k + t_j \geqslant 0$ and zero otherwise.

Assume that the smooth real coefficients are defined in $\overline{\Omega}$.

Let $x_0 \in \Gamma$. We assume that L' is elliptic at x_0. Usually (cf. [1] and [8] for instance), the SL condition at x_0 is defined via a local diffeomorphism sending a neighborhood of x_0 in Ω into a neighborhood of the origin in a half-plane. For ulterior computations, it is worth-while to take a special diffeomorphism which amounts to taking locally Cartesian coordinates x_1, x_2, respectively, tangent and (inward) normal to the boundary at x_0. We then consider only the principal parts of the equations and of the boundary conditions frozen at x_0. Next, we consider the corresponding boundary value problem obtained by formal tangential Fourier transform (i.e., $D_1 \to \xi_1$, with $\xi_1 \in \mathbb{R}$ and $u \to \widetilde{u}$) which amounts to the following algebraic conditions:

$$\begin{aligned} \widetilde{l'_{kj}}(x_0,\xi_1,D_2)\widetilde{u} &= 0 \quad \text{for} \quad x_2 > 0, \\ \widetilde{b'_{kj}}(x_0,\xi_1,D_2)\widetilde{u} &= \widetilde{g}_j \quad \text{for} \quad x_2 = 0, \end{aligned} \tag{3.3}$$

$j,k \in \{1,\dots,m\}$ (cf. [7, Section 3.2] for details, if necessary).

The problem (3.3) involves a system of ordinary differential equations with constant coefficients of the variable $x_2 \in \mathbb{R}^+$ and m boundary conditions at

$x_2 = 0$, whose solutions are classically a linear combination of terms of the form

$$\widetilde{u}(\xi_1, x_2) = \begin{cases} v e^{i\xi_2 x_2}, & v \in \mathbb{C}^3, \\ P(x_2) e^{i\xi_2 x_2}, & P \text{ is a polynomial} \\ & \text{in the case of Jordan block.} \end{cases} \tag{3.4}$$

Recalling that the system L is elliptic, it follows that the imaginary part of ξ_2 does not vanish. Furthermore, there are m solutions ξ_2 of $D(x_0, \xi_1, \xi_2) = 0$ with positive imaginary part that we denote ξ_2^+ (and m with negative imaginary part denoted by ξ_2^-).

We then try to solve (3.3) using only linear combinations of the m solutions of the form (3.4) for the m roots ξ_2^+ (i.e., exponentially decreasing towards the domain).

Definition 3.2. The SL *condition* is satisfied at $x_0 \in \Gamma$ if one of the following equivalent conditions holds:

1. The solution of the previous problem is defined uniquely.
2. Zero is the only solution of the homogeneous (i.e., with $g_j = 0$) previous problem.

Remark 3.5. The two conditions (which are equivalent) of the previous definition are clearly equivalent to the nonvanishing of the determinant of the corresponding algebraic "system."

Remark 3.6. The reason for defining the SL condition amounts to the possibility of solving the problem in a half-plane via tangential Fourier transform. The reason for not considering the ξ_2^- roots is that, for $x_2 > 0$, they should give exponentially growing Fourier transforms in $x_1 \to \xi_1$, which are not allowed in distribution theory (note that ξ_1 and ξ_2 are proportional since $D(\xi)$ is homogeneous).

The verification of the SL condition is often tricky. In some situations, we can use equivalent expressions which are simpler to treat. More precisely, define the function u by $u(x_1, x_2) = \widetilde{u}(\xi_1, x_2) e^{i|\xi_1| x_1}$, with $\widetilde{u}(\xi_1, x_2) = v e^{i\xi_2^+ x_2}$ (or expressed as exponential polynomial in the case of Jordan block), it is an exponentially decreasing function in the direction inwards the domain (as $x_2 \to +\infty$), it is also a periodic function in the tangential direction x_1 and it satisfies

$$\begin{aligned} \widetilde{l'_{kj}}(x_0, D_1, D_2) u &= 0 \quad \text{for} \quad x_2 > 0, \\ \widetilde{b'_{kj}}(x_0, D_1, D_2) u &= g_j \quad \text{for} \quad x_2 = 0, \end{aligned} \tag{3.5}$$

$j, k \in \{1, \ldots, m\}$.

The following proposition is very useful in the case where ellipticity is linked with positive energy integrals obtained by integrating by parts. For instance, we have the following assertion.

Proposition 3.1. *Consider the homogeneous problem associated with* (3.5) *(i.e., taking* $g_j = 0$*) for* $x_0 \in \Gamma$*. If any solution* u*, which is periodic in the tangential direction* x_1 *and exponentially decreasing in the direction* x_2 *inwards the domain, is zero, then the* SL *condition is satisfied.*

Remark 3.7. In order to have well-posed problems for elliptic systems, boundary conditions satisfying the SL condition should be prescribed at any points of the boundary. Their number is half the total order of the system.

Remark 3.8. The specific boundary conditions may differ from a point to another on the boundary. In particular, each connected component of the boundary may have its own set of boundary conditions. Otherwise, local changes of boundary conditions (as well as the non-smoothness of the boundary) induces local singularities.

3.3 Some Results for "Well-Posed" Elliptic Systems

Let us now consider a boundary value problem formed by an elliptic system with boundary conditions satisfying the SL condition. In what sense is this problem "well-behaved?" The obvious example of an eigenvalue problem, even for an equation shows that uniqueness is only ensured up to the kernel formed by the eigenvectors associated with the zero eigenvalue, whereas existence involves compatibility conditions (orthogonality to the kernel of the adjoint problem). The general results are those of Agmon, Douglis, and Nirenberg [1].

First, let us recall the definition of a Fredholm operator.

Definition 3.3. Let E and F be two Hilbert spaces and A an operator (closed with dense domain in E) from E into F. We say that A is a *Fredholm operator* if the following three conditions hold:

1. $\mathrm{Ker}(A)$ is of finite dimension,
2. $\mathrm{R}(A)$ is closed,
3. $\mathrm{R}(A)$ is of finite codimension.

The operator A is also said to be an *index operator*, the *index* is defined as $\dim \mathrm{Ker}(A) - \mathrm{codim}\, \mathrm{R}(A)$.

Let us consider an elliptic system of order $2m$ whose coefficients are smooth:

$$\begin{aligned} l_{kj}u_j &= f_k, \quad j,k \in \{1,\dots,l\} \text{ in } \Omega, \\ b_{hj}u_j &= g_h, \quad h \in \{1,\dots,m\} \text{ on } \partial\Omega, \end{aligned} \tag{3.6}$$

whose indices associated with unknowns, equations, and boundary conditions are t_j, s_j, and r_j respectively. Let ρ be a "big enough" real number, called *regularity index*. Consider the operator (3.6) as a linear operator from the space E to the space F defined by

$$E = \Pi_{j=1}^{l} H^{\rho+t_j}(\Omega),$$
$$F = \Pi_{j=1}^{l} H^{\rho-s_j}(\Omega) \times \Pi_{j=1}^{m} H^{\rho-r_j-\frac{1}{2}}(\partial\Omega). \tag{3.7}$$

The real ρ is chosen in order to give a sense to the traces which are involved, i.e., it is such that $\rho - r_j - 1/2 > 0$ for $j \in \{1, \ldots, m\}$.

The following assertion is the main result of the theory of Agmon, Douglis, and Nirenberg.

Theorem 3.1 (Agmon, Douglis, and Nirenberg [1]). *Let Ω be a bounded open set with smooth boundary Γ. Consider an elliptic system with boundary conditions satisfying the* SL *condition everywhere on Γ. Assume that the coefficients of the system are smooth and that u, f, and g satisfy* (3.6). *Then the following estimate holds:*

$$\|u\|_E \leqslant C(\|(f,g)\|_F + \|u\|_{(L^2(\Omega))^l}), \tag{3.8}$$

where C does not depend on u, f, and g. Moreover, the operator defined by (3.6) *from the space E to the space F, given by* (3.7), *is a Fredholm operator for all value of ρ such that $\rho - r_j - 1/2 > 0$ for $j \in \{1, \ldots, m\}$. Furthermore, the dimension of the kernel and the dimension of the subspace orthogonal to the range do not depend on ρ. The kernel is composed of smooth functions.*

Remark 3.9. The previous theorem means that, in general, the existence and uniqueness of the solution only hold up to a finite number of compatibility conditions for f and g and the existence of a solution holds up to a finite-dimensional kernel. More precise properties need specific properties of the system.

Remark 3.10. For all values of ρ the kernel formed by the eigenvectors corresponding to the eigenvalue 0 is of finite dimension and is composed of smooth functions, independent of ρ (in $\mathcal{C}^\infty(\overline{\Omega})$).

Remark 3.11. Denote by A the operator defined by (3.6) in the spaces E and F. Let us consider the case where $\dim \operatorname{Ker}(A) > 0$ and define the inverse B of A as a closed operator from $\mathrm{R}(A)$ to $E/\operatorname{Ker}(A)$. We have

$$\|\widetilde{u}\|_{E/\operatorname{Ker}(A)} \leqslant C\|(f,g)\|_F, \tag{3.9}$$

where $\widetilde{u}$ is an element of the equivalence class of u.

The element $\widetilde{u}$ can also be viewed as an element of the orthogonal of $\operatorname{Ker}(A)$ in E, which is identified with $E/\operatorname{Ker}(A)$. In such a case, there exists a unique $(\widetilde{u}, \widehat{u}) \in E/\operatorname{Ker}(A) \times \operatorname{Ker}(A)$ such that

$$u = \widetilde{u} + \widehat{u}.$$

Since Ker(A) is of finite dimension, all the norms are equivalent and we can choose for $\widehat{u}$ a norm in a space $H^{-\nu}$ with ν very big. Therefore, the inequality (3.8) can be rewritten as

$$\|u\|_E \leqslant C(\|\widetilde{u}\|_{E/\mathrm{Ker}(A)} + \|\widehat{u}\|_{H^{-\nu}}) \leqslant C(\|\widetilde{u}\|_{E/\mathrm{Ker}(A)} + \|u\|_{H^{-\nu}}), \quad (3.10)$$

for ν big enough such that $E \subset H^{-\nu}$. Recalling (3.9), we then deduce that

$$\|u\|_E \leqslant C(\|(f,g)\|_F + \|\widehat{u}\|_{H^{-\nu}}). \quad (3.11)$$

Moreover, the norm in $H^{-\nu}$ may be replaced by a seminorm provided that it is a norm on Ker(A).

Remark 3.12. In the case where dim Ker$(A) = 0$, the inverse B of the operator A is well defined on $R(A)$. It is a closed operator, hence it is bounded and the following estimate holds:

$$\|u\|_E \leqslant C\|(f,g)\|_F. \quad (3.12)$$

4 Study of Four Systems Involved in Shell Theory

In this section, we study four systems, called the *rigidity* system, the *membrane tension* system, the *membrane* system, and the *Koiter shell* system, which will appear in the sequel. We prove that these four systems satisfy the ellipticity condition, and we study some boundary conditions. Note that the boundary conditions may be different on Γ_0 and Γ_1 which are supposed to be disjoints.

Let us recall the situation: Ω is a connected bounded open set of $\mathbb{R}^2$ with $\mathcal{C}^\infty$ boundary $\Gamma = \Gamma_0 \cup \Gamma_1$ and $\Gamma_0 \cap \Gamma_1 = \varnothing$. The middle surface S of the shell is the image in $\mathcal{E}^3$ of Ω for the map

$$\varphi : (y^1, y^2) \in \overline{\Omega} \to \varphi(y) \in \mathcal{E}^3.$$

We assume that the ellipticity assumption of the surface holds:

$$b_{11}b_{22} - b_{12}^2 > 0 \quad \text{uniformly in} \quad \Omega.$$

4.1 The rigidity system

Let us begin with the *rigidity* system defined by $\gamma_{\alpha\beta}(u)$:

$$\begin{aligned}
\gamma_{11}(u) &:= \partial_1 u_1 - \Gamma_{11}^{\alpha} u_\alpha - b_{11} u_3, \\
\gamma_{22}(u) &:= \partial_2 u_2 - \Gamma_{22}^{\alpha} u_\alpha - b_{22} u_3, \\
\gamma_{12}(u) &:= \frac{1}{2}(\partial_2 u_1 + \partial_1 u_2) - \Gamma_{12}^{\alpha} u_\alpha - b_{12} u_3.
\end{aligned} \tag{4.1}$$

It is clear that u_α and u_3 play very different roles since u_α appears with derivatives, whereas u_3 only appears without. Therefore, take $(1,1,0)$ as the indices of the unknowns (u_1,u_2,u_3) and $(0,0,0)$ as equation indices in the order $(\gamma_{11},\gamma_{22},\gamma_{12})$. The principal system is obtained by substituting 0 for $\Gamma_{\lambda\mu}^{\alpha}$, but keeping $b_{\lambda\mu}$.

Lemma 4.1. *Under the ellipticity assumption of the surface* (2.16), *the rigidity system γ is elliptic of total order* 2 *in Ω.*

Proof. Substituting $-i\xi_\alpha$ for ∂_α in the principal system, we obtain a system whose determinant is $D(x,\xi) = 2b_{12}\xi_1\xi_2 - b_{22}\xi_1^2 - b_{11}\xi_2^2$. Hence, by the ellipticity assumption (2.16), for all $x \in \Omega$ we have $D(x,\xi) > 0$. □

4.1.1 Cauchy boundary conditions

It is classical that the Cauchy problem associated with an elliptic system is not well posed in the sense that it does not enjoy the existence, uniqueness, and stability of solutions. Nevertheless, the Cauchy problem associated with the rigidity system will be involved in the sequel, and we study it now. In particular, we need the following uniqueness assertion for solutions $u \in H^1 \times H^1 \times L^2$.

Lemma 4.2. *Under the ellipticity assumption of the surface (2.16), the system $\gamma_{\alpha\beta}(u) = 0$ in Ω with the boundary conditions $u_1 = u_2 = 0$ on a part of the boundary (of positive measure) admits a unique solution which is $u = 0$.*

Proof. Let us assume that $v \in H^1(\Omega) \times H^1(\Omega) \times H^2(\Omega)$ is such that $\gamma_{\alpha\beta}(v) = 0$ and $v_1 = v_2 = 0$ on a part of the boundary. By the ellipticity assumption (2.16), we know that $b_{11} \neq 0$ in $\overline{\Omega}$. We can eliminate v_3 from the first and third equations ($\gamma_{11}(v) = 0$ and $\gamma_{22}(v) = 0$) of the system γ. This yields the system of two equations for two unknowns (v_1, v_2):

$$\begin{aligned}
0 &= \partial_2 v_2 - \Gamma_{22}^{\alpha} v_\alpha - \frac{b_{22}}{b_{11}}(\partial_1 v_1 - \Gamma_{11}^{\alpha} v_\alpha), \\
0 &= \frac{1}{2}(\partial_2 v_1 + \partial_1 v_2) - \Gamma_{12}^{\alpha} v_\alpha - \frac{b_{12}}{b_{11}}(\partial_1 v_1 - \Gamma_{11}^{\alpha} v_\alpha).
\end{aligned} \tag{4.2}$$

The eliminated unknown being then given by

$$v_3 = \frac{1}{b_{11}}(\partial_1 v_1 - \Gamma_{11}^{\alpha} v_\alpha).$$

The problem then reduces to the uniqueness problem in the class $H^1(\Omega)$ of (v_1, v_2) satisfying

$$\begin{aligned} &\partial_1 v_1 - b_{11} v_3 = 0, \\ &\partial_2 v_2 - b_{22} v_3 = 0, \\ &\frac{1}{2}(\partial_2 v_1 + \partial_1 v_2) - b_{12} v_3 = 0, \end{aligned} \tag{4.3}$$

with $v_1 = v_2 = 0$ on a part of the boundary. This problem is more or less classical. Under analyticity hypotheses about the coefficients and the boundary, the uniqueness follows from the Holmgren local uniqueness theorem and analytic continuation (since u_1 and u_2 are, in this case, analytic inside Ω). Under the $\mathcal{C}^\infty$ hypotheses adopted here, the uniqueness follows from the theory of pseudoanalytic functions. There are two nearly equivalent theories of such functions attached to the names of L. Bers (cf., for example, the supplement of Chapter IV of [5], written by Bers himself) and I.N. Vekua [21].

Let (v_1, v_2) be a solution of (4.3) vanishing on a part Γ of the boundary. Let $(\widetilde{v}_1, \widetilde{v}_2)$ be an extension of (v_1, v_2) with values zero to an extended domain across Γ. Classically, $(\widetilde{v}_1, \widetilde{v}_2)$ satisfies the same system (4.3) on the extended domain and, according to the interior regularity theory for elliptic systems, is of class $\mathcal{C}^\infty$ inside it. The function $\widetilde{w} = \widetilde{v}_1 + i\widetilde{v}_2$ is pseudoanalytic, belongs to the class $\mathcal{C}^\infty$, and vanishes on the outer region of the extended domain. We then use either Theorem 3.5 of [21, p. 146], which gives directly the uniqueness, or the representation theorem of [5, p. 379]. In this case, $\widetilde{w}(z)$ admits the expression (here $z = x_1 + ix_2$):

$$\widetilde{w}(z) = e^{\delta(z)} f(z),$$

where $f(z)$ is analytic and $\delta(z)$ is continuous. Since $e^{\delta(z)}$ vanishes nowhere, the uniqueness follows. □

Remark 4.1. Strictly speaking, the evoked theorems of pseudoanalytic functions apply to systems with principal part of the canonical form

$$\begin{aligned} \partial_1 v_1 - \partial_2 v_2 &= \dots , \\ \partial_2 v_1 + \partial_1 v_2 &= \dots , \end{aligned} \tag{4.4}$$

so that the classical reduction to this form (cf., for example, [5, p. 169-170]) should be previously considered. But it is obvious that this does not modify the $\mathcal{C}^\infty$ regularity inside the domain.

Let us make several comments about this uniqueness result.

Remark 4.2. This result, known as the infinitesimal rigidity of the surface, does not depend on the curvilinear coordinates.

Remark 4.3. The key ingredients of the previous uniqueness result are a uniqueness theorem for the Cauchy problem for elliptic systems of two equations of order 1. It is not based upon a coercivity assumption for an elliptic system. But we know that the Cauchy problem for elliptic systems is precarious in the sense that it does not enjoy existence, uniqueness and stability of solutions. This means that such a system could lead to instability in the sense that there could exist v_1, v_2, v_3 very "big" in usual spaces such that $\gamma_{\alpha\beta}(v)$ are very "small."

4.1.2 Boundary value problems for the *rigidity* system

From now on, we consider the frame (O, a_1, a_2, a_3) to be orthonormal on the boundary and such that $u_t = (u_1, 0, 0)$ and $u_n = (0, u_2, 0)$, where u_t denotes the component of u in the tangential direction to the boundary and u_n is the component of u in the normal direction to the boundary and in the tangent plane. This point which is not absolutely necessary implies a special local parametrization.

Lemma 4.3. *The boundary condition $u_1 = g$ satisfies the* SL *condition for the system γ.*

Proof. We take as index of the boundary condition $r = -1$. Let x_0 belong to Γ. As explained in Section 3.2, using a partition of unity, local mappings, with axes y_1 tangential and y_2 inwards Γ, and dropping lower order differential terms, we obtain the new system

$$\text{For } y_2 > 0 \quad \begin{cases} \partial_1 u_1 - b_{11} u_3 = 0, \\ \partial_2 u_2 - b_{22} u_3 = 0, \\ \frac{1}{2}(\partial_2 u_1 + \partial_1 u_2) - b_{12} u_3 = 0. \end{cases} \tag{4.5}$$

We look for solutions which are exponentially decreasing as $y_2 \to +\infty$ of the form

$$u(y_1, y_2) = U e^{i\zeta y_2 + i\xi_1 y_1}, \ \xi_1 \in \mathbb{R} \setminus \{0\},$$

with $U = \begin{pmatrix} U_1 \\ U_2 \\ U_3 \end{pmatrix} \in \mathbb{C}^3$, $\mathrm{Im}(\zeta) > 0$. Substituting this solution into (4.5) and using the boundary condition, we have $U_1 = 0$. Consequently, $u_1 = 0$ everywhere and (4.5) gives also $u_2 = u_3 = 0$. □

Remark 4.4. Similarly to the proof of the previous result, we can prove that the following boundary conditions satisfy the SL condition:

1. $u_2 = g$ (take $r = -1$).
2. $u_3 = g$ (take $r = 0$).

Remark 4.5. Since Γ_0 and Γ_1 are disjoints and, by the previous statements, the boundary value problem

$$\begin{aligned} &\gamma_{\alpha\beta}(u) = 0 \ \text{ in } \ \Omega, \\ &u_2 = 0 \ \text{ on } \ \Gamma_0, \\ &u_3 = \widetilde{u} \ \text{ on } \ \Gamma_1. \end{aligned} \tag{4.6}$$

is "well posed" in the Agmon, Douglis, and Nirenberg sense. By Theorem 3.1 and Remark 3.9, together with the standard regularity theory for elliptic systems, it follows that u is of class C^∞ in $\Omega \cup \Gamma_0$ for any $\widetilde{u}$ (either smooth or not). Consequently, up to a kernel of finite dimension composed of smooth functions belonging to $\mathcal{C}^\infty(\overline{\Omega})^3$ (and, eventually, up to a compatibility condition (to belong to the range of the operator which is a closed subspace of finite codimension), the space $\{v,\ \gamma_{\alpha\beta}(v) = 0 \text{ in } \Omega,\ v_n = 0 \text{ on } \Gamma_0\}$ is isomorphic to the space $\mathcal{C}^\infty(\Gamma_1)$. The previous statements can be rephrased as follows: up to a finite-dimensional space composed of smooth functions, the space $\{v,\ \gamma_{\alpha\beta}(v) = 0 \text{ in } \Omega,\ v_n = 0 \text{ in } \Gamma_0\}$ is isomorphic to the space of traces on Γ_1:

$$\{\widetilde{v} \in C^\infty(\Gamma_1)\}, \tag{4.7}$$

the isomorphism is obtained by solving (4.5).

In the sequel, we shall consider indifferently the functions v (defined up to an additive element of the kernel) or their traces $\widetilde{v}$ on Γ_1.

4.2 The system of membrane tensions

Consider the *membrane tensions* system $\mathcal{T}$ of three equations with the three unknowns (T^{11}, T^{22}, T^{12}):

$$\begin{aligned} &-T^{11}_{|1} - T^{21}_{|2} = f^1, \\ &-T^{22}_{|2} - T^{21}_{|1} = f^2, \\ &-b_{11}T^{11} - 2b_{12}T^{12} - b_{22}T^{22} = f^3. \end{aligned} \tag{4.8}$$

It is apparent that the three unknowns play analagous roles. Concerning the equations, it is clear that the first and second ones are similar, but different from the third. Therefore, we consider $(1, 1, 0)$ as indices of equations and $(0, 0, 0)$ as indices of unknowns. The principal system $\mathcal{T}_P$ is obtained by replacing the covaraint differentiation $_{|\alpha}$ by the usual differentiation ∂_α (i.e., replacing $\Gamma^\lambda_{\alpha\beta}$ by zero). Proceeding in the same way as in the proof of Lemma 4.1, we obtain the following result.

Lemma 4.4. *Under the ellipticity assumption of the surface* (2.16), *the system $\mathcal{T}$ is elliptic of total order two.*

Remark 4.6. It is worth-while to study the Cauchy problem for the membrane tension system (4.8). This is done exactly as in Subsection 4.1.2 for the rigidity system. We eliminate one of the unknowns, T^{11} for instance and (4.8) reduces to an elliptic system of two first order equations in T^{12} and T^{22}. The Cauchy conditions are $T^{12} = T^{22} = 0$ on a part of the boundary. According to our special frame, this amounts to $T^{\alpha\beta}n_\beta = 0$. This Cauchy problem enjoys uniqueness, but not existence and stability in usual spaces.

Remark 4.7. The system of membrane tensions $\mathcal{T}$ (4.8) and the system of rigidity γ (4.1) are adjoint to each other. This is easily checked by covariant integration by parts on S. Indeed, neglecting boundary terms (we are only interested in the equations) and using (2.1), together with the symmetry of $T^{\alpha\beta}$, we have

$$\begin{aligned}\int_S T^{\alpha\beta}\gamma_{\alpha\beta}(u)\mathrm{d}s &= \int_S T^{\alpha\beta}\Big(\frac{1}{2}(u_{\alpha|\beta} + u_{\beta|\alpha}) - b_{\alpha\beta}u_3\Big)\mathrm{d}s \\ &= \int_S T^{\alpha\beta}\Big(u_{\alpha|\beta} - b_{\alpha\beta}u_3\Big)\mathrm{d}s \\ &= -\int_S \Big(T^{\alpha\beta}_{|\beta}u_\alpha + T^{\alpha\beta}b_{\alpha\beta}u_3\Big)\mathrm{d}s \\ &= \int_S \mathcal{T}(T)u\mathrm{d}s.\end{aligned}$$

4.3 The membrane system

By the *membrane* system we mean the system of three equations with three unknowns $u = (u_1, u_2, u_3)$ obtained from (4.8) when the tensions are written in terms of u, i.e.,

$$\begin{aligned}&- T^{11}_{|1}(u) - T^{21}_{|2}(u) = f^1, \\ &- T^{22}_{|2}(u) \quad T^{21}_{|1}(u) = f^2, \\ &- b_{11}T^{11}(u) - 2b_{12}T^{12}(u) - b_{22}T^{22}(u) = f^3,\end{aligned} \tag{4.9}$$

with

$$T^{\alpha\beta}(u) = A^{\alpha\beta\lambda\mu}\gamma_{\lambda\mu}(u) \tag{4.10}$$

and

$$T^{\alpha\beta}_{|k}(u) = \partial_k T^{\alpha\beta}(u) + \Gamma^\beta_{kn}T^{\alpha n}(u) + \Gamma^\alpha_{km}T^{\beta m}(u). \tag{4.11}$$

To prove the ellipticity of the *membrane* system, we replace it by another, equivalent one. Indeed, we take as unknowns u_1, u_2, u_3 and the supplementary auxiliary unknowns T^{11}, T^{22}, T^{12}. Inverting the matrix $A^{\alpha\beta\lambda\mu}$ in (4.10) and recalling the definition of γ, we obtain the following equivalent system:

$$\begin{aligned} &-T^{11}_{|1} - T^{21}_{|2} = f^1, \\ &-T^{22}_{|2} - T^{21}_{|1} = f^2, \\ &-b_{11}T^{11} - 2b_{12}T^{12} - b_{22}T^{22} = f^3, \end{aligned} \tag{4.12}$$

$$\begin{aligned} &u_{1|1} - b_{11}u_3 - C_{11\alpha\beta}T^{\alpha\beta} = 0, \\ &u_{2|2} - b_{22}u_3 - C_{22\alpha\beta}T^{\alpha\beta} = 0, \\ &\frac{1}{2}(u_{1|2} + u_{2|1}) - b_{12}u_3 - C_{12\alpha\beta}T^{\alpha\beta} = 0, \end{aligned} \tag{4.13}$$

where $C_{\alpha\beta\lambda\mu}$ are the *compliances* (the inverse matrix of $A^{\alpha\beta\lambda\mu}$). The system (4.12) and (4.13) is a system of six equations with the six unknowns $(T^{11}, T^{22}, T^{12}, u_1, u_2, u_3)$ (written in this order). We recognize the *membrane tension* system in (4.12) and the *rigidity* system in (4.13). Consider $(1,1,0,0,0,0)$ as indices of equations and $(0,0,0,1,1,0)$ as indices of unknowns. Then replacing the differentiation ∂_α by $-i\xi_\alpha$ and taking the determinant of the obtained system, we have a determinant of the form

$$\begin{vmatrix} D_{11} & 0 \\ D_{21} & D_{22} \end{vmatrix} = 0 = \left| D_{11} \right| \left| D_{22} \right|,$$

where $D_{\alpha\beta}$ are 3×3 matrices. Moreover, D_{11} and D_{22} are precisely those of the *membrane tension* system and the *rigidity* system respectively, and ellipticity follows. The same result is obviously obtained without using the auxiliary unknowns $T^{\alpha\beta}$. In fact, we have the following assertion.

Lemma 4.5. *Under the ellipticity assumption of the surface* (2.16), *the membrane system with indices* (*of unknowns and of equations*) $(1,1,0)$ $(1,1,0)$ *is elliptic of total order four.*

Let us now state boundary value problems which will be considered later on. Note that only two boundary conditions are considered on Γ_0.

Proposition 4.1. *The boundary value problem*

$$\begin{aligned} &-\partial_1 T^{11}(u) - \partial_2 T^{21}(u) = f^1, \\ &-\partial_2 T^{22}(u) - \partial_1 T^{21}(u) = f^2, \\ &-b_{11}T^{11}(u) - 2b_{12}T^{12}(u) - b_{22}T^{22}(u) = f^3, \\ &u_1 = u_2 = 0 \quad on \quad \Gamma_0, \\ &T^{\alpha\beta}(u)n_\alpha = 0 \quad on \quad \Gamma_1, \quad \beta \in \{1,2\}, \end{aligned} \tag{4.14}$$

with unknown u satisfies the SL *condition on Γ_0, but it does not on Γ_1.*

Remark 4.8. The partial differential boundary value problem (4.14) is formally associated with the variational problem (2.17) when $\varepsilon = 0$.

Proof. Let us fix $x_0 \in \Gamma$. According to the definition of the SL condition, we consider the homogeneous system with constant coefficients in which we only kept the principal terms, i.e., taking $\Gamma^\lambda_{\alpha\beta} = 0$, but $b_{\alpha\beta} \neq 0$ and $f^i = 0$.

After a change of coordinates with local mappings, still denoted by (x_1, x_2), we only have to consider solutions, which are exponentially decreasing in the direction inwards the domain (x_2), of the corresponding boundary value problem obtained by formal tangential Fourier transform. Denoting by $\widetilde{u}(\xi_1, x_2)$ such a solution, by periodicity, we can restrict the domain to the strip $B = (0, 2\pi/|\xi_1|) \times (0, +\infty)$ and we can consider the function

$$v(x_1, x_2) = e^{i\xi_1 x_1}\widetilde{u}(\xi_1, x_2), \tag{4.15}$$

which is periodic in the tangential direction x_1, decreasing as $x_2 \to +\infty$, and satisfies the homogeneous boundary condition associated with the principal part of (4.14). Recall that v satisfies the equations

$$\begin{aligned} &-\partial_1 T^{11}(v) - \partial_2 T^{21}(v) = 0, \\ &-\partial_2 T^{22}(v) - \partial_1 T^{21}(v) = 0, \\ &-b_{11}T^{11}(v) - 2b_{12}T^{12}(v) - b_{22}T^{22}(v) = 0. \end{aligned} \tag{4.16}$$

We multiply each line of (4.16) by the conjugate $\overline{v}_i$ and integrate by parts on the periodicity layer B. We see that, on the infinite boundary, the boundary integral vanishes because of the decreasing condition as $x_2 \to +\infty$. The boundary integral also vanishes on the lateral boundary (which is parallel to x_2) of the strip by the periodicity of v. Recalling the definition of T^{ij}, we obtain

$$\int_B A^{\alpha\beta\lambda\mu}\gamma_{\lambda\mu}(v)\gamma_{\alpha\beta}(\overline{v})\mathrm{d}x_1\mathrm{d}x_2 = 0, \tag{4.17}$$

where obviously all $\Gamma^\alpha_{\beta\gamma}$ vanish. Consequently, by the positivity property (2.10) of A, this yields

$$\int_B \Sigma_{\alpha\beta}|\gamma_{\alpha\beta}(\overline{v})|^2\mathrm{d}x_1\mathrm{d}x_2 = 0, \tag{4.18}$$

and then

$$\gamma_{\alpha\beta}(v) = 0 \quad \text{on} \quad B. \tag{4.19}$$

We have now to distinguish two cases.

If $x_0 \in \Gamma_0$, then reasoning as in Lemma 4.2 (or merely as in Lemma 4.3), we deduce that $v_1 = v_2 = v_3 = 0$, which means that the SL condition is satisfied on Γ_0.

Let now $x_0 \in \Gamma_1$ and

$$\gamma_{\alpha\beta}(v) = 0 \quad \text{on} \quad B. \tag{4.20}$$

By the definition (4.15) of v, this yields that $\widetilde{u}$ is a solution of the following system of ordinary differential equations of order 2:

$$\begin{aligned} &i\xi_1 \widetilde{u}_1 - b_{11}\widetilde{u}_3 = 0, \\ &\partial_2 \widetilde{u}_2 - b_{22}\widetilde{u}_3 = 0, \\ &\frac{1}{2}(\partial_2 \widetilde{u}_1 + i\xi_1 \widetilde{u}_2) - b_{12}\widetilde{u}_3 = 0. \end{aligned} \tag{4.21}$$

Since $b_{11} \neq 0$ and $b_{22} \neq 0$, this can be written as

$$\begin{aligned} \widetilde{u}_1 &= -i\frac{b_{11}}{\xi_1}\widetilde{u}_3, \\ \widetilde{u}_3 &= \frac{1}{b_{22}}\partial_2 \widetilde{u}_2, \\ b_{11}\partial_2^2 \widetilde{u}_2 &- 2ib_{12}\xi_1\partial_2\widetilde{u}_2 - b_{22}\xi_1^2\widetilde{u}_2 = 0. \end{aligned}$$

Recalling the ellipticity assumption (2.16), after an easy computation we find that there exists a complex solution $\widetilde{u}$ given by $\widetilde{u} = we^{\lambda_- x_2}$, where $w \neq 0$ and λ_- is the root with negative real part of

$$b_{11}\lambda^2 - 2ib_{12}\xi_1\lambda - b_{22}\xi_1^2 = 0.$$

This means that there exists nonzero v which is exponentially decreasing in the direction inwards the domain $v(\xi_1, x_2) = we^{i\xi_1 x_1}e^{\lambda_- x_2}$, with $\mathrm{Re}(\lambda_-) < 0$ such that $\gamma_{\alpha\beta}(v) = 0$ on B, and hence $T^{\alpha\beta}(v)n_\alpha = 0$ on Γ_1. Therefore, the SL condition is not satisfied on Γ_1. □

4.4 The Koiter shell system

The boundary value problem associated with the variational problem (2.17) with $\varepsilon > 0$ is classical and well posed (cf., for example, [4, 17]). It is elliptic of total order 8, and the boundary conditions satisfy the SL condition. The system of equations is obtained by integration by parts, which yields

$$\begin{aligned} &-T^{\alpha\gamma}_{|\alpha}(u) + \varepsilon^2 b^\gamma_\beta M^{\alpha\beta}_{|\alpha}(u) + \varepsilon^2\Big(b^\gamma_\alpha M^{\alpha\beta}(u)\Big)_{|\beta} = f^\gamma, \\ &-b_{\alpha\beta}T^{\alpha\beta}(u) - \varepsilon^2 M^{\alpha\beta}(u)_{|\alpha\beta} + \varepsilon^2 b^\gamma_\alpha b_{\gamma\beta}M^{\alpha\beta}(u) = f^3, \end{aligned} \tag{4.22}$$

where the flection moments $M^{\alpha\beta}$ were defined in (2.14) and (2.15). The boundary conditions on Γ_0 (supposed clamped) are

$$u_1 = u_2 = u_3 = \partial_n u_3 = 0 \text{ on } \Gamma_0, \tag{4.23}$$

while the *natural* boundary conditions on Γ_1 are in number of four, are not relevant (they are boundary terms obtained by integration by parts). We have the following assertion.

Proposition 4.2. *The boundary value problem associated with the variational problem* (2.17) *when* $\varepsilon > 0$ *considered as a system of three equations with the unknowns* u *is elliptic of total order* 8 *with indices* (1, 1, 2) *for the unknowns and the equations.*

5 A Sensitive Singular Perturbation Problem Arising in the Koiter Linear Shell Theory

Very few is known concerning elliptic problems with boundary conditions not satisfying the SL condition and there is no general theory concerning them to our knowledge. Linear shell theory is one physical theory where they are naturally involved.

5.1 Definition of the problem

Let us first recall the variational problem (2.17) we are interested in:

$$\begin{aligned} &\text{Find } u^\varepsilon \in V \text{ such that } \forall\, v \in V \\ &a(u^\varepsilon, v) + \varepsilon^2 b(u^\varepsilon, v) = \langle f, v \rangle, \end{aligned} \tag{5.1}$$

where $f \in V'$ is given, the brackets denote the duality between V' and V. More precisely, we consider the limit boundary partial differential system associated with (5.1) when $\varepsilon = 0$. This is the membrane system which, by Proposition 4.1, is elliptic, satisfies the SL on Γ_0, but does not on Γ_1.

5.2 Sensitive character

Let us now recall the definition of sensitive problem (cf. [7] and [16] for a more complete description). Let us comment a little on Proposition 4.1.

The SL condition is not satisfied on a free boundary when $\varepsilon = 0$ for the variational problem (5.1). Specifically, the membrane problem is of total order four for elliptic surfaces. The number of boundary conditions should be two. On a fixed boundary Γ_0, they are

$$u_1 = u_2 = 0. \tag{5.2}$$

Note that the trace of u_3 does not make sense within the membrane framework. The previous boundary conditions satisfy the SL condition. Oppositely, on the free boundary Γ_1 the conditions are

$$T^{\alpha\beta}(u)n_\beta = 0. \tag{5.3}$$

Let us admit that (4.14) has (in some sense) a solution u. Replacing it in the three equations (4.14) and in the boundary conditions on Γ_1 of (4.14), one fineds that the corresponding $T^{\alpha\beta}(u)$ satisfy the elliptic membrane tensions system with the Cauchy conditions on the part of Γ_1 of the boundary. Since this last problem has, in general, no solution in usual spaces, the membrane problem (4.14) cannot (in general) have solution in usual spaces. We shall see that the existence of a solution (as well as the convergence as $\varepsilon \to 0$) only holds in very abstract spaces (out of the distribution space).

On the other hand, the boundary condition (5.2) constitutes the Cauchy condition for the rigidity system $\gamma_{\alpha\beta}(u) = 0$. According to the uniqueness theorem for the elliptic Cauchy problem (cf. the proof of Lemma 4.2) an elliptic shell is inhibited (or geometrically rigid) provided that it is fixed (or clamped) on a part (or the whole) of the boundary. When the boundary is everywhere free, the shell is not inhibited. Coming back to the inhibited elliptic shells, we see that when the whole boundary is fixed, the membrane problem is classical (the boundary condition satisfies the SL condition). But, when a part of the boundary Γ_0 is fixed whereas another one Γ_1 is not, the boundary conditions satisfy the SL condition on Γ_0, but not on Γ_1. This problem is out of the classical theory of elliptic boundary value problems and is called *sensitive* for a reason which will be selfevident later.

Let us consider formally the variational formulation of the membrane problem (4.14) (i.e., with $\varepsilon = 0$):

$$\begin{aligned} &\text{Find } u \in V_a \text{ such that } \forall\, v \in V_a \\ &a(u, v) = \langle f, v \rangle, \end{aligned} \tag{5.4}$$

where V_a is the completion of the "Koiter space" V with the norm $\|v\|_a = a(v, v)^{1/2}$.

The fact that $\|v\|_a$ is a norm on V follows from Lemma 4.2.

At the present state, it should be noticed that the previous completion process is somewhat abstract and the elements of V_a are not necessarly distributions. Indeed, since the SL condition is not satisfied on Γ_1, we may construct corresponding solutions with $u \neq 0$ and $\gamma_{\alpha\beta}(u) = 0$ which are rapidly oscillating along Γ_1 and exponentially decreasing inwards Ω. This is only concerned with the higher order terms. When taking into account lower order terms (which are "small" for rapidly oscillating solutions), we see that we may have "large u" with "small $\gamma_{\alpha\beta}(u)$" (i.e., small $\Sigma_{\alpha,\beta}\|\gamma_{\alpha\beta}(u)\|_{L^2}$) and then small membrane energy. Accordingly, the dual V_a', where f must be taken for (5.4) to make sense, is "very small."

The above property originates the term "sensitive." The problem is unstable. Very small and smooth variations of f (even in $\mathcal{D}(\Omega)$) induce modifications of the solution which are large and singular (out of the distribution space).

5.3 Abstract convergence results as $\varepsilon \to 0$

In this subsection, we recall some abstract convergence results (in the norm of the specified spaces), see [3] and [7] for more details.

Recalling the problem we are studying, we know that the shell is geometrically rigid:

$$v \in V \text{ and } \gamma_{\alpha\beta}(v) = 0 \Longrightarrow v = 0. \tag{5.5}$$

Let A and B be the continuous operators from V into V' associated with the forms a and b by

$$\langle Au, v \rangle = a(u, v), \quad \langle Bu, v \rangle = b(u, v) \quad \forall\, u, v \in V, \tag{5.6}$$

so that Equation (5.1) becomes as follows:

$$Au^\varepsilon + \varepsilon^2 Bu^\varepsilon = f. \tag{5.7}$$

Lemma 5.1. *The operator A is injective and its range, $\mathcal{R}(A)$ is dense in V'.*

The proof is not difficult (cf.[7] if necessary).

It then appears that the operator A is a one-to-one mapping of V onto $\mathcal{R}(A)$, which is a dense subset of V'. Let us define a new norm by

$$\|v\|_{V_A} = \|Av\|_{V'}. \tag{5.8}$$

It is obvious that V is not complete for the previous norm. But A defines an isomorphism between V (with the norm V_A) and $\mathcal{R}(A)$ (with the norm V'). Automatically, A has an extension by continuity which is an isomorphism between the completions of both spaces. Denote by $\overline{A}$ the extended operator and by V_A the completion of V with the norm (5.8). Then $\overline{A}$ is an isomorphism between V_A and V' (which is the completion of $\mathcal{R}(A)$ with the norm of V'). Equation (5.7) may be written as

$$\overline{A}u^\varepsilon + \varepsilon^2 Bu^\varepsilon = f. \tag{5.9}$$

Remark 5.1. In order to pass to the limit as $\varepsilon \to 0$, the classical way consists in obtaining an a priori energy estimate of u^ε by taking the duality product of (5.9) with u^ε. But such a way needs a hypothesis of boundedness of the functional f with respect to the limit form a and this does not work for any $f \in V'$. In the general case, following an idea of Caillerie [3] (cf. also [7]), which consists in proving that the term $\varepsilon^2 Bu^\varepsilon$ tends to zero in V', one can pass this latter term to the right-hand side and show that it tends to f in V'. Then, using the fact that $\overline{A}$ is an isomorphism, it is possible to prove the existence of a limit of u^ε in V_A. Specifically, we have the following result.

Theorem 5.1. *There exists a unique element* u^0 *in* V_A *such that*

$$\overline{A}u^0 = f. \tag{5.10}$$

Moreover, the following strong convergence holds in V_A :

$$u^\varepsilon \to u^0 \quad as \quad \varepsilon \to 0, \tag{5.11}$$

where $u^\varepsilon \in V$ *is the solution of* (5.9).

The proof, which follows the trends outlined above, may be seen in [7].

Remark 5.2. It should be emphasized that Theorem 5.1 holds without special hypothesis on f (besides the obvious one $f \in V'$). The limit $u^0 \in V_A$ is the solution of the abstract problem (5.4), which is not a variational one. The classical variational theory of the limit needs a supplementary hypothesis on f : there exists $C > 0$ such that

$$\|\langle f, v\rangle\| \leqslant Ca(v,v)^{1/2} \quad \forall\, v \in V, \tag{5.12}$$

which is very restrictive in shell theory.

For the sake of completeness, let us give the elements of the classical limit theory under the assumption (5.12).

We first note that, in such a case, $a(v,v)^{1/2}$ defines a norm on V. Let V_a be the completion of V with respect to that norm (which should not be confused with V_A). We then note that (5.12) shows that f may be extended by continuity to an element of V_a'. We denote this extension by f again. It is obvious that the variational problem

$$\begin{aligned} &\text{Find } u^0 \in V_a \text{ such that } \forall\, v \in V_a \\ &a(u^0, v) = \langle f, v\rangle \end{aligned} \tag{5.13}$$

is well posed and has a unique solution. We then have the classical convergence result (cf., for example, [10] or even [18]).

Theorem 5.2. *Under the assumption* (5.12),

$$u^\varepsilon \to u^0 \quad \text{strongly in } V_a \text{ as } \varepsilon \to 0, \tag{5.14}$$

where u^ε *and* u^0 *are the solutions of* (5.1) *and* (5.10) *respectively.*

Let us now briefly recall the non-inhibited case where (5.5) does not hold. In such a situation, there is a convergence result towards a limit with vanishing membrane energy. More precisely, we define the kernel G of a by

$$G = \{v \in V;\ \gamma_{\alpha\beta}(v) = 0\} = \{v \in V;\ a(v,v) = 0\}. \tag{5.15}$$

Note that G is a Hilbert space with the norm of V. But since $a(v,v) = 0$ in G, we see that the norm of V in G is equivalent to $b(v,v)^{1/2}$. As a consequence, the problem

$$\begin{aligned} &\text{Find } v^0 \in G \text{ such that } \forall\, w \in G \\ &b(v^0, w) = \langle f, w \rangle \end{aligned} \tag{5.16}$$

is well posed and has a unique solution. Moreover, since the "limit form" a in (5.1) vanishes on G, it implies some kind of weakness in G. The solution will be very large, and we should define a new scaling in order to have a finite limit, $v^\varepsilon = \varepsilon^2 u^\varepsilon$, (5.1) becomes

$$\begin{aligned} &\text{Find } v^\varepsilon \in V \text{ such that } \forall\, w \in V \\ &\varepsilon^{-2} a(v^\varepsilon, w) + b(v^\varepsilon, w) = \langle f, w \rangle, \end{aligned} \tag{5.17}$$

we then have the following assertion (cf., for example, [14] for the proof).

Theorem 5.3. *Under the assumption $G \neq \varnothing$,*

$$v^\varepsilon \to v^0 \quad \textit{strongly in } \; V, \tag{5.18}$$

where v^ε and v^0 are the solutions of (5.17) *and* (5.16) *respectively.*

6 Heuristic Asymptotics in the Previous Problem

The aim of this section is the construction, in a heuristic way, of an approximate description of the solutions u^ε of the linear Koiter model for small values of ε. Indeed, coming back to the Koiter problem for $\varepsilon > 0$, in the sensitive case, the problem is not really to describe the limit problem (which, in general, has no solution in the distribution space; in particular, the space V_A (cf. (5.8)) where there is always a limit, is not a distribution space), but rather to give a good description of the solution u^ε for very small values of ε. This we shall try to do. We shall see that heuristic considerations allow us to construct a simplified model accounting for the main features of the problem.

To do so, we shall use the heuristic procedure of [7]. In this latter article, we addressed a model problem including a variational structure, somewhat analogous to the problem studied here, but simpler, as concerning an equation instead of a system. It is shown that the limit problem contains, in particular, an elliptic Cauchy problem. This problem was handled in both a rigorous (very abstract) framework and using a heuristic procedure for exhibiting the structure of the solutions with very small ε. The main difference is that, in the present work, we deal with systems instead of single equations.

We shall see that heuristic considerations involving minimization of energy allow us to reduce the problem to another on the boundary Γ_1. In that context, it is seen that the "pathological" operator A is represented by a smoothing operator S (i.e., sending any distribution to a $\mathcal{C}^\infty$ function), whereas the "classical" operator B is represented by a "classical" elliptic operator Q. Denoting by $s(x,\xi)$ and $q(x,\xi)$ the corresponding symbols (here, x is the arc on Γ_1), s is likely exponentially decreasing as $\xi \to \infty$, whereas q is algebraically growing. The action of $S+\varepsilon^2 Q$ on test functions is given by

$$(S+\varepsilon^2 Q)\theta(x) = (2\pi)^{-1}\int_{-\infty}^{+\infty} e^{i\xi x}[s(x,\xi)+\varepsilon^2 q(x,\xi)]\widetilde{\theta}(\xi)\mathrm{d}\xi. \tag{6.1}$$

It is then apparent that, when ε is small, the operator S is significant only for bounded values of ξ, whereas $\varepsilon^2 Q$ describes the behavior as $\xi \to \infty$. If $|\xi| \ll \log(1/\varepsilon)$, then the symbol of the operator $S+\varepsilon^2 Q$ is equal to $(1+o(1))s(x,\xi)$ and for $|\xi| \gg \log(1/\varepsilon)$, it is $(1+o(1))\varepsilon^2 q(x,\xi)$. The balance of S and $\varepsilon^2 Q$ is obtained for values of ξ such that

$$|\xi| \sim \log(1/\varepsilon). \tag{6.2}$$

This is the window of frequencies allowing a good description of the simultaneous influence of S and $\varepsilon^2 Q$, which is precisely our aim. Moreover, it is easily seen that the range of frequencies (6.2) is responsible for most of the contribution to the integral (6.1). This property is of great interest for the construction of the heuristic approximation. More precisely, the heuristics incorporate approximations for large $|\xi|$. This amounts to saying that only the most singular parts of the solutions are retained, or equivalently, that the approximate solutions are defined up to more regular terms. This is, for instance, the kind of approximation which is used in the construction of a parametrix. We also note that, as (6.2) involves "moderately large" values of $|\xi|$, the "general quality" of the approximation is not very good, as it is only accurate for very small values of ε.

Note that numerical computations [2] carried out with very reliable software (including an adapted mesh procedure) for the Koiter problem with very small values of ε agree with the overall trends of our heuristic procedure. It appears that most of the deformation consists in very large deformations along Γ_1 exponentially decreasing inwards Ω (then in good agreement with the "local lack of uniqueness" implied by the non-satisfied SL condition). Since ε decreases, the amplitude increases, whereas the wave length decreases very slowly, verifying fairly well (6.2). The paper [2] also contains numerical comparisons with the case where the shell is fixed all along its boundary, which is classical (as the SL condition is satisfied all along the boundary). The differences are drastic for small values of ε.

6.1 Introduction to the heuristic asymptotic

A first remark in the context described above is that sensitive problems may be considered as "intermediate" between "inhibited" and "non-inhibited." Indeed, "inhibited" means that $v \in V$ and $\gamma_{\alpha\beta}(v) = 0$ implies $v = 0$, whereas "non-inhibited" means that there are nonvanishing elements v of V such that $\gamma_{\alpha\beta}(v) = 0$. Strictly speaking, sensitive problems enter in the class "inhibited," but there are nonvanishing elements v of V with "very small" $\gamma_{\alpha\beta}(v)$.

In order to minimize the energy

$$a(v,v) + \varepsilon^2 b(v,v) - 2\langle f, v\rangle, \tag{6.3}$$

we may proceed as in non-inhibited problems. The solution with small ε "avoids" the (larger) membrane energy a, so that roughly speaking, solutions for small ε should have $\gamma_{\alpha\beta}(v)$ vanishing or at least very small with respect to v.

Obviously, it is impossible to impose the four boundary conditions (4.23) on Γ_0 with the "exact" system $\gamma_{\alpha\beta}(v) = 0$ as they imply $v = 0$.

Nevertheless, we shall see in Subsection 6.2.1 that it is possible to construct functions satisfying the two boundary conditions $u_n = u_t = 0$ on Γ_0 with the "non exact" system $\gamma_{\alpha\beta}(v) = 0$ in the sense that $\gamma_{\alpha\beta}(v)$ will be "very small" (i.e., $\Sigma_{\alpha,\beta}\|\gamma_{\alpha\beta}(v)\|_{L^2}$ will be very small). This will imply a "membrane boundary layer" in the vicinity of Γ_0 involving the bilinear form a. To this end, we first construct a set of functions v with only one vanishing component on Γ_0. Choosing (for instance) the normal component, we define

$$G^0 = \{v,\ \gamma_{\alpha\beta}(v) = 0 \text{ in } \Omega,\ v_2 = 0 \text{ on } \Gamma_0\}, \tag{6.4}$$

the regularity is not precisely stated as we shall later take the completion, we may consider C^∞ functions for instance. Note that v is a triplet of functions.

Recalling Remark 4.5, we know that up to a finite-dimensional space composed of smooth functions, the space G^0 is isomorphic to the space of traces on Γ_1:

$$\{w \in C^\infty(\Gamma_1)\} \tag{6.5}$$

the isomorphism is obtained by solving the problem

$$\begin{aligned} &\gamma_{\alpha\beta}(\widetilde{w}) = 0 \quad \text{in} \quad \Omega, \\ &\widetilde{w}_2 = 0 \quad \text{on} \quad \Gamma_0, \\ &\widetilde{w}_3 = w \quad \text{on} \quad \Gamma_1. \end{aligned} \tag{6.6}$$

In the sequel, when we will consider a function $\widetilde{w} \in G^0$, we will consider a function of the equivalence class for the quotient operation described in Remark 4.5. Moreover, we shall consider indifferently the functions $\widetilde{w}$ obtained after a quotient operation in $\overline{\Omega}$ (for the finite-dimensional space) or their traces w on Γ_1.

Moreover, the conditions $u_3 = \partial_n u_3 = 0$ on Γ_0 of (4.23) will be satisfied with the help of a "flection sublayer" involving the bilinear form b; its effect is not relevant (cf. Subsection 6.2.2).

According to the previous considerations, we shall consider the minimization problem on G^0 instead of on V. This modified problem obviously involves the a-energy and the $\varepsilon^2 b$-energy. A natural space for handling it should be the completion G of G^0 with the corresponding norm.

The fact that we may "neglect" the functions in the finite-dimensional space of smooth functions follows from the fact that we are interested in the singular part.

6.2 The boundary layer on Γ_0

Let $\widetilde{w}$ be in G^0 (cf. (6.4)), and let $\varepsilon > 0$ be fixed. The aim of this subsection is to build a modified function $\widetilde{w}^a$ of $\widetilde{w}$ in a narrow boundary layer of Γ_0 in order to satisfy the supplementary boundary conditions $\widetilde{w}_t = \widetilde{w}_3 = \partial_n \widetilde{w}_3 = 0$ on Γ_0.

The present problem is analogous to the "model problem" of [7] in the case of a singular perturbation, i.e., [7, Section 7.1.2]. Indeed, the membrane problem is of total order 4 allowing 2 boundary conditions ($\widetilde{w}_t = \widetilde{w}_n = 0$) on Γ_0, whereas the complete Koiter shell problem is of order 8, allowing 4 boundary conditions (we shall add $\widetilde{w}_3 = \partial_n \widetilde{w}_3 = 0$) on Γ_0. It appears that the first two conditions ($\widetilde{w}_t = \widetilde{w}_n = 0$) may be obtained from elements of G^0 by modifying them on account of a "membrane layer" which relies on the membrane system, of thickness of order $\frac{1}{\log(1/\varepsilon)}$ on Γ_0, whereas an irrelevant boundary layer will be considered in Subsection 6.2.2.

6.2.1 The membrane boundary layer on Γ_0

In this subsection, we proceed to modify the element $\widetilde{w}$ of G^0 in order to satisfy both conditions $u_1 = u_2 = 0$ on Γ_0.

Let $\widetilde{\Gamma}_0$ be a neighborhood of Γ_0 in $\mathbb{R}^2$ disjoint with Γ_1 and sufficiently narrow to be described by the curvilinear coordinates y_1 = arc of Γ_0 and y_2 = distance along the normal to Γ_0. Let $(\psi_j(y_1))_{j\in J}$ be a partition of the unity associated with Γ_0, and let $\eta \in C^\infty(\mathbb{R}_+; \mathbb{R}_+)$ be a cut-off function equal to 1 for small values of y_2.

The mappings θ_j defined by $\theta_j(y_1, y_2) = \psi_j(y_1)\eta(y_2)$, where y_2 is the (inward) normal coordinate along Γ_0, define a partition of unity in $\widetilde{\Gamma}_0$; in particular, for a given $\widetilde{w} \in G^0$, we have

$$\forall\ (y_1, y_2) \in \widetilde{\Gamma}_0,\ \widetilde{w}(y_1, y_2) = \Sigma_{j\in J}\theta_j(y_1, y_2)\widetilde{w}(y_1, y_2). \tag{6.7}$$

Let us fix j in J and y_2 such that $(y_1, y_2) \in \widetilde{\Gamma}_0$, the function $\theta_j(\cdot, y_2)\widetilde{w}(\cdot, y_2)$ has a compact support, we denote by $\widetilde{w}^j(\cdot, y_2)$ its extension by zero to $\mathbb{R}$ and by $\mathcal{F}(\widetilde{w}^j)$ the tangential Fourier transform, $y_1 \to \xi_1$, of $\widetilde{w}^j$.

Let us first exhibit the local structure of the Fourier transform of $\widetilde{w}^j$ close to Γ_0. Denoting by θ_j the multiplication operator by θ_j, recalling that the commutator of the operator γ associated with $\gamma_{\alpha\beta}$ and θ_j, denoted by $[\gamma, \theta_j]$, is a differential operator of lower order, taking the γ operator in the new coordinates (y_1, y_2) (which, according to our approximation close to Γ_0, has the same principal part), and using that $\widetilde{w} \in G^0$, we see that

$$\gamma_{\alpha\beta}(\widetilde{w}^j) + U_{\alpha\beta}(y, D)\widetilde{w}^j = 0 \quad \text{on} \quad \mathbb{R} \times (0, t) \tag{6.8}$$

for some $t > 0$, where $U_{\alpha\beta}$ is a differential operator of order less than the order of $\gamma_{\alpha\beta}$.

Now, according to the general trends of our boundary layer approximation, we can neglect the terms of lower order in (6.8) and proceed as in the construction of a parametrix (freezing coefficients, dropping lower order terms, solving such simpler equation via tangent Fourier transform, and gluing together the solutions for different j), so that (6.8) becomes

$$\gamma_{\alpha\beta}(\widetilde{w}^j) = 0 \quad \text{on} \quad \mathbb{R} \times (0, t). \tag{6.9}$$

The previous system can be rewritten as

$$\begin{aligned}
&\frac{\partial}{\partial y_1}\widetilde{w}_1^j - b_{11}\widetilde{w}_3^j = 0,\\
&\frac{\partial}{\partial y_2}\widetilde{w}_2^j - b_{22}\widetilde{w}_3^j = 0,\\
&\frac{1}{2}\Big(\frac{\partial}{\partial y_2}\widetilde{w}_1^j + \frac{\partial}{\partial y_1}\widetilde{w}_2^j\Big) - b_{12}\widetilde{w}_3^j = 0.
\end{aligned} \tag{6.10}$$

Taking the tangential Fourier transform denoted by $\mathcal{F}(\widetilde{w}^j)(\xi_1, y_2)$, we get

$$\Big(\widehat{\gamma}_0 + \widetilde{\gamma}_1\frac{\mathrm{d}}{\mathrm{d}y_2}\Big)\mathcal{F}(\widetilde{w}^j) = 0, \tag{6.11}$$

with

$$\widehat{\gamma}_0 = \begin{pmatrix} -i\xi_1 & 0 & -b_{11} \\ 0 & 0 & -b_{22} \\ 0 & -i\xi_1 & -2b_{12} \end{pmatrix} \quad \text{and} \quad \widetilde{\gamma}_1 = \begin{pmatrix} 0 & 0 & 0 \\ 0 & 1 & 0 \\ 1 & 0 & 0 \end{pmatrix}.$$

The general solution of the system (6.11) is

$$\mathcal{F}(\widetilde{w}^j)(\xi_1, y_2) = A\widetilde{w}_+ e^{\lambda_+(\xi_1)y_2} + B\widetilde{w}_- e^{\lambda_-(\xi_1)y_2}, \tag{6.12}$$

with

$$\widetilde{w}_+ = \begin{pmatrix} i\dfrac{\lambda_+(\xi_1)}{\xi_1}\dfrac{b_{11}}{b_{22}} \\ 1 \\ \dfrac{\lambda_+(\xi_1)}{b_{22}} \end{pmatrix}, \quad \widetilde{w}_- = \begin{pmatrix} i\dfrac{\lambda_-(\xi_1)}{\xi_1}\dfrac{b_{11}}{b_{22}} \\ 1 \\ \dfrac{\lambda_-(\xi_1)}{b_{22}} \end{pmatrix},$$

$$\lambda_+(\xi_1) = -i\xi_1\frac{b_{12}}{b_{11}} + \frac{|\xi_1|}{b_{11}}\sqrt{b_{11}b_{22} - b_{12}^2},$$

$$\lambda_-(\xi_1) = -i\xi_1\frac{b_{12}}{b_{11}} - \frac{|\xi_1|}{b_{11}}\sqrt{b_{11}b_{22} - b_{12}^2}.$$

Since $\widetilde{w} \in G_0$, it follows that $\mathcal{F}(\widetilde{w}_2^j)(\xi_1, 0) = 0$. Hence $A = -B$. Consequently,

$$\mathcal{F}(\widetilde{w}^j)(\xi_1, y_2) = \frac{b_{11}b_{22}}{2|\xi_1|\sqrt{b_{11}b_{22} - b_{12}^2}}\mathcal{F}(\widetilde{w}_3^j)(\xi_1, 0)\big(\widetilde{w}_+ e^{\lambda_+(\xi_1)y_2} - \widetilde{w}_- e^{\lambda_-(\xi_1)y_2}\big). \tag{6.13}$$

This expression exhibits the structure of (the Fourier transform of) $\widetilde{w}$ in a narrow neighborhood of Γ_0. It was expressed in terms of (the Fourier transform of) the trace of its third component on Γ_0, but this choice is arbitrary.

We now proceed to the modification of $\widetilde{w}^j$ in $\widetilde{w}^{ja}$ in a narrow boundary layer of Γ_0 in order to satisfy (always within our approximation) the equation coming from (4.22) for $\varepsilon = 0$ (this is the membrane boundary layer associated with the membrane system of Section 4.3). Using considerations similar to those leading to (6.8), this amounts to

$$\left(\widetilde{\gamma}^* \widetilde{A}_1 \widetilde{\gamma}\right)\widetilde{w}^{ja} + U(y, D)\widetilde{w}^{ja} = 0 \quad \text{on} \quad \mathbb{R} \times (0, t), \tag{6.14}$$

where U is a differential operator of lower order than four, $\widetilde{\gamma}^*$ denotes the operator

$$\widetilde{\gamma}^* = \begin{pmatrix} \partial_1 & 0 & -b_{11} \\ 0 & \partial_2 & -b_{22} \\ \partial_2 & \partial_1 & -2b_{12} \end{pmatrix},$$

and

$$\widetilde{A}_1 = \begin{pmatrix} A^{1111} & A^{1122} & A^{1112} \\ A^{2211} & A^{2222} & A^{2212} \\ A^{1211} & A^{1222} & A^{1212} \end{pmatrix}.$$

Therefore, dropping as before terms of lower order, we have

$$(\widetilde{\gamma}^* \widetilde{A}_1 \widetilde{\gamma})\widetilde{w}^{ja} = 0 \quad \text{on} \quad \mathbb{R} \times (0, t), \tag{6.15}$$

which can be rewritten as

$$\big((\widetilde{\gamma}_0^* - \widetilde{\gamma}_1^* \partial_2)\widetilde{A}_1(\widetilde{\gamma}_0 + \widetilde{\gamma}_1 \partial_2)\big)\widetilde{w}^{ja} = 0 \quad \text{on} \quad \mathbb{R} \times (0,t), \tag{6.16}$$

with $\widetilde{\gamma}^* = \overline{\widetilde{\gamma}}^T$ and

$$\widetilde{\gamma}_0 = \begin{pmatrix} \partial_1 & 0 & -b_{11} \\ 0 & 0 & -b_{22} \\ 0 & \partial_1 & -2b_{12} \end{pmatrix}.$$

Hence, taking the tangential Fourier transform, we look for solutions of the system

$$\Big(\Big(\overline{\widehat{\gamma}_0}^T - \widetilde{\gamma}_1^T \frac{\mathrm{d}}{\mathrm{d}y_2}\Big)\widetilde{A}_1\Big(\widehat{\gamma}_0 + \widetilde{\gamma}_1 \frac{\mathrm{d}}{\mathrm{d}y_2}\Big)\Big)\mathcal{F}(\widetilde{w}^{ja})(\xi_1, y_2) = 0, \tag{6.17}$$

with

$$\widehat{\gamma}_0 = \begin{pmatrix} -i\xi_1 & 0 & -b_{11} \\ 0 & 0 & -b_{22} \\ 0 & -i\xi_1 & -2b_{12} \end{pmatrix}.$$

At this moment, it is worth-while to compare (6.17) and (6.11). We see that the given function $\widetilde{w}^j$ (rather its Fourier transform) solves the "right half" of (6.17), i.e., the expression on the right of $\widetilde{A}_1$ in (6.17). Obviously, the "left half" accounts for the "adjoint part," coming with integration by parts from the bilinear form a (cf. (2.5)). Our aim in constructing the modified $\widetilde{w}^{ja}$ is to satisfy the conditions $\widetilde{w}_1^{ja} = \widetilde{w}_2^{ja} = 0$ on $y_2 = 0$, whereas for "large y_2" (in the sense of "out of the layer") the modified $\widetilde{w}^{ja}$ coincides (up to small terms) with the given $\widetilde{w}^j$. We now proceed to write down the general solution of (6.17) on account of its special structure.

For $\lambda \in \{\lambda_-(\xi_1), \lambda_+(\xi_2)\}$, let us consider the function k defined by

$$k(\xi_1, y_2) = (y_2 w + v)e^{\lambda y_2}, \tag{6.18}$$

where $w \in \{\widetilde{w}_-, \widetilde{w}_+\}$ is a solution of

$$\big(\widehat{\gamma}_0 + \lambda\widetilde{\gamma}_1\big)w = 0$$

and v is unknown. We then search for solutions of (6.17) under the form (6.18), i.e.,

$$\Big(\Big(\overline{\widehat{\gamma}_0}^T - \widetilde{\gamma}_1^T \frac{\mathrm{d}}{\mathrm{d}y_2}\Big)\widetilde{A}_1\Big(\widehat{\gamma}_0 + \widetilde{\gamma}_1 \frac{\mathrm{d}}{\mathrm{d}y_2}\Big)\Big)k(\xi_1, y_2) = 0. \tag{6.19}$$

We check that

$$\Big(\widehat{\gamma}_0 + \widetilde{\gamma}_1 \frac{\mathrm{d}}{\mathrm{d}y_2}\Big)\big(y_2 w + e^{\lambda y_2} v\big) = \big((\widehat{\gamma}_0 + \lambda\widetilde{\gamma}_1)v + \widetilde{\gamma}_1 w\big)e^{\lambda y_2}.$$

So that (6.19) becomes

$$\left(\left(\overline{\widehat{\gamma}_0}^T - \widetilde{\gamma}_1^T \frac{\mathrm{d}}{\mathrm{d}y_2}\right)\widetilde{A}_1\left(\widehat{\gamma}_0 + \widetilde{\gamma}_1 \frac{\mathrm{d}}{\mathrm{d}y_2}\right)\right)(y_2 w + v)e^{\lambda y_2}$$
$$= \left(\overline{\widehat{\gamma}_0}^T - \widetilde{\gamma}_1^T \frac{\mathrm{d}}{\mathrm{d}y_2}\right)\widetilde{A}_1\left((\widehat{\gamma}_0 + \lambda\widetilde{\gamma}_1)v + \widetilde{\gamma}_1 w\right)e^{\lambda y_2} = 0.$$

This amounts to saying that $\widetilde{A}_1\big((\widehat{\gamma}_0 + \lambda\widetilde{\gamma}_1)v + \widetilde{\gamma}_1 w\big)$ is an eigenvector of $\overline{\widehat{\gamma}_0}^T - \lambda\widetilde{\gamma}_1^T$ associated with the zero eigenvalue. Since dim Ker $\big(\overline{\widehat{\gamma}_0}^T - \lambda\widetilde{\gamma}_1^T\big) = 1$, denoting by u_0 a nonvanishing vector of Ker $\big(\overline{\widehat{\gamma}_0}^T - \lambda\widetilde{\gamma}_1^T\big)$, we see that v should satisfy

$$(\widehat{\gamma}_0 + \lambda\widetilde{\gamma}_1)v + \widetilde{\gamma}_1 w = \widetilde{A}_1^{-1}(\tau u_0) \quad \text{for some} \quad \tau \in \mathbb{C}. \tag{6.20}$$

According to the Fredholm alternative, a necessary and sufficient condition for the existence of such v is that

$$\widetilde{A}_1^{-1}(\tau u_0) - \widetilde{\gamma}_1 w \ \in (\mathrm{Vect}\ u_0)^{\perp}.$$

Since $\widetilde{A}_1$ is positive definite, we deduce that $(\widetilde{A}_1^{-1}u_0, u_0) > 0$ hence

$$\tau = \frac{(\widetilde{\gamma}_1 u, u_0)}{(\widetilde{A}_1^{-1}u_0, u_0)}$$

satisfies

$$(\tau\widetilde{A}_1^{-1}u_0 - \widetilde{\gamma}_1 w, u_0) = 0.$$

It follows that the vector $v \in \mathbb{C}^3$ exists and is unique (up to an additive and arbitrary eigenvector, which is irrelevant in the sequel). Consequently, k defined as above satisfies (6.19).

Repeating this argument twice (first for $\lambda_+(\xi_1)$, and then for $\lambda_-(\xi_1)$) and denoting by v_+ and v_- the corresponding vectors v, we see that

$$\mathcal{F}\big(\widetilde{w}^{ja}\big)(\xi_1, y_2) = C_1\widetilde{w}_e^{-\lambda_+(\xi_1)y_2} + C_2\widetilde{w}_- e^{-\lambda_-(\xi_1)y_2}$$
$$+ C_3\big(y_2\widetilde{w}_+ + v_+\big)e^{\lambda_+(\xi_1)y_2} + C_4\big(y_2\widetilde{w}_- + v_-\big)e^{\lambda_-(\xi_1)y_2}, \tag{6.21}$$

with arbitrary C_1, C_2, C_3, and C_4, is the general solution of (6.17).

We determine C_1, C_2, C_3, and C_4 in order to satisfy the boundary conditions $\widetilde{w}_1^{ja} = \partial_2\widetilde{w}_1^{ja} = 0$ at $y_2 = 0$ and the "matching condition" with $\widetilde{w}^j$, i.e., in the context of boundary layer theory (for large $|\xi_1|$), $\widetilde{w}^{ja}$ should become $\widetilde{w}^j$ out of the layer.

Let us now explain the process of matching the layer: out of the layer, we want $\widetilde{w}^{ja}$ to match with the given function $\widetilde{w}^j$. Since $|\xi_1| \gg 1$, we have $|\xi_1|y_2 \gg 1$ and

$$\frac{\sqrt{b_{11}b_{22} - b_{12}^2}}{b_{11}}|\xi_1|y_2 \gg 1,$$

which means that

$$y_2 \gg \frac{b_{11}}{\sqrt{b_{11}b_{22} - b_{12}^2}} \frac{1}{|\xi_1|}$$

(but we still impose that y_2 is small in order to be in a narrow layer of Γ_0 where (6.13) holds). This is perfectly consistent, as we will only use the functions for large $|\xi_1|$, hence the terms with coefficients C_2 and C_4 are "boundary layer terms" going to zero out of the layer (i.e., for $|y_2| \gg \mathcal{O}(\frac{1}{|\xi_1|})$).

The matching with (6.13) out of the layer then gives

$$C_3 = 0, \quad C_1 = \frac{b_{11}b_{22}}{2|\xi_1|\sqrt{b_{11}b_{22} - b_{12}^2}} \mathcal{F}(\widetilde{w}_3^j)(\xi_1, 0). \tag{6.22}$$

The two other constants C_2 and C_4 are determined by

$$\mathcal{F}(\widetilde{w}^{ja})_1(\xi_1, 0) = 0, \quad \mathcal{F}(\widetilde{w}^{ja})_2(\xi_1, 0) = 0,$$

which yields the existence of two constants α and β such that

$$C_2 = \alpha C_1, \quad C_4 = \beta C_1.$$

So that the modified solution is of the form

$$\mathcal{F}(\widetilde{w}^{ja})(\xi_1, y_2) = \frac{b_{11}b_{22}}{2|\xi_1|\sqrt{b_{11}b_{22} - b_{12}^2}} \Big(\widetilde{w}_+ e^{\lambda_+(\xi_1)y_2} + ((\alpha + \beta y_2)\widetilde{w}_- + \beta v_-)e^{\lambda_-(\xi_1)y_2}\Big)\mathcal{F}(\widetilde{w}_3^j)(\xi_1, 0). \tag{6.23}$$

The modification of the function $\widetilde{w}_j$ then consists in adding to it the inverse Fourier transform of

$$\frac{b_{11}b_{22}}{2|\xi_1|\sqrt{b_{11}b_{22} - b_{12}^2}} ((\alpha + 1 + \beta y_2)\widetilde{w}_- + \beta v_-)e^{\lambda_-(\xi_1)y_2}\mathcal{F}(\widetilde{w}_3^j)(\xi_1, 0). \tag{6.24}$$

In the sequel, we shall study the behavior of such an expression. The role of the constants α and β is not relevant, and we may assume, for instance, that $\alpha = -1$ and $\beta = 1$ (this amounts to change $\widetilde{w}_-$ and $\widetilde{v}_-$). As a result, the modification of the function $\widetilde{w}_j$ consists in adding to it the inverse Fourier transform of

$$\frac{b_{11}b_{22}}{2|\xi_1|\sqrt{b_{11}b_{22} - b_{12}^2}} (y_2\widetilde{w}_- + v_-)e^{\lambda_-(\xi_1)y_2}\mathcal{F}(\widetilde{w}_3^j)(\xi_1, 0). \tag{6.25}$$

More precisely, on account of considerations at the beginning of Section 6 (cf., in particular, (6.1) and (6.2)), the modification should only be effective for large $|\xi_1|$, accounting for "singular parts" of the solution. Moreover, in order to have $\widetilde{w}^a \in V$, we shall also impose $\widetilde{w}_1^{ja} = \partial_2 \widetilde{w}_1^{ja} = 0$ on Γ_0 (the other two conditions $\widetilde{w}_3^a = \partial_n \widetilde{w}_3^a = 0$ on Γ_0 will be addressed in Subsection 6.2.2).

To this end, we multiply the added term by a cut-off function avoiding low frequencies (it should be remembered that this is one of the typical devices in the construction of a parametrix). More precisely, on account of (6.2), we shall only keep frequencies of order more than or equal to $[\log(1/\varepsilon)]^{1/2}$, which preserve the useful region (6.2) and are large (then consistent with the fact that the modification is a layer). Moreover, in order to the modified function satisfy the boundary conditions, we must also take into account the low frequencies of the Fourier transform which we multiply by a smooth vector $\rho(y_2)$ such that $\rho_1(0) = \rho_2(0) = 0$ and $\rho(y_2) = 0$ for $y_2 > C$ for a certain C. The division into high and low frequencies is defined by a smooth function $H(z)$ equal to 1 for $|z| > 1$ and vanishing for $|z| < 1/2$, with $z = \frac{\xi}{[\log(1/\varepsilon)]^{1/2}}$. Finally, we define the function

$$h(\varepsilon, \xi, y_2) = \Big(1 - H\Big(\frac{\xi_1}{[\log(1/\varepsilon)]^{1/2}}\Big)\Big)\rho(y_2)$$

$$+ \frac{b_{11}b_{22}}{2|\xi_1|\sqrt{b_{11}b_{22} - b_{12}^2}}(y_2\widetilde{w}_- + v_-)e^{\lambda_-(\xi_1)y_2} H\Big(\frac{\xi}{[\log(1/\varepsilon)]^{1/2}}\Big), \tag{6.26}$$

which obviously has its first and second components vanishing for $y_2 = 0$. Now, we can modify the function $\widetilde{w}_j$ by

$$\delta\widetilde{w}_j \equiv \widetilde{w}_j^a - \widetilde{w}_j, \tag{6.27}$$

where $\delta\widetilde{w}_j$ is defined by its Fourier transform:

$$\mathcal{F}(\delta\widetilde{w}_j) = \mathcal{F}(\widetilde{w}_3^j)(\xi_1, 0)h(\varepsilon, \xi, y_2). \tag{6.28}$$

Remark 6.1. The constant C in the definition of $\rho(y_2)$ is chosen sufficiently small for this function to vanish out of the layer of Ω close to Γ_0 where the curvilinear coordinates y_1, y_2 operate. Rigorously speaking, the rest of the expression should also be multiplied by a cut-off function vanishing for $y_2 > C$, but this is practically not necessary since this part is exponentially small for large $|\xi_1|$.

Hence, summing over j and defining on Γ_0 the family (with parameter y_2) of pseudodifferential smoothing operators $\delta\sigma(\varepsilon, D_1, y_2)$ with symbol

$$\delta\sigma(\varepsilon, \xi_1, y_2) = \frac{|b_{11}|b_{22}}{2|\xi_1|\sqrt{b_{11}b_{22} - b_{12}^2}}(y_2\widetilde{w}_- + v_-)e^{\lambda_-(\xi_1)y_2}, \tag{6.29}$$

we see that the modification of the function $\widetilde{w}$

$$\delta\widetilde{w} = \widetilde{w}^a - \widetilde{w} \tag{6.30}$$

is precisely the action of $\delta\sigma(\varepsilon, D_1, y_2)$ on $\widetilde{w}_3^j(y_1, 0)$.

Once $\widetilde{w}^a$ is constructed, it is worth-while to compute its a-energy. This we proceed to do. More generally, we shall compute the form a for two functions $\widetilde{v}^a$ and $\widetilde{w}^a$.

Let us now compute the leading terms of the a-energy of the modified function $\widetilde{w}^a$.

Let $\widetilde{v}$ and $\widetilde{w}$ be two elements in G^0, and let $\widetilde{v}^a$ and $\widetilde{w}^a$ be the corresponding elements modified in the boundary layer. Since given $\widetilde{v}$ and $\widetilde{w}$ satisfy $\gamma_{\alpha\beta}(\widetilde{v}) = \gamma_{\alpha\beta}(\widetilde{w}) = 0$, the a-form is only concerned with the modification terms $\delta\widetilde{v}$ and $\delta\widetilde{w}$. Then, within our approximation, we have

$$a(\widetilde{v}^a, \widetilde{w}^a) = \int_{\Gamma_0} A^{\alpha\beta\lambda\mu} dy_1 \int_0^{+\infty} \gamma_{\alpha\beta}(\delta\widetilde{v}) \overline{\gamma_{\lambda\mu}(\delta\widetilde{w})} \mathrm{d}y_2, \tag{6.31}$$

where the integral in $\mathrm{d}y_2$ is only effective in the narrow layer. Using the partition of the unity θ_j and denoting, as before, by $\delta w_j(\cdot, y_2)$ the extension with value 0 to $\mathbb{R}$ of $\theta_j(\cdot, y_2)\delta w(\cdot, y_2)$, we have

$$a(\widetilde{v}^a, \widetilde{w}^a) = \Sigma_{j,k} \int_{\Gamma_0} A^{\alpha\beta\lambda\mu} dy_1 \int_0^{+\infty} \gamma_{\alpha\beta}(\delta\widetilde{v}_j) \overline{\gamma_{\lambda\mu}(\delta\widetilde{w}_k)} \mathrm{d}y_2. \tag{6.32}$$

Consequently, using the tangential Fourier transform $y_1 \to \xi_1$ and the Plancherel–Parseval theorem, dropping lower order terms (within our approximation, we only consider expressions with large $|\xi_1|$ which amounts to take $H = 1$ in (6.26)), we deduce that

$$\begin{aligned}
&a(\widetilde{v}^a, \widetilde{w}^a) \\
&= \Sigma_{j,k} \int_{-\infty}^{+\infty} \widetilde{A}_1 \mathrm{d}\xi_1 \int_0^{+\infty} \Big(\widehat{\gamma}_0 + \widetilde{\gamma}_1 \frac{\mathrm{d}}{\mathrm{d}y_2}\Big) \delta\sigma(\varepsilon, \xi, y_2) \mathcal{F}(\widetilde{v}_3^j)(\xi_1, 0)\big) \\
&\quad \times \overline{\Big(\widehat{\gamma}_0 + \widetilde{\gamma}_1 \frac{\mathrm{d}}{\mathrm{d}y_2}\Big) \delta\sigma(\varepsilon, \xi, y_2) \mathcal{F}(\widetilde{w}_3^k)(\xi_1, 0)\big)} \mathrm{d}y_2 \\
&= \Sigma_{j,k} \int_{-\infty}^{+\infty} \widetilde{A}_1 \mathrm{d}\xi_1 \int_0^{+\infty} \frac{b_{11}b_{22}}{2|\xi_1|\sqrt{b_{11}b_{22} - b_{12}^2}} \\
&\quad \times \Big((\widehat{\gamma}_0 + \lambda_- \widetilde{\gamma}_1) v_- + \widetilde{\gamma}_1 \widetilde{w}_- \Big) e^{\lambda_- y_2} \mathcal{F}(\widetilde{v}_3^j)(\xi_1, 0)\big) \\
&\quad \times \overline{\frac{b_{11}b_{22}}{2|\xi_1|\sqrt{b_{11}b_{22} - b_{12}^2}} \Big((\widehat{\gamma}_0 + \lambda_- \widetilde{\gamma}_1) v_- + \widetilde{\gamma}_1 \widetilde{w}_- \Big) e^{\lambda_- y_2} \mathcal{F}(\widetilde{w}_3^k)(\xi_1, 0)\big)} \mathrm{d}y_2.
\end{aligned}$$

Hence, on account of the definitions of $\widehat{\gamma}_0$, $\widetilde{\gamma}_1$, λ_- and $\widetilde{w}_-$ integrating in y_2, we know that

$$a(\widetilde{v}^a, \widetilde{w}^a) = \Sigma_{j,k} \int_{-\infty}^{+\infty} \theta|\xi_1| \mathcal{F}(\widetilde{v}^j)_{3|y_2=0} \overline{\mathcal{F}(\widetilde{w}^k)_{3|y_2=0}} h^2(\varepsilon, \xi, y_2) \mathrm{d}\xi_1, \quad (6.33)$$

with $\theta = \theta(A^{\alpha\beta\lambda\mu}, (v_-)_1(0), b_{\alpha\beta}, \mu_-)$, where $\mu_- = \frac{\lambda_(\xi_1)}{|\xi_1|}$ is independent of ξ_1.

The expression (6.33) only depends on $(\widetilde{v}^j)_{3|y_2=0}(y_1)$ and $(\widetilde{w}^k)_{3|y_2=0}(y_1)$, which are functions defined on Γ_0.

Remark 6.2. The important fact in (6.33) is the presence of $|\xi_1|$. This comes from $\int_0^{+\infty} e^{-\lambda_- y_2} \mathrm{d}y_2$ and analogous, on account that this integral is equal to $C/|\xi_1|$.

We now simplify this last expression by using a sesquilinear form involving pseudodifferential operators.

Then, defining the elliptic pseudodifferential operator $P(y_1, \frac{\partial}{\partial y_1})$ of order 1/2 with principal symbol

$$(\theta|\xi_1|)^{1/2} h(\varepsilon, \xi, y_2), \quad (6.34)$$

and summing over j and k, we obtain

$$a(\widetilde{v}^a, \widetilde{w}^a) = \int_{\Gamma_0} P\Big(\frac{\partial}{\partial s}\Big)(\widetilde{v}_3)_{|\Gamma_0} \overline{P\Big(\frac{\partial}{\partial s}\Big)(\widetilde{w}_3)_{|\Gamma_0}} \mathrm{d}s. \quad (6.35)$$

Remark 6.3. Since we only considered the principal terms for large $|\xi_1|$, we may define as well $P(\xi_1)$ by the symbol

$$P(\xi_1) = \theta(1 + |\xi_1|^2)^{1/4}. \quad (6.36)$$

The corresponding pseudodifferential operator is elliptic of order 1/2.

Remark 6.4. We shall use the definition (6.36), which is more pleasant than (6.34) since such P defines an isomorphism from $H^s(\Gamma_0)$ onto $H^{s+1/2}(\Gamma_0)$, $s \in \mathbb{R}$.

6.2.2 The flection sublayer on Γ_0

The structure of the flection sublayer, see the beginning of Section 6, accounting for the two new boundary conditions $\widetilde{w}_3 = \partial_n \widetilde{w}_3 = 0$ follows from classical issues in singular perturbation theory, as in [7, Section 7.1.2], [22], and [9]. It is mainly concerned with a drastic change of the normal component $\widetilde{w}_3$ (whereas the conditions on $\widetilde{w}_1$ and $\widetilde{w}_2$ are satisfied). The specific structure is analogous to the layer in [19].

The thickness is of order $\delta = \varepsilon^{1/2}$. This may be easily seen by taking into account only higher order terms in the membrane and the flection systems; eliminating $\widetilde{w}_1$ and $\widetilde{w}_2$, we obtain an equation for $\widetilde{w}_3$. The membrane terms are of order 4 and the flection terms are of order 8. In the layer, the derivatives of order n have an order of magnitude $\mathcal{O}(\frac{\widetilde{w}_3}{\delta^n})$. Since both membrane and flection terms are of the same order of magnitude in the layer, we thus have

$$\mathcal{O}\Big(\frac{\widetilde{w}_3}{\delta^4}\Big) = \varepsilon^2 \mathcal{O}\Big(\frac{\widetilde{w}_3}{\delta^8}\Big),$$

which furnishes $\delta = \mathcal{O}(\varepsilon^{1/2})$.

It is easily seen (as in [7, Section 7.1.2]) that the presence of this flection sublayer plays a negligible role in the asymptotic behavior. Indeed, proceeding as in the previous membrane layer, we see that the expression analogous to (6.35) has the form

$$\varepsilon^2 a_0(\widetilde{v}^a, \widetilde{w}^a) = \varepsilon^2 \int_{\Gamma_0} P_0\Big(\frac{\partial}{\partial s}\Big)(\widetilde{v})_{|\Gamma_0} \overline{P_0\Big(\frac{\partial}{\partial s}\Big)(\widetilde{w})_{|\Gamma_0}} \mathrm{d}s, \tag{6.37}$$

where P_0 is an operator of order 0. Going on to Subsection 6.4, the action of sublayer amounts to change $\mathcal{A}$ to $\mathcal{A} + \varepsilon^2 \mathcal{C}$, where $\mathcal{C}$ is a smoothing operator. Equivalently, we may change $\mathcal{B}$ to $\mathcal{B} + \mathcal{C}$ which is again a 3-order operator (as $\mathcal{C}$ is smoothing). The asymptotic behavior does not change. Equivalently, in (6.1), the effect of the sublayer is to change s to $s + \varepsilon^2 s_0$ where s_0 is a smoothing symbol, or q to $q + s_0$ which is again the symbol of an operator of order $2m > 0$.

For that reasons, the influence of the sublayer will no more be mentioned.

6.3 Formulation of the problem in the heuristic asymptotics

Presently, our aim is to formulate problem (5.1) on the space of u^a with $u \in G^0$. The forms $b(u, v)$ and $\langle f, v \rangle$ should be written within the framework of our formal asymptotics, for $\widetilde{u}^a$ and $\widetilde{v}^a$ obtained from u and v defined on Γ_1 by solving (6.6) and modifying $\widetilde{u}$ and $\widetilde{v}$ with the Γ_0-layer.

The computation of the b-energy form is exactly analogous to that of [7, Section 5.3]. It follows the ideas of the previous section in a much simpler situation. Since only the third component is involved in the higher order terms of the form b (cf. (2.15) and (2.2)), we have

$$b(\widetilde{u}^a, \widetilde{v}^a) \approx \int_{\Omega} B^{\alpha\beta\lambda\mu} \partial_{\alpha\beta} \widetilde{u}_3^a \partial_{\lambda\mu} \widetilde{v}_3^a \mathrm{d}\xi \mathrm{d}x. \tag{6.38}$$

Moreover, from (6.4)–(6.6) and according to our approximations analogous to the construction of a parametrix, $\widetilde{u}, \widetilde{v}$ are only significant in a narrow layer adjacent to Γ_1. The local structure is analogous to (6.12) where, obviously, the decreasing solution inwards the domain should be chosen. This gives the obvious local asymptotics

$$\widehat{\widehat{v}}_3(\xi, y) = \widehat{v}_3(\xi_1)e^{\lambda_-(\xi_1)y_2}, \tag{6.39}$$

where $\lambda_-(\xi_1)$ is proportional to $|\xi_1|$. After substitution (6.39) in (6.38) a computation analogous to that of Subsection 6.2.1 (but much easier) gives (using a partition of unity)

$$b(\widetilde{u}^a, \widetilde{v}^a) = \Sigma_{j,k} \int_{-\infty}^{+\infty} \zeta_{jk}(y_1)|\xi_1|^3 \widetilde{u}_3^j(\xi_1)\widetilde{v}_3^k(\xi_1)\mathrm{d}\xi_1,$$

where $\zeta_{jk}(y_1)$ are smooth positive functions on Γ_1 depending on the coefficients. The function $|\xi_1|^3$ comes obviously from the integrals in the normal direction of products of second order derivatives of functions of the form $e^{\lambda_-(\xi_1)y_2}$, with $\lambda_-(\xi_1)$ proportional to $|\xi_1|$.

Then, defining the pseudodifferential operator $Q(\frac{\partial}{\partial y_1})$ of order 3/2 with principal symbol

$$\sqrt{\zeta(y_1)|\xi_1|^3}, \tag{6.40}$$

we have within our approximation

$$\int_\Omega B^{\alpha\beta\lambda\mu}\partial_{\alpha\beta}u_3\partial_{\lambda\mu}v_3\mathrm{d}x = \int_{\Gamma_1} Q\Big(\frac{\partial}{\partial y_1}\Big)u\, Q\Big(\frac{\partial}{\partial y_1}\Big)v\mathrm{d}y_1. \tag{6.41}$$

We observe that the operator Q is only concerned with the trace on Γ_1 and y_1 which denotes its arc.

The formal asymptotic problem becomes

$$\begin{aligned}&\text{Find } \widetilde{u}^\varepsilon \in G \text{ such that } \forall\, \widetilde{v} \in G\\ &\int_{\Gamma_0} P\Big(\frac{\partial \widetilde{u}^\varepsilon}{\partial n}\Big)\overline{P\Big(\frac{\partial \widetilde{v}}{\partial n}\Big)}\mathrm{d}s + \varepsilon^2 \int_{\Gamma_1} Q(\widetilde{u}^\varepsilon)\, \overline{Q(\widetilde{v})}\mathrm{d}s = \langle f, w\rangle,\end{aligned} \tag{6.42}$$

where G is the completion of G^0 for the norm

$$\|\widetilde{v}\|_G^2 = \int_{\Gamma_0} \Big|P\Big(\frac{\partial v}{\partial n}\Big)\Big|^2 \mathrm{d}s + \int_{\Gamma_1} \Big|Q(v_3)\Big|^2 \mathrm{d}s$$

Remark 6.5. For $\varepsilon > 0$ (6.42) is a classical Lax–Milgram problem. The continuity and coerciveness follow from the ellipticity of the operators P and Q.

6.4 The formal asymptotics and its sensitive behavior

In the sequel, we denote

$$\alpha(\widetilde{v}^\varepsilon, \widetilde{w}) = \int_{\Gamma_0} P\Big(\frac{\partial \widetilde{v}^\varepsilon}{\partial n}\Big)\overline{P\Big(\frac{\partial \widetilde{w}}{\partial n}\Big)}\mathrm{d}s, \tag{6.43}$$

$$\beta(\widetilde{v}^\varepsilon, \widetilde{w}) = \int_{\Gamma_1} Q(\widetilde{v}^\varepsilon)\,\overline{Q(\widetilde{w})}\mathrm{d}s. \tag{6.44}$$

We observe that the problem (6.42) is again within the same abstract framework as the initial problem (2.17). Nevertheless, the context is different, as the nonlocal character of the new problem is apparent from the structure of the space G. Let us define the operators

$$\mathcal{A} \in \mathcal{L}(G, G'), \quad \mathcal{B} \in \mathcal{L}(G, G') \tag{6.45}$$

by

$$\alpha(v, w) = \langle \mathcal{A}v, w\rangle, \quad \beta(v, w) = \langle \mathcal{B}v, w\rangle. \tag{6.46}$$

Let $G_{\mathcal{A}}$ be the completion of G with the norm

$$\|v\|_{\mathcal{A}} = \|\mathcal{A}v\|_{G'}. \tag{6.47}$$

Denoting again by $\mathcal{A}$ its extension to $\mathcal{L}(G_{\mathcal{A}}, G')$, which is an isomorphism, we may rewrite (6.42) in the form

$$(\mathcal{A} + \varepsilon\mathcal{B})\widetilde{v}^\varepsilon = F, \tag{6.48}$$

where $F \in G'$ is defined by

$$\langle F, \widetilde{w}\rangle = \int_{\Omega} f\widetilde{w}\mathrm{d}x \quad \forall\, \widetilde{w} \in V. \tag{6.49}$$

It follows that

$$\widetilde{v}^\varepsilon \to \widetilde{v}^0 \text{ strongly in } G_{\mathcal{A}}, \tag{6.50}$$

where

$$\mathcal{A}\widetilde{v}^0 = F. \tag{6.51}$$

Reduction to a problem on Γ_1. In order to exhibit more clearly the unusual character of the problem, we shall now write (6.42) in another, equivalent form involving only the traces on Γ_1. Coming back to (6.6), let us define $\mathcal{R}_0$ as follows. For a given $w \in C^\infty(\Gamma_1)$ we solve (6.6) and obtain

$$\widetilde{w}_3 = \mathcal{R}_0 w. \tag{6.52}$$

By the regularity properties of the solution of (6.6), it follows that $\mathcal{R}_0 w$ is in $C^\infty(\Gamma_0)$. Moreover, we may take in (6.6) w in any $H^s(\Gamma_1)$, $s \in \mathbb{R}$, and the corresponding solution is of class C^∞ on Γ_0 and its neighborhood, so that $\mathcal{R}_0$ has an extension which is continuous from $H^s(\Gamma_1)$ to $C^\infty(\Gamma_0)$. We denote by $\mathcal{R}_0$ such an extension, so that

$$\mathcal{R}_0 \in \mathcal{L}(H^s(\Gamma_1), H^r(\Gamma_0)) \quad \forall\, s, r \in \mathbb{R}. \tag{6.53}$$

Then (6.42) may be written as a problem for the traces on Γ_1, where $F \in G'$ is defined by

$$\begin{aligned} &\text{Find } v^\varepsilon \in H^{3/2}(\Gamma_1) \text{ such that } \forall\, w \in H^{3/2}(\Gamma_1) \\ &\int_{\Gamma_0} P\Big(\frac{\partial}{\partial s}\Big)\mathcal{R}_0 v^\varepsilon \overline{P\Big(\frac{\partial}{\partial s}\Big)\mathcal{R}_0 w}\mathrm{d}s \\ &\qquad + \varepsilon^2 \int_{\Gamma_1} Q\Big(\frac{\partial}{\partial s}\Big) v^\varepsilon \overline{Q\Big(\frac{\partial}{\partial s}\Big) w}\mathrm{d}s = \int_\Omega F\widetilde{w}\mathrm{d}x, \end{aligned} \tag{6.54}$$

where the configuration space is obviously $H^{3/2}(\Gamma_1)$. The left-hand side with $\varepsilon > 0$ is continuous and coercive.

Remark 6.6. Coerciveness follows from the ellipticity of Q, as it is of order $3/2$. Strictly speaking, this only ensures coerciveness on the leading order terms, which may "forget" a finite-dimensional kernel. But this is controlled by $\mathcal{R}_0$ since it is a surjective operator. Indeed, $\mathcal{R}_0 v = 0$ implies $\gamma_{\alpha\beta}(\widetilde{v}) = 0$ with $\widetilde{v}_3 = \widetilde{v}_2 = 0$ on Γ_0, which implies $\widetilde{v} = 0$ (and then $v = 0$) using the uniqueness of the Cauchy problem for the rigidity system.

Here, $F \in H^{-3/2}(\Gamma_1)$ is defined for $f \in V'$ by

$$\langle F, w\rangle_{H^{-3/2}(\Gamma_1), H^{3/2}(\Gamma_1)} = \langle f, \widetilde{w}\rangle. \tag{6.55}$$

We note that, for instance, when the "loading" f is defined by a "force" F on Γ_1, this function is F in (6.54). Obviously, (6.54) may be written as

$$\Big(\mathcal{R}_0^* P^*\Big(\frac{\partial}{\partial s}\Big) P\Big(\frac{\partial}{\partial s}\Big)\mathcal{R}_0 + \varepsilon^2 Q^*\Big(\frac{\partial}{\partial s}\Big) Q\Big(\frac{\partial}{\partial s}\Big)\Big)\widetilde{v}^\varepsilon = F. \tag{6.56}$$

From (6.53) we see that $\mathcal{R}_0^*$ is also a smoothing operator, i.e.,

$$\mathcal{R}_0^* \in \mathcal{L}(H^{-r}(\Gamma_1), H^{-s}(\Gamma_0)) \quad \forall\, s, r \in \mathbb{R}. \tag{6.57}$$

Now, we define the new operators (but we use the same notation)

$$\mathcal{A} = \mathcal{R}_0^* P^* P \mathcal{R}_0 \ \in \mathcal{L}(H^s(\Gamma_1), H^r(\Gamma_0)) \quad \forall\, s, r \in \mathbb{R}, \tag{6.58}$$

$$\mathcal{B} = Q^* Q \ \in \mathcal{L}(H^{3/2}(\Gamma_1), H^{-3/2}(\Gamma_1)). \tag{6.59}$$

It is obvious that $\mathcal{B}$ is an elliptic pseudodifferential operator of order 3, whereas $\mathcal{A}$ is a smoothing (nonlocal) operator. Then (6.56) becomes

$$(\mathcal{A} + \varepsilon^2\mathcal{B})v^\varepsilon = F \ \text{ in } \ H^{-3/2}(\Gamma_1). \tag{6.60}$$

Once more, the problem (6.54) is within the general framework of (2.17), so that we can define the space $\mathcal{V} = H^{3/2}(\Gamma_1)$ and its completion $\mathcal{V}_A$ with the norm

$$\|v\|_{\mathcal{A}} = \|\mathcal{A}v\|_{H^{-3/2}(\Gamma_1)}. \tag{6.61}$$

Denoting similarly by $\mathcal{A}$ the continuous extension of $\mathcal{A}$, which is an isomorphism between $\mathcal{V}_{\mathcal{A}}$ and $\mathcal{V}'$, we obtain

$$u^\varepsilon \to u^0 \text{ strongly in } \mathcal{V}_{\mathcal{A}}, \tag{6.62}$$

where $u^0 \in \mathcal{V}_{\mathcal{A}}$ satisfies

$$\mathcal{A}u^0 = F. \tag{6.63}$$

It is obvious that this equation is uniquely solvable in $\mathcal{V}_{\mathcal{A}}$ for $F \in \mathcal{V}' = H^{-3/2}(\Gamma_1)$. But, the unusual character of this equation appears now clearly.

Proposition 6.1. *Let $F \in H^{-3/2}(\Gamma_1)$ and $F \notin C^\infty(\Gamma_1)$. Then the problem* (6.63) *has no u^0 solution in $\mathcal{D}'(\Gamma_1)$.*

Proof. If $u^0 \in \mathcal{D}'(\Gamma_0)$ was a solution of (6.63), as Γ_1 is compact, u^0 should be in some H^s, then recalling (6.58), we should have $\mathcal{A}u^0 \in C^\infty(\Gamma_0)$, which is not possible. Moreover, (6.60) is clearly of the form (6.1). □

References

1. Agmon, S., Douglis, A., Nirenberg, L.: Estimates near the boundary for solutions of elliptic partial differential equations satisfying general boundary conditions Part II. Commun. Pure. Appl. Math. **17**, 35–92 (1965)
2. Bechet, F., Millet, O., Sanchez-Palencia, E.: Singular perturbations generating complexification phenomena in elliptic shells. Comput. Mech. **43**, 207–221 (2009)
3. Caillerie, D.: Etude générale d'un type de problèmes raides et de perturbation singulière. Compt. Rend. Math. **323**, 835–840 (1996)
4. Ciarlet, P.G.: Mathematical Elasticity. III. Theory of Shells. North-Holland Publishing Co., Amsterdam (2000)
5. Courant, R., Hilbert, D.: Methods of Mathematical Physics. II. Interscience Publishers, Inc., New York (1962)
6. Douglis, A., Nirenberg, L.: Interior estimates for elliptic systems of partial differential equations. Commun. Pure. Appl. Math. **8**, 503–538 (1958)
7. Egorov, Yu.V., Meunier, N., Sanchez-Palencia, E.: Rigorous and heuristic treatment of certain sensitive singular perturbations. J. Math. Pures Appl. (9) **88**, no. 2, 123–147 (2007)
8. Egorov, Yu.V., Schulze, B.W.: Pseudo-differential Operators, Singularities and Applications. Birkhäuser, Basel (1997)
9. Georgescu, A.: Asymptotic Treatment of Differential Equations. Chapman Hall, London (1995)

10. Huet, D.: Phénomènes de perturbation singulière dans les problèmes aux limites. Ann. Inst. Fourier **10**, 61–150 (1960)
11. Kozlov, V., Maz'ya, V., Movchan, A.: Asymptotic Analysis of Fields in Multistructures. Clarendon Press, Oxford (1999)
12. Maz'ya, V., Nazarov, S., Plamenevskii, B.: Asymptotic Theory of Elliptic Boundary Value Problems in Singularly Perturbed Domains. I.II. Birkhäuser, Basel (2000)
13. Maz'ya, V., Slutskii, A.: Asymptotic analysis of the Navier–Stokes system in a plane domain with thin channels. Asymptotic Anal. **23**, no. 1, 59–89 (2000)
14. Meunier, N., Sanchez-Hubert, J., Sanchez-Palencia, E.: Various kinds of sensitive singular perturbations. Ann. Blaise Pascal **14**, no. 2, 199–242 (2007)
15. Meunier, N., Sanchez-Palencia, E.: Sensitive versus classical perturbation problem via Fourier transform. Math. Models Methods Appl. Sci. **16**, no. 11, 1783–1816, 2006.
16. Meunier, N., Sanchez-Palencia, E.: Integral approach to sensitive singular perturbations. [To appear]
17. Sanchez Hubert, J., Sanchez Palencia, E.: Vibration and Coupling of Continuous Systems. Asymptotic Methods. Springer, Berlin etc. (1989)
18. Sanchez-Hubert, J., Sanchez-Palencia, É.: Coques élastiques minces. Propriétés asymptotiques. Masson, Paris (1997)
19. Sanchez-Palencia, E.: On internal and boundary layers with unbounded energy in thin shell theory. Elliptic case. Asymptotic Anal. **36**, no. 2, 169–185 (2003)
20. Taylor, M.E.: Pseudodifferential Operators. Princeton Univ. Press, Princeton, NJ (1981)
21. Vekua, I.L.: Generalised Analytic Functions. Pergamon Press, London (1962)
22. Vishik, M.I., Lyusternik, L.A.: Regular degeneration and boundary layer for linear differential equations with small parameter (Russian). Usp. Mat. Nauk **12**, no. 5, 3–122 (1957)

On the Existence of Positive Solutions of Semilinear Elliptic Inequalities on Riemannian Manifolds

Alexander Grigor'yan and Vladimir A. Kondratiev

Abstract We consider elliptic inequalities of type $\Delta u + u^\sigma \leqslant 0$ on geodesically complete Riemannian manifolds and prove sharp sufficient conditions in terms of capacities and volumes for the nonexistence of positive solutions.

1 Introduction and the Main Results

Let M be a smooth connected Riemannian manifold. Consider the differential inequality on M

$$\operatorname{div}(A(x)\nabla u) + V(x)u^\sigma \leqslant 0, \tag{1.1}$$

where ∇ and div are the Riemannian gradient and divergence respectively, $u = u(x)$ is an unknown positive function on M, $\sigma > 1$ is a given constant, V is a given positive measurable function on M, and A is a given measurable tensor field on M such that $A(x)$ is a nonnegative definite symmetric operator in the tangent space T_xM. The inequality (1.1) is understood in a weak sense to be explained below.

We are concerned with the question when (1.1) has no positive solution u on M. This question in the setting of Euclidean spaces has a long history, starting with the pioneering work of Gidas and Spruck [3]. We refer the reader to [10] for the survey of this problem. Let us cite only one result in this direction, which already exhibits the phenomenon that the answer depends on the interplay of all the data, including the geometry of M and the value

Alexander Grigor'yan
Department of Mathematics, University of Bielefeld, 33501 Bielefeld, Germany
e-mail: alexander.grigorian@googlemail.com

Vladimir A. Kondratiev
Department of Mathematics, Moscow State University, 119992 Moscow, Russia
e-mail: vla-kondratiev@yandex.ru

A. Laptev (ed.), *Around the Research of Vladimir Maz'ya II: Partial Differential Equations*, 203
International Mathematical Series 12, DOI 10.1007/978-1-4419-1343-2_8,

of σ. Indeed, it is known that the following inequality in $\mathbb{R}^n$, $n > 2$,

$$\Delta u + u^\sigma \leqslant 0$$

has no positive solution if and only if $\sigma \leqslant \frac{n}{n-2}$. If $n \leqslant 2$, then there is no positive solution for any σ.

The previously developed methods for investigation of the above question include such advanced tools as Harnack inequalities and estimates of fundamental solutions. Here, we adopt another approach that originates from [8] and that uses only very basic tools as capacities and volumes. This enables us to replace a traditional assumption on $A(x)$ to be positive definite, by the nonnegative definiteness.

In the rest of this section, we state the main results: the capacity tests and the volume test. The former are proved in Section 2, and the latter in Section 4. In Sections 3 and 5, we give examples showing the sharpness of the above tests.

Let us explain in what sense we understand (1.1). Let μ be the Riemannian measure on M. All the spaces $L^p(M)$ will be considered with respect to μ. Recall that $W^1(M)$ is the Sobolev space defined by

$$W^1(M) = \left\{ f \in L^2(M) : |\nabla f| \in L^2(M) \right\},$$

where ∇f is the weak gradient of f. Let $W_c^1(M)$ be a subspace of $W^1(M)$ that consists of functions with compact support.

Similarly, define a local Sobolev space $W^1_{\mathrm{loc}}(M)$ by

$$W^1_{\mathrm{loc}}(M) = \left\{ f \in L^2_{\mathrm{loc}}(M) : |\nabla f| \in L^2_{\mathrm{loc}}(M) \right\}.$$

Definition. A function u on M is called a *positive (weak) solution* of the inequality (1.1) on M if u is a positive function from $W^1_{\mathrm{loc}}(M)$ such that $\frac{1}{u} \in L^\infty_{\mathrm{loc}}(M)$ and for any nonnegative function $\psi \in W_c^1(M)$ the following inequality holds:

$$-\int_M (A(x)\nabla u, \nabla\psi)\, d\mu + \int_M V(x) u^\sigma \psi d\mu \leqslant 0, \tag{1.2}$$

where $(\cdot,\cdot)$ is the inner product in T_xM given by the Riemannian metric.

To ensure the finiteness of the integrals in (1.2), we assume henceforth that the function $x \mapsto \|A(x)\|$ is locally bounded, where $\|A(x)\|$ is the norm of the operator $A(x)$ in T_xM, i.e., the maximal eigenvalue of the operator $A(x)$. Indeed, since $K := \operatorname{supp}\psi$ is compact, we have

$$\begin{aligned}\int_M |(A(x)\nabla u, \nabla\psi)|\, d\mu &= \int_K |(A(x)\nabla u, \nabla\psi)|\, d\mu \\ &\leqslant \operatorname{ess\,sup}_{x\in K} \|A(x)\|\, \|\nabla u\|_{L^2(K)} \|\nabla\psi\|_{L^2(K)} < \infty.\end{aligned}$$

The second integral in (1.2) is then finite due to (1.2).

Consider the following quadratic form in the tangent space $T_x M$:

$$(\xi, \eta)_A := (A(x)\xi, \eta)$$

and the corresponding seminorm

$$|\xi|_A := (A(x)\xi, \xi)^{1/2}.$$

In particular, for any function $f \in W^1_{\mathrm{loc}}(M)$ we have

$$|\nabla f|_A = (A(x)\nabla f, \nabla f)^{1/2}.$$

Definition. Let A and V be as above. Fix two constants $p > 0$ and $q \geqslant 0$. For any precompact set $K \subset M$ we define the *capacity* $\mathrm{cap}_{p,q}(K)$ as follows:

$$\mathrm{cap}_{p,q}(K) := \inf_{\varphi \in \mathcal{T}(K)} \int_M |\nabla \varphi|_A^p \, V^{-q} d\mu,$$

where $T(K)$ is the class of test functions defined by

$$\mathcal{T}(K) = \left\{ \varphi \in \mathrm{Lip}_c(M) : 0 \leqslant \varphi \leqslant 1 \text{ and } \varphi \equiv 1 \text{ in a neighborhood of } \overline{K} \right\},$$

and $\mathrm{Lip}_c(M)$ is the class of Lipschitz functions on M with compact support.

If $q = 0$, we write

$$\mathrm{cap}_p(K) \equiv \mathrm{cap}_{p,0}(K) = \inf_{\varphi \in \mathcal{T}(K)} \int_M |\nabla \varphi|_A^p \, d\mu,$$

so that $\mathrm{cap}_p(K)$ is independent of $V(x)$. It is well known that if $\mathrm{cap}_2(K) = 0$ for any compact set (or for some compact set with nonempty interior), then any positive solution of the inequality

$$\mathrm{div}\,(A(x)\nabla u) \leqslant 0 \tag{1.3}$$

must be constant (cf., for example, [5]). Since any positive solution of (1.1) satisfies (1.3), in this setting we obtain $u \equiv \mathrm{const}$, which implies $u \equiv 0$ by (1.1). Hence the condition $\mathrm{cap}_2(K) = 0$ implies the absence of a positive solution of (1.1) for any potential $V(x)$ and any σ.

We state now more subtle conditions in terms of higher capacities that take into account also σ and $V(x)$. Fix some $\sigma > 1$ in (1.1) and set

$$p = \frac{2\sigma}{\sigma - 1}, \quad q = \frac{1}{\sigma - 1}. \tag{1.4}$$

Theorem 1.1. *If for some compact set $K \subset M$ with nonempty interior the following condition is satisfied:*

$$\operatorname{cap}_{p-2\varepsilon,q-\varepsilon}(K) = o\left(\varepsilon^{p/2}\right) \quad \text{as } \varepsilon \to 0+, \tag{1.5}$$

then (1.1) *has no positive solution on M.*

Theorem 1.2. *If for some compact set $K \subset M$ with nonempty interior and some $\varepsilon \in (0,q]$ the following condition is satisfied:*

$$\operatorname{cap}_{p-2\varepsilon,q-\varepsilon}(K) = 0, \tag{1.6}$$

then (1.1) *has no positive solution on M.*

Let $d(x,y)$ be the geodesic distance on M, and let $B(x,r)$ be the open geodesic ball of radius r centered at $x \in M$. Assume further that M is geodesically complete, which is equivalent to the relative compactness of all geodesic balls in M. For any $\varepsilon \geqslant 0$ consider a measure ν_ε on M defined by

$$d\nu_\varepsilon = \|A\|^{p/2-\varepsilon}\, V^{-(q-\varepsilon)} d\mu,$$

where p and q are the same as in (1.4).

Theorem 1.3. *Let M be a geodesically complete Riemannian manifold. Assume that for some $x_0 \in M$, $C > 0$, and $\varkappa < q$ the following inequality holds:*

$$\nu_\varepsilon\left(B\left(x_0,r\right)\right) \leqslant C r^{p+C\varepsilon} \log^{\varkappa} r \tag{1.7}$$

for all large enough r and all small enough $\varepsilon > 0$. Then (1.1) *has no positive solution.*

As we will see in Section 5, the restriction $\varkappa < q$ is sharp. More precisely, in the case $\varkappa > q$, there is a counterexample with a positive solution. The borderline case $\varkappa = q$ requires further investigation.

Note that

$$d\nu_0 = \|A\|^{p/2}\, V^{-q} d\mu$$

and

$$d\nu_\varepsilon = \left(\frac{V}{\|A\|}\right)^{\varepsilon} d\nu_0$$

(although the latter makes sense only if $\|A(x)\| > 0$ a.e.). It is clear that the condition (1.7) holds provided that

$$\nu_0\left(B\left(x_0,r\right)\right) \leqslant C r^{p}\left(\log r\right)^{\varkappa} \tag{1.8}$$

and

$$\frac{V}{\|A\|}(x) \leqslant C\left(1 + d\left(x,x_0\right)\right)^{C} \tag{1.9}$$

for some $C > 0$.

Example. Let $A = \mathrm{id}$ and $V \equiv 1$ so that the inequality (1.1) becomes

$$\Delta u + u^\sigma \leqslant 0. \tag{1.10}$$

The condition (1.7) is equivalent to

$$\mu\left(B\left(x_0, r\right)\right) \leqslant C r^p \log^\varkappa r, \tag{1.11}$$

where p and $\varkappa$ are related to σ as above, i.e.,

$$p = \frac{2\sigma}{\sigma - 1} \quad \text{and} \quad \varkappa < \frac{1}{\sigma - 1}.$$

Assume now that (1.11) is given with some nonnegative p and $\varkappa$, and determine for which $\sigma > 1$ the inequality (1.1) has no positive solutions. Set

$$p_\sigma = \frac{2\sigma}{\sigma - 1}, \quad \varkappa_\sigma = \frac{1}{\sigma - 1}.$$

If either $p < p_\sigma$ or $p = p_\sigma$ and $\varkappa < \varkappa_\sigma$, then (1.11) implies

$$\mu\left(B\left(x_0, r\right)\right) \leqslant C r^{p_\sigma} \log^{\varkappa_\sigma - \varepsilon} r$$

for some $\varepsilon > 0$, whence the absence of positive solutions of (1.10) follows. In terms of σ, the above conditions are satisfied in any of the three cases:

1. $p \leqslant 2$, $\sigma > 1$, $\varkappa$ is any,
2. $p > 2$, $\sigma < \frac{p}{p-2}$, $\varkappa$ is any,
3. $p > 2$, $\sigma = \frac{p}{p-2}$, and $\varkappa < \frac{p}{2} - 1$.

For example, if $M = \mathbb{R}^n$, then (1.11) holds with $p = n$ and $\varkappa = 0$. If $n \leqslant 2$, then (1.10) has no positive solutions for any $\sigma > 1$ (in fact, for any real σ), and if $n > 2$, then (1.10) has no positive solutions for $\sigma \leqslant \frac{n}{n-2}$, as it was already mentioned above.

2 Proof of the Capacity Tests

Here, we prove Theorems 1.1 and 1.2, using the approach of Kurta [8]. Let u be a positive solution of (1.1). We first obtain some estimates of u without using specific hypotheses of Theorems 1.1 and 1.2. Fix some function $\varphi \in \mathrm{Lip}_c(M)$ such that $0 \leqslant \varphi \leqslant 1$, constants $0 < t \leqslant 1$, $s \geqslant 2$, and take in (1.2) the test function $\psi = u^{-t}\varphi^s$. It is clear that ψ has compact support and is bounded due to the local boundedness of u^{-1}. We have

$$\nabla\psi = -tu^{-t-1}\varphi^s \nabla u + su^{-t}\varphi^{s-1}\nabla\varphi,$$

whence it is clear that $|\nabla\psi| \in L^2(M)$ and, consequently, $\psi \in W_c^1(M)$. From (1.2) we obtain

$$t\int_M |\nabla u|_A^2 u^{-t-1}\varphi^s d\mu + \int_M u^{\sigma-t}\varphi^s V d\mu \leqslant s\int_M (\nabla u, \nabla\varphi)_A u^{-t}\varphi^{s-1}d\mu. \quad (2.1)$$

By the Cauchy–Schwarz inequality, we estimate the right-hand side of (2.1) as follows:

$$\begin{aligned} s\int_M (\nabla u, \nabla\varphi)_A u^{-t}\varphi^{s-1}d\mu &= \int_M \left(\sqrt{t}u^{-\frac{t+1}{2}}\varphi^{s/2}\nabla u, \frac{s}{\sqrt{t}}u^{-\frac{t-1}{2}}\varphi^{s/2-1}\nabla\varphi\right)_A d\mu \\ &\leqslant \frac{t}{2}\int_M |\nabla u|_A^2 u^{-t-1}\varphi^s d\mu + \frac{s^2}{2t}\int_M |\nabla\varphi|_A^2 u^{1-t}\varphi^{s-2}d\mu. \end{aligned}$$

Substituting into (2.1) and cancelling out the half of the first term in (2.1), we obtain

$$\frac{t}{2}\int_M |\nabla u|_A^2 u^{-t-1}\varphi^s d\mu + \int_M u^{\sigma-t}\varphi^s V d\mu \leqslant \frac{s^2}{2t}\int_M |\nabla\varphi|_A^2 u^{1-t}\varphi^{s-2}d\mu. \quad (2.2)$$

In what follows, assume that $0 < t < 1$ and set

$$\alpha = \frac{\sigma - t}{1 - t}, \quad \beta = \frac{\sigma - t}{\sigma - 1}$$

so that α and β are Hölder conjugate. Applying the Young inequality in the form

$$\int fg\,d\mu \leqslant \int |f|^\alpha d\mu + \int |g|^\beta d\mu,$$

we estimate the right-hand side of (2.2) as follows:

$$\begin{aligned} \frac{s^2}{2t}\int_M |\nabla\varphi|_A^2 u^{1-t}\varphi^{s-2}d\mu &= \frac{1}{2}\int_M \left[u^{1-t}\varphi^{\frac{s}{\alpha}}V^{\frac{1}{\alpha}}\right]\left[\frac{s^2}{t}|\nabla\varphi|_A^2\varphi^{\frac{s}{\beta}-2}V^{-\frac{1}{\alpha}}\right]d\mu \\ &\leqslant \frac{1}{2}\int_M u^{\sigma-t}\varphi^s V d\mu + \frac{1}{2}\left(\frac{s^2}{t}\right)^{\frac{\sigma-t}{\sigma-1}}\int_M |\nabla\varphi|_A^{2\frac{\sigma-t}{\sigma-1}}\varphi^{s-2\frac{\sigma-t}{\sigma-1}}V^{-\frac{1-t}{\sigma-1}}d\mu. \end{aligned}$$

Now, we substitute this estimate into (2.2), using also that

$$\left(\frac{s^2}{t}\right)^{\frac{\sigma-t}{\sigma-1}} \leqslant \left(\frac{s^2}{t}\right)^{\frac{\sigma}{\sigma-1}}$$

and

$$\varphi^{s-2\frac{\sigma-t}{\sigma-1}} \leqslant 1$$

provided that

$$s > \frac{2\sigma}{\sigma-1},$$

which will be assumed in the sequel. Noticing that a half of the middle term in (2.2) cancels out and multiplying by 2, we obtain

$$t\int\limits_M |\nabla u|_A^2 u^{-t-1}\varphi^s d\mu + \int\limits_M u^{\sigma-t}\varphi^s V d\mu \leqslant \left(\frac{s^2}{t}\right)^{\frac{\sigma}{\sigma-1}} \int\limits_M |\nabla\varphi|_A^{2\frac{\sigma-t}{\sigma-1}} V^{-\frac{1-t}{\sigma-1}} d\mu. \tag{2.3}$$

Proof of Theorem 1.1. Let K be a compact set from (1.5), and let φ be a test function from the class $\mathcal{T}(K)$. Applying (2.3) with this function φ and taking the infimum in φ on the right-hand side, we obtain

$$\int_K u^{\sigma-t} V d\mu \leqslant \left(\frac{s^2}{t}\right)^{\frac{\sigma}{\sigma-1}} \operatorname{cap}_{2\frac{\sigma-t}{\sigma-1},\frac{1-t}{\sigma-1}}(K) = c_{s,\sigma}\varepsilon^{-p/2}\operatorname{cap}_{p-2\varepsilon,q-\varepsilon}(K), \tag{2.4}$$

where $\varepsilon = t/(\sigma-1)$. Letting $\varepsilon \to 0$ and using the hypothesis (1.5), we see that the right-hand side here goes to 0, whence

$$\int_K u^\sigma V d\mu = 0,$$

which contradicts the positivity of u and V. □

Proof of Theorem 1.2. Let K be a compact set from (1.6). If $0 < \varepsilon < q$, then we set $t = \varepsilon(\sigma-1)$ so that $0 < t < 1$. Then the right-hand side of (2.4) vanishes due to (1.6), whence we again obtain a contradiction.

If $\varepsilon = q$, then $t = 1$ and (2.3), (2.4) do not apply. In this case, the condition (1.6) becomes $\operatorname{cap}_2(K) = 0$, which implies that any positive solution of the inequality (1.3) is constant. Hence (1.1) has no positive solution. Alternatively, from (2.2) with $s = 2$ we obtain

$$\int_M u^{\sigma-1}\varphi^2 V d\mu \leqslant 2\int_M |\nabla\varphi|_A^2 d\mu. \tag{2.5}$$

The hypothesis $\operatorname{cap}_2(K) = 0$ implies that the infimum of the right-hand side of (2.5) over all $\varphi \in \mathcal{T}(K)$ is equal to 0, which completes the proof. □

The condition (1.6) of Theorem 1.2 can be replaced by the following assumption: there is a constant $C > 0$ such that for any compact set $K \subset M$,

$$\operatorname{cap}_{p-2\varepsilon,q-\varepsilon}(K) \leqslant C. \tag{2.6}$$

Indeed, using certain properties of capacities (cf. [6, Lemma 2.5]), it is possible to show that (2.6) implies (1.6).

3 Examples to the Capacity Test

In this section, $M = \mathbb{R}^n$, $n > 2$, μ is the Lebesgue measure, $A(x) = (a_{ij}(x))$, where $a_{ij} \in L^\infty(\mathbb{R}^n)$, and $V(x) \equiv 1$. Set $\sigma = \frac{n}{n-2}$ which is the critical exponent for the problem (1.1). Let B_R be the Euclidean ball of radius R centered at the origin. Let us use the following expression for the Euclidean capacity (cf. [1, 9]): for any $s \in (1, n)$

$$\inf_{\varphi \in \mathcal{I}(B_R)} \int_{\mathbb{R}^n} |\nabla \varphi|^s \, d\mu = (n-s)^{s-1} \frac{\omega_n}{(s-1)^s} R^{n-s},$$

where ω_n is the area of the unit sphere in $\mathbb{R}^n$. Since for the above value of σ we have $p = n$ and $\|A\|$ is uniformly bounded, for $s = p - 2\varepsilon = n - 2\varepsilon$ we obtain

$$\begin{aligned}
\operatorname{cap}_{p-2\varepsilon,q-\varepsilon}(B_R) &\leqslant C \inf_{\varphi \in \mathcal{I}(B_R)} \int_{\mathbb{R}^n} |\nabla \varphi|^{n-2\varepsilon} \, d\mu \\
&= C\,(2\varepsilon)^{n-2\varepsilon-1} \frac{\omega_n}{(n-2\varepsilon-1)^{n-2\varepsilon}} R^{2\varepsilon} \\
&= O(\varepsilon^n) \quad \text{as } \varepsilon \to 0.
\end{aligned}$$

The condition (1.5), i.e.,

$$\operatorname{cap}_{p-2\varepsilon,q-\varepsilon}(B_R) = o(\varepsilon^{n/2}),$$

is obviously satisfied, and, by Theorem 1.1, we see that (1.1) has no positive solution. This result was previously known for positive definite matrices $A(x)$.

Let us show that one cannot set $\varepsilon = 0$ in Theorem 1.1, i.e., the condition

$$\operatorname{cap}_{p,q}(K) = 0$$

does not necessarily imply the nonexistence of positive solutions. Before we can state an example supporting this claim, let us cite the following theorem of Atkinson.

Proposition 3.1 (Atkinson [2]). *Let $\sigma > 1$ be a constant, and let $\beta(x)$ be a continuous function on $(x_0, +\infty)$ such that*

$$\int^\infty x\,|\beta(x)|\,dx < \infty. \tag{3.1}$$

Then there exists a positive solution $y(x)$ to the differential equation

$$y'' + \beta(x)y^{\sigma} = 0$$

in an interval $(x_1, +\infty)$ with a large enough x_1 such that

$$y(x) \to 1 \quad \text{and} \quad y'(x) \to 0 \quad \text{as} \quad x \to +\infty.$$

We use the following generalization of Proposition 3.1.

Proposition 3.2. *Let $\alpha(x)$ be a positive C^1-function on $(x_0, +\infty)$ such that*

$$\int^{\infty} \frac{dx}{\alpha(x)} < \infty. \tag{3.2}$$

Define the function $\gamma(x)$ on $(x_0, +\infty)$ by

$$\gamma(x) = \int_x^{\infty} \frac{ds}{\alpha(s)}.$$

Let $\beta(x)$ be a continuous function on $(x_0, +\infty)$ such that

$$\int^{\infty} \gamma(x)^{\sigma} |\beta(x)| \, dx < \infty. \tag{3.3}$$

Then the differential equation

$$(\alpha(x)y')' + \beta(x) y^{\sigma} = 0 \tag{3.4}$$

has a positive solution $y(x)$ on an interval $(x_1, +\infty)$ for large enough x_1 such that

$$y(x) \sim \gamma(x) \quad \text{as } x \to +\infty.$$

Proof. Introducing an independent variable $z = \frac{1}{\gamma(x)}$ and a function $u(z) = y(x)z$, we find, by the chain rule, that

$$\frac{d^2u}{dz^2} + \widetilde{\beta}(z) u^{\alpha} = \alpha\gamma^3 \frac{d}{dx}\left(\alpha \frac{dy}{dx}\right) + \frac{\widetilde{\beta}}{\gamma^{\sigma}} y^{\sigma},$$

so that (3.4) is equivalent to the equation

$$\frac{d^2u}{dz^2} + \widetilde{\beta}(z) u^{\alpha} = 0$$

with $\widetilde{\beta}(z) = \alpha\gamma^{\sigma+3}\beta$. By Proposition 3.1, this equation has a positive solution in a neighborhood of $+\infty$ provided that

$$\int^{\infty} z \left|\widetilde{\beta}(z)\right| dz < \infty. \tag{3.5}$$

By (3.2), $z \to \infty$ is equivalent to $x \to \infty$. Since

$$dz = -\frac{\gamma'}{\gamma^2} dx = \frac{1}{\alpha\gamma^2} dx,$$

the condition (3.5) becomes

$$\int^{\infty} \frac{1}{\gamma} |\beta(x)| \alpha\gamma^{\sigma+3} \frac{1}{\alpha\gamma^2} dx < \infty,$$

which coincides with (3.3). Finally, by Proposition 3.1, there is a solution $u(z) \sim 1$ as $z \to \infty$, which implies $y(x) \sim \gamma(x)$ as $x \to \infty$. □

Our purpose here is to construct in $\mathbb{R}^n$ a positive solution $u(x)$ of the inequality

$$\sum_{i=1}^{n} \frac{\partial}{\partial x_i} \left(a(r) \frac{\partial u}{\partial x_i} \right) + u^{\sigma} \leqslant 0, \tag{3.6}$$

where $\sigma = \frac{n}{n-2}$, r is the polar radius in $\mathbb{R}^n$, and the function $a(r)$ is a positive constant for small r and

$$a(r) = \log^k r \quad \text{for large} \quad r,$$

where k can be any constant such that

$$k > \frac{n-2}{n}. \tag{3.7}$$

Since $p = n$ and $V \equiv 1$, the corresponding capacity is given by

$$\operatorname{cap}_{p,q}(K) = \operatorname{cap}_n(K) = \inf_{\varphi \in \mathcal{T}(K)} \int_{\mathbb{R}^n} a^{n/2}(r) |\nabla\varphi|^n dx.$$

Evaluation of this capacity by the variational method shows that for any ball B_R centered at the origin

$$\operatorname{cap}_n(B_R) = c_n \left(\int_R^{\infty} \frac{dr}{\left(a^{n/2}(r) r^{n-1}\right)^{\frac{1}{n-1}}} \right)^{1-n},$$

where $c_n > 0$. Hence $\operatorname{cap}_n(B_R) = 0$ if and only if

$$\frac{nk}{2(n-1)} \leqslant 1. \tag{3.8}$$

It is clear that there is k such that both conditions (3.7) and (3.8) are satisfied. With this k, we obtain an example, where $\operatorname{cap}_{p,q}(K) = 0$ for any compact set K, whereas the inequality (3.6) has a positive solution.

We construct such a solution as a function of r only, so we write $u = u(r)$. Writing u' and a' for the derivative in r and using that $\frac{\partial x_i}{\partial r} = \frac{x_i}{r}$, one easily obtains

$$\sum_{i=1}^{n} \frac{\partial}{\partial x_i}\left(a(r)\frac{\partial u}{\partial x_i}\right) = au'' + a'u' + \frac{(n-1)a}{r}u'$$
$$= r^{1-n}\left(a(r)r^{n-1}u'\right)'.$$

Hence (3.6) is equivalent to

$$\left(a(r)r^{n-1}u'\right)' + r^{n-1}u^{\alpha} \leqslant 0. \tag{3.9}$$

The condition (3.2) of Proposition (3.2) is obviously satisfied. The function $\gamma(r)$ is given for large r by

$$\gamma(r) = \int_r^{\infty} \frac{ds}{s^{n-1}a(s)} = \int_r^{\infty} \frac{ds}{s^{n-1}\log^k s} \simeq r^{-(n-2)}\log^{-k} r.$$

The condition (3.3) with $\beta(r) = r^{n-1}$ is satisfied provided that

$$\int^{\infty} r^{-\sigma(n-2)}\log^{-\sigma k} r^{n-1}dr = \int^{\infty} \frac{dr}{r\log^{\alpha k} r} < \infty,$$

which is exactly the case where $k > \frac{1}{\sigma}$, which is the same as (3.7). By Proposition 3.2, there is a positive solution $u(r)$ of (3.9) in some interval $[r_0, +\infty)$ such that

$$u(r) \sim \gamma(r) \simeq r^{-(n-2)}\log^{-k} r \quad \text{as} \quad r \to \infty,$$

in particular, $u(r) \to 0$ as $r \to \infty$. By increasing r_0 if necessary, we can assume that $u'(r_0) < 0$. For small values of r, namely for $r \leqslant \xi$ where ξ will be specified later on, the function $a(r)$ will be defined to be a constant whose value will also be determined later.

So far consider the linear equation

$$v'' + \frac{n-1}{r}v' + \varepsilon v = 0$$

that has a solution $v(r)$ with the initial conditions

$$v(0) = 1, \quad v'(0) = 0.$$

This solution is positive and decreasing for $r < r_\varepsilon$ for some positive r_ε and vanishes at r_ε; moreover, $r_\varepsilon \to \infty$ as $\varepsilon \to 0$. Since $0 < v \leqslant 1$ in $(0, r_\varepsilon)$, it follows that v is a positive solution in $(0, r_\varepsilon)$ of the inequality

$$v'' + \frac{n-1}{r}v' + \varepsilon v^{\sigma} \leqslant 0. \tag{3.10}$$

Choose ε so small that $r_\varepsilon > r_0$ and

$$\frac{v'}{v}(r_0) > \frac{u'}{u}(r_0). \tag{3.11}$$

This is possible to achieve because for small enough ε the function $v(r)$ is nearly constant 1 up to $2r_0$ and $v'(r_0)$ can be made arbitrarily close to 0 (although negative), whereas $u'(r_0) < 0$ by construction.

Compare the functions $u(r)$ and $v(r)$ in the interval $[r_0, r_\varepsilon)$. Set

$$c = \inf_{r\in[r_0,r_\varepsilon)} \frac{u(r)}{v(r)}.$$

Since $u(r)/v(r) \to \infty$ as $r \to r_\varepsilon$, the value c is attained at some point, say $\xi \in [r_0, r_\varepsilon)$. We claim that $\xi > r_0$. Indeed, at $r = r_0$, we have by (3.11)

$$\left(\frac{u}{v}\right)'(r_0) = \frac{u'v - uv'}{v^2}(r_0) < 0,$$

so that $u(r)/v(r)$ takes smaller values for some $r > r_0$. Hence the minimum point ξ is contained in an open interval (r_0, r_ε), and, at this point, we have

$$\left(\frac{u}{v}\right)'(\xi) = 0.$$

It follows that

$$u(\xi) = cv(\xi) \quad \text{and} \quad u'(\xi) = cv'(\xi). \tag{3.12}$$

Now, we extend/redefine the function $u(r)$ for $r < \xi$ by setting $u(r) = cv(r)$. From (3.10) it follows that u satisfies in $(0, \xi]$ the inequality

$$u'' + \frac{n-1}{r}u' + \frac{\varepsilon}{c^{\sigma-1}}u^{\alpha} \leqslant 0.$$

Hence, setting $a(r) \equiv c^{\sigma-1}/\varepsilon$ in $[0, \xi]$, we see that u satisfies (3.6) for $r \leqslant \xi$. Since u satisfies (3.6) also for $r \geqslant \xi$ and u is differentiable at ξ by (3.12), we find that u is a weak solution of (3.6) in $\mathbb{R}^n$.

4 Proof of the Volume Test

Here, we prove Theorem 1.3 by using Theorem 1.1. Using the obvious inequality

$$|\nabla\varphi|_A \leqslant \|A\|^{1/2} |\nabla\varphi|, \tag{4.1}$$

where $|\nabla\varphi|$ is the Riemannian length of the gradient $\nabla\varphi$, and setting $\varepsilon = \frac{t}{\sigma-1}$ in (2.3), we see that the integral on the right-hand side of (2.3) can be estimated as follows:

$$\int_M |\nabla\varphi|_A^{p-2\varepsilon} V^{-(q-\varepsilon)} d\mu \leqslant \int_M |\nabla\varphi|^{p-2\varepsilon} \|A\|^{p/2-\varepsilon} V^{-(q-\varepsilon)} d\mu = \int_M |\nabla\varphi|^{p-2\varepsilon} d\nu_\varepsilon. \tag{4.2}$$

Next we apply the following result: for any Radon measure ν on a complete Riemannian manifold, any $s > 1$, and any ball $B_r = B(x_0, r)$

$$\inf_{\varphi\in\mathcal{T}(B_r,M)} \int_M |\nabla\varphi|^s d\nu \leqslant C_s \left(\int_r^\infty \left(\frac{\rho}{\nu(B_\rho)} \right)^{\frac{1}{s-1}} d\rho \right)^{1-s} \tag{4.3}$$

(cf. [4, 5, 7] and [9, Section 2.2.2, Lemma 1]). The constant C_s is locally uniformly bounded in the interval $s \in (1, +\infty)$. The range of s that we are interested in is $s \approx p$, so that we can assume that C_s is uniformly bounded from above independently of s.

Applying (4.3) with $\nu = \nu_\varepsilon$ and $s = p - 2\varepsilon$ and combining with (4.2), we obtain

$$\operatorname{cap}_{p-2\varepsilon,q-\varepsilon}(B_r) \leqslant C \left(\int_r^\infty \left(\frac{\rho}{\nu_\varepsilon(B_\rho)} \right)^{\frac{1}{p-1-2\varepsilon}} d\rho \right)^{1+2\varepsilon-p}.$$

The condition (1.5) of Theorem 1.1 will be satisfied provided that

$$\lim_{\varepsilon\to 0} \varepsilon^{\frac{p}{2(p-1)}} \int_r^\infty \left(\frac{\rho}{\nu_\varepsilon(B_\rho)} \right)^{\frac{1}{p-1-2\varepsilon}} d\rho = \infty.$$

In the view of the hypothesis (1.7), it suffices to show that

$$\lim_{\varepsilon\to 0} \varepsilon^{\frac{p}{2(p-1)}} \int_r^\infty \rho^{-\frac{p-1+C\varepsilon}{p-1-2\varepsilon}} (\log\rho)^{-\frac{\varkappa}{p-1-2\varepsilon}} d\rho = \infty, \tag{4.4}$$

where r can be assumed large enough (but fixed). Making change $\rho = e^t$ and setting $\delta = \frac{(C+2)\varepsilon}{p-1-2\varepsilon}$, we find that the integral in (4.4) is equal to

$$\int_{\log r}^\infty \exp(-\delta t)\, t^{-\frac{\varkappa}{p-1-2\varepsilon}} dt = \delta^{\frac{\varkappa}{p-1-2\varepsilon}-1} \int_{\delta\log r}^\infty \exp(-\tau)\, \tau^{-\frac{\varkappa}{p-1-2\varepsilon}} d\tau. \tag{4.5}$$

As $\varepsilon \to 0$, the right-hand side of (4.5) is of the order

$$\text{const}\ \varepsilon^{\frac{\varkappa}{p-1}-1},$$

where const is a positive constant. Hence the expression under the limit in (4.4) is of the order

$$\varepsilon^{\frac{p}{2(p-1)}+\frac{\varkappa}{p-1}-1}.$$

By hypothesis, we have $\varkappa < q = p/2 - 1$, which implies that the exponent of ε here is negative, which proves (4.4).

Remark. Assume that M is geodesically complete and consider the measure

$$d\nu = \|A\| d\mu.$$

It is clear that

$$\operatorname{cap}_2(K) = \inf_{\varphi \in \mathcal{T}(K)} \int |\nabla\varphi|_A^2 \, d\mu \leqslant \inf_{\varphi \in \mathcal{T}(K)} \int_M |\nabla\varphi|^2 \, d\nu.$$

Using the estimate (4.3), we find that if

$$\int^{\infty} \frac{rdr}{\nu\left(B_r\right)} = \infty, \tag{4.6}$$

then $\operatorname{cap}_2(K) = 0$. As was noted in Section 1, the latter implies that (1.1) has no positive solution regardless of V and σ. The condition (4.6) is satisfied if, for example,

$$\nu\left(B_r\right) \leqslant Cr^2 \quad \text{for all large } r. \tag{4.7}$$

5 Examples to the Volume Test

In this section, $M = \mathbb{R}^n$,

$$V(x) \simeq r^{-\alpha_1} \log^{-\alpha_2} r \quad \text{and} \quad \|A(x)\| \simeq r^{\beta_1} \log^{\beta_2} r \tag{5.1}$$

as $r := |x| \to \infty$, where α_i, β_i are real constants.

If $\beta_1 < 2 - n$, then it is easy to verify that the condition (4.7) of Remark 4 is satisfied. Hence there is no positive solution of (1.1) for any σ and V.

In the sequel, we assume that

$$\beta_1 + n - 2 > 0.$$

For the functions $V(x)$ and $\|A(x)\|$ from (5.1) the condition (1.9) is obviously satisfied. Hence the hypothesis (1.7) of Theorem 1.3 can be replaced by (1.8). Let us estimate $\nu_0\left(B_R\right)$, where B_R is the ball of radius R centered at the origin. For large R we have

$$\nu_0\left(B_R\right) \simeq \int_2^R \left(r^{\beta_1} \log^{\beta_2} r\right)^{p/2} \left(r^{\alpha_1} \log^{\alpha_2} r\right)^q r^{n-1} dr$$

$$= \int_2^R r^{\frac{\alpha_1+\beta_1\sigma}{\sigma-1}+n-1} (\log r)^{\frac{\alpha_2+\beta_2\sigma}{\sigma-1}} dr \leqslant CR^{\frac{\alpha_1+\beta_1\sigma}{\sigma-1}+n} (\log R)^{\frac{\alpha_2+\beta_2\sigma}{\sigma-1}} .$$

The condition (1.8) is satisfied in the two cases (in all cases $\sigma > 1$):

1) either

$$\frac{\alpha_1+\beta_1\sigma}{\sigma-1} + n < \frac{2\sigma}{\sigma-1}$$

2) or

$$\frac{\alpha_1+\beta_1\sigma}{\sigma-1} + n = \frac{2\sigma}{\sigma-1} \quad \text{and} \quad \frac{\alpha_2+\beta_2\sigma}{\sigma-1} < \frac{1}{\sigma-1}.$$

Solving these inequalities, we find that (1.8) is satisfied and hence (1.1) has no positive solutions provided that one of the following two cases takes place:

1) $\sigma < \sigma^* := \frac{n-\alpha_1}{\beta_1+n-2}$.

2) $\sigma = \sigma^*$ and $\alpha_2 + \beta_2\sigma < 1$.

Assuming that $\sigma^* > 1$, let us show that, in the opposite case

$$\sigma = \sigma^*, \quad \alpha_2 + \beta_2\sigma > 1 \tag{5.2}$$

a positive solution of (1.1) does exist, which will show the sharpness of the volume test of Theorem 1.3.

The construction uses Proposition 3.2 and is similar to the example in Section 3. Assuming (5.2), we construct a positive solution in $\mathbb{R}^n$ of the inequality

$$\sum_{i=1}^n \frac{\partial}{\partial x_i}\left(a(r)\frac{\partial u}{\partial x_i}\right) + V(r)u^\sigma \leqslant 0, \tag{5.3}$$

where $r = |x|$,

$$a(r) = r^{\beta_1}\log^{\beta_2} r \quad \text{and} \quad V(r) = r^{-\alpha_1}\log^{-\alpha_2} r \quad \text{for large } r.$$

In the polar coordinates, the inequality (5.3) becomes

$$\left(a(r) r^{n-1}u'\right)' + r^{n-1}V(r)u^\alpha \leqslant 0. \tag{5.4}$$

The condition (3.2) of Proposition 3.2 becomes

$$\int^\infty \frac{dr}{r^{\beta_1+n-1}\log^{\beta_2} r} < \infty,$$

which is true due to $\beta_1 + n - 2 > 0$. Setting

$$\gamma(r) = \int_r^\infty \frac{ds}{a(s)s^{n-1}} = \int^\infty \frac{ds}{s^{\beta_1+n-1}\log^{\beta_2} s} \simeq r^{-(\beta_1+n-2)}\log^{-\beta_2} r,$$

we see that the condition (3.3) of Proposition 3.2 is equivalent to

$$\int^{\infty} \left(r^{-(\beta_1+n-2)} \log^{-\beta_2} r\right)^{\sigma} r^{n-1} \left(r^{-\alpha_1} \log^{-\alpha_2} r\right) dr < \infty,$$

which, by $\sigma = \frac{n-\alpha_1}{\beta_1+n-2}$, is equivalent to

$$\int^{\infty} r^{-1} \log^{-(\alpha_2+\sigma\beta_2)} r dr < \infty.$$

The latter is obviously satisfied due to $\alpha_2 + \sigma\beta_2 > 1$. By Proposition 3.2, we conclude that (5.4) has a positive solution $u(r)$ in a neighborhood of $+\infty$ such that

$$u(r) \simeq r^{-(\beta_1+n-2)} \log^{-\beta_2} r \quad \text{as} \quad r \to \infty.$$

Arguing further as in Section 3, one extends this function to be a solution of (5.3) on $\mathbb{R}^n$.

Similarly, one can show the existence of a positive solution of (5.3) in the case $\sigma > \sigma^*$.

Acknowledgments. Both authors were partially supported by SFB 701 of the German Research Council.

References

1. Adams, R.A.: Sobolev spaces. Academic Press, New York etc. (1975)
2. Atkinson, F.V.: On second order nonlinear oscillations. Pacif. J. Math. **5**, no. 1, 643–647 (1955)
3. Gidas, B., Spruck, J.: Global and local behavior of positive solutions of nonlinear elliptic equations. Commun. Pure Appl. Math. **34**, 525–598 (1981)
4. Grigor'yan, A.: On the existence of positive fundamental solution of the Laplace equation on Riemannian manifolds (Russian). Mat. Sb. **128**, no. 3, 354–363 (1985); English transl.: Math. USSR Sb. **56**, 349–358 (1987)
5. Grigor'yan, A.: Analytic and geometric background of recurrence and non-explosion of the Brownian motion on Riemannian manifolds. Bull. Am. Math. Soc. **36**, 135–249 (1999)
6. Grigor'yan, A., Netrusov, Yu., Yau, S.-T.: Eigenvalues of elliptic operators and geometric applications. In: Eigenvalues of Laplacians and Other Geometric Operators. Surveys in Differential Geometry. IX, pp. 147–218 (2004)
7. Holopainen, I.: Volume growth, Green's functions and parabolicity of ends. Duke Math. J. **97**, no. 2, 319–346 (1999)
8. Kurta, V.V.: On the absence of positive solutions to semilinear elliptic equations (Russian). Tr. Mat. Inst. Steklova **227**, 162–169 (1999); English transl.: Proc. Steklov Inst. Math. **227**, no. 4, 155–162 (1999)
9. Maz'ya, V.G.: Sobolev Spaces. Springer, Berlin etc. (1985)
10. Mitidieri, E., Pokhozhaev, S.I.: Apriori estimates and the absence of solutions of nonlinear partial differential equations and inequalities (Russian). Tr. Mat. Inst. Steklova **234**, 1–384 (2001); English transl.: Proc. Steklov Inst. Math. **234**, no. 3, 1–362 (2001)

Recurrence Relations for Orthogonal Polynomials and Algebraicity of Solutions of the Dirichlet Problem

Dmitry Khavinson and Nikos Stylianopoulos

Dedicated to Professor Vladimir Maz'ya in recognition of his substantial contribution to the subject of Sobolev spaces

Abstract We show that any finite-term recurrence relation for planar orthogonal polynomials in a domain implies that the domain must be an ellipse. Our proof relies on Schwarz function techniques and on elementary properties of functions in Sobolev spaces.

1 Introduction

Let Ω be a bounded simply connected domain in the complex plane, and let $\{p_n\}_{n=0}^{\infty}$ denote the sequence of *Bergman orthogonal polynomials* of Ω,

$$p_n(z) = \gamma_n z^n + \gamma_{n-1} z^{n-1} + \cdots, \quad \gamma_n > 0, \quad n = 0, 1, 2, \ldots,$$

of polynomials which are orthonormal with respect to the inner product

$$\langle f, g \rangle := \int_\Omega f(z)\overline{g(z)}dA(z),$$

where dA is the area measure. The associated L^2-norm is defined by

$$\|f\|_{L^2(\Omega)} := \langle f, f \rangle^{\frac{1}{2}} = \left\{ \int_\Omega |f(z)|^2 dA(z) \right\}^{\frac{1}{2}}.$$

Dmitry Khavinson
Department of Mathematics and Statistics, University of South Florida, 4202 E. Fowler Avenue, PHY114, Tampa, FL 33620-5700, USA, e-mail: dkhavins@cas.usf.edu

Nikos Stylianopoulos
Department of Mathematics and Statistics, University of Cyprus, P.O. Box 20537, 1678 Nicosia, Cyprus, e-mail: nikos@ucy.ac.cy

A. Laptev (ed.), *Around the Research of Vladimir Maz'ya II: Partial Differential Equations*, International Mathematical Series 12, DOI 10.1007/978-1-4419-1343-2_9,

Let $\Omega' := \overline{\mathbb{C}} \setminus \overline{\Omega}$ denote the complement (in $\overline{\mathbb{C}}$) of $\overline{\Omega}$, and let Φ denote the conformal map $\Omega' \to \mathbb{D}' := \{w : |w| > 1\}$ normalized so that, near infinity,

$$\Phi(z) = \gamma z + \gamma_0 + \frac{\gamma_1}{z} + \frac{\gamma_2}{z^2} + \cdots, \quad \gamma > 0. \tag{1.1}$$

Note that the constant $1/\gamma$ gives the (logarithmic) *capacity* $\mathrm{cap}(\Gamma)$ of the boundary Γ of Ω (cf., for example, [19, 20]). The inverse conformal map $\Psi := \Phi^{-1} : \mathbb{D}' \to \Omega'$ has a Laurent series expansion of the form

$$\Psi(w) = bw + b_0 + \frac{b_1}{w} + \frac{b_2}{w^2} + \cdots, \tag{1.2}$$

valid for $|w| > 1$, where $b = 1/\gamma = \mathrm{cap}(\Gamma)$.

It is well known that orthogonal polynomials with respect to any measure μ on the real line do satisfy a three-term recurrence relation (cf., for example, [20]). By contrast, polynomials orthogonal with respect to the area measure, or the arc-length measure, in the complex plane $\mathbb{C}$, do not favor recurrence relations. To this end, Lempert [12] produced examples of several, rather special domains, where the associated orthogonal polynomials do not satisfy ANY finite-term recurrence relation. Putinar and the second author noted [14] that the fact that "the Bergman polynomials of Ω satisfy a finite-term recurrence relation" is, actually, equivalent to the fact that "any Dirichlet problem in Ω, with polynomial data, possesses a polynomial solution." The latter is the hypothesis of the so-called Khavinson–Shapiro conjecture [11] which states that *only ellipses* (*or ellipsoids in higher dimensions*) *have this property.* This conjecture has attracted some attention and the reader is referred to [2, 15, 8, 10] and the references therein for results reporting on the recent progress in that direction. In [14], the authors showed that if the Bergman polynomials of a simply connected domain Ω satisfy *any* finite-term recurrence relation and, in addition, the (necessarily algebraic) boundary of Ω, $\partial\Omega \subseteq \{P(x,y) = 0, P \text{ is a polynomial}\}$ satisfies the condition

(B) the set $\{P = 0\}$ is bounded in $\mathbb{C}$,

then Ω is an ellipse and the recurrence relation must be a three-term relation.

The main point of this paper is to remove the assumption (B). We do this, however, by assuming a finite-term recurrence of constant width, rather than one of variable width, as was the case in [14]. More precisely, we show that if the Bergman orthogonal polynomials of Ω satisfy an $(N+1)$-term recurrence relation, with N a positive integer, then Ω is an ellipse and $N = 2$.

Yet, in order for our argument to work, it is not enough to assume that Ω is merely simply connected, though a C^2-smooth Jordan boundary curve is sufficient. It remains an open question whether our results hold for any simply connected domain. We strongly believe so, but we were not able to extend our proof to that case.

2 The Main Results

Let Ω be a bounded simply connected planar domain. Consider the Bergman space $L_a^2(\Omega)$ associated with Ω. It is the Hilbert space of functions analytic and square integrable in Ω. In this paper, we assume that the boundary Γ of Ω is a Jordan curve. Under this assumption, the Bergman polynomials $\{p_n\}_{n=0}^{\infty}$ of Ω form a complete orthonormal system in $L_a^2(\Omega)$ (cf., for example, [9] for weaker assumptions on Γ regarding completeness in $L_a^2(\Omega)$).

A standard way to construct the Bergman polynomials is by means of the Gram–Schmidt process. This is a linear algorithm that computes the sequence of the orthonormal polynomials recursively, by using as data the entrances of the complex moment matrix $H := [\mu_{m,n}]_{m,n=0}^{\infty}$ of Ω:

$$\mu_{m,n} := \int_{\Omega} z^m \overline{z}^n dA(z). \tag{2.1}$$

It turns out that the complex moment matrix H alone suffices to determine the (unique) sequence of Bergman polynomials of Ω, and this determination is done, for each p_n, in a finite number of steps and by using a finite section of the moment matrix. (For more details regarding the general question of uniqueness properties of complex moments see [5].)

It is clear that for all $n = 0, 1, 2, \ldots$

$$zp_n(z) = \sum_{k=0}^{n+1} a_{k,n} p_k(z), \quad n = 0, 1, \ldots, \tag{2.2}$$

where the Fourier coefficients $a_{k,n}$ are given by $a_{k,n} = \langle zp_n, p_k \rangle$. Then

$$\sum_{k=0}^{n+1} |a_{k,n}|^2 < \infty, \quad n = 0, 1, \ldots.$$

The coefficients $a_{k,n}$ constitute the entrances of an infinite *upper Hessenberg matrix* M. This matrix is closely related to the multiplication operator by z (the *Bergman shift* operator) $T_z : L_a^2(\Omega) \to L_a^2(\Omega)$, defined by $(T_z f)(z) = zf(z)$, in the sense that T_z can be represented with respect to the basis $\{p_n\}_{n=0}^{\infty}$ by M. Note that T_z is linear and bounded on $L_a^2(\Omega)$.

Definition 2.1. We say that the Bergman polynomials $\{p_n\}_{n=0}^{\infty}$ satisfy an $(N+1)$-*term recurrence relation* for some fixed positive integer N if for any $n \geqslant N-1$,

$$zp_n(z) = a_{n+1,n}p_{n+1}(z) + a_{n,n}p_n(z) + \ldots + a_{n-N+1,n}p_{n-N+1}(z). \tag{2.3}$$

If the Bergman polynomials satisfy an $(N+1)$-term recurrence relation then one easily sees (cf. [14]) that the adjoint operator T_z^* of the Bergman shift, and its multiples, increase the degree of a polynomial $p(z)$ subject to the constraint

$$\deg[(T_z^*)^m p] \leqslant m(N-1) + \deg p, \quad m \in \mathbb{N}. \tag{2.4}$$

This follows easily from the fact that T_z^* can be represented with respect to the basis $\{p_n\}_{n=0}^{\infty}$ by the adjoint matrix M^* of M which, in this case, has a lower Hessenberg and banded form of constant width $N+1$.

The next result confirms the Khavinson–Shapiro conjecture (cf. [11, 2, 15]) under an additional assumption on the degree of the polynomial solution to the Dirichlet problem.

Theorem 2.1. *Let Ω be a bounded simply connected domain in $\mathbb{C}$ with a C^2-smooth Jordan boundary Γ. Assume that there exists a positive integer $N := N(\Omega)$ with the property that the Dirichlet problem*

$$\Delta u = 0 \quad in \quad \Omega, \qquad u = \overline{z}^m z^n \quad on \quad \Gamma, \tag{2.5}$$

has a polynomial solution of analytic degree $\leqslant m(N-1)+n$ (in z) and of conjugate analytic degree $\leqslant n(N-1)+m$ (in $\overline{z}$), for all positive integers m and n. Then Ω is an ellipse and $N=2$.

Remark 2.1. Considering the polynomial $p(x,y) = \overline{z}z (= x^2+y^2)$ in (2.5), it is easy to see that, under the assumptions of Theorem 2.1, the boundary curve Γ must be a part of the zero set of an algebraic polynomial and hence a piecewise analytic curve.

It is well known that the Bergman polynomials of an ellipse satisfy a three-term recurrence relation. In fact, it is easy to check that they are suitably normalized Chebyshev polynomials of the 2nd kind. The associated Hessenberg matrix in this case is triangular and goes by the name of *Jacobi matrix*. The following theorem states that this is the only possible case for an $(N+1)$-recurrence to occur.

Theorem 2.2. *With Ω and Γ as in Theorem 2.1, assume that the Bergman orthogonal polynomials for Ω satisfy an $(N+1)$-term recurrence relation, with some $N \geqslant 2$. Then Ω is an ellipse and hence $N=2$.*

We note that the conclusion of Theorem 2.2 for polynomials orthogonal with respect to the arc-length measure that satisfy a three-term recurrence relation (i.e., under the assumption $N=2$) goes back to Duren [7]. A similar result, as that of [7], but for polynomials orthogonal with respect to the harmonic measure on Γ, was established in [6].

Theorem 2.1 becomes an easy consequence of Theorem 2.2, after we establish the equivalence between the assumptions of the two theorems. This latter task was essentially done in [14, Theorem 1] under a more general definition for recurrences and thus without specific reference to the degree of the polynomial solution of (2.5). For our purposes here, however, we require the following explicit version of Theorem 1 of [14].

Proposition 2.1. *Let Ω be a bounded simply connected domain in $\mathbb{C}$ with a C^2-smooth Jordan boundary Γ. Then there exists a positive integer $N := N(\Omega)$ such that for all positive integers m and n the Dirichlet problem* (2.5) *with polynomial data $\overline{z}^m z^n$ has a polynomial solution of degree $\leqslant m(N-1)+n$ in z and $\leqslant n(N-1)+m$ in $\overline{z}$ if and only if the Bergman orthogonal polynomials for Ω satisfy an $(N+1)$-term recurrence relation.*

The following result, which gives the ratio asymptotics for the Bergman polynomials, is needed in establishing Theorem 2.2. Its proof is a simple consequence of the strong asymptotics for Bergman polynomials over domains with smooth boundaries, established by Suetin [19, Theorem 1.2] and, thus, we omit it.

Lemma 2.1. *Assume that Ω is a bounded simply connected domain in $\mathbb{C}$ with a C^2-smooth Jordan boundary Γ. Let $\{p_n\}_{n=0}^{\infty}$ denote the sequence of Bergman polynomials of Ω. Then*

$$\lim_{n\to\infty} \frac{p_{n+1}(z)}{p_n(z)} = \Phi(z), \quad z \in \overline{\Omega'}. \tag{2.6}$$

We note, in passing, that strong asymptotics for Bergman polynomials were first derived by Carleman [3] under the assumption that Γ is analytic.

Remark 2.2. For Ω simply connected and bounded, a well-known result by Fejér asserts that the zeros of $p_n(z)$ $(n \in \mathbb{N})$ are contained in $Co(\overline{\Omega})$, where $Co(\overline{\Omega})$ denotes the convex hull of $\overline{\Omega}$. Under the additional assumption on Γ in Lemma 2.1, from [19, Theorem 1.2] it follows that there exists a positive integer n_0 such that the sequence $\{p_n\}_{n=n_0}^{\infty}$ has no zeros in Ω'.

Remark 2.3. Lemma 2.1 is precisely the reason we need to assume the C^2-smoothness of Γ in Theorem 2.1. Although we were not able to extend the ratio asymptotics to more general sets, we believe that (2.6) holds for arbitrary domains Ω, such that $\Gamma = \partial\Omega = \partial\Omega'$ is a continuum.

3 Proofs

Proof of Proposition 2.1. Fix two positive integers m, n and assume that the Bergman orthogonal polynomials for Ω satisfy an $(N+1)$-term recurrence relation. Then, in view of (2.4),

$$(T_z^*)^m z^{n-1} = q(z),$$

where q is a polynomial of degree $\leqslant m(N-1)+n-1$. Therefore,

$$\overline{z}^m z^{n-1} = q(z) + h(z), \quad z \in \Omega,$$

where $h \in L^2(\Omega) \ominus L_a^2(\Omega)$. Let $Q(z)$ be a polynomial satisfying $Q' = q$. According to the so-called Khavin lemma (cf., for example, [18, p. 26]), $h = \partial g$ with g in the Sobolev space $W_0^{1,2}(\Omega)$. Integrating against ∂, we find

$$\overline{z}^m z^n = Q(z) + g(z) + \overline{f(z)}, \quad z \in \Omega, \tag{3.1}$$

where $f \in L_a^2(\Omega)$. Since Γ is smooth, it follows that $g = 0$, a.e. on $\Gamma = \partial\Omega$ and thus

$$\overline{z}^m z^n = Q(z) + \overline{f(z)} \quad \text{for a. e.} \quad z \in \Gamma.$$

Moreover, from (3.1), the Poincaré inequality and the smoothness of Γ, we infer easily (cf., for example, [1]) that f, in fact, belongs to the Hardy space $H^2(\Omega)$. (For the most up to date theory of Sobolev spaces we refer the reader to [13].) Similarly, we have

$$\overline{z}^n z^m = G(z) + \overline{f_1(z)} \quad \text{for a. e.} \quad z \in \Gamma,$$

where G is a polynomial of degree $\leqslant n(N-1)+m$ and $f_1 \in L_a^2(\Omega) \cap H^2(\Omega)$. Hence $Q(z) + \overline{f(z)} = \overline{G(z)} + f_1(z)$, $z \in \Omega$ and, since Ω is simply connected, we infer that $Q(z) = f_1(z) + c$ and $G(z) = f(z) + \overline{c}$, $z \in \Omega$, for some constant c. Hence the Dirichlet problem in Ω with data $\overline{z}^m z^n$ has a polynomial solution whose analytic degree (in z) is $\leqslant m(N-1)+n$ and its antianalytic degree (in $\overline{z}$) is $\leqslant n(N-1)+m$.

For the converse, assume that the Dirichlet problem for Ω with data $\overline{z}^m z^n$ has a polynomial solution $u(z) = Q(z) + \overline{G(z)}$, where Q and G are complex polynomials with $\deg(Q) \leqslant m(N-1)+n$.

Let $h(z)$ be a bounded analytic function in $\overline{\Omega}$. Then, by the Stokes and Cauchy theorems, for $n \geqslant 1$ we obtain

$$\begin{aligned}\langle (T_z^*)^m z^{n-1}, h\rangle &= \langle \overline{z}^m z^{n-1}, h\rangle = \int_\Omega \overline{z}^m z^{n-1}\overline{h(z)}\, dA(z)\\ &= -\frac{1}{2ni}\int_\Gamma \overline{\zeta^m}\zeta^n\overline{h(\zeta)}\, d\overline{\zeta} = \frac{1}{n}\int_\Omega Q'(z)\overline{h(z)}\, dA(z) = \langle q, h\rangle,\end{aligned}$$

where $q(z) := Q'(z)/n$. This implies

$$(T_z^*)^m z^{n-1} = q(z),$$

where $\deg(q) \leqslant m(N-1)+n-1$, and hence the finite-term recurrence relation for the Bergman polynomials. □

Proof of Theorem 2.2. Assume that the Bergman polynomials of Ω satisfy the recurrence relation (2.3) for some $N \geqslant 2$. Then from Proposition 2.1 and Remark 2.1 we see that Γ must be piecewise analytic.

Now we argue as in [7, p. 314]. For the moment, we assume that each of the $N+1$ sequences of the Fourier coefficients

$$\alpha_n^{(1)} := a_{n+1,n}, \quad \alpha_n^{(2)} := a_{n,n}, \quad \ldots, \quad \alpha_n^{(N+1)} := a_{n-N+1,n},\ n \in \mathbb{N}, \tag{3.2}$$

is bounded, and then proceed as follows:

(i) divide both sides of (2.3) by $p_n(z)$ (for $z \in \mathbb{C} \setminus Co(\overline{\Omega})$);

(ii) take the limit as $n \to \infty$, $n \in \Lambda$, on both sides of the resulting equation, where Λ is an appropriate subsequence of $\mathbb{N}$ chosen so that each sequence in (3.2) tends to a finite limit;

(iii) note that

$$\frac{p_{n-k}}{p_n} = \frac{p_{n-1}}{p_n}\frac{p_{n-2}}{p_{n-1}} \cdots \frac{p_{n-k}}{p_{n-k+1}}, \quad k \leqslant N-1;$$

(iv) apply Lemma 2.1.

The above steps yield that the inverse exterior conformal map $\Psi : \mathbb{D}' \to \Omega'$ has a finite Laurent expansion of the form

$$\Psi(w) = bw + b_0 + \frac{b_1}{w} + \frac{b_2}{w^2} + \cdots + \frac{b_{N-1}}{w^{N-1}}, \quad |w| > 1. \tag{3.3}$$

To verify that all the sequences in (3.2) are bounded, one simply has to apply the Cauchy–Schwarz inequality, for $j = 1, 2, \ldots, N+1$ and $n \geqslant N-1$:

$$\begin{aligned} |\alpha_n^{(j)}| = |a_{n+2-j,n}| &= \left| \int_\Omega \overline{z} p_{n+2-j}(z)\overline{p_n(z)} dA(z) \right| \\ &\leqslant \|z\|_\infty \, \|p_{n+2-j}\|_{L^2(\Omega)} \, \|p_n\|_{L^2(\Omega)} = \|z\|_\infty, \end{aligned}$$

where $\|\cdot\|_\infty$ stands for the sup-norm on $\overline{\Omega}$.

From (3.3) it follows that Ψ is a rational function. This implies that Ω' is an unbounded quadrature domain. Hence the associated Schwarz function $S(z)$ with $S(z) = \overline{z}$ on $\Gamma := \partial\Omega = \partial\Omega'$ has a meromorphic extension to Ω', i.e.,

$$S(z) = r(z) + \sum_{j=1}^{M} \sum_{l=1}^{k_j} \frac{c_{j,l}}{(z - z_j)^l} + f(z), \tag{3.4}$$

where $z_j \in \Omega' \setminus \{\infty\}$, $k_j \in \mathbb{N}$, $r(z)$ is a polynomial of degree d, and $f(z)$ is analytic and bounded in Ω' (cf., for example, [17]).

We first show that all the constants $c_{j,l}$, $j = 1, \ldots, M$, $l = 1, \ldots, k_j$, in (3.4) vanish.

Let $P(z) := \prod_{j=1}^{M}(z - z_j)^{k_j}$. Consider the Dirichlet problem (2.5) with data $\overline{z}P(z)$. Our hypothesis and Proposition 2.1 imply that there exist analytic polynomials $h(z)$ and $g(z)$ such that

$$\overline{z}P(z) = g(z) + \overline{h(z)}, \quad z \in \Gamma. \tag{3.5}$$

(Note that $\deg(h) \geqslant 1$; otherwise, on Γ, $\overline{z} = S(z)$ is equal to a rational function and $\Gamma = \partial\Omega$ is a circle, according to a well-known theorem of Davis [4, p. 104].) Let $R(z) = \overline{S(z)}$ be the anticonformal reflection about Γ. It is obvious that, by (3.4), $R(z)$ extends to Ω' and has poles at ∞ and $\{z_j\}_{j=1}^{M}$. From (3.4) and (3.5) we see that, on Γ,

$$g(z) + \overline{h(R(z))} = r(z)\,P(z) + F(z), \tag{3.6}$$

where F(z) is analytic in $\Omega' \setminus \{\infty\}$ and it may have a pole of order at most $\sum_{j=1}^{M} k_j$ at ∞. Since both sides of (3.6) are analytic functions of z, (3.6) holds on any path originating on Γ along which $S(z)$ continues analytically.

Now, let γ be any path in $\Omega' \setminus \{\infty\}$ joining Γ to a given pole z_j and avoiding all other poles. Then the right-hand side of (3.6) stays bounded on γ and so does $g(z)$, while $\overline{h(R(z))} \to \infty$ at z_j because $|R(z)| \to \infty$ at z_j and $h(z)$ is a (nonconstant) polynomial. This is a contradiction and therefore $S(z)$ can have no finite poles in Ω', i.e.,

$$S(z) = r(z) + f(z), \quad z \in \Omega', \tag{3.7}$$

where $f(z)$ is analytic in Ω' (including ∞).

Consider now the Dirichlet problem (2.5) with data $z\overline{z} = |z|^2$. In view of our hypothesis and Proposition 2.1, we can find a polynomial $g(z)$ of degree $k \geqslant 2$ (since if $k \leqslant 1$, then Γ is obviously a circle and $N = 1$) such that for $z \in \Gamma$

$$2\mathrm{Re}\{g(z)\} = g(z) + \overline{g(z)} = |z|^2 = zS(z) \tag{3.8}$$

or, by (3.7),

$$g(z) + \overline{g(R(z))} = z\,r(z) + zf(z), \quad z \in \Gamma. \tag{3.9}$$

Consider a path γ in $\Omega' \setminus \{\infty\}$ joining Γ to ∞. Since both sides of (3.9) are analytic in Ω', (3.9) holds along γ. Yet, near ∞ we have $|g(z)| \sim |z|^k$, $|\overline{g(R(z))}| \sim |z|^{dk}$, and the right-hand side of (3.9) behaves as $O(|z|^{d+1})$, This can only be possible if $dk = d + 1$, i.e., since $k \geqslant 2$, only if $d = 1$ and $k = 2$. From this it follows that Ω is an ellipse and $N = 2$. However, it may be worthwhile to point out the following observation as well. Thus,

$$S(z) = c\,z + f(z), \quad z \in \Omega', \tag{3.10}$$

with f analytic in Ω'.

But this implies right away that Ω' is a null-quadrature domain (cf. [17] and [18]). Indeed, using the Green and Cauchy theorems, for any number $m \geqslant 3$ we have

$$\begin{aligned}\int_{\Omega'} \frac{1}{z^m} dA(z) &= -\frac{1}{2i}\int_{\Gamma} \frac{1}{z^m}\overline{z}dz = -\frac{1}{2i}\int_{\Gamma} \frac{1}{z^m} S(z)dz \\ &= -\frac{1}{2i}\int_{|z|=\rho} \frac{1}{z^m}\{c\,z + f(z)\}dz = 0\end{aligned}$$

for large enough ρ since f is analytic in Ω' (including ∞). From this, in view of [16, Theorem 1], we infer that Ω' must be the exterior of an ellipse. Hence Ω is an ellipse and thus $N = 2$. □

4 Concluding Remarks

We finish with a number of remarks.

(i) As we have pointed out above, the main place where the C^2-regularity of the boundary $\Gamma = \partial\Omega$ is needed was the application of the strong asymptotics for Bergman polynomials of Suetin [19], that yield Lemma 2.1. Moreover, it is clear from the proof of Theorem 2.2 that we only need (2.6) to hold on a continuum subset of Ω', in a neighborhood of ∞. It looks quite plausible that, in this weaker form, (2.6) holds for arbitrary, bounded Jordan domains. Yet, we have not been able to derive it for such general domains or find a pertinent result in the literature.

(ii) For the most updated account on the status of the Khavinson–Shapiro conjecture in its full generality, mostly due to the work of Render [15], we refer the reader to the recent survey [10].

(iii) For a quite different approach, regarding singularities of solutions of the Dirichlet problem in $\mathbb{R}^2$, we refer the reader to [2] (cf. also [8]).

(iv) The finiteness of only the first column of the adjoint matrix M^* associated with T_z^* is not sufficient to yield Theorems 2.1–2.2 or Proposition 2.1 (unless, of course, $a_{0,n} = 0$, $n \geqslant 2$; cf. [14, Prop. 1]). For example, take Γ to be the bounded component of $\{x^2 + y^2 - 1 + \epsilon(x^3 - 3xy^2) = 0\}$, where $\epsilon > 0$ is small enough so that Γ is a perturbation of the unit circle. Then the quadratic data $z\overline{z}$ are matched on Γ by a cubic harmonic polynomial, despite the fact that Γ is not an ellipse.

(v) The assumption that Ω is simply connected is not really necessary. As is seen from the arguments of [2], the hypothesis in the Khavinson–Shapiro conjecture implies that Ω is simply connected.

(vi) The condition that a finite-term recurrence relation (of some constant width $N+1$), satisfied by the Bergman polynomials of Ω, is stronger than the hypothesis of the Khavinson–Shapiro conjecture for Ω. This is so because the Khavinson–Shapiro conjecture does not involve any assumption on the degree of the polynomial solution. *Thus, a full proof of the conjecture is still amiss.*

(vii) If the hypothesis of the Khavinson–Shapiro conjecture is satisfied then, clearly, $\Gamma = \partial\Omega$ is algebraic and hence piecewise analytic. Yet, in order to be able to use the ratio asymptotics for the Bergman polynomials, as they have been obtained by Suetin [19], we must eliminate the possibility that Γ has cusps. Perhaps, whenever the hypothesis of the Khavinson–Shapiro conjecture holds (cf. Proposition 2.1) the cusps cannot occur a priori. We were not able to prove this either. We note however that it is possible to have cusped curves on which a quadratic matches a harmonic polynomial, for example, $y^2 = x^3 - 3y^2x$.

Acknowledgment. Part of this work was carried out during authors' respective visits to the University of Cyprus and the University of South Florida. The authors also acknowledge, gratefully, support by the National Science Foundation (D.K.) and the University of Cyprus Research Committee (N.S.).

References

1. Adams, R.A., Fournier, J.J.F.: Sobolev Spaces. Elsevier, Amsterdam (2003)
2. Bell, S.R., Ebenfelt, P., Khavinson, D., Shapiro, H.S.: On the classical Dirichlet problem in the plane with rational data. J. Anal. Math. **100**, 157–190 (2006)
3. Carleman, T.: Über die approximation analytisher funktionen durch lineare aggregate von vorgegebenen potenzen. Ark. Mat., Astr. Fys. **17**, no. 9, 215–244 (1923)
4. Davis, P.: The Schwarz Function and its Applications. MAA, Buffalo, NY (1974)
5. Davis, P., Pollak, H.: On the analytic continuation of mapping functions. Trans. Am. Math. Soc. **87**, 198–225 (1958)
6. Dovgoshei, A.A.: Three-term recurrence relation for polynomials orthogonal with respect to harmonic measure. Ukrainian Math. J. **53**, 167–177 (2001)
7. Duren, P.: Polynomials orthogonal over a curve. Michigan Math. J. **12**, 313–316 (1965)
8. Ebenfelt, P., Khavinson, D., Shapiro, H.S.: Algebraic aspects of the Dirichlet problem. In: Quadrature Domains and Their Applications, pp. 151–172, Birkhäuser, Basel (2005)
9. Hedberg, L.I.: Approximation in the mean by analytic functions. Trans. Am. Math. Soc. **163**, 157–171 (1972)
10. Khavinson, D., Lundberg, E.: The search for singularities of solutions to the Dirichlet problem: recent developments. Preprint. http://shell.cas.usf.edu/~dkhavins/publications.html.
11. Khavinson, D., Shapiro, H.S.: Dirichlet's problem when the data is an entire function. Bull. London Math. Soc. **24**, no. 5, 456–468 (1992)
12. Lempert, L.: Recursion for orthogonal polynomials on complex domains. In: Fourier Analysis and Approximation Theory (Proc. Colloq., Budapest, 1976), Vol. II, pp. 481–494. North-Holland, Amsterdam (1978)
13. Maz'ya, V.G.: Sobolev Spaces. Springer, Berlin etc. (1985)
14. Putinar, M., Stylianopoulos, N.: Finite-term relations for planar orthogonal polynomials. Complex Anal. Oper. Theory **1**, no. 3, 447–456 (2007)
15. Render, H.: Real Bargmann spaces, Fischer decompositions, and sets of uniqueness for polyharmonic functions. Duke Math. J. **142**, no. 2, 313–352 (2008)
16. Sakai, M.: Null quadrature domains. J. Anal. Math. **40**, 144–154 (1981)
17. Shapiro, H.S.: Unbounded quadrature domains. In: Complex analysis, I. Proc. Spec. Year, College Park/Md., 1985–86, Lect. Notes Math. **1275**, 287–331 (1987)
18. Shapiro, H.S.: The Schwarz Function and its Generalization to Higher Dimensions. John Wiley & Sons Inc., New York (1992)
19. Suetin, P.K.: Polynomials Orthogonal over a Region and Bieberbach Polynomials. Am. Math. Soc., Providence, R.I. (1974)
20. Szegő, G.: Orthogonal Polynomials. Am. Math. Soc., Providence, R.I. (1975)

On First Neumann Eigenvalue Bounds for Conformal Metrics

Gerasim Kokarev and Nikolai Nadirashvili

Abstract We prove that on any orientable surface with nonempty boundary there exists a conformal class of Riemannian metrics whose first Neumann eigenvalues satisfy the Hersch isoperimetric inequality.

1 Introduction

Let M be an orientable compact surface with boundary. For a Riemannian metric g on M we denote by

$$0 = \lambda_0(g) < \lambda_1(g) \leqslant \ldots \leqslant \lambda_k(g) \leqslant \ldots$$

the Neumann eigenvalues of the Laplace operator $-\Delta_g$. The result of Li and Yau [5] says that if M has zero genus, then for any metric g the following inequality holds:

$$\lambda_1(g)\, Vol_g(M) \leqslant 8\pi, \tag{1.1}$$

where $Vol_g(M)$ stands for the volume of M with respect to g. This is a version of an earlier result by Hersch [4], who proved the same inequality for closed Riemannian surfaces of zero genus. The inequality above is sharp in the sense that for any M there exists a sequence of metrics for which the left-hand side tends to 8π. These metrics can be obtained by excision of small spherical caps from the round sphere. It is a simple exercise to show that the bound

Gerasim Kokarev
School of Mathematics, The University of Edinburgh, King's Buildings, Mayfield Road, Edinburgh EH9 3JZ, UK, e-mail: G.Kokarev@ed.ac.uk

Nikolai Nadirashvili
Laboratoire d'Analyse, Topologie, Probabilités UMR 6632, Centre de Mathématiques et Informatique, Université de Provence, 39 rue F. Joliot-Curie, 13453 Marseille Cedex 13, France, e-mail: nicolas@cmi.univ-mrs.fr

A. Laptev (ed.), *Around the Research of Vladimir Maz'ya II: Partial Differential Equations*, International Mathematical Series 12, DOI 10.1007/978-1-4419-1343-2_10,

above does not hold for Riemannian surfaces of higher genus. However, there are upper bounds via quantities depending on the genus of M (cf. [5]).

The purpose of this note is to show that the inequality (1.1) still holds for surfaces of arbitrary genus within certain conformal classes of Riemannian metrics. More precisely, we prove the following statement.

Theorem. *Let M be an orientable compact surface with nonempty boundary. Then there exists a metric g_* on M such that for any conformal metric $g = \phi \cdot g_*$ the first Neumann eigenvalue of any subdomain $\Omega \subseteq M$ satisfies the estimate*

$$\lambda_1(g)\, Vol_g(\Omega) \leqslant 8\pi,$$

where $Vol_g(\Omega)$ is the total volume of (Ω, g).

Mention that the statement of the theorem continues to hold for conformal metrics $g = \phi \cdot g_*$ defined only in the interior of M and allowed to approach infinity or zero near the boundary. The topology induced by the former metrics makes M into a noncompact surface, and in this case we consider compact subdomains $\Omega \subset M$. The topology induced by metrics with zeroes on the boundary corresponds to shrinking these parts of the boundary to points.

The theorem demonstrates a sharp contrast with the behavior of first eigenvalues in conformal classes on *closed* Riemannian surfaces. More precisely, Colbois and El Soufi [1] prove the following "opposite" statement: for any conformal class c on a closed Riemannian surface M the inequality

$$\sup_{g \in c} \lambda_1(g)\, Vol_g(M) \geqslant 8\pi \tag{1.2}$$

holds. The second author raised a question whether the equality in (1.2) occurs only when M has zero genus. More generally, it is interesting to know for which conformal classes (on surfaces with or without boundary) the quantity $\sup_{g \in c} \lambda_1(g)\, Vol_g(M)$ equals 8π.

Finally, we would like to mention the contribution to the field made by V. Maz'ya. He was one of the first to study the Neumann problem for domains with nonregular boundaries in the 1960's. His paper [6] on the subject, in particular, contains estimates for the first Neumann eigenvalue in terms of the so-called isocapacitory inequalities.

2 Preliminaries

2.1 Notation

Let M be a smooth surface. Recall that for a Riemannian metric g on M the Laplace operator $-\Delta_g$ in local coordinates (x^i), $1 \leqslant i \leqslant 2$, has the form

$$-\Delta_g = -\frac{1}{\sqrt{|g|}}\frac{\partial}{\partial x^i}\left(\sqrt{|g|}g^{ij}\frac{\partial}{\partial x^j}\right),$$

where (g_{ij}) are components of the metric g, (g^{ij}) is the inverse tensor, and $|g|$ stands for $\det(g_{ij})$. Above we use the summation convention for the repeated indices. For a subdomain $\Omega \subseteq M$ with a nonempty boundary Σ, we denote by

$$0 = \lambda_0(g) < \lambda_1(g) \leqslant \ldots \leqslant \lambda_k(g) \leqslant \ldots$$

its Neumann eigenvalues. These are real numbers for which the equation

$$(\Delta_g + \lambda_k(g))u = 0, \qquad \left.\frac{\partial u}{\partial \nu}\right|_{\Sigma} = 0, \tag{2.1}$$

has a nontrivial solution on Ω; here ν stands for the outer normal direction along Σ. The solutions of Equation (2.1) are called eigenfunctions, and their collection over all eigenvalues forms a complete orthogonal basis in $L^2(\Omega)$. Recall that by variational characterization the eigenvalue $\lambda_k(g)$ is the infimum of the Rayleigh quotient

$$\mathrm{R}_g(u) = \left(\int_\Omega |\nabla u|^2\, dVol_g\right) \Big/ \left(\int_\Omega u^2 dVol_g\right)$$

over the set of all non-zero $W^{1,2}$-smooth functions u that are L_2-orthogonal to the eigenfunctions for λ_0, λ_1, ..., λ_{k-1}.

For a map $\varphi : M \to S^2$ by the *energy* of φ we mean the quantity

$$E(\varphi) = \sum_i \int_M \left|\nabla(x^i \circ \varphi)\right|^2 dVol_g,$$

where S^2 is regarded as the unit sphere in $\mathbf{R}^3$, and (x^i), $i = 1, 2, 3$, are the Euclidean coordinates. The important observation is that the energy $E(\varphi)$ is invariant with respect to a conformal change of the metric on M. If the domain M is a closed Riemann surface and the map φ is holomorphic, then the energy satisfies the following identity:

$$E(\varphi) = 8\pi \cdot \deg(\varphi), \tag{2.2}$$

where $\deg(\varphi)$ is the degree of φ (cf. [2]).

2.2 Reformulation of the theorem

Our main technical tool is an extension of the first eigenvalue to a more general class of conformal metrics. To explain the approach, we start with the following definition.

Definition. Let c be a conformal class of smooth Riemannian metrics on a surface M with or without boundary. For any finite Borel measure μ on M by the *first eigenvalue* $\lambda_1(\mu, c)$ of the pair (μ, c) we mean the infimum of the Rayleigh type quotient

$$\mathrm{R}_c(\mu, u) = \left(\int_M |\nabla u|^2 \, dVol_{g_*}\right) \Big/ \left(\int_M u^2 d\mu\right)$$

over all non-zero smooth functions such that $\displaystyle\int u d\mu = 0$. The metric g_* in the Rayleigh quotient above is assumed to be in c; by the conformal invariance of the numerator, the quantity $\mathrm{R}_c(\mu, u)$ does not depend on the choice of such a metric.

It is straightforward to see that the first eigenvalue $\lambda_1(\mu, c)$ is finite for any finite measure whose support is not a single point. The natural space of test functions for the Rayleigh quotient above is $L_2(M, \mu) \cap L_2^1(M, \, Vol_{g_*})$. The second space in the intersection is formed by distributions on M whose first derivatives are in $L_2(M, \, Vol_{g_*})$ (cf. [7]). Following classical terminology, we call the minimizers for the Rayleigh quotient $\mathrm{R}_c(\mu, \cdot)$ *first eigenfunctions.* They satisfy the following integral identity

$$\int_M \langle \nabla u, \nabla \varphi \rangle \, dVol_{g_*} = \lambda_1(\mu, c) \int_M u \cdot \varphi d\mu$$

for any smooth function φ and, in particular, form a linear space(cf. Proposition 5 in Section 4).

Recall that a measure μ is called *absolutely continuous* with respect to Vol_{g_*} if it has the form

$$\mu(X) = \int_X \phi \cdot dVol_{g_*},$$

where $X \subseteq M$ is a Borel set, and ϕ is a nonnegative L_1-function. Such measures correspond to metrics $\phi \cdot g_*$, thus, allowing them to be singular, degenerate, and have a nontrivial pole set. Below we call a measure μ *admissible* if it is absolutely continuous and its density function ϕ is bounded.

The theorem in Section 1 is a consequence of the following statement.

Theorem*. *Let M be an orientable compact surface with nonempty boundary. Then there exists a conformal class c of smooth Riemannian metrics on M such that for any admissible measure μ the first eigenvalue satisfies the estimate*

$$\lambda_1(\mu, c)\mu(M) \leqslant 8\pi.$$

2.3 The Yang-Yau estimate revisited

Our method uses the following version of the result by Yang and Yau [8, p. 58].

Proposition 1. *Let M be a closed Riemann surface, and let c be the conformal class induced by the complex structure. Suppose that M admits a holomorphic map $\varphi : M \to S^2$ of degree d. Then for any finite absolutely continuous measure μ on M the first eigenvalue satisfies the estimate*

$$\lambda_1(\mu, c)\mu(M) \leqslant 8\pi d.$$

Proof. First, Hersch's lemma holds for absolutely continuous measures (cf. [4, 5]). More precisely, there exists a conformal transformation s of S^2 such that

$$\int_M (x^i \circ s \circ \varphi) d\mu = 0 \qquad \text{for any } i = 1, 2, 3,$$

where x^i are the coordinate functions in $\mathbf{R}^3$. Using $(x^i \circ s \circ \varphi)$'s as test functions for the Rayleigh quotient, we obtain

$$\lambda_1(\mu, c) \int_M (x^i \circ s \circ \varphi)^2 d\mu \leqslant \int_M \left|\nabla(x^i \circ s \circ \varphi)\right|^2 dVol_{g_*}.$$

Summing up these inequalities over all i's and using the identity $\sum (x^i)^2 = 1$ on the unit sphere, we see that

$$\lambda_1(\mu, c)\mu(M) \leqslant \sum_i \int_M \left|\nabla(x^i \circ s \circ \varphi)\right|^2 dVol_{g_*}.$$

The right-hand side here is the energy of the map $(s \circ \varphi)$, which by formula (2.2) equals $8\pi d$. □

As a consequence, we obtain a version of the Hersch isoperimetric inequality for finite absolutely continuous measures on the sphere S^2. The estimates of Li and Yau [5] for the first eigenvalue via the conformal volume carry over our setting also.

3 The Proof

The proof is divided into a collection of lemmas. We start with the case where the boundary of M has at most two connected components; the general case can be reduced to the latter. First, we introduce more notation.

By $\widetilde{M}$ we denote below the doubling of M, i.e., the closed surface obtained by identifying the boundaries of two copies of M;

$$\widetilde{M} = M_0 \cup M_1, \qquad M_0 \cap M_1 = \Sigma,$$

where M_0 and M_1 are diffeomorphic to M, and Σ stands for their boundaries. Further, by r we denote the *reflection* with respect to Σ, i.e., the diffeomorphism of $\widetilde{M}$ that swaps M_0 and M_1, and is identical on Σ.

Lemma 2. *For any orientable surface M whose boundary has at most two connected components there exists a hyperelliptic complex structure on its doubling $\widetilde{M}$ (that makes $\widetilde{M}$ into a branched double cover over S^2) such that the reflection r is anti-holomorphic.*

Proof. Let γ be the genus of M. First, suppose that the boundary Σ has two connected components. Then the doubling $\widetilde{M}$ is a closed orientable surface of genus $2\gamma + 1$. To prove the statement, we describe a hyperelliptic model for $\widetilde{M}$ invariant under the reflection r (for the notation cf. [3]).

Take two copies of S^2, and let Σ_1 and Σ_2 be big circles on them respectively. On each sphere we make $2\gamma + 2$ "cuts" such that:

- the "cuts" do not intersect Σ_1 (or Σ_2 respectively);
- the "cuts" are invariant with respect to the reflection over the hyperplane through Σ_1 (or through Σ_2 respectively).

Now $\widetilde{M}$ is constructed by gluing the "north bank" of each "cut" on the first copy with the "south bank" of the corresponding "cut" on the second copy. These glued copies are called sheets of $\widetilde{M}$, and the holomorphic covering map is given by the natural projection onto one of the sheets. The reflections over the hyperplanes through Σ_1 and Σ_2 give rise to the conformal reflection r of $\widetilde{M}$.

In the case where M has a connected boundary, the doubling surface $\widetilde{M}$ has genus 2γ. The hyperelliptic model for it can be constructed in a similar fashion. More precisely, now we make $2\gamma + 1$ "cuts" on each copy of S^2. As above they should be made symmetrically with respect to the big circles and such that only one "cut" intersects the circle transversally. The doubling $\widetilde{M}$ is obtained by gluing these copies. □

Let $\widetilde{g}$ be a reflection invariant metric on $\widetilde{M}$ from the conformal class determined by the hyperelliptic complex structure. To obtain such a metric, one can start with an arbitrary metric $\widetilde{h}$ (form this conformal class), and set $\widetilde{g}$ to be $(\widetilde{h} + r^*\widetilde{h})$. By g_* we denote its restriction onto M.

Now we prove the theorem when the boundary of M has at most two connected components.

Lemma 3. *Let M be an orientable compact surface whose boundary has at most two connected components. Denote by g_* the metric on M obtained by restricting $\widetilde{g}$, and by c its conformal class. Then for any admissible measure μ the first eigenvalue satisfies the inequality*

$$\lambda_1(\mu, c)\mu(M) \leqslant 8\pi.$$

Proof. Let μ be a finite measure on M, and let $\widetilde{\mu}$ be its reflection invariant extension to the doubling surface $\widetilde{M}$. By $\widetilde{c}$ we also denote the conformal class of the reflection invariant metric $\widetilde{g}$. First, we claim that $\lambda_1(\widetilde{\mu},\widetilde{c})$ coincides with $\lambda_1(\mu,c)$. Indeed, for any smooth test function u for the Rayleigh quotient on M, its reflection invariant extension $\widetilde{u}$ to $\widetilde{M}$ can be taken as a test function for the Rayleigh quotient on $\widetilde{M}$. Thus, we have

$$\lambda_1(\widetilde{\mu},\widetilde{c}) \leqslant \mathrm{R}_{\widetilde{c}}(\widetilde{\mu},\widetilde{u}) = \mathrm{R}_c(\mu,u).$$

Taking the infimum over all test functions u, we conclude that $\lambda_1(\widetilde{\mu},\widetilde{c})$ is not greater than $\lambda_1(\mu,c)$.

To prove the inverse, we start with a first eigenfunction v for the measure $\widetilde{\mu}$ on $\widetilde{M}$. By Proposition 6, such an eigenfunction exists and, by Proposition 5, its symmetric part

$$\widetilde{u} = (v + v\circ r)/2$$

is also a first eigenfunction. Denote by u the restriction of $\widetilde{u}$ onto M. Clearly, it satisfies the orthogonality hypothesis $\displaystyle\int u d\mu = 0$ and can be used as a test function for the Rayleigh quotient on M. Thus, we obtain the inverse inequality

$$\lambda_1(\mu,c) \leqslant \mathrm{R}_c(\mu,u) = \mathrm{R}_{\widetilde{c}}(\widetilde{\mu},\widetilde{u}) = \lambda_1(\widetilde{\mu},\widetilde{c}).$$

Now since the total mass $\widetilde{\mu}(\widetilde{M})$ equals $2\mu(M)$, the statement follows by the combination of Lemma 2 with the Yang–Yau estimate (Proposition 1). □

Finally, we consider the case where the boundary of M has more than two connected components.

Lemma 4. *Let M be an orientable compact surface whose boundary has $\ell > 2$ connected components. Then there exists a conformal class c of smooth Riemannian metrics on M such that for any admissible measure μ the first eigenvalue satisfies the inequality*

$$\lambda_1(\mu,c)\mu(M) \leqslant 8\pi.$$

Proof. Denote by $\overline{M}$ a surface obtained from M by gluing $(\ell-1)$ (or $(\ell-2)$) disks along its boundary; the result is a compact surface whose boundary is connected (or has two connected components). Let $\overline{c}$ be a conformal class on $\overline{M}$ which satisfies the conclusions of Lemma 3, and let c be the conformal class on M obtained by restricting the metrics from $\overline{c}$. For any admissible measure μ on M, we denote be $\overline{\mu}$ its extension by zero to $\overline{M}$. Then for any test function $\overline{u}$ for the Rayleigh quotient $\mathrm{R}_{\overline{c}}(\overline{\mu},\cdot)$ on $\overline{M}$, its restriction u to M satisfies the orthogonality hypothesis $\displaystyle\int u d\mu = 0$. Using the latter as a test function for $\mathrm{R}_c(\mu,\cdot)$, we obtain

$$\lambda_1(\mu,c) \leqslant \mathrm{R}_c(\mu,u) \leqslant \mathrm{R}_{\overline{c}}(\overline{\mu},\overline{u}).$$

The second inequality here follows from the fact that the denominators in the corresponding Rayleigh quotients coincide, while the numerator of the first is not greater than the one of the second. Taking the infimum over all test functions $\overline{u}$, we conclude that $\lambda_1(\mu, c)$ is not greater than $\lambda_1(\overline{\mu}, \overline{c})$. Since the total masses of μ and $\overline{\mu}$ coincide, we finally obtain

$$\lambda_1(\mu, c)\mu(M) \leqslant \lambda_1(\overline{\mu}, \overline{c})\overline{\mu}(\overline{M}) \leqslant 8\pi,$$

where the last inequality follows by Lemma 3. □

4 Appendix: Auxiliary Propositions

The purpose of this section is to give more details on the properties of minimizers (first eigenfunctions) for the Rayleigh type quotient $\mathrm{R}_c(\mu, u)$. Below M denotes a smooth compact surface (with or without boundary) equipped with a conformal class c of smooth Riemannian metrics and a finite Borel measure μ.

Proposition 5. *Suppose that the first eigenvalue $\lambda_1(\mu, c)$ of a given pair (μ, c) on M is positive and finite. Then a nontrivial test function u for the Rayleigh quotient $\mathrm{R}_c(\mu, \cdot)$ is its minimizer if and only if it satisfies the integral identity*

$$\int_M \langle \nabla u, \nabla \varphi \rangle dVol_{g_*} = \lambda_1(\mu, c) \int_M u \cdot \varphi d\mu \tag{4.1}$$

for any smooth function φ.

Proof. First, suppose that u satisfies the integral identity (4.1). Setting φ to be a non-zero constant, we obtain $\int u d\mu = 0$. Further approximating u by smooth functions with respect to the norm in $L_2(M, \mu) \cap L_2^1(M, Vol_{g_*})$, we conclude that

$$\int_M |\nabla u|^2 \, dVol_{g_*} = \lambda_1(\mu, c) \int_M u^2 d\mu.$$

Thus, we see that u is indeed a minimizer.

Conversely, suppose that u minimizes the Rayleigh quotient. For a smooth zero mean value function φ consider the family $u_t = u + t\varphi$. Since u is a minimizer, the derivative of $\mathrm{R}_c(\mu, u_t)$ with respect to t vanishes at $t = 0$. This shows that Equation (4.1) holds for any zero mean value φ. Now the claim follows from the observation that equation (4.1) holds trivially for constant functions φ. □

Proposition 6. *Suppose that the measure μ of a given pair (μ, c) on M is admissible. Then there exists a minimizer for the Rayleigh quotient $\mathrm{R}_c(\mu, \cdot)$, and the first eigenvalue $\lambda_1(\mu, c)$ is positive.*

Proof. Let u_n be a minimizing sequence of smooth test functions,

$$\int_M u_n^2 d\mu = 1, \qquad \int_M |\nabla u_n|^2 \, dVol_{g_*} \longrightarrow \lambda_1(\mu, c) \quad \text{as} \quad n \to +\infty.$$

Denote by $\overline{u}_n$ the zero mean value part of u_n with respect to the measure Vol_{g_*}. By the Poincaré inequality the sequence $\overline{u}_n$ is bounded in $W^{1,2}(M, Vol_{g_*})$ and, hence, contains a subsequence (also denoted by $\overline{u}_n$) converging weakly in $W^{1,2}(M, Vol_{g_*})$ and strongly in $L_2(M, Vol_{g_*})$ to some function $\overline{u}$. Since the measure μ is admissible, the sequence $\overline{u}_n$ also converges to $\overline{u}$ in $L_2(M, \mu)$. Using the orthogonality hypothesis $\int u_n d\mu = 0$, one further concludes that the original test functions u_n converge to $u = \overline{u} + C$ in $L_2(M, \mu)$, where C is an appropriate constant. In particular, the limit function u satisfies the identities

$$\int_M u^2 d\mu = 1, \qquad \int_M u d\mu = 0.$$

By the lower semi-continuity of the Dirichlet energy, we further obtain

$$\begin{aligned}\int_M |\nabla u|^2 \, dVol_{g_*} &= \int_M |\nabla \overline{u}|^2 \, dVol_{g_*} \\ &\leqslant \liminf \int_M |\nabla \overline{u}_n|^2 \, dVol_{g_*} = \lambda_1(\mu, c).\end{aligned}$$

Thus, we see that the function u is a minimizer for the Rayleigh quotient. Due to the hypothesis $\int u d\mu = 0$ it can not be constant and, hence, the first eigenvalue $\lambda_1(\mu, c)$ is positive. □

References

1. Colbois, B., El Soufi, A.: Extremal eigenvalues of the Laplacian in a conformal class of metrics: the 'conformal spectrum'. Ann. Global Anal. Geom. **24**, 337–349 (2003)
2. Eells, J., Lemaire, L.: Selected Topics in Harmonic Maps. Am. Math. Soc., Providence, RI (1983)
3. Farkas, H. M., Kra, I.: Riemann Surfaces. Springer, New York (1992)
4. Hersch, J.: Quatre propriétés isopérimétriques de membranes sphériques homogènes. C. R. Acad. Sci. Paris Sér. A-B **270**, A1645–A1648 (1970)
5. Li, P., Yau, S.-T.: A new conformal invariant and its applications to the Willmore conjecture and the first eigenvalue of compact surfaces. Invent. Math. **69**, 269–291 (1982)

6. Maz'ya, V.: On Neumann's problem in domains with nonregular boundaries Russian. Sib. Mat. Zh. **9**, 1322–1350 (1968); English transl.: Sib. Math. J. **9**, 990–1012 (1968)
7. Maz'ya, V.G.: Sobolev Spaces. Springer, Berlin etc. (1985)
8. Yang, P., Yau, S.-T.: Eigenvalues of the Laplacian of compact Riemann surfaces and minimal submanifolds. Ann. Scuola Norm. Sup. Pisa Cl. Sci. (4) **7**, 55–63 (1980)

Necessary Condition for the Regularity of a Boundary Point for Porous Medium Equations with Coefficients of Kato Class

Vitali Liskevich and Igor I. Skrypnik

Abstract We prove the necessity of the Wiener test for the regularity of a boundary point for a wide class of porous medium type equations with lower order terms in the structure conditions. The coefficients corresponding to the lower order terms are assumed to be in the Kato class, which generalizes known results.

1 Introduction and the Main Results

The capacitary Wiener test for the regularity of a boundary point with respect to the Dirichlet problem for the Laplace equation [19] has been extended to various classes of elliptic and parabolic equations. Littman, Stampacchia, Weinberger [10] showed that the same test applies to uniformly elliptic second order divergence type equations with bounded measurable coefficients. Degenerate linear elliptic equations were studied by Fabes, Jerison, Kenig [4]. For elliptic equations extensions also include nonlinear p-Laplace type equations (sufficiency was established by Maz'ya [12], extended by Gariepy and Ziemer [5] to general classes, and necessity proved by Kilpeläinen and Malý [6]), nondivergence type equations (Bauman [1]), fully nonlinear equations (Labutin [7]), higher order equations (Maz'ya [14]) and the quest continues (cf. also [13, 11] for problems and historical notes).

Vitali Liskevich
Department of Mathematics, University of Swansea, Swansea, SA2 8PP, UK
e-mail: `v.a.liskevich@swansea.ac.uk`

Igor I. Skrypnik
Institute of Applied Mathematics and Mechanics, Donetsk, Ukraine
e-mail: `iskrypnik@iamm.donbass.com`

A. Laptev (ed.), *Around the Research of Vladimir Maz'ya II: Partial Differential Equations*, International Mathematical Series 12, DOI 10.1007/978-1-4419-1343-2_11,

The Wiener condition in the linear parabolic case has been obtained by Landis [9], Lanconelli [8], Evans, Gariepy [3], for quasilinear parabolic equations in [20, 21, 16] under some restrictions on the structure conditions.

In this paper, we are interested in the boundary regularity problem for generalized porous medium type quasilinear equations with measurable coefficients. The interior regularity for equations of such a type was studied in [2] with $L^{p,q}$ type conditions on the structure coefficients, and the boundary regularity in [21], [16] for a particular case of structure conditions ($m = 1$, see below). General porous medium type quasilinear equations with structure coefficients in $L^{p,q}$ classes were studied in [17]. Here we give a generalization of the above results to the case of coefficients satisfying the Kato class condition (see the precise definition below). The Kato class turns out to be the optimal class for a number of qualitative properties of elliptic and parabolic equations, such as standard estimates of fundamental solutions, continuity of weak solutions, the Harnack inequality (cf., for example, [15]). We show that the Kato class condition on the structure coefficients for the general quasilinear porous medium equation preserves the classical Wiener condition for the regularity of a boundary point. The result seems to be new even in the linear case. We adapt the Kilpeläinen–Malý method [6] which was originally used for elliptic equations and applied in the parabolic case in [18].

Let Ω be a domain in $\mathbb{R}^n$, $0 < T < \infty$. Set $\Omega_T = \Omega \times (0,T)$. In this paper we study solutions to the equation

$$u_t - \operatorname{div} \mathbf{A}(x,t,u,\nabla u) = a_0(x,t,u,\nabla u) \tag{1.1}$$

in a neighborhood of the parabolic boundary of Ω_T.

Throughout the paper, we suppose that $m > \frac{n-2}{n}$, and the functions $\mathbf{A} : \Omega \times \mathbb{R}^+ \times \mathbb{R} \times \mathbb{R}^n \to \mathbb{R}^n$ and $a_0 : \Omega \times \mathbb{R}^+ \times \mathbb{R} \to \mathbb{R}^n$ are such that $\mathbf{A}(\cdot,\cdot,u,\zeta), a_0(\cdot,\cdot,u,\zeta)$ are Lebesgue measurable for all $u \in \mathbb{R}, \zeta \in \mathbb{R}^N$, and $\mathbf{A}(x,t,\cdot,\cdot), a_0(x,t,\cdot,\cdot)$ are continuous for almost all $(x,t) \in \Omega_T$.

We also assume that the following structure conditions are satisfied:

$$\mathbf{A}(x,t,u,\zeta)\zeta \geqslant c_1|u|^{m-1}|\zeta|^2, \quad x \in \Omega, t > 0,\, u \in \mathbb{R},\, \zeta \in \mathbb{R}^n, \tag{1.2}$$

$$|\mathbf{A}(x,t,u,\zeta)| \leqslant c_2|u|^{m-1}|\zeta| + g_1(x)|u|^m + f_1(x), \tag{1.3}$$

$$(\mathbf{A}(x,t,u,\zeta) - \mathbf{A}(x,t,u,\zeta')) \cdot (\zeta - \zeta') > 0, \quad \zeta \neq \zeta' \in \mathbb{R}^n, \tag{1.4}$$

$$|a_0(x,t,u,\zeta)| \leqslant \nu(x)|u|^{m-1}|\zeta| + g_2(x)|u|^m + f_2(x), \tag{1.5}$$

with some positive constants c_1 and c_2 and nonnegative functions $f_i(x)$, $g_i(x)$, $i = 1,2$, and $\nu(x)$.

We introduce the Kato class

$$K := \left\{ g \in L^1(\Omega) \lim_{R\to 0} \sup_{x\in\Omega} \int_0^R \left\{ \frac{1}{r^{n-2}} \int_{B_r(x)\cap\Omega} |g(z)|dz \right\} \frac{dr}{r} = 0 \right\}, \tag{1.6}$$

where $B_r(x) = \{z \in \Omega : |z - x| < r\}$. It is easy to see that this condition is equivalent to the standard (in the linear theory) Kato class condition [15].

We also introduce the class

$$\widetilde{K} := \left\{ g \in L^1(\Omega)\ \lim_{R\to 0} \sup_{x\in\Omega} \int_0^R \left\{ \frac{1}{r^{n-2}} \int_{B_r(x)\cap\Omega} |g(z)|dz \right\}^{\frac{1}{2}} \frac{dr}{r} = 0 \right\}, \tag{1.7}$$

We impose the following condition on the structure coefficients:

$$g_1^2 + f_1^2 \in \widetilde{K}, \quad g_2 + f_2 + \nu^2 \in K. \tag{1.8}$$

Recall that $u \in V(\Omega_T) := C(0, T; L^2(\Omega)) \cap \{u : |u|^{\frac{m-1}{2}} |\nabla u| \in L^2(\Omega_T)\}$ is a weak solution to (1.1) if

$$\int_\Omega u(x,\tau)\varphi(x,\tau)dx + \int_o^\tau \int_\Omega \{-u\varphi_t + \mathbf{A}(x,t,u,\nabla u)\nabla\varphi - a_0(x,t,u,\nabla u)\varphi\}\, dxdt = 0 \tag{1.9}$$

for $\varphi(x,t) = \xi(x,t)\psi(x,t)$, where $\psi \in V(\Omega_T)$ is such that $\dfrac{\partial\psi}{\partial t} \in L^2(\Omega_T)$, $\xi \in C^\infty(\overline{\Omega_T})$ vanishes in a neighborhood of $(0,0)$, and $\tau \in (0,T)$.

Without loss of generality, we can assume that $\dfrac{\partial u}{\partial t} \in L^2(\Omega_T)$; otherwise, we can pass to the Steklov average of u.

We use the notation $B = B_r(x_0)$ and $Q = B \times (t_0 - r^2, t_0 + r^2)$.

A point $(x_0, t_0) \in S_T \equiv \partial\Omega \times (0,T)$ is said to be *regular* for Equation (1.1) if for any solution u to (1.1) satisfying $\varphi(u - f) \in L^2(0,T; \overset{\circ}{W}{}^{1,2}(\Omega))$ with $f \in C(\overline{\Omega}_T) \cap W^{1,2}(\Omega_T)$ and $\varphi \in C^1(\mathbb{R}^{n+1})$, $\varphi(x,t) = 1$ in a neighborhood of (x_0, t_0) we have

$$\lim_{r\to 0} \operatorname{ess\,sup}_{(x,t)\in\Omega_T\cap Q} u(x,t) = \lim_{r\to 0} \operatorname{ess\,inf}_{(x,t)\in\Omega_T\cap Q} u(x,t) = f(x_0,t_0). \tag{1.10}$$

The main result of the paper is formulated as follows.

Theorem 1.1. *Let u be a bounded weak solution to (1.1) in Ω_T. Assume that the conditions (1.2)–(1.5), and (1.8) are fulfilled. If a point $(x_0, t_0) \in S_T$ is regular for (1.1), then*

$$\int_0^1 \frac{C(B_r(x_0) \setminus \Omega)}{r^{n-1}} dr = \infty, \tag{1.11}$$

where $C(E)$ denotes the Newtonian capacity of a set $E \subset \mathbb{R}^n$.

Remark 1.2. Under the assumption that the solution u is bounded, the conditions (1.2), (1.3), and (1.5) can be written in the form

$$\mathbf{A}(x,t,u,\zeta)\zeta \geqslant c_1|u|^{m-1}|\zeta|^2, \tag{1.12}$$

$$|\mathbf{A}(x,t,u,\zeta)| \leqslant c_2|u|^{m-1}|\zeta| + f_1(x), \tag{1.13}$$

$$|a_0(x,t,u)| \leqslant \nu(x)|u|^{m-1}|\zeta| + f_2(x), \tag{1.14}$$

The rest of the note contains the proof of the above theorem. Assuming that

$$\int_0^{1/2} \frac{C(B_r(x_0)\setminus\Omega)}{r^{n-1}}dr < \infty \tag{1.15}$$

we prove that the point $(x_0,t_0)\in S_T$ is irregular.

In the sequel γ denotes a constant whose value is of no importance and which may vary from line to line.

2 Auxiliary Estimates

Let l and r be positive numbers such that

$$\frac{1}{4} \leqslant l \leqslant M := \operatorname{ess\,sup}_{\Omega_T}|u(x,t)|, \quad (t_0-r^2,t_0+r^2)\subset(0,T).$$

Let $\xi,\eta : B\to\mathbb{R}$ and $\zeta:\mathbb{R}^n\to\mathbb{R}$ be such that $\xi,\eta\in W^{1,2}(B)$, $\xi\zeta\in \overset{\circ}{W}{}^{1,2}(B)$

$$\mathbf{1}_{B_{r/2}(x_0)} \leqslant \xi \leqslant \mathbf{1}_{B_r(x_0)}, \quad |\nabla\xi|\leqslant\gamma r^{-1},$$

$$0\leqslant\zeta\leqslant 1, \quad \zeta(x)=0, \quad x\in\mathbb{R}^n\setminus\Omega,$$

$$0\leqslant\eta\leqslant 1, \quad (1-\eta)(1-\zeta)=0.$$

Denote $L=Q\cap\{(x,t):u(x,t)>l\}$, $L(\tau)=\{(x,t)\in L: t=\tau\}$, $E=L\cap\{x:\eta(x)<1\}$, and $F=L\cap\{x:\eta(x)=1\}$.

Let $\theta\in C^1(\mathbb{R})$ be such that

$$\mathbf{1}_{(t_0-\frac{4}{9}r^2,t_0+\frac{4}{9}r^2)} \leqslant \theta \leqslant \mathbf{1}_{(t_0-r^2,t_0+r^2)}.$$

We set

$$\lambda=\frac{1}{n},\ \rho(\lambda)=\frac{2}{1-\lambda}$$

and

$$w(x,t)=\left[\frac{1}{\delta}\int_l^{u(x,t)}\left(1+\frac{s-l}{\delta}\right)^{-\frac{1+\lambda}{2}}ds\right]_+.$$

Lemma 2.1. *Let the assumptions of Theorem* 1.1 *hold. Then for any* $\varepsilon>0$, $l\in[\frac{1}{4},M]$, *and* $k>2$

$$\operatorname{ess\,sup}_t \int_{L(t)\cap\{u>l+\varepsilon\delta\}} \frac{u-l}{\delta}[\xi(x)\zeta(x)\theta(t)]^k dx$$
$$+ \iint_L |\nabla w|^2[\xi(x)\zeta(x)\theta(t)]^k dx\,dt$$
$$\leqslant \frac{\gamma(\varepsilon)}{r^2}\iint_E \left(1+\frac{u-l}{\delta}\right)^{1+\lambda}[\xi\theta]^{k-2}dx\,dt + \frac{\gamma(\varepsilon)}{\delta}r^2\int_B |\nabla\zeta|^2 dx$$
$$+ \gamma(\varepsilon)\delta^{-2}r^2\int_B f_1^2 dx\,dt + \gamma(\varepsilon)\delta^{-1}r^2\int_B (\nu^2+f_2)dx\,dt. \tag{2.1}$$

Proof. We test (1.1) by

$$u\left[\int_l^u \left(1+\frac{s-l}{\delta}\right)^{-1-\lambda} ds\right]_+ [\xi\zeta\theta]^k.$$

Then we have

$$\operatorname{ess\,sup}_t \int_{L(t)} \left\{\int_l^{u(x,t)} v\int_l^v \left(1+\frac{s-l}{\delta}\right)^{-1-\lambda} ds\,dv\right\}[\xi(x)\zeta(x)\theta(t)]^k dx$$
$$+ \iint_L \left\{u^{m-1}\int_l^{u(x,t)}\left(1+\frac{s-l}{\delta}\right)^{-1-\lambda} ds + u^m\left(1+\frac{u-l}{\delta}\right)^{-1-\lambda}\right\}$$
$$\times |\nabla u|^2[\xi\zeta\theta]^k dx\,dt$$
$$\leqslant \gamma\iint_L \left\{\int_l^{u(x,t)} v\int_l^v \left(1+\frac{s-l}{\delta}\right)^{-1-\lambda} ds\,dv\right\}|\theta_t|[\xi\zeta\theta]^{k-1}dx\,dt$$
$$+ \gamma\iint_L u\int_l^{u(x,t)}\left(1+\frac{s-l}{\delta}\right)^{-1-\lambda} ds\left\{u^{m-1}|\nabla u| + f_1\right\}|\nabla(\xi\zeta)|$$
$$\times [\xi\zeta\theta]^{k-1}dx\,dt$$
$$+ \gamma\iint_L u\int_l^{u(x,t)}\left(1+\frac{s-l}{\delta}\right)^{-1-\lambda} ds\left\{\nu u^{m-1}|\nabla u| + f_2\right\}[\xi\zeta\theta]^k dx\,dt. \tag{2.2}$$

By the conditions on l,

$$\frac{1}{4} \leqslant u(x,t) \leqslant M, \quad \text{for } (x,t)\in L. \tag{2.3}$$

It is obvious that

$$\int_l^u \left(1+\frac{s-l}{\delta}\right)^{-1-\lambda} ds \leqslant \frac{\delta}{\lambda} \tag{2.4}$$

and

$$\int_l^u v \int_l^v \left(1+\frac{s-l}{\delta}\right)^{-1-\lambda} ds\, dv \leqslant \gamma\delta u(u-l). \tag{2.5}$$

We also note that for $u > l + \varepsilon\delta$

$$\int_l^u v \int_l^v \left(1+\frac{s-l}{\delta}\right)^{-1-\lambda} ds$$
$$\geqslant \frac{1}{2} u \int_l^{\frac{u+l}{2}} \left(1+\frac{s-l}{\delta}\right)^{-1-\lambda} (u-s) ds \geqslant \gamma(\varepsilon)\delta u(u-l), \tag{2.6}$$

and

$$\gamma(\varepsilon)^{-1}\left(\frac{u-l}{\delta}\right)^{\frac{1-\lambda}{2}} \leqslant w \leqslant \gamma(\varepsilon)\left(\frac{u-l}{\delta}\right)^{\frac{1-\lambda}{2}}. \tag{2.7}$$

Hence from (2.2) we obtain

$$\begin{aligned}
&\delta \operatorname{ess\,sup}_t \int_{L(t)} (u-l)[\xi(x)\zeta(x)\theta(t)]^k dx \\
&+ \iint_L \left\{ \int_l^u \left(1+\frac{s-l}{\delta}\right)^{-1-\lambda} ds + \left(1+\frac{u-l}{\delta}\right)^{-1-\lambda} \right\} \\
&\quad \times |\nabla u|^2 [\xi\zeta\theta]^k dx\, dt \\
&\leqslant \frac{\gamma(\varepsilon)\delta^2}{r^2} \iint_L \frac{u-l}{\delta} [\xi\zeta\theta]^{k-1} dx\, dt + \gamma(\varepsilon) \iint_L \int_l^u \left(1+\frac{s-l}{\delta}\right)^{-1-\lambda} ds \\
&\quad \times |\nabla u||\nabla(\xi\zeta)|[\xi\zeta\theta]^{k-1} dx\, dt \\
&+ \gamma(\varepsilon) \iint_L u \int_l^u \left(1+\frac{s-l}{\delta}\right)^{-1-\lambda} ds \nu |\nabla u| [\xi\zeta\theta]^k dx\, dt \\
&+ \gamma(\varepsilon) \iint_L \int_l^u \left(1+\frac{s-l}{\delta}\right)^{-1-\lambda} ds\, f_1 |\nabla(\xi\zeta)|[\xi\zeta\theta]^{k-1} dx\, dt \\
&+ \gamma(\varepsilon) \iint_L \int_l^u \left(1+\frac{s-l}{\delta}\right)^{-1-\lambda} ds\, f_2 [\xi\zeta\theta]^k dx\, dt \\
&:= \sum_{i=1}^5 I_i(L).
\end{aligned} \tag{2.8}$$

By (2.3) and (2.4), we have

$$I_5(L) \leqslant \gamma(\varepsilon)\delta r^2 \int_B f_2 dx. \tag{2.9}$$

The remaining integrals on the left-hand side of (2.8) are represented as the sums of integrals over E and F.

First, we estimate the integrals over E where $\zeta(x) = 1$. We have

$$I_2(E) \leqslant \frac{1}{8}\iint_L \left(1+\frac{u-l}{\delta}\right)^{-1-\lambda} |\nabla u|^2[\xi\zeta\theta]^k dx\, dt + \frac{\gamma(\varepsilon)\delta^2}{r^2}\iint_E \left(1+\frac{u-l}{\delta}\right)^{1+\lambda} [\xi\theta]^{k-2} dx\, dt, \tag{2.10}$$

$$I_3(E) \leqslant \frac{1}{8}\iint_L \int_l^u \left(\frac{s-l}{\delta}\right)^{-1-\lambda} ds |\nabla u|^2[\xi\zeta\theta]^k dx\, dt + \gamma(\varepsilon)\iint_E u^2 \int_l^u \left(\frac{s-l}{\delta}\right)^{-1-\lambda} ds\, \nu^2 [\xi\zeta\theta]^k dx\, dt$$

$$\leqslant \frac{1}{8}\iint_L \int_l^u \left(\frac{s-l}{\delta}\right)^{-1-\lambda} ds |\nabla u|^2[\xi\zeta\theta]^k dx\, dt + \gamma(\varepsilon)\delta r^2 \int_B \nu^2 dx, \tag{2.11}$$

$$I_4(E) \leqslant \frac{\gamma(\varepsilon)\delta^2}{r^2}\iint_E \left(1+\frac{u-l}{\delta}\right)^{1+\lambda} [\xi\theta]^{k-2} dx\, dt + \gamma(\varepsilon) r^2 \int_B f_1^2 dx. \tag{2.12}$$

To estimate the corresponding integrals over F, we test (1.1) by $(u - l)_+[\xi\zeta\theta]^k$. We obtain

$$\operatorname{ess\,sup}_t \int_{L(t)} (u-l)^2[\xi\zeta\theta]^k dx + \iint_L |\nabla u|^2 [\xi\zeta\theta]^k dx\, dt \leqslant \frac{\gamma}{r^2}\iint_L [\xi\zeta\theta]^{k-1} dx\, dt + \gamma \iint_L |\nabla(\xi\zeta)|^2 [\xi\zeta\theta]^{k-2} dx\, dt + \gamma \iint_L h_1 [\xi\zeta\theta]^{k-2} dx\, dt, \tag{2.13}$$

where $h = f_1^2 + f_2 + \nu^2$. From (1.11) and the Poincaré inequality it follows that

$$\operatorname{ess\,sup}_t \int_{L(t)} (u-l)^2[\xi\zeta\theta]^k dx + \iint_L |\nabla u|^2 [\xi\zeta\theta]^k dx\, dt \leqslant \gamma r^2 \left(\int_B |\nabla\zeta|^2 dx + \int_B h_1 dx\right). \tag{2.14}$$

Hence, by (2.14),

$$I_1(F) + I_2(F) + I_3(F) + I_4(F) \leqslant \gamma(\varepsilon) r^2 \delta \left(\int_B |\nabla\zeta|^2 dx + \int_B h_1 dx\right). \tag{2.15}$$

The required (2.1) follows from (2.8)–(2.15) and (2.3), (2.7). □

Let $R_0 \in (0,1)$ be small enough. We set $r_j = \frac{R_0}{2^j}$, $j = 1, 2, \dots$, $B_j = B_{r_j}(x_0)$, $Q_j = B_j \times (t_0 - r_j^2, t_0 + r_j^2)$. Let $\xi_j \in C_0^1(B_j)$ be such that $\xi_j = 1$ on B_{j+1}, and let $\theta \in C_0^1(t_0 - r_j^2, t_0 + r_j^2)$ be such that $\theta = 1$ on $(t_0 - \frac{4}{9}r_j^2, t_0 + \frac{4}{9}r_j^2)$. Also let $g_j \in C_0^\infty(B_1(x_0))$ be such that $g_j = 1$ on $B_j \setminus \Omega$ and

$$\int_{B_1(x_0)} |\nabla g_j|^2 dx \leqslant \gamma C(B_j \setminus \Omega). \tag{2.16}$$

We set $g_j' = \min\{1, [g_j]_+\}$, $\eta_j = \min\{1, 3g_j' + 3g_{j-1}'\}$, and $\zeta_j = \min\{1, (2 - 3g_j')_+\}$.

Define a sequence of positive numbers $(l_j)_{j \in \mathbb{N}}$ by $l_0 = \frac{1}{4}$ and for $j \geqslant 1$

$$a = \frac{1}{r_j^{n+2}} \iint_{L_j} \left(\frac{u(x,t) - l_j}{l_{j+1} - l_j} \right)^{1+\lambda} [\xi_j(x)\zeta_j(x)\theta_j(t)]^{k-2} dx\, dt, \tag{2.17}$$

where $L_j = \Omega_T \cap \{(x,t) : u(x,t) > l_j\}$ and a is a positive constant which will be chosen later depending on n, m, c_0, c_1, and M.

We set

$$\delta_j = l_{j+1} - l_j. \tag{2.18}$$

The following lemma is the key of the Kilpeläinen–Malý method [6].

Lemma 2.2. *Let the assumptions of Theorem 1.1 hold. Then for every $j \geqslant 1$ the following inequality holds:*

$$\delta_j \leqslant \frac{1}{2}\delta_{j-1} + \gamma \frac{C(B_{j-1} \setminus \Omega)}{r_j^{n-2}} + \gamma \frac{1}{r_j^{n-2}} \int_{B_j} (\nu^2 + f_2) dx + \gamma \left(\frac{1}{r_j^{n-2}} \int_{B_j} f_1^2 dx\, dt \right)^{\frac{1}{2}}. \tag{2.19}$$

Proof. We fix $j \geqslant 1$. We can assume that

$$\delta_j > \frac{1}{2}\delta_{j-1} \tag{2.20}$$

since otherwise the assertion is obvious. Decompose L_j as $L_J = L_j' \cup L_j''$, where

$$L_j = L_j \cap \left\{ (x,t) : \frac{u(x,t) - l_j}{\delta_j} < \varepsilon \right\},$$

$$L_j'' = L_j \cap \left\{ (x,t) : \frac{u(x,t) - l_j}{\delta_j} > \varepsilon \right\},$$

$\varepsilon \in (0,1)$. We have

$$\frac{1}{r_j^{n+2}} \iint_{L_j'} \left(\frac{u-l_j}{\delta_j}\right)^{1+\lambda} [\xi_j\zeta_j\theta_j]^{k-2} dx\,dt$$
$$\leqslant \gamma \frac{\varepsilon^{1+\lambda}}{r_j^{n+2}} \left\{ \iint_{E_j} dx\,dt + \iint_{F_j} [\xi_j\zeta_j\theta_j]^{k-2} dx\,dt \right\}. \tag{2.21}$$

Since $\zeta_{j-1}(x) = \xi_{j-1}(x) = \theta_{j-1}(t) = 1$ for $(x,t) \in E_j$, it follows that

$$\iint_{E_j} dx\,dt \leqslant \iint_{E_j} \left(\frac{u-l_{j-1}}{\delta_{j-1}}\right)^{1+\lambda} [\xi_{j-1}\zeta_{j-1}\theta_{j-1}]^{k-2} dx\,dt \leqslant \gamma r_j^{n+2} a. \tag{2.22}$$

By the Poincaré inequality and (2.16)

$$\iint_{F_j} [\xi_j\zeta_j\theta_j]^{k-2} dx\,dt \leqslant \gamma r_j^4 \left(C(B_j \setminus \Omega) + C(B_{j-1} \setminus \Omega)\right). \tag{2.23}$$

Let w_j be defined by (2.1) with $l = l_j$, $\delta = \delta_j$. Then

$$w_j(x,t) \geqslant \gamma(\varepsilon) \left(\frac{u(x,t)-l_j}{\delta_j}\right)^{\frac{1-\lambda}{2}}, \quad (x,t) \in L_j''. \tag{2.24}$$

By the above inequality and the Sobolev embedding theorem, we find

$$\frac{1}{r_j^{n+2}} \iint_{L_j''} \left(\frac{u-l_j}{\delta_j}\right)^{1+\lambda} [\xi_j\zeta_j\theta_j]^{k-2} dx\,dt$$
$$\leqslant \frac{\gamma}{r_j^{n+2}} \iint_{L_j''} w_j^{2\frac{1+\lambda}{1-\lambda}} [\xi_j\zeta_j\theta_j]^{k-2} dx\,dt$$
$$\leqslant \frac{\gamma}{r_j^{n+2}} \left\{ \operatorname{ess\,sup}_t \iint_{L_j(t)} w_j^{\rho(\lambda)} [\xi_j\zeta_j\theta_j]^k dx \right\}^{2/n}$$
$$\times \iint_{L_j} \left| \nabla \left[w_j (\xi_j\zeta_j\theta_j)^{\frac{k-2}{2}} \right] \right|^2 dx\,dt. \tag{2.25}$$

By Lemma 2.1 and the fact that $\frac{1}{4} \leqslant u(x,t) \leqslant M$ for $(x,t) \in L_j$, from (2.25) we find

$$\frac{1}{r_j^{n+2}} \iint_{L_j''} \left(\frac{u-l_j}{\delta_j}\right)^{1+\lambda} [\xi_j\zeta_j\theta_j]^{k-2} dx\,dt$$
$$\leqslant \left\{ \frac{1}{r_j^{n+2}} \iint_{E_j} \left(1 + \frac{u-l_j}{\delta_j}\right)^{1+\lambda} [\xi_j\zeta_j\theta_j]^{k-4} dx\,dt + \frac{1}{\delta_j} r_j^{2-n} C(B_j \setminus \Omega) \right)$$

$$+\frac{1}{\delta_j^2} r_j^{2-n} \int_{B_j} f_1^2 dx\, dt + \frac{1}{\delta_j} r_j^{2-n} \int_{B_j} (\nu^2 + f_2) dx\, dt \Bigg\}^{1+2/n}. \tag{2.26}$$

We estimate

$$\begin{aligned} &\frac{1}{r_j^{n+2}} \iint_{E_j} \left(1 + \frac{u - l_j}{\delta_j}\right)^{1+\lambda} [\xi_j \zeta_j \theta_j]^{k-4} dx\, dt \\ &\leqslant \frac{\gamma}{r_j^{n+2}} \iint_{E_j'} dx\, dt + \frac{\gamma(\varepsilon)}{r_j^{n+2}} \iint_{E_j''} \left(\frac{u - l_j}{\delta_j}\right)^{1+\lambda} [\xi_{j-1} \zeta_{j-1} \theta_{j-1}]^{k-2} dx\, dt \\ &\leqslant \gamma a, \end{aligned} \tag{2.27}$$

where the first term on the right-hand side of (2.27) was estimated as in (2.22), whereas for the second term we used (2.20) and the fact that $\xi_{j-1}(x) = \zeta_{j-1}(x) = \theta_{j-1}(t) = 1$ for $(x,t) \in L_j$.

From (2.21)–(2.27) we obtain

$$\begin{aligned} a &\leqslant \gamma \varepsilon^{1+\lambda} a + \frac{\gamma}{\delta_j} r_j^{2-n} C(B_{j-1} \setminus \Omega) + \gamma r_j^{2-n} C(B_{j-1} \setminus \Omega) \\ &+ \gamma(\varepsilon) \Bigg\{ a + \frac{1}{\delta_j} r_j^{2-n} C(B_{j-1} \setminus \Omega) + r_j^{2-n} C(B_{j-1} \setminus \Omega) \\ &+ \frac{1}{\delta_j^2} r_j^{2-n} \int_{B_j} f_1^2 dx + \frac{1}{\delta_j} r_j^{2-n} \int_{B_j} (\nu^2 + f_2) dx \Bigg\}^{1+2/n}. \end{aligned} \tag{2.28}$$

We fix R_0 such that

$$\int_0^{2R_0} \frac{C(B_r(x_0) \setminus \Omega)}{r^{n-1}} dr \leqslant \gamma^{-1} a,$$

which is possible due to (1.15). We fix ε by the condition $\gamma \varepsilon^{1+\lambda} = 1/8$ and a by the condition $\gamma(\varepsilon) a^{2/n} = 1/8$. Then (2.28) implies (2.19). □

3 Proof of Theorem 1.1

Let r_j, B_j, ξ_j, ζ_j, θ_j, and g_j be the same as in the previous section. Summing up the inequality (2.19) with respect to j from 1 to J, we find

$$l_J \leqslant \frac{1}{4} + \frac{1}{2}\delta_0 + \gamma \sum_{j=1}^{\infty} \frac{C(B_{j-1} \setminus \Omega)}{r_j^{n-2}}$$

$$+\gamma\sum_{j=1}^{\infty}\left(\frac{1}{r_j^{n-2}}\int_{B_j} f_1^2 dx\right)^{\frac{1}{2}} dx\,dt + \gamma\sum_{j=1}^{\infty}\int_{B_j}(f_2+\nu^2)dx. \qquad (3.1)$$

This implies that the sequence $(l_j)_{j\in\mathbb{N}}$ is convergent. Let $l = \lim_{j\to\infty} l_j$. From (2.17) we find

$$\delta_0 \leqslant \gamma\left(R_0^{-n}\operatorname{ess\,sup}_t \int_{\Omega} u dx\right)^{\frac{1}{1+\lambda}}. \qquad (3.2)$$

Denote $Q'_{j+1} = \{B_{j+1}\cap\Omega\}\times(t_0 - r_{j+1}^2, t_0 + r_{j+1}^2)$, $G_{j+1} = \{(x,t)\in Q'_{j+1} : g_j(x) > 1/3\}$, and $Q''_{j+1} = Q'_{j+1}\setminus G_{j+1}$. Define a function w_ε by the formula

$$w_\varepsilon(x,t) = \left(\frac{u(x,t)-l-\varepsilon}{\varepsilon}\right)^{\frac{1-\lambda}{2}}, \qquad \varepsilon\in(0,1).$$

If $u(x,t)\geqslant l+2\varepsilon$, then $w_\varepsilon(x,t)\geqslant\gamma(\varepsilon)$. By the definition of the parabolic capacity, from Lemma 2.1 we find

$$\begin{aligned}
& r_j^{-n}\Gamma(Q''_{j+1}\cap\{u>l+2\varepsilon\}) \\
& \leqslant \gamma(\varepsilon) r_j^{-n}\left\{\operatorname{ess\,sup}_t \int_{\Omega} w_\varepsilon^2[\xi_j\zeta_j\theta_j]^k dx\,dt + \iint_{\Omega_T}\left|\nabla\left(w_\varepsilon[\xi_j\zeta_j\theta_j]^{k/2}\right)^2 dx\,dt\right.\right\} \\
& \leqslant \gamma(\varepsilon)\left\{\frac{1}{r_j^{2+n}}\iint_{E_j}\left(1+\frac{u-l-\varepsilon}{\varepsilon}\right)^{1+\lambda}[\xi_j\zeta_j\theta_j]^{k-2} dx\,dt\right. \\
& \left. + r_j^{2-n} C(B_{j-1}\setminus\Omega) + \left(r_j^{2-n}\int_{B_j} f_1^2 dx\,dt\right)^{\frac{1}{2}} + r_j^{2-n}\int_{B_j}(\nu^2+f_2)dx\,dt\right\}.
\end{aligned} \qquad (3.3)$$

Summing up the above inequalities, for sufficiently large J we find

$$\begin{aligned}
& \int_0^{r_J}\frac{\Gamma(Q'_r\cap\{u>l+\varepsilon\})}{r^{n+1}} dr \\
& \leqslant \gamma(\varepsilon)\delta_0 + \gamma(\varepsilon)\int_0^{r_J}\frac{C(B_r\setminus\Omega)}{r^{n-1}} dr + \gamma(\varepsilon)\int_0^{r_J}\left(\frac{1}{r^{n-2}}\int_{B_r} f_1^2 dx\,dt\right)^{\frac{1}{2}}\frac{dr}{r} \\
& + \gamma(\varepsilon)\int_0^{r_J}\frac{1}{r^{n-2}}\int_{B_r}(\nu^2+f_2)dx\,dt\,\frac{dr}{r}.
\end{aligned} \qquad (3.4)$$

We choose $f\in C_0^\infty(B_\varkappa(x_0))$ such that

$$0\leqslant f\leqslant 1, \quad f(x)=1 \text{ for } x\in B_\varkappa(x_0), \quad |\nabla f|\leqslant 2\varkappa^{-1}. \qquad (3.5)$$

Let $g \in C_0^\infty(0,T)$ be such that $0 \leqslant g \leqslant 1$, $g(t) = 1$ for $t \in (\frac{t_0}{2}, \frac{t_0+T}{2})$, and $|g'(t)| \leqslant \gamma(t_0)$.

We consider a solution u to Equation (1.1) such that

$$u(x,t) = f(x)g(t), \quad (x,t) \in S_T, \tag{3.6}$$

$$u(x,0) = 0, \quad x \in \Omega. \tag{3.7}$$

The existence of a solution follows from (1.4) in accordance with the theory of monotone operators.

By (3.5), it is easy to verify that

$$\operatorname{ess\,sup}_t \int_\Omega u^2 dx \leqslant \gamma(t_0)\varkappa^{n-2}. \tag{3.8}$$

Indeed, testing (1.1) by $\varphi = u - fg$, we find

$$\int_\Omega u^2 dx - \iint_{\Omega_T} u_t f g dx\, dt + \iint_{\Omega_T} |u|^{m-1}|\nabla u|^2 dx\, dt$$
$$\leqslant \gamma \iint_{\Omega_T} |\nabla f| g(|u|^{m-1}|\nabla u| + f_1) dx\, dt + \gamma \iint_{\Omega_T} fg(\nu|u|^{m-1}|\nabla u| + f_2) dx\, dt.$$

By the Young inequality,

$$\int_\Omega u^2 dx \leqslant \gamma \int_\Omega f^2 g^2 dx + \gamma \iint_{\Omega_T} |u|\, f|g_t| dx\, dt + \gamma \iint_{\Omega_T} |u|^{m-1}|\nabla f|^2 g^2 dx\, dt$$
$$+ \gamma \iint_{\Omega_T} |\nabla f| g\, f_1 dx\, dt + \gamma \iint_{\Omega_T} \nu^2 f^2 g^2 |u|^{m-1} dx\, dt + \gamma \iint_{\Omega_T} fg\, f_2 dx\, dt.$$

Then

$$\int_\Omega f^2 g^2 dx \leqslant \gamma \varkappa^n,$$
$$\int_{\Omega_T} |u| f |g_t| dx\, dt \leqslant \gamma(t_0, M)\varkappa^n,$$
$$\int_{\Omega_T} |u|^{m-1}|\nabla f|^2 g^2 dx\, dt \leqslant \gamma(T, M)\varkappa^{n-2}.$$

By the Cauchy–Schwarz inequality and the definition of the Kato class,

$$\iint_{\Omega_T} |\nabla f| g f_1 dx\, dt \leqslant \gamma(T)\varkappa^{n-2},$$
$$\iint_{\Omega_T} \nu^2 f^2 g^2 |u|^{m-1} dx\, dt \leqslant \gamma(T, M)\varkappa^{n-2},$$
$$\iint_{\Omega_T} fg f_2 dx\, dt \leqslant \gamma(T)\varkappa^{n-2}.$$

Hence

$$\int_{\Omega} u^2 dx \leqslant \gamma(t_0, T, M)\varkappa^{n-2}.$$

From (3.1) and (3.2) we find

$$l \leqslant \frac{1}{4} + \gamma(t_0)\left(\frac{\varkappa^{\frac{n-2}{2}}}{R_0}\right)^{\frac{1}{1+\lambda}} + \gamma(t_0)\int_0^{R_0} \frac{C(B_r(x_0)\setminus\Omega)}{r^{n-1}}dr$$

$$+ \gamma(t_0)\int_0^{R_0}\left(\frac{1}{r^{n-2}}\int_{B_r(x_0)} f_1^2 dx\right)^{\frac{1}{2}}\frac{dr}{r}$$

$$+ \gamma(t_0)\int_0^{R_0}\frac{1}{r^{n-2}}\int_{B_r(x_0)}(\nu^2 + f_2)dx\frac{dr}{r}. \tag{3.9}$$

Choosing a sufficiently small R_0 and then $\varkappa$, we get $l \leqslant 3/4$.

From (3.4) it follows that for $l \geqslant 7/8$

$$\int_0^{1/2}\frac{\Gamma(Q'_r \cap \{u > l\})}{r^{n+1}}dr < \infty. \tag{3.10}$$

By the Poincaré inequality,

$$\text{mes } E \leqslant \gamma r^2 \Gamma(E), \text{for } E \subset B_{R_0}(x_0).$$

Hence

$$\liminf_{r\to 0}\{r^{-n-2}\,\text{mes}\,[Q'_r \cap \{u > l\} = 0. \tag{3.11}$$

This implies that

$$\liminf_{r\to 0} \operatorname{ess\,inf}_{(x,t)\in Q'_r\cap\Omega_T} u(x,t) \leqslant 7/8, \tag{3.12}$$

which violates (1.10) and proves that the point (x_0, t_0) is irregular. □

Acknowledgments. The authors would like to acknowledge the support of the Royal Society through the International Joint Project Grant 2006/R1.

References

1. P. Bauman, P.: A Wiener test for nondivergence structure, second-order elliptic equations. Indiana Univ. Math. J. **34:4**, 825–844 (1985)
2. DiBenedetto, E.: Continuity of weak solutions to a general porous medium equation. Indiana Univ. Math. J. **32**, 83–118 (1983)
3. Evans, L.C., Gariepy, R.: Wiener's criterion for the heat equation. Arch. Ration. Mech. Anal. **78**, 293–314 (1982)

4. Fabes, E., Jerison, D., Kenig, C.: The Wiener test for degenerate elliptic equations. Ann. Inst. Fourier (Grenoble) **32**, 151–182 (1982)
5. Gariepy, R., Ziemer, W.: A regularity condition at the boundary for solutions of quasilinear elliptic equations. Arch. Ration. Mech. Anal. **67**, 25–39 (1977)
6. Kilpeläinen, T., Malý, J.: The Wiener test and potential estimates for quasilinear elliptic equations. Acta Math. **172**, 137–161 (1994)
7. Labutin, D.: Potential estimates for a class of fully nonlinear elliptic equations. Duke Math. J. **111**, 1–49 (2002)
8. Lanconelli, E.: Sul problema di Dirichlet per l'equazione del calore. Ann. Mat. Pura Appl. **106**, 11–38 (1975)
9. Landis, E. M.: Necessary and sufficient conditions for the regularity of a boundary point for the Dirichlet problem for the heat equation (Russian). Dokl. Akad. Nauk SSSR **185**, 517–520 (1969)
10. Littman, W., Stampacchia, G., Weinberger, H.F.: Regular points for elliptic equations with discontinuous coefficients. Ann. Scuola Norm. Sup. Pisa **17**, no. 3, 43–77 (1963)
11. Malý, J., Ziemmer, W.: Fine Regularity of Solutions of Elliptic Partial Differential Equations. Am. Math. Soc., Providence, RI (1997)
12. Maz'ya, V.G.: The continuity at a boundary point of solutions of quasilinear equations (Russian). Vestnik Leningrad. Univ. **25**, no. 13, 42–55 (1970). Correction. ibid **27:1**, 160 (1972); English transl.: Vestnik Leningrad Univ. Math. **3**, 225–242 (1976)
13. Maz'ya, V.G.: Unsolved problems connected with the Wiener criterion. Proc. Sympos. Pure Math. **60**, 199–208 (1997)
14. Maz'ya, V.G.: The Wiener test for higher order elliptic equations. Duke Math. J. **115**, 479–512 (2002)
15. Simon, B.: Schrödinger semigroups. Bull. Am. Math. Soc. (N.S.) **7**, 447–526 (1982)
16. Skrypnik, I.V.: A necessary condition for the regularity of a boundary point for a quasilinear parabolic equation (Russian). Mat. Sb. **183**, 3–22 (1992); English transl.: Russian Acad. Sci. Sb. Math. **76**, no. 2, 265–281 (1993)
17. Skrypnik, I.I.: On regularity of a boundary point for porous medium equation (Russian). Tr. Inst. Prikl. Mat. Mekh. **4**, 171–178 (1999)
18. Skrypnik, I.I.: On the Wiener criterion for quasilinear degenerate parabolic equations (Russian). Dokl. Akad. Nauk **398**, no. 4, 458–461 (2004); English transl.: Dokl. Math. **70**, No 2, 743–746 (2004)
19. Wiener, N.: The Dirichlet problem. J. Math. Phys. Mass. Inst. Tech. **3**, 127–146 (1924); Reprint: Wiener, N.: Collected Works with Commentaries. Vol. 1, pp. 394–413. MIT Press, Cambridge (1976)
20. Ziemer, W.: Behavior at the boundary of solutions of quasilinear parabolic equations. J. Differ. Equ. **35**, 291–305 (1980)
21. Ziemer, W.: Interior and boundary continuity of weak solutions of degenerate parabolic equations. Trans. Am. Math. Soc. **271**, 733–748 (1982)

The Problem of Steady Flow over a Two-Dimensional Bottom Obstacle

Oleg Motygin and Nikolay Kuznetsov

Abstract The linear boundary value problem describing a steady flow over a two-dimensional obstacle (bottom protrusion) is considered. This is a mixed problem for a harmonic function in an indented strip of constant width at infinity, where asymmetric conditions are imposed on the gradient. Under rather general assumptions on the obstacle, the existence of a unique solution is proved for all values of the nonnegative parameter (the reciprocal of the Froude number squared) of the problem, except possibly for a sequence of values that tends from above to the critical value.

1 Introduction

The paper [16] by Vainberg and Maz'ya was one of the pioneering works concerned with rigorous treatment of the linearized problem describing a steady flow bounded above by a free surface and containing an immersed two-dimensional body. The approach proposed in [16] for the case of a totally immersed body was based on the integral-equation technique. With its help, the theorem on the unique solvability was proved for all nonnegative values of the parameter (cf. ν in the relation (1.3) below), except possibly for a finite number of values. This approach was developed further in [7], where

Oleg Motygin
Laboratory for Mathematical Modelling of Wave Phenomena, Institute for Problems in Mechanical Engineering, Russian Academy of Sciences, V.O., Bol'shoy pr. 61, 199178 St. Petersburg, Russia
e-mail: o.v.motygin@gmail.com

Nikolay Kuznetsov
Laboratory for Mathematical Modelling of Wave Phenomena, Institute for Problems in Mechanical Engineering, Russian Academy of Sciences, V.O., Bol'shoy pr. 61, 199178 St. Petersburg, Russia
e-mail: nikolay.g.kuznetsov@gmail.com

A. Laptev (ed.), *Around the Research of Vladimir Maz'ya II: Partial Differential Equations*, International Mathematical Series 12, DOI 10.1007/978-1-4419-1343-2_12,

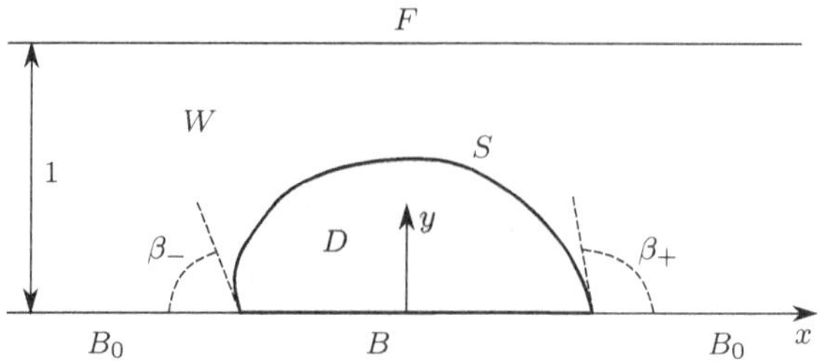

Fig. 1 A sketch of geometry of the problem.

a more complicated problem of the flow about a surface-piercing body was investigated. These and other results on this topic are summarized in the book [8].

In the recent paper [14], the case of a rectangular bottom obstacle was considered using an alternative variational technique, which was first applied to the problem of a stream in the presence of a dock [13]. Then the case of a more general surface-piercing obstacle was investigated in [12]. The unique-solvability theorem obtained in the latter paper has an advantage over that in [8], being valid for all subcritical streams and obstacles of rather general geometry, whereas the relevant theorem in [8] is valid only for a circular cylinder in infinitely deep water. On the other hand, the variational method, as it is used in [14], gives an unduly weak result in the case of a bottom obstacle; not only the obstacle geometry is restricted to rectangles, but there is a condition connecting the depth of submergence with the parameter values for which the unique solvability holds.

The aim of the present paper is to show that the classical integral-equation technique (cf. [8, Section 7.2]) yields the existence of a unique solution for a wide class of bottom obstacles under a rather natural restriction on the parameter values; in particular, the theorem proved in [14] turns out to be a special case of our result. Furthermore, the problem of bottom obstacle arises as one of the limiting cases to be investigated when studying the problem of a two-layer flow about an interface-piercing body.

1.1 Statement of the problem

Let the water layer have a constant depth outside the bottom obstacle which is a two-dimensional protrusion of finite width. The flow is assumed to be two-dimensional and orthogonal to the obstacle generators. Cartesian coordinates are chosen so that the x-axis coincides with the bottom outside the obstacle (cf. B_0 in Fig. 1), and the y-axis is directed opposite to gravity. Without loss of generality, only nondimensional variables and parameters are used for the description of flow, and so the layer depth is taken to be equal to one outside the obstacle. Moreover, the direction of flow is taken opposite to the x-axis, and the flow velocity is supposed to be constant at infinity upstream, i.e., as $x \to +\infty$. Cross-sections of the water domain, obstacle *etc.* are shown in Fig. 1 (in what follows, we omit the word "cross-section" speaking about these sets). The origin is taken on the interval B, along which the obstacle domain D is adjacent to the x-axis, and

$$W = L \setminus \overline{D}, \quad F = \{(x, y) : x \in \mathbb{R}, y = 1\}, \qquad S = \partial D \setminus \overline{B}$$

are the water domain, the free surface of water, and the wetted surface of obstacle, respectively. Here, $L = \{(x, y) : x \in \mathbb{R}, 0 < y < 1\}$ and $\overline{S}$ is supposed to be a closed C^2-curve. The right (P_+) and left (P_-) end-points of B are assumed to be corner points of ∂D and ∂W. The directed into W angle enclosed between B_0 and the unilateral tangent to S at $P_\pm$ is denoted by $\beta_\pm \neq 0, \pi$.

The presence of an obstacle in the water flow induces the velocity field described by the gradient of a velocity potential u. For determining u defined in $\overline{W}$, we have the following boundary-value problem:

$$\nabla^2 u = 0 \quad \text{in } W, \tag{1.1}$$

$$u_{xx} + \nu u_y = 0 \quad \text{on } F, \tag{1.2}$$

$$u_y = 0 \quad \text{on } B_0, \tag{1.3}$$

$$\frac{\partial u}{\partial n} = f \quad \text{on } S, \tag{1.4}$$

$$\sup_{W \setminus E} |\nabla u| < \infty, \qquad \lim_{x \to +\infty} |\nabla u(x, y)| = 0 \text{ uniformly in } y \in [0, 1], \tag{1.5}$$

$$\int_{W \cap E} |\nabla u|^2 \, dx \, dy < \infty. \tag{1.6}$$

Here, $\nu > 0$ is a nondimensional parameter equal to gh/U^2, and the dimensional quantities in this fraction are as follows: the acceleration due to gravity g; the constant velocity U and the constant depth h, characterizing the undisturbed flow at infinity upstream. (In particular, h is used for defining nondimensional coordinates (x, y).) The unit normal $\mathbf{n}$ is directed into W, and $f \in C^{0,\alpha}(\overline{S})$ in the Neumann condition (1.4) ($f = Un_x$ in the simplest case of rigid obstacle). We denote by E an arbitrary compact set such that P_+ and P_- are its inner points.

It is clear that an arbitrary constant term added to u does not violate the validity of the conditions (1.1)–(1.6).

1.2 The main result

It is clear that the dispersion equation corresponding to the problem (1.1)–(1.6) is as follows (cf. [8, Section 6.3.1]):

$$Q_0(k) = 0, \quad \text{where} \quad Q_0(k) = k - \nu \tanh k. \tag{1.7}$$

If $\nu \neq 1$, then this equation has at most one positive zero; no zeros for $\nu \in (0,1)$ and exactly one zero $\lambda_0 \in (0,+\infty)$, which increases with $\nu > 1$.

Since Q_0 has a triple zero at $k = 0$ when $\nu = 1$, the Green function of the problem is not welldefined in this *critical* case. Therefore, it is natural to consider the problem (1.1)–(1.6) only for noncritical values of the parameter ν, in which case the Green function has the following representation (cf. [10, Section 3] for the definition and details):

$$G(z,\zeta) = -\frac{1}{2\pi}\Big\{\log|z-\zeta| + w(z,\zeta) + r.v.\int_0^{+\infty}\big[\nu\cosh k(y+\eta-1) \\ + k\sinh k(y+\eta-1) + (\nu+k)\,\mathrm{e}^{-k}\cosh k(y-\eta)\big]\frac{\cos k(x-\xi)}{k\,Q_0(k)\cosh k}\,\mathrm{d}k\Big\}. \tag{1.8}$$

Here, $z = x + \mathrm{i}y$, $\zeta = \xi + \mathrm{i}\eta$ (for the sake of brevity the complex notation is used for points in the (x,y)-plane), and

$$w(z,\zeta) = \frac{\pi\nu}{1-\nu}(x-\xi) + H(\nu-1)\frac{2\pi\nu\cosh\lambda_0 y\cosh\lambda_0\eta}{\lambda_0(\cosh^2\lambda_0-\nu)}\sin\lambda_0(x-\xi),$$

where H is the Heaviside function. The integral in (1.8), which has a singularity of the second order at $k = 0$, is regularized as follows:

$$r.v.\int_0^{+\infty}\frac{f(k)}{k^2}\,\mathrm{d}k = \int_0^{+\infty}\left[\frac{f(k) - f(0) - k\,\mathrm{e}^{-k}f'(0)}{k^2}\right]\mathrm{d}k.$$

One can verify directly that the expressions used for the regularization do not depend on x and y. In the *subcritical* regime $\nu > 1$, the integral $r.v.\int_0^{+\infty}$ is also understood as the Cauchy principal value at $k = \lambda_0$.

If $|\xi| <$ const, then the Green function has the following asymptotic behavior at infinity:

$$G(z,\zeta) = -\frac{H(-x)}{\pi}w(z,\zeta) + \psi_\pm(x,y) \quad \text{as } \pm x \to +\infty, \tag{1.9}$$

where $\psi_\pm = O(|x|^{-1})$ and $|\nabla\psi_\pm| = O(|x|^{-2})$. The factor $H(\nu-1)$ in the second term of w shows that a source generates no waves at infinity downstream in the *supercritical* regime $\nu \in (0,1)$. The linear term in w means that in any regime a source changes the rate of flow and the flow level at infinity downstream.

As in [7] (cf. also [8, Section 8.1.4]), the Green function allows us to seek a solution u of the problem (1.1)–(1.6) in the form of a simple layer potential

$$(\mathcal{U}\mu)(z) = \int_S \mu(\zeta)\, G(z;\zeta)\,\mathrm{d}s_\zeta, \quad z \in \overline{L}.$$

Here, $\mu \in C^{0,\alpha}(S) \cap C_\varkappa(S)$, where $\alpha \in (0,1)$ is the Hölder exponent and $C_\varkappa(S)$ denotes the Banach space of functions which are continuous on S, with the finite norm

$$\|\mu\|_\varkappa = \sup\{|y|^{1-\varkappa}\,|\mu(z)| : z \in S\}, \quad \varkappa \in (0,1). \tag{1.10}$$

It is obvious that $\mathcal{U}\mu$ satisfies the relations (1.1)–(1.3). Formula (1.9) implies that the conditions (1.5) also hold for $\mathcal{U}\mu$. Using results from [5, Chapter 11, Section 3], one obtains more properties when $\mu \in C_\varkappa(S)$, namely:

$$\int_{R_d} |\nabla \mathcal{U}\mu|^2\,\mathrm{d}x\,\mathrm{d}y < \infty,$$

where $R_d = \{|x| < d,\ 0 < y < 1\}$, and $\mathcal{U}\mu$ is bounded in $\overline{R_d}$.

The boundedness of the Dirichlet integral over R_d yields that the condition (1.6) is true for $\mathcal{U}\mu$, and also allows us to conclude that $\mathcal{U}\mu$ belongs to $C^1(\overline{W})$ and $C^1(\overline{D})$ (cf., for example, [18] for the proof). However, the normal derivative of $\mathcal{U}\mu$ is discontinuous across S (cf., for example, [3, Section 2.4] for the well-known results of the potential theory). Since $\mu \in C^{0,\alpha}(S)$, the potential $\mathcal{U}\mu$ has derivatives which are Hölder continuous on $W \cup S$ and $D \cup S$ with exponent α. Moreover, the normal derivative has the limit

$$\frac{\partial \mathcal{U}\mu}{\partial n_\pm}(z) = \mp\,\mu(z) + (T\mu)(z) \quad \text{on } S, \tag{1.11}$$

which is uniform for z belonging to any compact subset of S; the subscript $+$ ($-$) denotes the derivative on the side directed to W (D). The operator

$$(T\mu)(z) = 2\int_S \mu(\zeta)\frac{\partial G}{\partial n_z}(z;\zeta)\,\mathrm{d}s_\zeta \tag{1.12}$$

understood as improper integral, defines a Hölder continuous function on S. Thus, one has to determine μ from the integral equation

$$-\,\mu(z) + (T\mu)(z) = 2\,f(z), \quad z \in S, \tag{1.13}$$

in order to obtain $\mathcal{U}\mu$ satisfying the condition (1.4). It is important that the operator T is not compact in $C_\varkappa(S)$ (cf. [8, Section 8.1.4]).

Now, we are in a position to formulate our main result.

Theorem 1.1. *Let $f \in C^{0,\alpha}(\overline{S})$ for some $\alpha \in (0,1)$. Then for every positive $\nu \neq 1$, except possibly for a sequence tending to one from above, the problem* (1.1)–(1.6) *has a unique* (*up to an additive constant*) *solution $u = \mathcal{U}\mu$. Here, μ is a unique solution of the integral equation* (1.13) *in the space $C_\varkappa(S)$ for the values of $\varkappa$ belonging to the interval*

$$\left(0, \min_{\pm} \{1 + |1 - 2\beta_\pm/\pi|\}^{-1}\right). \tag{1.14}$$

Remark 1.1. If $S \notin C^2$, but consists of $n+1$ regular C^2-arcs having angle joints at points $P_1, \dots, P_n \in S$, then the Neumann condition (1.4) and Equation (1.13) remain valid for $z \in S \setminus \{P_1, \dots, P_n\}$.

Denoting by $\beta_1, \dots, \beta_n$ the directed to W angles formed by unilateral tangents to S at $P_1, \dots, P_n$, we modify the definition of $C_\varkappa(S)$ so that the weight $|y|^{1-\varkappa}$ (cf. formula (1.10)), which is essential near $P_\pm$, is replaced by $[\mathrm{dist}(z, P_k)]^{1-\varkappa}$ near every point $P_k \in \{P_1, \dots, P_n\}$. Then changing $\min_\pm \{1 + |1 - 2\beta_\pm/\pi|\}^{-1}$ to

$$\min\left\{\min_{\pm} \{1 + |1 - 2\beta_\pm/\pi|\}^{-1}, \min_{1,\dots,n} \{1 + |1 - \beta_n/\pi|\}^{-1}\right\} \tag{1.15}$$

in the formulation of Theorem 1.1, we obtain the assertion that is true for this more general geometry. In particular, this assertion is true for a rectangular bottom obstacle considered in [14].

2 Auxiliary Assertions

2.1 Asymptotics at infinity

The following lemma is similar to the asymptotic results in [8, Sections 7.2 and 8.2], but here both types of behavior (for subcritical and supercritical regimes) are represented by a single formula.

Lemma 2.1. *A solution of the problem* (1.1)–(1.6) *has the following asymptotics as $\pm x \to +\infty$:*

$$u(x,y) = H(-x)\{\mathcal{Q}x + H(\nu - 1)\cosh\lambda_0 y\,(\mathcal{A}\sin\lambda_0 x + \mathcal{B}\cos\lambda_0 x)\} \\ + C_\pm + \varphi_\pm(x,y). \tag{2.1}$$

Here,

$$\varphi_\pm(x,y) = O(|x|^{-1}) \quad \textit{and} \quad |\nabla\varphi_\pm(x,y)| = O(|x|^{-2}),$$

and

$$C_+ - C_- = \frac{\nu}{1-\nu} \int_S \left(x \frac{\partial u}{\partial n} - u\, n_x \right) \mathrm{d}s.$$

Furthermore, the coefficients in (2.1) *are as follows:*

$$\mathcal{Q} = \frac{\nu}{1-\nu} \int_S \frac{\partial u}{\partial n} \, \mathrm{d}s,$$

$$\mathcal{A} = \frac{2\nu}{\lambda_0(\nu - \cosh^2 \lambda_0)} \int_S \Big\{ u \frac{\partial}{\partial n} [\cosh \lambda_0 (y+1) \cos \lambda_0 x] - \frac{\partial u}{\partial n} \cosh \lambda_0 (y+1) \cos \lambda_0 x \Big\} \, \mathrm{d}s, \tag{2.2}$$

$\mathcal{B}$ *has the same form as* $\mathcal{A}$, *but* $\cos \lambda_0 x$ *must be changed to* $-\sin \lambda_0 x$.

The proof of this lemma literally repeats that of the asymptotic proposition in [8, Section 8.1.2], and we leave it to the reader.

Remark 2.1. If $\mathcal{Q} \neq 0$ in the asymptotic formula (2.1), then this means (like for a source), that the presence of an obstacle causes the induced rate of flow at infinity downstream depending on f through (2.2) and (1.4). However, this rate of flow vanishes when the orthogonality condition $\int_S f \, \mathrm{d}s = 0$ holds.

2.2 Uniqueness theorem for the supercritical regime

In [9], an approach based on use of stream function was applied for proving the uniqueness of a solution in the case of an obstacle totally immersed in a supercritical flow. Here, we show that this approach, which also proved to be useful for a surface-piercing obstacle [6], works for a bottom obstacle as well.

Lemma 2.2. *Let* u *be a solution of the homogeneous problem* (1.1)–(1.6) (*i.e.,* f *vanishes identically*). *If* $\nu \in (0,1)$, *then* $u \equiv \text{const}$ *in* $\overline{W}$.

Proof. Let v be a conjugate to u harmonic function in W. From the results obtained in [18] it follows that u and v belong to $C^1(\overline{W})$. Then the Cauchy–Riemann equations and the homogeneous Neumann condition on $S \cup B_0$ yield that

$$v = c_0 \quad \text{on } \overline{S \cup B_0}, \quad c_0 = \text{const}. \tag{2.3}$$

Furthermore, the Cauchy–Riemann equations and the condition (1.2) imply that

$$v_y - \nu v = c_1 \quad \text{on } F, \quad c_1 = \text{const}. \tag{2.4}$$

Since $\lim_{x\to+\infty} |\nabla v(x,y)| = 0$ by the second condition (1.5), form the relation (2.4) it follows that

$$v(x,1) \to -\nu^{-1} c_1 \quad \text{as } x \to +\infty.$$

Using the second condition (1.5) again, we get

$$c_0 + \nu^{-1} c_1 = \lim_{x\to+\infty} [v(x,0) - v(x,1)] = - \lim_{x\to+\infty} \int_0^1 v_y(x,y)\, \mathrm{d}y = 0.$$

Now, the proper choice of the additive constant for v gives

$$v = 0 \quad \text{on } \overline{S \cup B_0}, \qquad v_y - \nu v = 0 \quad \text{on } F, \tag{2.5}$$

in view of (2.3).

Since $\nu \in (0,1)$ and f vanishes identically, the expression in braces in formula (2.1) is equal to zero, which guarantees that the Dirichlet integral

$$\int_W |\nabla u|^2 \, \mathrm{d}x \, \mathrm{d}y = \int_W |\nabla v|^2 \, \mathrm{d}x \, \mathrm{d}y$$

is finite, and so the Green formula gives

$$\int_W |\nabla v|^2 \, \mathrm{d}x\mathrm{d}y = \int_{\partial W} v \frac{\partial v}{\partial n} \, \mathrm{d}s = \nu \int_F v^2 \, \mathrm{d}x. \tag{2.6}$$

Here, $\mathbf{n}$ is the exterior normal on ∂W in the second integral, and the last equality is a consequence of the relations (2.5).

Extending v to $\overline{D}$ by zero, we have

$$v(x,1) = \int_0^1 v_y(x,y)\, \mathrm{d}y \quad \text{for all } x \in \mathbb{R}.$$

Applying the Schwarz inequality to the left-hand side of the last equality squared and integrating with respect to x, we get

$$\int_F v^2 \, \mathrm{d}x \leqslant \int_W |v_y|^2 \, \mathrm{d}x \, \mathrm{d}y \leqslant \int_W |\nabla v|^2 \, \mathrm{d}x \, \mathrm{d}y.$$

Comparing this with (2.6), we obtain

$$(1-\nu) \int_F v^2 \, \mathrm{d}x \leqslant 0,$$

and so v vanishes on F. Therefore, $v \equiv 0$ in $\overline{W}$, from which the required assertion follows. □

2.3 The Fredholm alternative for the integral equation

Since the operator T is not compact in $C_\varkappa(S)$, the validity of the Fredholm alternative for the integral equation (1.13) is guaranteed in this space when $|T| < 1$, where $|\cdot|$ is the essential norm of an operator (cf., for example, [8, Section 2.1.3]).

Lemma 2.3. *The estimate*

$$|T| < \max_{\pm} \frac{\sin \varkappa |\pi - 2\beta_\pm|}{\sin \varkappa \pi}$$

holds for the essential norm of the integral operator (1.12) *in the space* $C_\varkappa(S)$, $\varkappa \in (0,1)$.

Proof. Let us consider the following equivalent representation of the Green function (cf. [8, Section 6.3.2]):

$$G(z,\zeta) = -\frac{1}{2\pi}\left\{ \log|z-\zeta| + \log|z-\overline{\zeta}| + w(z,\zeta) - \frac{2\pi\nu}{1-\nu}(x-\xi) \right.$$
$$\left. + 2\,p.v. \int_0^{+\infty} \left[\frac{(k+\nu)\,\cos k(x-\xi)\cosh ky \cosh k\eta}{k\,\mathrm{e}^k\,Q_0(k)\cosh k} + \frac{\nu}{(\nu-1)k^2} - \frac{\mathrm{e}^{-k}}{k} \right] \mathrm{d}k \right\}.$$

Here, the integral is understood as the Cauchy principal value at $k = \lambda_0$ when $\nu > 1$, but the integrand has a finite limit as $k \to 0$ because the regularizing term is added to the integrand.

This form of the Green function allows us to write

$$(\mathcal{U}\mu)(z) = \frac{1}{2\pi}\int_{S\cup S'} \mu(\zeta)\log|z-\zeta|\,\mathrm{d}s_\zeta + \int_S \mu(\zeta)\,G'(z;\zeta)\,\mathrm{d}s_\zeta, \quad z \in \overline{L}.$$

Here, the density μ is extended to $S' = \{z \in (-\infty,+\infty)\times(-1,0) : \overline{z} \in S\}$ as an even in y function and

$$G'(z,\zeta) = G(z,\zeta) + \frac{1}{2\pi}\left(\log|z-\zeta| + \log|z-\overline{\zeta}|\right).$$

From the last representation of $\mathcal{U}\mu$ we get

$$(T\mu)(z) = \frac{1}{\pi}\int_{S\cup S'} \mu(\zeta)\frac{\partial \log|z-\zeta|}{\partial n_z}\,\mathrm{d}s_\zeta + 2\int_S \mu(\zeta)\frac{\partial G'}{\partial n_z}(z;\zeta)\,\mathrm{d}s_\zeta,$$

and the operator corresponding to the second term on the right-hand side is compact in $C_\varkappa(S)$. The operator defined by the first term was investigated by Carleman [2] (cf. also [8, Section 8.1.4]), and his calculations give the required estimate of $|T|$. □

Combining the estimate of $|T|$ obtained in Lemma 2.3 and the condition (1.14) for $\varkappa$, we get $|T| < 1$, which allows us to formulate

Corollary 2.1. *Let* $\varkappa$ *belong to the interval* (1.14). *Then the Fredholm alternative is valid for the integral equation* (1.13) *in the space* $C_\varkappa(S)$.

Remark 2.2. Let S be of the type described in Remark 1.1, and let $C_\varkappa(S)$ be the space from that remark with the modified value of $\varkappa$. Combining Lemma 2.3 and calculations of Carleman [2], one obtains $|T|$ in the new space $C_\varkappa(S)$ is less than expression (1.15), and so $|T| < 1$. Hence the Fredholm alternative is valid for the integral equation (1.13) in this space.

2.4 The limit problem of flow under rigid lid

Formally letting $\nu \to \infty$ in the condition (1.2), we obtain the limit problem describing flow over bottom obstacle when the upper boundary is the rigid lid instead of the free surface. To distinguish a solution of the limit problem from u, solving the problem (1.1)–(1.6), we denote by u_0 the former. Thus, u_0 satisfies the relations (1.1), (1.3)–(1.6), and the Neumann condition

$$\frac{\partial u_0}{\partial y} = 0 \quad \text{when } y = 1. \tag{2.7}$$

The Green function of the latter problem is obtained in [17] and has the form:

$$G_0(z,\zeta) = -\frac{1}{2\pi} \log\Big| \Big(1 - \mathrm{e}^{-\pi|x-\xi| + \mathrm{i}\pi(y+\eta-2)}\Big) \times \Big(1 - \mathrm{e}^{-\pi|x-\xi| + \mathrm{i}\pi(y-\eta)}\Big)\Big| + \frac{1}{2}|x-\xi| + \frac{1}{2}(\xi - x). \tag{2.8}$$

Its asymptotic behavior as $|x| \to \infty$ and $|\zeta| \leqslant c < \infty$ is as follows:

$$G_0(z,\zeta) = H(-x)(x-\xi) + O\big(\mathrm{e}^{-\pi|x|}\big).$$

Therefore, the following formula is analogous to (2.1):

$$u_0(z) = c_\pm - H(-x)\, x \int_S \frac{\partial u_0}{\partial n}\, \mathrm{d}s + O\big(\mathrm{e}^{-\pi|x|}\big) \quad \text{as } \pm x \to +\infty. \tag{2.9}$$

Here, c_+ and c_- are constants such that

$$c_+ - c_- = \int_S \left(u_0\, n_x - x\, \frac{\partial u_0}{\partial n} \right) \mathrm{d}s.$$

The following lemma concerns the unique solvability of the problem about flow under rigid lid.

Lemma 2.4. *For every* $f \in C^{(0,\alpha)}(\overline{S})$, $\alpha \in (0,1)$, *the problem* (1.1), (1.3)–(1.6), *and* (2.7) *has a unique (up to an additive constant) solution*

$$u_0(z) = \int_S \mu(\zeta)\, G_0(z;\zeta)\, \mathrm{d}s_\zeta, \quad z \in \overline{L}, \tag{2.10}$$

where μ *satisfies the uniquely solvable integral equation*

$$-\mu(z) + 2\int_S \mu(\zeta) \frac{\partial G_0}{\partial n_z}(z,\zeta)\, \mathrm{d}s_\zeta = 2\, f(z) \tag{2.11}$$

and belongs to $C_\varkappa(S)$ *with* $\varkappa$ *from the interval* (1.14).

Proof. For proving the uniqueness we have to show that $u_0 \equiv \text{const}$ on $\overline{W}$ when f vanishes identically. Using the homogeneous condition (2.7) in formula (2.9), we get

$$\int_W |\nabla u_0|^2\, \mathrm{d}x\, \mathrm{d}y < \infty,$$

where the condition (1.6) is also taken into account. Therefore, the Green formula is applicable and it shows that the Dirichlet integral vanishes, which gives the required result.

Seeking u_0 in the form of the simple layer potential (2.10), one arrives at the integral equation (2.11) in the same way as Equation (1.13) was obtained for the problem (1.1)–(1.6). The Fredholm alternative holds for Equation (2.11) because the corresponding integral operator has the essential norm in $C_\varkappa(S)$ strictly less than one when $\varkappa$ belongs to the interval (1.14). Proofs of these assertions are similar to those in Subsection 2.3. Thus, Equation (2.11) has a unique solution in $C_\varkappa(S)$, when the homogeneous equation

$$-\mu(z) + 2\int_S \mu(\zeta) \frac{\partial G_0}{\partial n_z}(z,\zeta)\, \mathrm{d}s_\zeta = 0 \tag{2.12}$$

has only a trivial solution.

Indeed, assuming that μ_0 is a nontrivial solution of (2.12), we consider the following simple layer potential

$$\mathcal{U}_0(z) = \int_S \mu_0(\zeta)\, G_0(z;\zeta)\, \mathrm{d}s_\zeta, \quad z \in \overline{L}.$$

The normal derivative $\partial \mathcal{U}_0/\partial n_+$ vanishes on S in view of Equation (2.12), and so $\mathcal{U}_0$ is a solution of the homogeneous problem (1.1), (1.3)–(1.6), and (2.7). Hence $\mathcal{U}_0 \equiv c$ on $\overline{W}$, where c is a constant as was demonstrated above.

Now, we apply the Green formula to $\mathcal{U}_0 - c$ in D, thus obtaining

$$0 = \int_D (\mathcal{U}_0 - c)\nabla^2(\mathcal{U}_0 - c)\, \mathrm{d}x\mathrm{d}y = \int_D |\nabla \mathcal{U}_0|^2\, \mathrm{d}x\mathrm{d}y$$

because $\mathcal{U}_0 - c$ vanishes on S and $(\mathcal{U}_0 - c)_y$ vanishes on B. Therefore, $\mathcal{U}_0$ is constant both in $\overline{D}$ and $\overline{W}$, and so μ_0 vanishes identically on S being proportional to

$$\partial\mathcal{U}_0/\partial n_- - \partial\mathcal{U}_0/\partial n_+$$

as follows from the formula analogous to (1.11). □

2.5 Inequality

The following inequality will be used in Subsection 3.1.

Lemma 2.5. *The following inequality holds:*

$$\Delta(k, \lambda_0) \geqslant 1 \quad \textit{for } k \geqslant 0 \textit{ and } \lambda_0 \geqslant 1 \tag{2.13}$$

(note that $\nu > 1$ in this case), where

$$\Delta(k, \lambda_0) = \frac{(k - \nu \tanh k)[k + \lambda_0/(\nu - 1)]}{k(k - \lambda_0)} \quad \textit{for } k \neq \lambda_0,$$

and this function is extended continuously to $k = \lambda_0$.

Proof. It is easy to see that

$$\Delta(\lambda_0, \lambda_0) = \frac{\lambda_0}{\lambda_0 - \tanh \lambda_0}\left(1 - \frac{2\lambda_0}{\sinh 2\lambda_0}\right) > 1 \quad \text{for } \lambda_0 \geqslant 1.$$

If $k < \lambda_0$, then for proving (2.13) it suffices to show that the function

$$\begin{aligned} g(k, \lambda_0) &= [\Delta(k, \lambda_0) - 1]\,(k - \lambda_0)/\lambda_0 \\ &= 1 + \frac{\tanh \lambda_0}{\lambda_0 - \tanh \lambda_0} - \left[1 + \frac{\lambda_0 \tanh \lambda_0}{k(\lambda_0 - \tanh \lambda_0)}\right]\frac{\tanh k}{\tanh \lambda_0} \end{aligned}$$

is nonpositive for $k \in [0, \lambda_0)$; the last equality is based on the definition of $\lambda_0 \geqslant 1$. The derivative

$$\frac{\partial g}{\partial k}(k, \lambda_0) = \frac{\lambda_0 \tanh \lambda_0 \sinh k \cosh k - k^2(\lambda_0 - \tanh \lambda_0) - k\lambda_0 \tanh \lambda_0}{k^2(\lambda_0 - \tanh \lambda_0) \tanh \lambda_0 \cosh^2 k} \tag{2.14}$$

vanishes if and only if $f_1(k) = f_2(k, \lambda_0)$, where

$$f_1(k) = \frac{\sinh k \cosh k}{k} \quad \text{and} \quad f_2(k, \lambda_0) = 1 + \frac{k(\lambda_0 - \tanh \lambda_0)}{\lambda_0 \tanh \lambda_0}.$$

It is easy to check that

$$f_1''(k) = \mathrm{e}^{-2k}\big[(2k-1)\mathrm{e}^{4k} + 2k + 1\big]/4k^2 > 0 \quad \text{for } k > 0,$$

and so f_1 is a convex monotonically increasing function for $k > 0$ with the horizontal tangent at $k = 0$. Furthermore, f_2 is linear in k with the slope $(\lambda_0 - \tanh\lambda_0)/(\lambda_0 \tanh\lambda_0) > 0$, and so for every $\lambda_0 \geqslant 1$ the partial derivative (2.14) vanishes only once for $k > 0$ because $f_1(0) = f_2(0, \lambda_0) = 1$. Moreover, the zero of $(\partial g/\partial k)(k, \lambda_0)$ belongs to the interval $(0, \lambda_0)$ on the k-axis since $g(0, \lambda_0) = g(\lambda_0, \lambda_0) = 0$. Finally, noting that the numerator of (2.14) is positive for $k = \lambda_0$ being equal to $\lambda_0(\sinh\lambda_0)^2 - \lambda_0^3$, we get $(\partial g/\partial k)(\lambda_0, \lambda_0) > 0$. Thus, the obtained properties of $g(k, \lambda_0)$ show that this function is nonpositive for $k \in [0, \lambda_0)$, which is what we set out to prove.

In order to prove (2.13) for $k > \lambda_0$, we note that

$$g(k, \lambda_0) \to \frac{\lambda_0 \tanh\lambda_0 - \lambda_0 + \tanh\lambda_0}{\tanh\lambda_0(\lambda_0 - \tanh\lambda_0)} \quad \text{as } k \to +\infty.$$

Here, the numerator is equal to $(\mathrm{e}^{2\lambda_0} - 2\lambda_0 - 1)/(\mathrm{e}^{2\lambda_0} + 1)$, and so one easily obtains $g(+\infty, \lambda_0) > 0$ when $\lambda_0 \geqslant 1$. Moreover, we have $g(k, \lambda_0) > 0$ for $k \in (\lambda_0, \lambda_0 + \varepsilon)$, where $\varepsilon > 0$. If $g(k, \lambda_0)$ attains negative values for some $k \in (\lambda_0, +\infty)$, then this function must have at least two extrema on this interval. However, it was proved above that the partial derivative (2.14) vanishes only once, and so $g(k, \lambda_0) > 0$, which gives the required inequality (2.13) for all $k > \lambda_0$ and $\lambda_0 \geqslant 1$. □

3 Proof of Theorem 1.1

If $\nu \in (0, 1)$, then the result is a consequence of Lemma 2.2. In order to show this, one has to apply the considerations used for proving Lemma 2.4.

3.1 The existence of a solution for $\nu > 1$

The existence proof for the subcritical case follows guidelines proposed in [16] (cf. also [8, Section 8.1.4]) and is based on the fact that the operator T depends analytically on ν in a complex neighborhood of the half-axis $(1, +\infty)$.

According to Lemma 2.4, the operator $-I + T_0$ is invertible in $C_\varkappa(S)$ when $\varkappa$ belongs to the interval (1.14). Here, I is the identity operator and

$$(T_0\mu)(z) = 2\int_S \mu(\zeta) \frac{\partial G_0}{\partial n_z}(z, \zeta)\, \mathrm{d}s_\zeta$$

is the operator that appears in Equation (2.11). If $\|T - T_0; C_\varkappa(S)\|$ is sufficiently small for $\varkappa$ belonging to the interval (1.14) and all values of ν exceeding some number greater than one, then the operator $-I+T$ is also invertible in $C_\varkappa(S)$ for these values of ν and $\varkappa$. Moreover, Corollary 2.1 guarantees that the Fredholm alternative holds for Equation (1.13) in $C_\varkappa(S)$ for such values of $\varkappa$. Then the theorem on invertibility of an operator function analytically depending on a parameter (cf. [15] for the theorem in a Banach space and [4, Section 5] for the better-known result in a Hilbert space) yields that Equation (1.13) is uniquely solvable in $C_\varkappa(S)$ for all $\nu \in (1, +\infty)$, except possibly for a sequence of isolated values. Since the equation is uniquely solvable for large ν, the sequence of exceptional values (if exists) tends to one.

These considerations prove the assertion of Theorem 1.1 about the existence of a solution to the problem (1.1)–(1.6) and the representation of this solution as $\mathcal{U}\mu$ provided that the smallness of $\|T - T_0; C_\varkappa(S)\|$ is established. Thus, we turn to estimating this norm for large values of ν (or large values of λ_0, what is the same in view of $\nu = \lambda_0 \tanh \lambda_0$), for which purpose it is essential to evaluate $\|\nabla_{x,y}[G(z,\zeta) - G_0(z,\zeta)]; C(\overline{S} \times \overline{S})\|$.

Another representation of the Green function (2.8) (cf. [10, Appendix A])

$$G_0(z,\zeta) = -\frac{1}{2\pi}\Big\{ \log|z-\zeta| - r.v. \int_0^{+\infty} \frac{[\cosh k(y+\eta-1) + \mathrm{e}^{-k}\cosh k(y-\eta)] \cos k(x-\xi)\,\mathrm{d}k}{k \sinh k} \Big\}$$

allows us to write

$$G(z,\zeta) - G_0(z,\zeta) = G_\mathrm{s}(z,\zeta) + G_\mathrm{c}(z,\zeta), \tag{3.1}$$

where

$$G_\mathrm{s}(z,\zeta) = -\frac{1}{4\pi}\, r.v. \int_0^\infty \left[\frac{(\nu-k)\,\mathrm{e}^k}{k\,Q_0(k)\cosh k} + \frac{\mathrm{e}^k}{\sinh k} \right] \frac{\mathrm{e}^{-k(y+\eta)}\cos k(x-\xi)\,\mathrm{d}k}{k}.$$

The estimates obtained in [10] give the following asymptotic formula for the second term:

$$\|\nabla G_\mathrm{c}; C(\overline{S} \times \overline{S})\| = O\left(\nu^{-1}\right) \quad \text{as } \nu \to +\infty. \tag{3.2}$$

To show that G_s has the same asymptotic behavior, i.e.,

$$\|\nabla G_\mathrm{s}; C(\overline{S} \times \overline{S})\| = O\left(\nu^{-1}\right) \quad \text{as } \nu \to +\infty, \tag{3.3}$$

we write

$$\frac{\partial G_\mathrm{s}}{\partial x}(z,\zeta) = \frac{1}{4\pi} \int_0^\infty J(k,\lambda_0)\,\mathrm{e}^{-k(y+\eta+1)} \sin k(x-\xi)\,\mathrm{d}k + I_x^{(1)} + I_x^{(2)}, \tag{3.4}$$

$$\frac{\partial G_s}{\partial y}(z,\zeta) = -\frac{1}{4\pi}\int_0^\infty J(k,\lambda_0)\,\mathrm{e}^{-k(y+\eta+1)}\cos k(x-\xi)\,\mathrm{d}k + I_y^{(1)} + I_y^{(2)}. \quad (3.5)$$

Here,

$$J(k,\lambda_0) = \frac{(\nu-k)\mathrm{e}^{2k}}{k\,Q_0(k)\cosh k} + \frac{\mathrm{e}^{2k}}{\sinh k} + \frac{1}{k(\nu-1)} + \frac{(\nu-\lambda_0)\mathrm{e}^{2\lambda_0}}{(\lambda_0-k)\,Q_0'(\lambda_0)\cosh\lambda_0} \quad (3.6)$$

is chosen so that it is a real-analytic function of k for $k > -\varepsilon$, where ε is a certain positive number. Furthermore,

$$Q_0'(\lambda_0) = Q_0'(k)\big|_{k=\lambda_0} = 1 - \nu + \nu^{-1}\lambda_0^2$$

and

$$I_x^{(1)} = \frac{1}{4\pi(1-\nu)}\int_0^\infty \frac{\mathrm{e}^{-k(y+\eta+1)}\sin k(x-\xi)}{k}\,\mathrm{d}k,$$

$$I_y^{(1)} = -\frac{1}{4\pi(1-\nu)}\int_0^\infty \frac{\mathrm{e}^{-k(y+\eta+1)}\cos k(x-\xi) - \mathrm{e}^{-k}}{k}\,\mathrm{d}k,$$

$$I_x^{(2)} = \frac{(\nu-\lambda_0)\,\mathrm{e}^{2\lambda_0}}{4\pi\,Q_0'(\lambda_0)\cosh\lambda_0}\,p.v.\int_0^\infty \frac{\mathrm{e}^{-k(y+\eta+1)}\sin k(x-\xi)}{k-\lambda_0}\,\mathrm{d}k,$$

$$I_y^{(2)} = -\frac{(\nu-\lambda_0)\,\mathrm{e}^{2\lambda_0}}{4\pi\,Q_0'(\lambda_0)\cosh\lambda_0}\,p.v.\int_0^\infty \frac{\mathrm{e}^{-k(y+\eta+1)}\cos k(x-\xi)}{k-\lambda_0}\,\mathrm{d}k.$$

It is clear that $\|I_x^{(1)}; C(\overline{S}\times\overline{S})\| = O(\nu^{-1})$ as $\nu\to\infty$, and the same is true for $I_y^{(1)}$ because the integral in the expression for this function is equal to $-\log|Z|$, where $Z = \mathrm{i}\big[x-\xi+\mathrm{i}(y+\eta+1)\big]$ and $|Z|\neq 0$ when $z,\zeta\in\overline{S}$. Furthermore,

$$I_x^{(2)} = -\frac{(\nu-\lambda_0)\,\mathrm{e}^{2\lambda_0}}{4\pi\,Q_0'(\lambda_0)\cosh\lambda_0}\,\mathrm{Im}\big\{\mathrm{e}^{\lambda_0 Z}\,\mathrm{Ei}(-\lambda_0 Z)\big\},$$

$$I_y^{(2)} = \frac{(\nu-\lambda_0)\,\mathrm{e}^{2\lambda_0}}{4\pi\,Q_0'(\lambda_0)\cosh\lambda_0}\,\mathrm{Re}\big\{\mathrm{e}^{\lambda_0 Z}\,\mathrm{Ei}(-\lambda_0 Z)\big\}, \quad (3.7)$$

where Ei denotes the exponential integral (cf., for example, [1, Section 5.1]). The well-known asymptotics of this function is as follows:

$$\mathrm{e}^z\,\mathrm{Ei}(-z) = \log z + O(1) \quad \text{as } |z|\to 0.$$

Moreover, we have

$$\mathrm{e}^z\,\mathrm{Ei}(-z) = -\mathrm{i}\pi\mathrm{e}^z\,\mathrm{sgn}\big(\mathrm{Im}\,z\big) + r(z), \quad \text{where } |\,r(z)| \leqslant 3|z|^{-1}$$

(cf. [11] for the proof). Therefore, $\mathrm{e}^z \,\mathrm{Ei}(-z) - \log z/(1+|z|)$ is bounded in the half-plane $\operatorname{Re} z \geqslant 0$, and so

$$|\mathrm{e}^z \,\mathrm{Ei}(-z)| \leqslant |\log z|/(1+|z|) + c \leqslant |\log z| + c,$$

where c is a constant. Hence

$$\begin{aligned}\left|\mathrm{e}^{\lambda_0 Z}\,\mathrm{Ei}(-\lambda_0 Z)\right| &\leqslant \left|\log\left[\mathrm{i}\lambda_0(x-\xi+\mathrm{i}(y+\eta+1))\right]\right| + c\\ &\leqslant |\log\lambda_0| + \left|\log\left[\mathrm{i}(x-\xi)-(y+\eta+1)\right]\right| + c,\end{aligned}$$

and the argument of the second logarithm is bounded and separated from zero when $z,\zeta \in \overline{S}$. Therefore, we arrive at the following estimate:

$$\|I^{(2)}_{x,y}; C(\overline{S}\times\overline{S})\| = O\big(\mathrm{e}^{-\nu}\log\nu\big) \quad \text{as } \nu\to\infty.$$

Indeed, the dispersion equation (1.7) implies that $\nu - \lambda_0 = O\big(\nu\,\mathrm{e}^{-2\nu}\big)$ as $\nu\to\infty$ and

$$Q_0'(\lambda_0) \to 1 \quad \text{as } \lambda_0\to\infty. \tag{3.8}$$

Therefore, the fraction in (3.7) is $O\big(\mathrm{e}^{-\nu}\big)$, where the last relation is taken into account.

To complete the proof of (3.3), it remains to consider each of the integrals standing first on the right-hand sides of formulas (3.4) and (3.5). For this purpose we express $J(k,\lambda_0)$ given by formula (3.6) in terms of $\varDelta(k,\lambda_0)$ (cf. Lemma 2.5, where this function is defined) and $\beta(k) = \mathrm{e}^{-k}\cosh k$:

$$\begin{aligned}&J(k,\lambda_0)\\ &= \frac{(\nu-k)\mathrm{e}^k\left[k+\lambda_0/(\nu-1)\right]}{k(k-\lambda_0)\,\beta(k)\,\varDelta(k,\lambda_0)} + \frac{\mathrm{e}^{2k}}{\sinh k} + \frac{1}{k(\nu-1)} + \frac{(\nu-\lambda_0)\mathrm{e}^{\lambda_0}}{(\lambda_0-k)\,Q'(\lambda_0)\,\beta(\lambda_0)}\\ &= \frac{R(k,\lambda_0)}{2\,P(k,\lambda_0)\,\beta(\lambda_0)\,Q_0'(\lambda_0)\lambda_0(1-\mathrm{e}^{-2\lambda_0})^2\left[1-\lambda_0^{-1}+(1+\lambda_0^{-1})\mathrm{e}^{-2\lambda_0}\right]^2},\end{aligned}$$

where the formula $\nu = \lambda_0/\tanh\lambda_0$ is applied for justifying the last equality. The following notation is used:

$$P(k,\lambda_0) = \alpha(k)\,\beta(k)\,\varDelta(k,\lambda_0), \quad \alpha(k) = (1+k)k^{-1}\mathrm{e}^{-k}\sinh k,$$

and

$$R(k,\lambda_0) = r_1(k,\lambda_0) + \lambda_0^{-1}\,R_0(k,\lambda_0), \tag{3.9}$$

where

$$r_1(k,\lambda_0) = \frac{\lambda_0\big(\lambda_0\mathrm{e}^{-\lambda_0} - k\mathrm{e}^{-k}\big)}{\lambda_0 - k} \quad \text{and} \quad R_0(k,\lambda_0) = \sum_{j=0}^{10} c_j\mathrm{e}^{-j\lambda_0}.$$

Here, the coefficients $c_0, c_1, \ldots, c_{10}$ are functions of $b = b(k) = \mathrm{e}^{-k}$,

$$s_1 = s_1(k) = \frac{\mathrm{e}^{-k} - 1}{k}, \quad s_2 = s_2(k) = \frac{\mathrm{e}^{-k} - 1 + k}{k^2},$$

and $h_i = h_i(k, \lambda_0) = \mathrm{e}^{-\lambda_0} s_i(k - \lambda_0)$, $i = 1, 2$, and these coefficients have the following form obtained with the help of the computer algebra system *Maxima*:

$$\begin{aligned} c_0 = 4^{-1}\big\{&4(\lambda_0 s_1 - \lambda_0 + 1)s_2 + \big[2(b+1)^2\lambda_0 + b^2 + 2b + 3\big]s_1^2 \\ &+ (b+3)\lambda_0 s_1^3 + \big[b^2(b+1)(\lambda_0+3) + (b+5)(1-\lambda_0)\big]s_1 \\ &- 2\big[(b^3 - b + 2)h_1 + 4b\big]\lambda_0 + 2b(1-b^2)h_1 + 4b\big\}, \end{aligned}$$

$$\begin{aligned} c_1 = 2^{-1}\big\{&(2(b+1)^2(\lambda_0 - 1)\lambda_0 s_1^2 + 4(b+1)(\lambda_0 - 1)\big(b^2\lambda_0 - \lambda_0 + b^2\big)s_1 \\ &+ 4(b^2-1)h_1^2\lambda_0^4 + 2b^3 h_1\lambda_0^3 - 4(b^2-1)(h_1+2b)\lambda_0^2 h_1 + (b^2-1)(4b-1)h_1\big\}, \end{aligned}$$

$$\begin{aligned} c_2 = -4^{-1}\big\{&4\big[(4\lambda_0+1)\lambda_0 s_1 - 4\lambda_0^2 + 3\lambda_0 + 3\big]s_2 + (b+3)(4\lambda_0+1)\lambda_0 s_1^3 \\ &+ [8(b+1)^2\lambda_0^2 + (4\lambda_0+3)(b^2+2b+3)]s_1^2 + 8(1-b^2)h_1 + 12b \\ &+ \big[4(b^3+b^2-b-5)\lambda_0^2 + (b^3+b^2)(11\lambda_0+3) + (5b+17)(1+\lambda_0)\big]s_1 \\ &+ 16(h_2 - bh_1^2 + h_1)\lambda_0^4 + 4\big[4bh_1^2 - 9b - 4h_2 + 2(-b^3+2b^2+b-5)h_1\big]\lambda_0^2 \\ &+ 4\big[4b - (b^2+1)h_1\big]\lambda_0^3 + \big[8b - 4(b^2-1)(2b+1)h_1\big]\lambda_0\big\}, \end{aligned}$$

$$\begin{aligned} c_3 = {}&(b+1)^2\lambda_0(\lambda_0+1)s_1^2 + 2(b+1)\big[2\lambda_0^2(b^2-1) - \lambda_0 + 2b^2 + 1\big]s_1 \\ &+ \big[4(b^2-1)h_1^2 + b(b^2+1)h_1\big]\lambda_0^3 + \big[2(-b^3+b^2+2b-1)h_1 - 4h_1^2\big]\lambda_0^2 \\ &+ 4b^2h_1^2\lambda_0^4 - (b^2-1)(3b-2)\lambda_0 h_1 + 4b(1-b^2)h_1, \end{aligned}$$

$$\begin{aligned} c_4 = 2^{-1}\big\{&4\big[\lambda_0(2\lambda_0-1)s_1 - 2\lambda_0^2 + 7\lambda_0 + 1\big]s_2 + \lambda_0(b+3)(2\lambda_0-1)s_1^3 \\ &- 8(h_2 - 2bh_1^2 + h_1)\lambda_0^4 + \big[4(b^2+2b+4)\lambda_0 + b^2 + 2b + 3\big]s_1^2 \\ &+ \big[2(b-3)\lambda_0^2 + (11b+39)\lambda_0 - b^2(1+b)(2\lambda_0^2 - 5\lambda_0 + 1) + b + 5\big]s_1 \\ &+ 2\big[-8h_2 + 8bh_1^2 + (b^2-9)h_1 - 4b\big]\lambda_0^3 - 4(2h_2 + b^2h_1 + 5b)\lambda_0^2 \\ &+ 2\big[(b^3 - 3b^2 - b + 5)h_1 + 16b\big]\lambda_0 + 2(b-4)(b^2-1)h_1 + 4b\big\}, \end{aligned}$$

$$\begin{aligned} c_5 = {}&(b+1)^2(1-\lambda_0)\lambda_0 s_1^2 + 2(b+1)(b^2\lambda_0^2 - \lambda_0^2 - \lambda_0 - b^2 - 2)s_1 \\ &+ 2(b^2+3)\lambda_0^4 h_1^2 + \big[4(b^2+1)h_1^2 + b(1-b^2)h_1\big]\lambda_0^3 \\ &+ 2\big[(b^2-1)h_1^2 + bh_1\big]\lambda_0^2 + (3b^3 + b^2 - b - 1)\lambda_0 h_1 + (b^2-1)(2b+1)h_1, \end{aligned}$$

$$\begin{aligned} c_6 = 2^{-1}\big\{&4\big[\lambda_0(2\lambda_0+1)s_1 - 2\lambda_0^2 - 7\lambda_0 + 1\big]s_2 + \lambda_0(b+3)(2\lambda_0+1)s_1^3 \\ &+ \big[4(b+1)^2\lambda_0^2 + (1-6\lambda_0)(b^2+2b+3)\big]s_1^2 + 8(h_2 + bh_1^2 + h_1)\lambda_0^4 \end{aligned}$$

$$+ \big[-2(b+5)\lambda_0^2 - (5b+33)\lambda_0 + b^2(1+b)(2\lambda_0^2 + 3\lambda_0 + 1) + 3b + 7\big] s_1$$
$$+ 2\big[8bh_1^2 - (b^2+1)h_1 + 4b\big]\lambda_0^3 - 4\big[2h_2 - 2bh_1^2 + (b^3 - b + 3)h_1 + 5b\big]\lambda_0^2$$
$$+ \big[2(1-b^2)(2b-3)h_1 - 32b\big]\lambda_0 + 4(b^2-1)h_1 + 4b\big\},$$

$$c_7 = -\lambda_0(b+1)^2(\lambda_0+1)s_1^2 + 4h_1^2\lambda_0^4 - \big[b(b^2+1)h_1 - 8h_1^2\big]\lambda_0^3$$
$$- 2\big[(b^3+b^2-1)h_1 - 2h_1^2\big]\lambda_0^2 - \lambda_0(b^2-1)(b+2)h_1 + 2(b+1)(\lambda_0+1)s_1,$$

$$c_8 = -4^{-1}\big\{4\big[\lambda_0(4\lambda_0-1)s_1 - 4\lambda_0^2 - 3\lambda_0 + 3\big]s_2 + \lambda_0(b+3)(4\lambda_0-1)s_1^3$$
$$+ \big[4(b-3)\lambda_0^2 + (5b-7)\lambda_0 - b^2(1+b)(4\lambda_0^2 + 5\lambda_0 - 1) + 3b + 15\big]s_1$$
$$+ \big[-2(3b^2+6b+7)\lambda_0 + 3b^2 + 6b + 9\big]s_1^2 - 16(h_2+h_1)\lambda_0^4$$
$$- 4\big[8h_2 - (b^2-8)h_1 + 4b\big]\lambda_0^3 - 4\big[4h_2 + 2(2-b^2)h_1 + 9b\big]\lambda_0^2$$
$$+ 2\big[b(b^2+2b-1)h_1 - 4b\big]\lambda_0 + 2b(b^2-1)h_1 + 12b\big\},$$

$$c_9 = -2^{-1}\big\{2b\lambda_0^2(\lambda_0+2)h_1 + (b^2+2b-1)\lambda_0 h_1 + (b^2-1)h_1\big\},$$

$$c_{10} = 4^{-1}\big\{4(1+\lambda_0 - \lambda_0 s_1)s_2 - (b+3)\lambda_0 s_1^3 + (b^2+2b+3)s_1^2$$
$$+ (b^3+b^2-b+3)(\lambda_0+1)s_1 + 4b(\lambda_0+1)^2\big\}.$$

It is easy to check (cf. [10]) that the function $(-1)^i\, s_i(k)$ is positive, continuous, and monotonically decreasing for $k \geqslant 0$, and so $|s_i(k)| \leqslant |s_i(0)| = 1/i$, $i = 1, 2$. This implies that $|h_i(k, \lambda_0)| \leqslant \mathrm{e}^{-\lambda_0}\,|s_i(-\lambda_0)| \leqslant \lambda_0^{-i}$ for $\lambda_0 > 0$ and $k \geqslant 0$. In view of these estimates, the above formulas for $c_1, \ldots, c_{10}$ show that $\sum_{j=1}^{10} c_j \mathrm{e}^{-j\lambda_0}$ decays exponentially as $\lambda_0 \to \infty$ because the factor $\mathrm{e}^{-j\lambda_0}$ is present in every term of the sum.

Furthermore, the above estimates show that the coefficient c_0 grows as a linear function of λ_0, and so

$$\sup_{k>0} |R_0(k, \lambda_0)| = O(\lambda_0) \quad \text{as } \lambda_0 \to \infty. \tag{3.10}$$

Estimating $r_1(k, \lambda_0)$ is straightforward, but rather tedious (cf. the proof of Lemma 2.5) and leads to the conclusion that

$$\sup_{k>0,\, \lambda_0>1} |r_1(k, \lambda_0)| \leqslant \text{const} < \infty,$$

which combined with the relations (3.9) and (3.10) gives

$$\sup_{k>0} |R(k, \lambda_0)| = O(1) \quad \text{as} \quad \lambda_0 \to \infty.$$

From Lemma 2.5 and formulas for α and β it follows that

$$\inf_{k>0} |P(k, \lambda_0)| \geqslant 1/4.$$

This inequality combined with the previous relation and (3.8) gives

$$\sup_{k>0} |J(k,\lambda_0)| = O(\lambda_0^{-1}) \quad \text{as } \lambda_0 \to \infty.$$

Here, the representation of J as a single fraction is used. Therefore, the estimate

$$\left| \int_0^\infty J(k,\lambda_0) \mathrm{e}^{-k(y+\eta+1)} \begin{Bmatrix} \sin \\ \cos \end{Bmatrix} k(x-\xi) \, \mathrm{d}k \right| \leqslant \frac{\sup_{k>0} |J(k,\lambda_0)|}{y+\eta+1} = O(\lambda_0^{-1}) \quad \text{as } \lambda_0 \to \infty$$

holds for $z, \zeta \in \overline{S}$.

Now, we use this relation and the estimates derived for $I^{(i)}_{x,y}$ $(i = 1, 2)$ in formulas (3.4) and (3.5), thus obtaining (3.3), which together with (3.2) and (3.1) yields

$$\|\nabla(G - G_0); C(\overline{S} \times \overline{S})\| = O(\lambda_0^{-1}) \quad \text{as } \lambda_0 \to \infty.$$

Then the same estimate is true for $\|T - T_0; C(\overline{S})\|$, and, in view of the inequality $\|T - T_0; C(\overline{S})\| \geqslant \|T - T_0; C_\varkappa(S)\|$, we arrive at the required relation

$$\|T - T_0; C_\varkappa(S)\| \to 0 \quad \text{as } \lambda_0 \to \infty.$$

This completes the proof of the existence and representation assertions of Theorem 1.1.

3.2 The uniqueness of a solution for $\nu > 1$

Let us consider the problem describing the flow running over the obstacle $\overline{D}$ along the x-axis. A solution $u'(x,y)$ of the new problem must satisfy the same relations (1.1)–(1.3), (1.6), and the first condition (1.5). The second condition (1.5) must be changed to

$$\lim_{x \to -\infty} |\nabla u'(x,y)| = 0 \text{ uniformly in } y \in [0,1]$$

and the right-hand side term f in the condition (1.4) must be changed to f' as well; for example, we have $\partial u'/\partial n = -U n_x$ on S in the case of rigid obstacle, i.e., the sign of this term is opposite in comparison with that in the condition for u.

In the same way as in [8, Section 7.1.3], one finds that the Green function for the new problem has the form

$$G'(z,\zeta) = G(z,\zeta) + \pi^{-1}\, w(z,\zeta),$$

where G is given by formula (1.8). If $|\xi| <$ const, then the following asymptotic formula holds

$$G'(z,\zeta) = \frac{H(x)}{\pi}\, w(z,\zeta) + \psi'_\pm(x,y) \quad \text{as } \pm x \to +\infty,$$

where $\psi'_\pm = O(|x|^{-1})$ and $|\nabla \psi'_\pm| = O(|x|^{-2})$.

Analogous to Lemma 2.1, if $\int_S f'\, \mathrm{d}s = 0$ (only such solutions are considered below), then u' has the following asymptotic behavior as $\pm x \to +\infty$:

$$u'(x,y) = H(x) \cosh \lambda_0 y \,(\mathcal{A}' \sin \lambda_0 x + \mathcal{B}' \cos \lambda_0 x) + C'_\pm + \varphi'_\pm(x,y). \quad (3.11)$$

Here, $\varphi'_\pm(x,y) = O\big(|x|^{-1}\big)$, $|\nabla \varphi'_\pm(x,y)| = O\big(|x|^{-2}\big)$, and the formulas for $C'_+ - C'_-$, $\mathcal{A}'$, and $\mathcal{B}'$ are similar to those in Lemma 2.1, but with u changed to u'.

Let us turn to the question of existence for u'. Note that the problem for u' coincides with that for u after the variable change $x \mapsto -x$, but for the obstacle reflected about the y-axis. Therefore, the results proved in Subsection 3.1 about the solvability of the problem (1.1)–(1.6) remain valid for the new statement. Thus, u' does exist for all $\nu > 1$ with a possible exception of values that belong to a sequence tending to one. Of course, the latter exceptional set might distinguish from that for the problem (1.1)–(1.6).

Now, let $\mathcal{S}$ be the subset of the half-axis $\nu > 1$ such that if $\nu \in \mathcal{S}$, then both u and u' do exist. It is clear that $\mathcal{S}$ coincides with the half-axis $\nu > 1$, except possibly for a sequence of values tending to one from above. In the same way as the problems describing opposite flows about a totally immersed body (they are considered in [8, Section 7.1.3]), the present problems for u and u' are "adjoint" to each other in the following sense.

Let u and u' be solutions of the problems describing flows running over the obstacle $\overline{D}$ in the opposite directions. If they correspond to the same $\nu \in \mathcal{S}$ and to f and f' such that

$$\int_S f\, \mathrm{d}s = \int_S f'\, \mathrm{d}s = 0,$$

then

$$\int_S u \frac{\partial u'}{\partial n}\, \mathrm{d}s = \int_S u' \frac{\partial u}{\partial n}\, \mathrm{d}s. \quad (3.12)$$

The proof of this assertion literally repeats that of Lemma 5.1 in [10] and essentially uses the asymptotic formulas (2.1) and (3.11).

To show the uniqueness for the problem (1.1)–(1.6), we assume that $\nu \in \mathcal{S}$ and u is a solution of this problem satisfying the homogeneous condition (1.4). Then (3.12) takes the form

$$\int_S u f' \, ds = 0,$$

where an arbitrary function orthogonal to constants stands as f', and so $u = \text{const}$ on S. Since $\partial u/\partial n = 0$ on S, applying the uniqueness theorem for the harmonic Cauchy problem, we get $u = \text{const}$ in W, which completes the proof.

Acknowledgments. O. V. Motygin thanks INTAS for the financial support through the YSF grant no. 06-1000014-6343, and School of Mathematics, Loughborough University, for hospitality.

References

1. Abramowitz, M., Stegun, I.A.: Handbook of Mathematical Functions. Dover, New York (1965)
2. Carleman, T.: Über das Neumann–Poincarésche Problem für ein Gebiet mit Ecken. Ålmqvist & Wiksell, Uppsala (1916)
3. Colton, D., Kress, R.: Integral Equation Methods in Scattering Theory. Wiley-Intersci., New York etc. (1983)
4. Gohberg, I., Krein, M.: Introduction to the Theory of Linear Nonselfadjoint Operators in Hilbert Space. Am. Math. Soc., Providence, RI (1969)
5. Kantorovich, L.V., Akilov, G.P.: Functional Analysis. Pergamon, Oxford (1982)
6. Kuznetsov, N.: On uniqueness and solvability in the linearized two-dimensional problem of a supercritical stream about a surface-piercing body. Proc. Roy. Soc. Lond. A **450**, 233–253 (1995) doi: 10.1098/rspa.1995.0083
7. Kuznetsov, N.G., Maz'ya, V.G.: On unique solvability of the plane Neumann–Kelvin problem (Russian). Mat. Sb. **135**, 440–462 (1988); English transl.: Math. USSR Sb. **63**, 425–446 (1989) doi:10.1070/sm1989v063n02abeh003283
8. Kuznetsov, N., Maz'ya, V., Vainberg, B.: Linear Water Waves: A Mathematical Approach. Cambridge Univ. Press, Cambridge (2002)
9. Lahalle, D.: Calcul des efforts sur un profil portant d'hydroptere par couplage eléments finis – représentation intégrale. ENSTA Rapport de Recherche. **187** (1984)
10. Motygin, O.V.: Uniqueness and solvability in the linearized two-dimensional problem of a body in a finite depth subcritical stream. Euro. J. Appl. Math. **10**, 141–156 (1999) doi: 10.1017/s0956792599003691
11. Motygin, O.V., McIver, P.: A uniqueness criterion for linear problems of wave-body interaction. IMA J. Appl. Math. **68**, 229–250 (2003) doi: 10.1093/imamat/68.3.229
12. Pagani, C.D.: The Neumann–Kelvin problem revisited. Appl. Anal. **85**, 277–292 (2006) doi: 10.1080/00036810500276589
13. Pierotti, D.: On unique solvability and regularity in the linearized two-dimensional wave-resistance problem. Quart. Appl. Math. **61**, 639–655 (2003)
14. Pierotti, D., Simioni, P.: The steady two-dimensional flow over a rectangular obstacle lying on the bottom. J. Math. Anal. Appl. **342**, 1467–1480 (2008) doi: 10.1016/j.jmaa.2008.01.020

15. Trofimov, V.P.: The root subspaces of operators that depend analytically on a parameter (Russian). Mat. Issledov. **3**(9), 117–125 (1968)
16. Vainberg, B.R., Maz'ya, V.G.: On the plane problem of the motion of a body immersed in a fluid (Russian). Tr. Mosk. Mat. O-va **28**, 35–56 (1973); English transl.: Trans. Moscow Math. Soc. **28**, 33–55 (1973)
17. Werner, P.: A Green's function approach to the potential flow around obstacles in a two-dimensional channel. In: Kleinman, R. et al. (eds), Direct and Inverse Boundary Value Problems. Methoden und Verfahren der math. physik. **37**, Peter Lang, Frankfurt am Main etc. (1991)
18. Wigley, N.M.: Mixed boundary value problems in plane domains with corners. Math. Z. **115**, 33–52 (1970) doi: 10.1007/bf01109747

Well Posedness and Asymptotic Expansion of Solution of Stokes Equation Set in a Thin Cylindrical Elastic Tube

Grigory P. Panasenko and Ruxandra Stavre

Abstract We extend the previous results for an interaction problem between a viscous fluid and an elastic structure to a three-dimensional case. We consider a nonstationary, axisymmetric, creeping flow of a viscous incompressible fluid through a long and narrow cylindrical elastic tube. The creeping flow is described by the Stokes equations, and for the wall displacement we consider the Koiter equation. The well posedness of the problem is proved by means of its variational formulation. We perform an asymptotic analysis of the problem with respect to two small parameters, for the periodic case. The small error between the exact solution and the asymptotic one justifies our asymptotic expansions.

1 Introduction

Problems involving the interaction between a fluid and a deformable structure have been studied extensively in the last years, due to their applications in many areas such as: engineering, biomechanics, biology, hydroelasticity etc. ([3, 2, 5] are some examples of works dealing with the variational study for such problems).

Mathematical modeling and numerical simulation for this type of problems allow better understanding of phenomena involved in vascular diseases.

Grigory P. Panasenko
Laboratory of Mathematics of the University of Saint-Etienne (LaMUSE EA3989), University Jean Monnet, 23, rue Dr Paul Michelon 42023 Saint-Etienne, France
e-mail: Grigory.Panasenko@univ-st-etienne.fr

Ruxandra Stavre
Institute of Mathematics "Simion Stoilow", Romanian Academy, P.O. Box 1-764, 014700 Bucharest, Romania
e-mail: Ruxandra.Stavre@imar.ro

A. Laptev (ed.), *Around the Research of Vladimir Maz'ya II: Partial Differential Equations*, International Mathematical Series 12, DOI 10.1007/978-1-4419-1343-2_13,

In order to model the blood flow through an artery or some vessel diseases, a few years ago we proposed an asymptotic approach for a fluid-structure interaction problem. We began, in [12], with the nonstationary viscous flow in a thin rectangle with highly rigid elastic walls, when at the ends of the flow domain periodicity conditions are set. The fluid flow was simulated by the Stokes equations and the elastic wall behavior for the transversal displacement was described by the Sophie Germain equation. With respect to two small dependent parameters, we constructed an asymptotic solution, justified by a theorem on the error estimates.

Within the same framework, we continued the asymptotic study of the fluid-structure interaction problem with the nonperiodic case. To obtain an asymptotic solution satisfying the same boundary conditions as the exact one, we introduce some boundary layer correctors. The construction of the boundary layer functions, the proof of their properties and the justification of the proposed asymptotic solution can be found in [13].

A numerical simulation for the nonperiodic case was performed in [11]. This approach confirms, from the numerical point of view, the boundary layer formation in a neighborhood of the ends of the elastic channel.

The next step in our asymptotic study for the fluid-structure interaction problem was to consider a fluid with variable viscosity. In this way, we were able to describe better some blood and vessel diseases. In addition to the previous case, we had to introduce further correctors, corresponding to the variable viscosity. The statement of the problem in this case, the construction of the asymptotic solution, the study of the problems for the correctors corresponding to the variable viscosity and the justification of the asymptotic solution can be found in [14].

The purpose of the present paper is to extend the previous results (obtained in 2D) to a three-dimensional case.

We consider a nonstationary, axisymmetric, creeping flow of a viscous incompressible fluid through a long and narrow cylindrical elastic tube. The creeping flow is described by the Stokes equations. We neglect the longitudinal displacement of the elastic wall and we consider for the radial displacement the Koiter equation, which is slightly different with respect to the model used in the previous papers.

We perform an asymptotic analysis of the problem with respect to two small parameters. We begin the 3D study with the periodic case. The first small parameter, ε, is defined as the ratio of the radius of the right cylinder to the period of the flow; the second one, δ, is the ratio of the linear density to the stiffness of the wall. For various ratios of these two small parameters, an asymptotic expansion of a periodic solution is constructed when the parameter δ is taken of the form $\delta = \varepsilon^\gamma$, with $\gamma \in \mathbb{N}$. The asymptotic expansion is different for three following cases: $\gamma > 3$ (very rigid wall), $\gamma < 3$ (soft wall) and $\gamma = 3$ (critical case). In each of these cases, the leading term is calculated and compared with the leading term obtained in the two-dimensional case, in [12]. As expected, the case $\gamma > 3$ is close to the rigid wall problem, while in the other two cases the solution is quite different. In the critical case, we obtain for the wall displacement a nonstandard sixth order in space parabolic equation. In the case of the soft wall $\gamma < 3$, there is an important dilation of the channel under a very small pressure.

In order to obtain the error between the exact and asymptotic solutions which justifies the asymptotic expansion, we prove some auxiliary results on existence, uniqueness, regularity of the exact solution and we establish some *a priori* estimates. The variational study of the problem is more complicated in the 3D axisymmetric case than in the two-dimensional case since we deal with weighted function spaces (cf. [7]).

We consider the periodicity condition at the ends of the tube. This condition is set in order not to complicate the asymptotic analysis by the consideration of the boundary layers. However, the construction of the boundary layer correctors could be developed in the same way as in [13], applying the results of the stabilization of solution at infinity in unbounded domains (cf. [4, 6, 8]).

We introduce the asymptotic solution, defined as in the 2D case. The asymptotic expansion is rigorously justified by the error obtained between the exact and the asymptotic solutions.

An asymptotic analysis of a viscous quasi-static flow through a narrow cylindrical elastic tube was also performed in [1]. The authors have shown that the error between the solution of the axisymmetric Stokes equations coupled with the Navier equations for the wall displacement and the main term of the asymptotic solution is of order ε^2 in the interior of the domain and of order $\varepsilon^{3/2}$ near the boundaries.

Asymptotic analysis of the flow in a channel is the first step for studying flows in so called tubular structures and lattice-like structures. In particular, the Navier–Stokes equations set in such structures with Dirichlet condition at the boundary were considered in [10] and [9]. Junctions of a "massive" 3D body with thin branches were studied in the book [6], mainly for the elasticity equation.

In the present paper, we consider the Koiter equation for the elastic wall displacement and the rigidity of the wall is the second parameter, that is great. This differs our paper from [1] (as well as the construction of the complete asymptotic expansion of the solution).

2 Notation and Problem Statement

We study the nonstationary, creeping flow of a viscous incompressible fluid through a thin right cylinder with elastic lateral wall, representing a small artery. We consider a periodic flow, the domain of motion being given by

$$\Omega_\varepsilon = \{x \in \mathbb{R}^3 : \ x = (r\cos\theta, r\sin\theta, z), r \in (0, \varepsilon), \theta \in [0, 2\pi), z \in (0, 1)\}, \quad (2.1)$$

where ε is a small parameter representing the radius of the cylinder.

We study the axisymmetric flow; hence the velocity and the pressure of the fluid satisfy in $D_\varepsilon \times (0,T)$ the Stokes equations in cylindrical coordinates:

$$\rho_f \frac{\partial u_z}{\partial t} - \mu\Big(\frac{\partial^2 u_z}{\partial z^2} + \frac{\partial^2 u_z}{\partial r^2} + \frac{1}{r}\frac{\partial u_z}{\partial r}\Big) + \frac{\partial p}{\partial z} = f_z,$$
$$\rho_f \frac{\partial u_r}{\partial t} - \mu\Big(\frac{\partial^2 u_r}{\partial z^2} + \frac{\partial^2 u_r}{\partial r^2} + \frac{1}{r}\frac{\partial u_r}{\partial r} - \frac{1}{r^2}u_r\Big) + \frac{\partial p}{\partial r} = f_r, \tag{2.2}$$
$$\frac{\partial u_z}{\partial z} + \frac{\partial u_r}{\partial r} + \frac{1}{r}u_r = 0,$$

where ρ_f and μ are the density and the viscosity of the fluid respectively, $\mathbf{f} = (f_z, f_r)$ is the exterior force applied to the fluid, $T > 0$ is an arbitrary given constant, and D_ε is the section of the flow domain, defined as follows:

$$D_\varepsilon = \{(z,r) : \; z \in (0,1),\, r \in (0,\varepsilon)\}. \tag{2.3}$$

The fluid interacts with the elastic part of the boundary of D_ε, $\Gamma_\varepsilon = \{(z,\varepsilon) : \; z \in (0,1)\}$, producing a displacement of this structure. We neglect the displacement in Oz direction and for the radial displacement we use the Koiter model.

Computing the action of the viscous fluid on the elastic wall, we are leaded for the radial displacement d to the following equation:

$$\rho h \frac{\partial^2 d}{\partial t^2} + \frac{h^3 E}{12(1-\sigma^2)}\frac{\partial^4 d}{\partial z^4} - \frac{\sigma}{6(1-\sigma^2)}\frac{h^3 E}{\varepsilon^2}\frac{\partial^2 d}{\partial z^2} + \frac{hE}{\varepsilon^2(1-\sigma^2)}\Big(1 + \frac{h^2}{12\varepsilon^2}\Big)d$$
$$+\overline{\nu}\frac{\partial^5 d}{\partial z^4 \partial t} = p/_{r=\varepsilon} + \frac{2\mu}{\varepsilon}u_r/_{r=\varepsilon} + g \;\text{ in }\; (0,1)\times(0,T), \tag{2.4}$$

where ρ is density of the elastic membrane, h is thickness, σ is the Poisson coefficient, $\overline{\nu}$ is the viscosity coefficient, E is the Young modulus, and $g\mathbf{e}_r$ is the exterior force applied on the elastic structure. A viscous type term $\overline{\nu}\dfrac{\partial^5 d}{\partial z^4 \partial t}$ was added to ensure more regularity for the unknowns.

Usually, the Young modulus has a very big value (E is of order $10^4 - 10^6$ Pa) and this value becomes more important if the elastic medium is more rigid.

We assume that the characteristic longitudinal space scale for vessels is of order of cm. Hence we scale every derivative in z by the factor 10^2. Equation (2.4) becomes

$$a\frac{\partial^2 d}{\partial t^2} + \frac{b}{\delta}\frac{\partial^4 d}{\partial z^4} - \frac{c}{\delta}\frac{\partial^2 d}{\partial z^2} + \frac{f}{\delta}d + \nu\frac{\partial^5 d}{\partial z^4 \partial t}$$
$$= p/_{r=\varepsilon} + \frac{2\mu}{\varepsilon}u_r/_{r=\varepsilon} + g \;\text{ in }\; (0,1)\times(0,T), \tag{2.5}$$

where

$$a = \rho h, \quad \frac{b}{\delta} = \frac{10^8 h^3 E}{12(1-\sigma^2)}, \quad \frac{c}{\delta} = \frac{10^4 \sigma}{6(1-\sigma^2)} \frac{h^3 E}{\varepsilon^2},$$

$$\frac{f}{\delta} = \frac{hE}{\varepsilon^2(1-\sigma^2)}\Big(1 + \frac{h^2}{12\varepsilon^2}\Big), \quad \nu = 10^8 \overline{\nu}.$$

In our asymptotic analysis, we take a, b, c, f, ν, and all the coefficients appearing in (2.2) after scaling as $\mathcal{O}(1)$.

We study this problem for $t \in [0, T]$, with T an arbitrary positive constant and we assume that the membrane is not very elastic so that the displacement of the boundary is small enough. Consequently, at each time t, we can consider with a good approximation the fluid flow equations in the initial configuration. The coupled system modeling the physical problem described before can be written in the following form:

$$\begin{aligned}
&\rho_f \frac{\partial u_z}{\partial t} - \mu\Big(\frac{\partial^2 u_z}{\partial z^2} + \frac{\partial^2 u_z}{\partial r^2} + \frac{1}{r}\frac{\partial u_z}{\partial r}\Big) \\
&\qquad\qquad + \frac{\partial p}{\partial z} = f_z \text{ in } D_\varepsilon \times (0,T), \\
&\rho_f \frac{\partial u_r}{\partial t} - \mu\Big(\frac{\partial^2 u_r}{\partial z^2} + \frac{\partial^2 u_r}{\partial r^2} + \frac{1}{r}\frac{\partial u_r}{\partial r} - \frac{1}{r^2} u_r\Big) \\
&\qquad\qquad + \frac{\partial p}{\partial r} = f_r \text{ in } D_\varepsilon \times (0,T), \\
&\frac{\partial u_z}{\partial z} + \frac{\partial u_r}{\partial r} + \frac{1}{r} u_r = 0 \text{ in } D_\varepsilon \times (0,T), \\
&a \frac{\partial^2 d}{\partial t^2} + \frac{b}{\delta}\frac{\partial^4 d}{\partial z^4} - \frac{c}{\delta}\frac{\partial^2 d}{\partial z^2} + \frac{f}{\delta} d + \nu \frac{\partial^5 d}{\partial z^4 \partial t} \\
&\quad = p/_{r=\varepsilon} + \frac{2\mu}{\varepsilon} u_r/_{r=\varepsilon} + g \text{ in } (0,1) \times (0,T), \\
&u_z, u_r, p, d \text{ 1- periodic in } z, \\
&u_r = 0 \text{ on } r = 0, \\
&\mathbf{u} = \frac{\partial d}{\partial t}\mathbf{e}_r \text{ on } r = \varepsilon, \\
&\mathbf{u}(z,r,0) = \mathbf{0}, \\
&d(z,0) = \frac{\partial d}{\partial t}(z,0) = 0.
\end{aligned} \tag{2.6}$$

The compatibility condition for (2.6) is given by

$$\int_0^1 d(z,t)\mathrm{d}z = 0 \text{ in } (0,T). \tag{2.7}$$

Remark 2.1. In addition to the boundary conditions $(2.6)_{6,7}$, the axisymmetry of the motion gives also:

$$\frac{\partial u_z}{\partial r} = 0 \quad \text{on} \quad r = 0.$$

More precisely, if u_z, u_r, and p are smooth functions satisfying $(2.6)_{1,2,3}$, then

$$\begin{aligned} &\lim_{r\to 0} \frac{\partial u_z}{\partial r} = 0, \\ &\lim_{r\to 0} \Big(\frac{1}{r}\frac{\partial u_r}{\partial r} - \frac{1}{r^2}u_r\Big) = \text{finite}, \\ &\lim_{r\to 0} \frac{u_r}{r} = \text{finite}. \end{aligned} \tag{2.8}$$

Indeed, supposing that $\lim_{r\to 0} \frac{\partial u_z}{\partial r} \neq 0$, it follows that $\lim_{r\to 0} \frac{1}{r}\frac{\partial u_z}{\partial r}$ is infinite, which leads to a contradiction if we pass to the limit in $(2.6)_1$ with $r \to 0$.

The other two relations of (2.8) are easy to obtain.

3 Variational Analysis of the Problem

This section deals with the weak formulation of the physical problem. This formulation allows us to obtain the existence, uniqueness, and regularity results and some *a priori* estimates, which are necessary to justify the asymptotic expansion introduced in the next section. The variational setting for the 3D axisymmetric problem is more complicated than that for the two-dimensional case, since the function spaces we are dealing with are less regular than the usual ones.

In order to obtain the weak formulation of the problem, we introduce the function spaces

$$\begin{aligned} &(L^2_r(D_\varepsilon))^2 = \Big\{\boldsymbol{\varphi} : D_\varepsilon \mapsto \mathbb{R}^2 : \int_{D_\varepsilon} r\boldsymbol{\varphi}^2(z,r)\mathrm{d}z\mathrm{d}r < \infty\Big\}, \\ &(H^1_r(D_\varepsilon))^2 = \Big\{\boldsymbol{\varphi} \in (L^2_r(D_\varepsilon))^2 : \int_{D_\varepsilon} r|\nabla\boldsymbol{\varphi}|^2(z,r)\mathrm{d}z\mathrm{d}r < \infty\Big\}, \\ &\mathcal{V}^\varepsilon = \Big\{\boldsymbol{\varphi} \in (\mathcal{D}(D_\varepsilon))^2 : \frac{\partial\varphi_z}{\partial z} + \frac{\partial\varphi_r}{\partial r} + \frac{1}{r}\varphi_r = 0\Big\}, \\ &V^\varepsilon_{\text{per}} = \Big\{\boldsymbol{\varphi} \in (H^1_r(D_\varepsilon))^2 : \frac{\partial\varphi_z}{\partial z} + \frac{\partial\varphi_r}{\partial r} + \frac{1}{r}\varphi_r = 0; \\ &\qquad \varphi_r = 0 \quad \text{on} \quad r = 0;\ \varphi_z = 0 \quad \text{on} \quad \Gamma_\varepsilon; \boldsymbol{\varphi} \text{ 1-periodic in } z\Big\}, \end{aligned}$$

$$\widetilde{V}^{\varepsilon}_{\text{per}} = \{\varphi \in V^{\varepsilon}_{\text{per}} : \varphi_r = 0 \text{ on } \Gamma_\varepsilon\},$$
$$\mathcal{V}^{\varepsilon}_{c} = \Big\{\varphi \in (\mathcal{D}(D_\varepsilon))^2 : \frac{\partial \varphi_z}{\partial z} + \frac{\partial \varphi_r}{\partial r} = 0\Big\},$$
$$V^{\varepsilon}_{c,per} = \Big\{\varphi \in (H^1_r(D_\varepsilon))^2 : \frac{\partial \varphi_z}{\partial z} + \frac{\partial \varphi_r}{\partial r} = 0; \varphi_r = 0 \text{ on } r = 0; \\ \varphi_z = 0 \text{ on } \Gamma_\varepsilon;\ \varphi \text{ 1-periodic in } z\Big\},$$

$$B_{0,\text{per}} = \Big\{b \in H^2(0,1) : \int_0^1 b(z)\mathrm{d}z = 0;\ b \text{ 1-periodic in } z\Big\}.$$

Remark 3.1. The space $V^{\varepsilon}_{\text{per}}$ is defined as the closure in the H^1_r norm of the space of smooth periodic in z vector-valued functions (φ_r, φ_z) with φ_r vanishing in some neighborhood of the line $r = 0$ and with φ_z vanishing in some neighborhood of the line $r = \varepsilon$ and satisfying the relation

$$\frac{\partial \varphi_z}{\partial z} + \frac{\partial \varphi_r}{\partial r} + \frac{1}{r}\varphi_r = 0.$$

$\mathcal{D}(D_\varepsilon)$ is the space of smooth functions with compact support.

We have the following result.

Proposition 3.1. *The norms generated by the scalar products*

$$(\varphi, \psi)_{(H^1_r(D_\varepsilon))^2} = \int_{D_\varepsilon} r(\varphi \cdot \psi + \nabla\varphi \cdot \nabla\psi)\mathrm{d}z\mathrm{d}r,$$

$$(\varphi, \psi)_{\widetilde{V}} = \int_{D_\varepsilon} r\Big(\frac{\partial \varphi_z}{\partial z}\frac{\partial \psi_z}{\partial z} + \frac{\partial \varphi_z}{\partial r}\frac{\partial \psi_z}{\partial r} + \frac{\partial \varphi_r}{\partial z}\frac{\partial \psi_r}{\partial z} + \frac{\partial \varphi_r}{\partial r}\frac{\partial \psi_r}{\partial r} + \frac{\varphi_r}{r}\frac{\psi_r}{r}\Big)\mathrm{d}z\mathrm{d}r$$

are equivalent on $\widetilde{V}^{\varepsilon}_{\text{per}}$.

Proof. Since $\varphi_z = 0$ on Γ_ε, we can apply the Poincaré inequality on Ω_ε, which gives

$$\int_{\Omega_\varepsilon} \varphi_z^2 \mathrm{d}x\mathrm{d}y\mathrm{d}z \leqslant \frac{1}{2}\varepsilon^2 \int_{\Omega_\varepsilon} \Big(\Big(\frac{\partial \varphi_z}{\partial x}\Big)^2 + \Big(\frac{\partial \varphi_z}{\partial y}\Big)^2 + \Big(\frac{\partial \varphi_z}{\partial z}\Big)^2\Big)\mathrm{d}x\mathrm{d}y\mathrm{d}z.$$

By changing the coordinates and using the axisymmetry property, it follows that

$$\int_{D_\varepsilon} r\varphi_z^2 \mathrm{d}z\mathrm{d}r \leqslant \frac{1}{2}\varepsilon^2 \int_{D_\varepsilon} r\Big(\Big(\frac{\partial \varphi_z}{\partial z}\Big)^2 + \Big(\frac{\partial \varphi_z}{\partial r}\Big)^2\Big)\mathrm{d}z\mathrm{d}r.$$

Hence

$$\|\varphi\|^2_{(H^1_r(D_\varepsilon))^2} \leqslant \int_{D_\varepsilon} r\Big(\varphi_r^2 + \Big(1 + \frac{1}{2}\varepsilon^2\Big)\Big(\frac{\partial \varphi_z}{\partial z}\Big)^2 + \Big(1 + \frac{1}{2}\varepsilon^2\Big)\Big(\frac{\partial \varphi_z}{\partial r}\Big)^2 \\ + \Big(\frac{\partial \varphi_r}{\partial z}\Big)^2 + \Big(\frac{\partial \varphi_r}{\partial r}\Big)^2\Big)\mathrm{d}z\mathrm{d}r \leqslant (1 + \frac{1}{2}\varepsilon^2)\|\varphi\|^2_{\widetilde{V}}.$$

The converse inequality is obtained by replacing $\frac{1}{r}\varphi_r$ by $-\frac{\partial \varphi_z}{\partial z} - \frac{\partial \varphi_r}{\partial r}$. □

Taking for the data the regularity

(H1) $\mathbf{f} \in L^2(0,T;(L^2_r(D_\varepsilon))^2)$, $\mathbf{f}$ 1-periodic in z, $g \in L^2(0,T;B_{0,\mathrm{per}})$,

we consider the variational problem

Find $(\mathbf{u}, d) \in L^2(0,T;V^\varepsilon_{\mathrm{per}}) \times H^1(0,T;B_{0,\mathrm{per}})$,
with $(\mathbf{u}', d'') \in L^2(0,T;(V^\varepsilon_{c,per})') \times L^2(0,T;(B_{0,\mathrm{per}})')$,
such that the following two relations are satisfied a. e. in $(0,T)$:

$$
\begin{aligned}
&\rho_f \frac{d}{dt}\int_{D_\varepsilon} r\mathbf{u}\cdot\mathbf{v} + \mu \int_{D_\varepsilon} r\Big(\frac{\partial u_z}{\partial z}\frac{\partial v_z}{\partial z} + \frac{\partial u_z}{\partial r}\frac{\partial v_z}{\partial r} + \frac{\partial u_r}{\partial z}\frac{\partial v_r}{\partial z} \\
&\quad + \Big(\frac{\partial u_r}{\partial r} - \frac{u_r}{r}\Big)\Big(\frac{\partial v_r}{\partial r} - \frac{v_r}{r}\Big)\Big) + a\varepsilon \frac{d}{dt}\int_0^1 \frac{\partial d}{\partial t}\beta + \frac{b\varepsilon}{\delta}\int_0^1 \frac{\partial^2 d}{\partial z^2}\beta'' \\
&\quad + \frac{c\varepsilon}{\delta}\int_0^1 \frac{\partial d}{\partial z}\beta' + \frac{f\varepsilon}{\delta}\int_0^1 d\beta + \nu\varepsilon \frac{d}{dt}\int_0^1 \frac{\partial^2 d}{\partial z^2}\beta'' \\
&= \int_{D_\varepsilon} r\mathbf{f}\cdot\mathbf{v} + \varepsilon\int_0^1 g\beta \quad \forall \mathbf{v} \in V^\varepsilon_{\mathrm{per}},\ \forall \beta \in B_{0,\mathrm{per}}, v_r = \beta \ \text{ on } \ \Gamma_\varepsilon, \\
&\quad u_r = \frac{\partial d}{\partial t} \ \text{on } \Gamma_\varepsilon, \\
&\mathbf{u}(z,r,0) = \mathbf{0}, \quad d(z,0) = \frac{\partial d}{\partial t}(z,0) = 0.
\end{aligned}
\tag{3.1}
$$

The relation of 3.1 holds *a.e.* in $(0,T)$ due to the $L^2(0,T)$ regularity of all its terms which do not contain the time derivative of the integral.

Here,

(H2) rho_f, μ, a, b, ν are positive constants and c, $f \geqslant 0$.

The next result gives the existence, uniqueness, and regularity of a solution of the variational problem (3.1) and the existence of a solution of (2.6).

Theorem 3.1. *Let* (H1), (H2) *hold. Then*

(a) *the problem* (3.1) *has a unique solution* $(\mathbf{u}, d)$, *with the regularity* $\mathbf{u}' \in L^2(0,T;(L^2_r(D_\varepsilon))^2)$, $d'' \in L^2((0,1)\times(0,T))$;

(b) *there exists a unique function* $p \in L^2(0,T;H^1_r(D_\varepsilon))$, *1-periodic in* z, *such that* $(\mathbf{u}, p, d)$ *satisfies the system* (2.6) *a. e. in* $D_\varepsilon \times (0,T)$.

Proof. We begin by obtaining the uniqueness of $(\mathbf{u}, d)$. Let $(\mathbf{u}_i, d_i)$, $i = 1,2$, be two solutions for (3.1). Subtracting the relations (3.1) for $i = 1,2$, denoting $(\mathbf{u}, d) = (\mathbf{u}_1 - \mathbf{u}_2, d_1 - d_2)$, and taking $(\mathbf{v}, \beta) = \Big(\mathbf{u}(t)\frac{\partial d}{\partial t}(t)\Big)$ for a test function, we find

$$\frac{\rho_f}{2}\frac{d}{dt}\int_{D_\varepsilon} r\mathbf{u}^2 + \mu \int_{D_\varepsilon} r\Big(\Big(\frac{\partial u_z}{\partial z}\Big)^2 + \Big(\frac{\partial u_z}{\partial r}\Big)^2 + \Big(\frac{\partial u_r}{\partial z}\Big)^2$$

$$+ \Big(\frac{\partial u_r}{\partial r} - \frac{u_r}{r}\Big)^2\Big) + \frac{a\varepsilon}{2}\frac{d}{dt}\int_0^1 \Big(\frac{\partial d}{\partial t}\Big)^2 + \frac{b\varepsilon}{2\delta}\frac{d}{dt}\int_0^1 \Big(\frac{\partial^2 d}{\partial z^2}\Big)^2$$

$$+ \frac{c\varepsilon}{2\delta}\frac{d}{dt}\int_0^1 \Big(\frac{\partial d}{\partial z}\Big)^2 + \frac{f\varepsilon}{2\delta}\frac{d}{dt}\int_0^1 d^2 + \nu\varepsilon\int_0^1 \Big(\frac{\partial^3 d}{\partial z^2 \partial t}\Big)^2 = 0.$$

The integration from 0 to t of the previous equality, together with the initial conditions, yields the uniqueness of $(\mathbf{u}, d)$.

The existence and regularity of $\mathbf{u}$ and d are obtained by the Galerkin method. Moreover, this method provides the *a priori* estimates necessary for justification of our asymptotic approximations.

Let $\{\beta_j\}_{j\in\mathbb{N}}$ be a basis for the space $B_{0,\mathrm{per}}$. In the sequel, we construct a basis for $\widetilde{V}^\varepsilon_{\mathrm{per}}$.

For any $\mathbf{f} \in (L^2_r(D_\varepsilon))^2$ we consider the problem (in the sense of variational formulation)

$$\begin{aligned}
&\text{Find } (\boldsymbol{\psi}, q) \in \widetilde{V}^\varepsilon_{\mathrm{per}} \times L^2_r(D_\varepsilon) \text{ which satisfies a. e. in } D_\varepsilon\\
&-\mu\Delta\boldsymbol{\psi} + \nabla q = \mathbf{f},\\
&q \text{ 1-periodic in } z,
\end{aligned} \tag{3.2}$$

with

$$\Delta\boldsymbol{\psi} = \Big(\frac{\partial^2\psi_z}{\partial z^2} + \frac{\partial^2\psi_z}{\partial r^2} + \frac{1}{r}\frac{\partial\psi_z}{\partial r}\Big)\mathbf{e}_z + \Big(\frac{\partial^2\psi_r}{\partial z^2} + \frac{\partial^2\psi_r}{\partial r^2} + \frac{1}{r}\frac{\partial\psi_r}{\partial r} - \frac{1}{r^2}\psi_r\Big)\mathbf{e}_r.$$

The uniqueness of the function $\boldsymbol{\psi}$ allows us to define the operator $T \in \mathcal{L}((L^2_r(D_\varepsilon))^2)$, $T\mathbf{f} = \boldsymbol{\psi}$. A standard computation shows that T is compact and selfadjoint and, by a classical theorem for separable Hilbert spaces, the space $(L^2_r(D_\varepsilon))^2$ has a basis $\{\boldsymbol{\psi}_k\}_{k\in\mathbb{N}}$ where $\boldsymbol{\psi}_k$ is the eigenfunction of T corresponding to the eigenvalue $1/\lambda_k$ with $\lambda_k > 0$. So, $\boldsymbol{\psi}_k$ satisfies the relation

$$T\boldsymbol{\psi}_k = \frac{1}{\lambda_k}\boldsymbol{\psi}_k,$$

i.e., for $\mathbf{f} = \boldsymbol{\psi}_k$ and $\boldsymbol{\psi} = \frac{1}{\lambda_k}\boldsymbol{\psi}_k$, the problem (3.2) takes the form

$$\begin{aligned}
&\boldsymbol{\psi}_k \in \widetilde{V}^\varepsilon_{\mathrm{per}},\\
&\mu(\boldsymbol{\psi}_k, \boldsymbol{\varphi})_{\widetilde{V}} = \lambda_k(\boldsymbol{\psi}_k, \boldsymbol{\varphi})_{(L^2_r(D_\varepsilon))^2}, \quad \forall\boldsymbol{\varphi} \in \widetilde{V}^\varepsilon_{\mathrm{per}}.
\end{aligned} \tag{3.3}$$

Let us prove that this basis is also a basis for $\widetilde{V}^\varepsilon_{\mathrm{per}}$.

We define the space $S = \overline{\{\boldsymbol{\psi}_1, \ldots, \boldsymbol{\psi}_n, \ldots\}}^{\|\ \|_{\tilde{V}}}$. The construction of a basis for $\widetilde{V}^{\varepsilon}_{\text{per}}$ is achieved if we prove that $S = \widetilde{V}^{\varepsilon}_{\text{per}}$. Since S is a closed subspace of $\widetilde{V}^{\varepsilon}_{\text{per}}$, we have $\widetilde{V}^{\varepsilon}_{\text{per}} = S \bigoplus S^{\perp}$. Let $\boldsymbol{\varphi}$ be an element of $S^{\perp}$. Taking it for a test function in (3.3), we find

$$(\boldsymbol{\psi}_k, \boldsymbol{\varphi})_{(L^2_r(D_\varepsilon))^2} = 0 \quad \forall k \in \mathbb{N}.$$

This yields

$$(\mathbf{v}, \boldsymbol{\varphi})_{(L^2_r(D_\varepsilon))^2} = 0 \quad \forall \mathbf{v} \in (L^2_r(D_\varepsilon))^2,$$

i.e., $S^{\perp} = \{\mathbf{0}\}$.

For any element of the basis $\{\beta_j\}_{j\in\mathbb{N}}$ we consider the problem

$$\begin{aligned}
& -\mu \Delta \boldsymbol{\varphi}_j + \nabla p_j = \mathbf{0}, \quad \text{in } D_\varepsilon, \\
& \frac{\partial \varphi_{j,z}}{\partial z} + \frac{\partial \varphi_{j,r}}{\partial r} + \frac{1}{r}\varphi_{j,r} = 0 \ \text{ in } D_\varepsilon, \\
& \boldsymbol{\varphi}_j = \mathbf{0} \ \text{ on } \partial D_\varepsilon \backslash \Gamma_\varepsilon, \\
& \boldsymbol{\varphi}_j = \beta_j \mathbf{e}_r \ \text{ on } \Gamma_\varepsilon.
\end{aligned} \tag{3.4}$$

By the uniqueness of $\boldsymbol{\varphi}_j$, $j \in \mathbb{N}$, we can define new functions for $m, n \in \mathbb{N}$

$$\begin{aligned}
& d_n(z,t) = \sum_{j=1}^{n} b_j(t)\ \beta_j(z), \\
& \mathbf{u}_n^m(z,r,t) = \sum_{i=1}^{m} a_i(t)\ \boldsymbol{\psi}_i(z,r) + \sum_{j=1}^{n} \dot{b}_j(t)\ \boldsymbol{\varphi}_j(z,r),
\end{aligned} \tag{3.5}$$

where a_i and b_j are unknown functions satisfying $a_i(0) = b_j(0) = \dot{b}_j(0) = 0$. These functions are determined by solving the problem

$$\begin{aligned}
& \rho_f \int_{D_\varepsilon} r \frac{\partial \mathbf{u}_n^m}{\partial t} \cdot \boldsymbol{\psi}_i + \mu I(\mathbf{u}_n^m, \boldsymbol{\psi}_i) = \int_{D_\varepsilon} r\mathbf{f} \cdot \boldsymbol{\psi}_i, \quad i = 1, \ldots, m, \\
& \rho_f \int_{D_\varepsilon} r \frac{\partial \mathbf{u}_n^m}{\partial t} \cdot \boldsymbol{\varphi}_j + \mu I(\mathbf{u}_n^m, \boldsymbol{\varphi}_j) + a\varepsilon \int_0^1 \frac{\partial^2 d_n}{\partial t^2}\beta_j + \frac{b\varepsilon}{\delta}\int_0^1 \frac{\partial^2 d_n}{\partial z^2}\beta_j'' \\
& + \frac{c\varepsilon}{\delta}\int_0^1 \frac{\partial d_n}{\partial z}\beta_j' + \frac{f\varepsilon}{\delta}\int_0^1 d_n \beta_j + \nu\varepsilon \int_0^1 \frac{\partial^3 d_n}{\partial z^2 \partial t}\beta_j'' \\
& = \int_{D_\varepsilon} r\mathbf{f} \cdot \boldsymbol{\varphi}_j + \varepsilon \int_0^1 g\beta_j, \ j = 1, \ldots, n,
\end{aligned} \tag{3.6}$$

$$\mathbf{u}_n^m(z,r,0) = \mathbf{0}, \qquad d_n(z,0) = \frac{\partial d_n}{\partial t}(z,0) = 0,$$

where

$$I(\varphi,\psi)=\int_{D_\varepsilon} r\Big(\frac{\partial\varphi_z}{\partial z}\frac{\partial\psi_z}{\partial z}+\frac{\partial\varphi_z}{\partial r}\frac{\partial\psi_z}{\partial r}+\frac{\partial\varphi_r}{\partial z}\frac{\partial\psi_r}{\partial z}+\Big(\frac{\partial\varphi_r}{\partial r}-\frac{\varphi_r}{r}\Big)\Big(\frac{\partial\psi_r}{\partial r}-\frac{\psi_r}{r}\Big)\Big).$$

We note that the relation

$$\mathbf{u}_n^m(z,\varepsilon,t)=\frac{\partial d_n}{\partial t}(z,t)$$

follows from $\boldsymbol{\psi}_i=\mathbf{0}$ on Γ_ε and $(3.4)_4$.

In the sequel, taking into account the obvious relations

$$I(\boldsymbol{\varphi},\boldsymbol{\psi})=(\boldsymbol{\varphi},\boldsymbol{\psi})_{\widetilde V}\quad \forall\,\boldsymbol{\varphi},\boldsymbol{\psi}\in\widetilde V^\varepsilon_{\rm per}\ \text{ or }\ \boldsymbol{\varphi}\in V^\varepsilon_{\rm per},\ \boldsymbol{\psi}\in\widetilde V^\varepsilon_{\rm per},$$

$$I(\boldsymbol{\varphi},\boldsymbol{\psi})=(\boldsymbol{\varphi},\boldsymbol{\psi})_{\widetilde V}-\int_{\Gamma_\varepsilon}\varphi_r\psi_r\mathrm{d}z\quad\forall\,\boldsymbol{\varphi},\boldsymbol{\psi}\in V^\varepsilon_{\rm per}$$

and the construction of the basis $\{\boldsymbol{\psi}_k\}_{k\in\mathbb{N}}$, we obtain the following differential system for the unknown functions a_i and b_j:

$$\begin{aligned}
&\rho_f\dot a_i(t)+\lambda_i a_i(t)+\rho_f\sum_{k=1}^n p_{ki}\,\ddot b_k(t)+\mu\sum_{k=1}^n r_{ki}\dot b_k(t)=\int_{D_\varepsilon} r\mathbf{f}\cdot\boldsymbol{\psi}_i,\\
&\rho_f\sum_{k=1}^m p_{jk}\dot a_k(t)+\mu\sum_{k=1}^m r_{jk}a_k(t)+\sum_{k=1}^n(\rho_f s_{jk}+a\varepsilon\delta_{jk})\ddot b_k(t)\\
&\quad+\sum_{k=1}^n(\mu t_{jk}+\nu\varepsilon v_{jk})\dot b_k(t)+\sum_{k=1}^n\Big(\frac{b\varepsilon}{\delta}v_{jk}+\frac{c\varepsilon}{\delta}w_{jk}+\frac{f\varepsilon}{\delta}\delta_{jk}\Big)b_k(t)\\
&\quad=\int_{D_\varepsilon} r\mathbf{f}\cdot\boldsymbol{\varphi}_j+\varepsilon\int_0^1 g\beta_j,\\
&a_i(0)=b_j(0)=\dot b_j(0)=0,\quad i=1,\dots,m,\ j=1,\dots,n,
\end{aligned}\tag{3.7}$$

where

$$\begin{aligned}
p_{kl}&=\int_{D_\varepsilon} r\boldsymbol{\varphi}_k\cdot\boldsymbol{\psi}_l,\ r_{kl}=I(\boldsymbol{\varphi}_k,\boldsymbol{\psi}_l),\ s_{kl}\\
&=\int_{D_\varepsilon} r\boldsymbol{\varphi}_k\cdot\boldsymbol{\varphi}_l,\ t_{kl}=I(\boldsymbol{\varphi}_k,\boldsymbol{\varphi}_l),\ v_{kl}=\int_0^1\beta_k''\beta_l'',\ w_{kl}=\int_0^1\beta_k'\beta_l'.
\end{aligned}$$

The existence and uniqueness of a solution of (3.7) follow from the classical results for ordinary differential systems. Hence the functions a_i and b_j are uniquely determined.

We obtain the first estimates as follows. Multiply $(3.6)_1$ by $a_i(t)$ and add the equalities for $i=1,\dots,m$. Then multiply $(3.6)_2$ by $\dot b_j(t)$, add the equalities for $j=1,\dots,n$ and add the previous two equalities. By the definition (3.5),

$$\rho_f \int_{D_\varepsilon} r\frac{\partial \mathbf{u}_n^m}{\partial t}\cdot \mathbf{u}_n^m + \mu I(\mathbf{u}_n^m, \mathbf{u}_n^m) + a\varepsilon \int_0^1 \frac{\partial^2 d_n}{\partial t^2}\frac{\partial d_n}{\partial t} + \frac{b\varepsilon}{\delta}\int_0^1 \frac{\partial^2 d_n}{\partial z^2}\frac{\partial^3 d_n}{\partial z^2 \partial t}$$

$$+ \frac{c\varepsilon}{\delta}\int_0^1 \frac{\partial d_n}{\partial z}\frac{\partial^2 d_n}{\partial z \partial t} + \frac{f\varepsilon}{\delta}\int_0^1 d_n \frac{\partial d_n}{\partial t} + \nu\varepsilon \int_0^1 \left(\frac{\partial^3 d_n}{\partial z^2 \partial t}\right)^2$$

$$= \int_{D_\varepsilon} r\mathbf{f}\cdot \mathbf{u}_n^m + \varepsilon\int_0^1 g\frac{\partial d_n}{\partial t}.$$

Since $d_n(t) \in B_{0,\text{per}}$, from the previous equality it remains to obtain the estimates

$$\|\mathbf{u}_n^m\|_{L^\infty(0,T;(L_r^2(D_\varepsilon))^2)} \leqslant C\big(\|\mathbf{f}\|_{L^2(0,T;(L_r^2(D_\varepsilon))^2)} + \|g\|_{L^2((0,1)\times(0,T))}\big),$$

$$\left(\left\|\frac{\partial u_{n,z}^m}{\partial z}\right\|^2_{L^2(0,T;L_r^2(D_\varepsilon))} + \left\|\frac{\partial u_{n,z}^m}{\partial r}\right\|^2_{L^2(0,T;L_r^2(D_\varepsilon))} + \left\|\frac{\partial u_{n,r}^m}{\partial z}\right\|^2_{L^2(0,T;L_r^2(D_\varepsilon))}\right.$$
$$\left. + \left\|\frac{\partial u_{n,r}^m}{\partial r}\right\|^2_{L^2(0,T;L_r^2(D_\varepsilon))} + \left\|\frac{u_{n,r}^m}{r}\right\|^2_{L^2(0,T;(L_r^2 D_\varepsilon))}\right)^{1/2}$$
$$\leqslant C\big(\|\mathbf{f}\|_{L^2(0,T;(L_r^2(D_\varepsilon))^2)} + \|g\|_{L^2((0,1)\times(0,T))}\big),$$

$$\varepsilon^{1/2}\left\|\frac{\partial d_n}{\partial t}\right\|_{L^\infty(0,T;L^2(0,1))} \leqslant C\big(\|\mathbf{f}\|_{L^2(0,T;(L_r^2(D_\varepsilon))^2)} + \|g\|_{L^2((0,1)\times(0,T))}\big),$$

$$\left(\frac{\varepsilon}{\delta}\right)^{1/2}\left\|\frac{\partial^2 d_n}{\partial z^2}\right\|_{L^\infty(0,T;L^2(0,1))} \leqslant C\big(\|\mathbf{f}\|_{(L^2(0,T;(L_r^2(D_\varepsilon))^2)} + \|g\|_{L^2((0,1)\times(0,T))}\big),$$

$$\varepsilon^{1/2}\left\|\frac{\partial^3 d_n}{\partial z^2 \partial t}\right\|_{L^2((0,1)\times(0,T))} \leqslant C\big(\|\mathbf{f}\|_{L^2(0,T;(L_r^2(D_\varepsilon))^2)} + \|g\|_{L^2((0,1)\times(0,T))}\big). \tag{3.8}$$

Hereinafter, C stands for a constant independent on ε. The other estimates given by the previous inequality are a consequence of $(3.8)_4$ and the classical Poincaré inequality, applied for $G = (0,1)\times(0,T)$:

$$\|u\|_{L^2(G)} \leqslant \tilde{c}\|\nabla u\|_{(L^2(G))^\alpha} \quad \forall\, u \in \{\varphi \in H^1(G) : \int_G \varphi(x)\mathrm{d}x = 0\}, \tag{3.9}$$

In the sequel, the viscous term, added in the Koiter equation, allow us to obtain the regularity stated in the theorem for the functions $\mathbf{u}$ and d. This regularity is necessary for obtaining $p(t) \in H_r^1(D_\varepsilon)$, property which gives sense to the first term on the right-hand side of $(2.6)_4$.

We multiply $(3.6)_1$ by $\dot{a}_i(t)$ and add the equalities for $i = 1,\ldots,m$. We multiply $(3.6)_2$ by $\ddot{b}_j(t)$ and add the equalities for $j = 1,\ldots,n$ and then add the two equalities. This yields

$$\rho_f \int_{D_\varepsilon} r\Big(\frac{\partial \mathbf{u}_n^m}{\partial t}\Big)^2 + \frac{\mu}{2}\frac{d}{dt} I(\mathbf{u}_n^m, \mathbf{u}_n^m) + a\varepsilon \int_0^1 \Big(\frac{\partial^2 d_n}{\partial t^2}\Big)^2 + \frac{b\varepsilon}{\delta}\int_0^1 \frac{\partial^2 d_n}{\partial z^2}\frac{\partial^4 d_n}{\partial z^2 \partial t^2}$$

$$+ \frac{c\varepsilon}{\delta}\int_0^1 \frac{\partial d_n}{\partial z}\frac{\partial^3 d_n}{\partial z \partial t^2} + \frac{f\varepsilon}{\delta}\int_0^1 d_n \frac{\partial^2 d_n}{\partial t^2} + \frac{\nu\varepsilon}{2}\frac{d}{dt}\int_0^1 \Big(\frac{\partial^3 d_n}{\partial z^2 \partial t}\Big)^2$$

$$= \int_{D_\varepsilon} r\mathbf{f}\cdot \frac{\mathbf{u}_n^m}{\partial t} + \varepsilon \int_0^1 g \frac{\partial^2 d_n}{\partial t^2}. \tag{3.10}$$

Integrating by parts T_4, T_5, and T_6 on the left-hand side of (3.10), integrating from 0 to t, and using the initial conditions, we get

$$\rho_f \Big\| \frac{\partial \mathbf{u}_n^m}{\partial t}\Big\|^2_{L^2(0,T;(L^2_r(D_\varepsilon))^2)} + \frac{\mu}{2} I(\mathbf{u}_n^m, \mathbf{u}_n^m)(t) + a\varepsilon \Big\|\frac{\partial^2 d_n}{\partial t^2}\Big\|^2_{L^2((0,1)\times(0,T))}$$

$$+ \frac{\nu\varepsilon}{2}\Big\|\frac{\partial^3 d_n}{\partial z^2 \partial t}\Big\|^2_{L^\infty(0,T;L^2(0,1))}$$

$$= -\frac{b\varepsilon}{\delta}\int_0^1 \frac{\partial^2 d_n}{\partial z^2}\frac{\partial^3 d_n}{\partial z^2 \partial t} + \frac{b\varepsilon}{\delta}\Big\|\frac{\partial^3 d_n}{\partial z^2 \partial t}\Big\|^2_{L^2((0,1)\times(0,T))}$$

$$- \frac{c\varepsilon}{\delta}\int_0^1 \frac{\partial d_n}{\partial z}\frac{\partial^2 d_n}{\partial z \partial t} + \frac{c\varepsilon}{\delta}\Big\|\frac{\partial^2 d_n}{\partial z \partial t}\Big\|^2_{L^2((0,1)\times(0,T))}$$

$$- \frac{f\varepsilon}{\delta}\int_0^1 d_n \frac{\partial d_n}{\partial t} + \frac{f\varepsilon}{\delta}\Big\|\frac{\partial d_n}{\partial t}\Big\|^2_{L^2((0,1)\times(0,T))} + \int_0^t \int_{D_\varepsilon} r\mathbf{f}\cdot \frac{\partial \mathbf{u}_n^m}{\partial t}$$

$$+ \varepsilon \int_0^t \int_0^1 g \frac{\partial^2 d_n}{\partial t^2}.$$

Majorating the right-hand side of the above equality and using (3.8), we find

$$\frac{\rho_f}{2}\Big\| \frac{\partial \mathbf{u}_n^m}{\partial t}\Big\|^2_{L^2(0,T;(L^2_r(D_\varepsilon))^2)} + \frac{\mu}{2} I(\mathbf{u}_n^m, \mathbf{u}_n^m)(t)$$

$$+ \frac{a\varepsilon}{2}\Big\|\frac{\partial^2 d_n}{\partial t^2}\Big\|^2_{L^2((0,1)\times(0,T))} + \frac{\nu\varepsilon}{2}\Big\|\frac{\partial^3 d_n}{\partial z^2 \partial t}\Big\|^2_{L^\infty(0,T;L^2(0,1))}$$

$$\leqslant \frac{\alpha}{\delta}\big(\|\mathbf{f}\|^2_{L^2(0,T;(L^2_r(D_\varepsilon))^2)} + \|g\|^2_{L^2((0,1)\times(0,T))}\big)$$

$$+ \frac{2c\varepsilon}{\delta}\Big\|\frac{\partial^2 d_n}{\partial z \partial t}\Big\|^2_{L^2((0,1)\times(0,T))}. \tag{3.11}$$

For majorating the last term of (3.11), we again use (3.9) and $(3.8)_5$. We obtain the second estimates given by

$$\left\|\frac{\partial \mathbf{u}_n^m}{\partial t}\right\|_{L^2(0,T;(L_r^2(D_\varepsilon))^2)} \leqslant \frac{C}{\delta^{1/2}}\big(\|\mathbf{f}\|_{L^2(0,T;(L_r^2(D_\varepsilon))^2)} + \|g\|_{L^2((0,1)\times(0,T))}\big),$$

$$\left(\left\|\frac{\partial u_{n,z}^m}{\partial z}\right\|^2_{L^\infty(0,T;(L_r^2 D_\varepsilon))} + \left\|\frac{\partial u_{n,z}^m}{\partial r}\right\|^2_{L^\infty(0,T;L_r^2(D_\varepsilon))} + \left\|\frac{\partial u_{n,r}^m}{\partial z}\right\|^2_{L^\infty(0,T;L_r^2(D_\varepsilon))}\right.$$

$$\left. + \left\|\frac{\partial u_{n,r}^m}{\partial r}\right\|^2_{L^\infty(0,T;L_r^2(D_\varepsilon))} + \left\|\frac{u_{n,r}^m}{r}\right\|^2_{L^\infty(0,T;L_r^2(D_\varepsilon))}\right)^{1/2}$$

$$\leqslant \frac{C}{\delta^{1/2}}\big(\|\mathbf{f}\|_{L^2(0,T;(L_r^2(D_\varepsilon))^2)} + \|g\|_{L^2((0,1)\times(0,T))}\big), \tag{3.12}$$

$$\varepsilon^{1/2}\left\|\frac{\partial^2 d_n}{\partial t^2}\right\|_{L^2((0,1)\times(0,T))} \leqslant \frac{C}{\delta^{1/2}}\big(\|\mathbf{f}\|_{L^2(0,T;(L_r^2(D_\varepsilon))^2)} + \|g\|_{L^2((0,1)\times(0,T))}\big),$$

$$\varepsilon^{1/2}\left\|\frac{\partial^3 d_n}{\partial z^2 \partial t}\right\|_{L^\infty(0,T;L^2(0,1))} \leqslant \frac{C}{\delta^{1/2}}\big(\|\mathbf{f}\|_{L^2(0,T;(L_r^2(D_\varepsilon))^2)} + \|g\|_{L^2((0,1)\times(0,T))}\big).$$

Using the previous estimates, we can pass to the limit in (3.6) and obtain the existence of a pair $(\mathbf{u}, d)$ with the desired regularity for the functions $\mathbf{u}$ and d satisfying

$$\rho_f \int_{D_\varepsilon} r\frac{\partial \mathbf{u}}{\partial t}\cdot \boldsymbol{\psi} + \mu I(\mathbf{u}, \boldsymbol{\psi}) = \int_{D_\varepsilon} r\mathbf{f}\cdot\boldsymbol{\psi} \quad \forall\, \boldsymbol{\psi} \in \mathcal{V}^\varepsilon. \tag{3.13}$$

The previous equality can also be written as

$$\Big\langle \rho_f \frac{\partial \mathbf{u}}{\partial t} - \mu\Delta\mathbf{u} - \mathbf{f}, r\boldsymbol{\psi}\Big\rangle = 0 \quad \forall\, \boldsymbol{\psi} \in \mathcal{V}^\varepsilon. \tag{3.14}$$

Since $\boldsymbol{\psi} \in \mathcal{V}^\varepsilon$ if and only if $r\boldsymbol{\psi} \in \mathcal{V}_c^\varepsilon$, it follows that we can apply the De Rham theorem and obtain the existence of a distribution p satisfying

$$\rho_f \frac{\partial \mathbf{u}}{\partial t} - \mu\Delta\mathbf{u} - \mathbf{f} = -\nabla p \;\text{ in }\; \mathcal{D}'(D_\varepsilon \times (0,T)). \tag{3.15}$$

From $(3.12)_1$ we obtain for $\dfrac{\partial \mathbf{u}}{\partial t}$ the regularity $L^2(0,T;(L_r^2(D_\varepsilon))^2)$; moreover, taking into account the regularity of $\mathbf{f}$, we also have $\Delta\mathbf{u} \in L^2(0,T;(L_r^2(D_\varepsilon))^2)$. As a consequence of these regularities, (3.15) gives $p \in L^2(0,T;H_r^1(D_\varepsilon))$.

The next step in the proof of the theorem is to obtain the 1-periodicity of p with respect to z. For this purpose, we multiply (3.15) by $r\boldsymbol{\psi}$, with $\boldsymbol{\psi} \in \widetilde{V}_{\text{per}}^\varepsilon$ and integrate over D_ε. By the above-established regularity, we can write

$$\rho_f \int_{D_\varepsilon} r\frac{\partial \mathbf{u}}{\partial t}\cdot\boldsymbol{\psi} + \mu(\mathbf{u}, \boldsymbol{\psi})_{\widetilde{V}}$$

$$- \int_0^\varepsilon r(p(0,r,t) - p(1,r,t))\psi_z(0,r,t)\mathrm{d}r = \int_{D_\varepsilon} r\mathbf{f}\cdot\boldsymbol{\psi}. \tag{3.16}$$

We take $\psi \in \widetilde{V}^{\varepsilon}_{\text{per}}$ in (3.13) and compute (3.13)–(3.16) with this test function. Taking into account the properties of I, we find

$$\int_0^{\varepsilon} r(p(0,r,t) - p(1,r,t))\psi_z(0,r,t)\mathrm{d}r = 0 \quad \forall\, \psi \in \widetilde{V}^{\varepsilon}_{\text{per}}.$$

This equality yields the 1-periodicity of p with respect to z.

In order to obtain the existence result for (3.1), we compute

$$\int_{D_\varepsilon} (3.15) \cdot r\boldsymbol{\varphi}_j$$

for $\boldsymbol{\varphi}_j$, the unique function satisfying (3.4).

Using again the properties of I, we find

$$\rho_f \int_{D_\varepsilon} r\frac{\partial \mathbf{u}}{\partial t} \cdot \boldsymbol{\varphi}_j + \mu(\mathbf{u}, \boldsymbol{\varphi}_j)_{\widetilde{V}} + \mu \int_0^1 u_r/_{r=\varepsilon}\beta_j \mathrm{d}z$$
$$+ \varepsilon \int_0^1 p_{/r=\varepsilon}\beta_j \mathrm{d}z = \int_{D_\varepsilon} r\mathbf{f} \cdot \boldsymbol{\varphi}_j. \tag{3.17}$$

Passing to the limit in $(3.6)_2$ as $m, n \to \infty$, we get

$$\rho_f \int_{D_\varepsilon} r\frac{\partial \mathbf{u}}{\partial t} \cdot \boldsymbol{\varphi}_j + \mu I(\mathbf{u}, \boldsymbol{\varphi}_j) + a\varepsilon \int_0^1 \frac{\partial^2 d}{\partial t^2}\beta_j + \frac{b\varepsilon}{\delta}\int_0^1 \frac{\partial^2 d}{\partial z^2}\beta_j''$$
$$+ \frac{c\varepsilon}{\delta}\int_0^1 \frac{\partial d}{\partial z}\beta_j' + \frac{f\varepsilon}{\delta}\int_0^1 d\beta_j + \nu\varepsilon \int_0^1 \frac{\partial^3 d}{\partial z^2 \partial t}\beta_j''$$
$$= \int_{D_\varepsilon} r\mathbf{f} \cdot \boldsymbol{\varphi}_j + \varepsilon \int_0^1 g\beta_j. \tag{3.18}$$

Computing (3.18)–(3.17) and using the fact that $\{\beta_j\}_{j\in\mathbb{N}}$ is a basis, we obtain

$$a\int_0^1 \frac{\partial^2 d}{\partial t^2}\beta + \frac{b}{\delta}\int_0^1 \frac{\partial^2 d}{\partial z^2}\beta'' + \frac{c}{\delta}\int_0^1 \frac{\partial d}{\partial z}\beta' + \frac{f}{\delta}\int_0^1 d\beta + \nu\int_0^1 \frac{\partial^3 d}{\partial z^2 \partial t}\beta''$$
$$= \frac{2\mu}{\varepsilon}\int_0^1 u_r/_{r=\varepsilon}\beta + \int_0^1 p_{/r=\varepsilon}\beta + \int_0^1 g\beta \quad \forall\, \beta \in B_{0,\text{per}}. \tag{3.19}$$

The fact that the pair $(\mathbf{u}, d)$ verifies $(3.1)_1$ is a consequence of the following computation: $\int_{D_\varepsilon}(3.15) \cdot r\boldsymbol{\psi} + \varepsilon(3.19)$ for any $\boldsymbol{\psi} \in V^{\varepsilon}_{\text{per}}$, $\psi_r = \beta$ on Γ_ε.

From (3.1), (3.15), and (3.19) it is obvious that $(\mathbf{u}, p+h, d)$ satisfies Equations $(2.6)_{1,2,4}$ a.e. in $D_\varepsilon \times (0,T)$, where $h = h(t)$ is an arbitrary function.

The last step of the proof is the uniqueness of pressure. Let us suppose that $p + h_1$ and $p + h_2$ are two functions satisfying (3.15) and $(2.6)_4$. Since

$(2.6)_4$ uniquely gives the average value of the pressure, the uniqueness of this function is obtained and the proof is complete. □

For saying that the variational problem (3.1) represents the weak formulation of the system (2.6), it remains to prove the following assertion.

Proposition 3.2. *If* $\mathbf{u}$, p, *and* d *are smooth functions satisfying the coupled system* (2.6) *in a classical sense, then* $(\mathbf{u}, d)$ *is a solution of the variational problem* (3.1).

Proof. The assertion is obtained as follows. We compute

$$\int_{D_\varepsilon} ((2.6)_1 r\psi_z + (2.6)_2 r\psi_r$$

for any $\psi \in V^\varepsilon_{\text{per}}$, $\psi_r = \beta$ on Γ_ε, $\beta \in B_{0,\text{per}}$. We compute

$$\varepsilon \int_0^1 (2.6)_4 \beta$$

and add the two obtained equalities. Then we obtain the relation $(3.1)_1$ and the proof is complete. □

Remark 3.2. If we consider $(\mathbf{f}_1, g_1)$ and $(\mathbf{f}_2, g_2)$ two given data, we obtain for the difference of the corresponding solutions, $(\mathbf{u}_1 - \mathbf{u}_2, d_1 - d_2)$ some *a priori* estimates, given by (3.8) and (3.12).

As a consequence of this theorem, we can also obtain some estimates for the pressure.

Corollary 3.1. *Let* $(\mathbf{u}, p, d)$ *be a unique solution of* (2.6) *with the regularity given by Theorem* 3.1. *Then*

$$\begin{aligned} &\|\nabla p\|_{L^2(0,T;(L^2_r(D_\varepsilon))^2)} \\ &\quad \leqslant \frac{C}{\delta^{1/2}} \left(\|\mathbf{f}\|_{L^2(0,T;(L^2_r(D_\varepsilon))^2)} + \|g\|_{L^2((0,1)\times(0,T))} \right). \end{aligned} \tag{3.20}$$

Proof. From $(3.12)_1$ and (3.14) we get

$$\begin{aligned} &\|\Delta \mathbf{u}\|_{L^2(0,T;(L^2_r(D_\varepsilon))^2)} \\ &\quad \leqslant \frac{C}{\delta^{1/2}} \left(\|\mathbf{f}\|_{L^2(0,T;(L^2_r(D_\varepsilon))^2)} + \|g\|_{L^2((0,1)\times(0,T))} \right). \end{aligned} \tag{3.21}$$

Taking into account the regularity of p, established in Theorem 3.1, the relations (3.15), $(3.12)_1$ and (3.21) give the estimate (3.20) for the pressure. □

Corollary 3.2. *Under the assumptions of Theorem* 3.1, *the following estimates hold:*

$$\|\mathbf{u}\|_{L^\infty(0,T;(L_r^2(D_\varepsilon))^2)} \leqslant C\big(\|\mathbf{f}\|_{L^2(0,T;(L_r^2(D_\varepsilon))^2)} + \|g\|_{L^2((0,1)\times(0,T))}\big),$$

$$\left(\left\|\frac{\partial u_z}{\partial z}\right\|^2_{L^2(0,T;L_r^2(D_\varepsilon))} + \left\|\frac{\partial u_z}{\partial r}\right\|^2_{L^2(0,T;L_r^2(D_\varepsilon))} + \left\|\frac{\partial u_r}{\partial z}\right\|^2_{L^2(0,T;L_r^2(D_\varepsilon))}\right.$$

$$\left. + \left\|\frac{\partial u_r}{\partial r}\right\|^2_{L^2(0,T;L_r^2(D_\varepsilon))} + \left\|\frac{u_r}{r}\right\|^2_{L^2(0,T;(L_r^2 D_\varepsilon))}\right)^{1/2}$$

$$\leqslant C\big(\|\mathbf{f}\|_{L^2(0,T;(L_r^2(D_\varepsilon))^2)} + \|g\|_{L^2((0,1)\times(0,T))}\big), \tag{3.22}$$

$$\varepsilon^{1/2}\left\|\frac{\partial d}{\partial t}\right\|_{L^\infty(0,T;L^2(0,1))} \leqslant C\big(\|\mathbf{f}\|_{L^2(0,T;(L_r^2(D_\varepsilon))^2)} + \|g\|_{L^2((0,1)\times(0,T))}\big),$$

$$\left(\frac{\varepsilon}{\delta}\right)^{1/2}\left\|\frac{\partial^2 d}{\partial z^2}\right\|_{L^\infty(0,T;L^2(0,1))} \leqslant C\big(\|\mathbf{f}\|_{(L^2(0,T;(L_r^2(D_\varepsilon))^2)} + \|g\|_{L^2((0,1)\times(0,T))}\big),$$

$$\varepsilon^{1/2}\left\|\frac{\partial^3 d}{\partial z^2 \partial t}\right\|_{L^2((0,1)\times(0,T))} \leqslant C\big(\|\mathbf{f}\|_{L^2(0,T;(L_r^2(D_\varepsilon))^2)} + \|g\|_{L^2((0,1)\times(0,T))}\big)$$

and

$$\left\|\frac{\partial \mathbf{u}}{\partial t}\right\|_{L^2(0,T;(L_r^2(D_\varepsilon))^2)} \leqslant \frac{C}{\delta^{1/2}}\big(\|\mathbf{f}\|_{L^2(0,T;(L_r^2(D_\varepsilon))^2)} + \|g\|_{L^2((0,1)\times(0,T))}\big),$$

$$\left(\left\|\frac{\partial u_z}{\partial z}\right\|^2_{L^\infty(0,T;(L_r^2 D_\varepsilon))} + \left\|\frac{\partial u_z}{\partial r}\right\|^2_{L^\infty(0,T;L_r^2(D_\varepsilon))} + \left\|\frac{\partial u_r}{\partial z}\right\|^2_{L^\infty(0,T;L_r^2(D_\varepsilon))}\right.$$

$$\left. + \left\|\frac{\partial u_r}{\partial r}\right\|^2_{L^\infty(0,T;L_r^2(D_\varepsilon))} + \left\|\frac{u_r}{r}\right\|^2_{L^\infty(0,T;L_r^2(D_\varepsilon))}\right)^{1/2}$$

$$\leqslant \frac{C}{\delta^{1/2}}\big(\|\mathbf{f}\|_{L^2(0,T;(L_r^2(D_\varepsilon))^2)} + \|g\|_{L^2((0,1)\times(0,T))}\big), \tag{3.23}$$

$$\varepsilon^{1/2}\left\|\frac{\partial^2 d}{\partial t^2}\right\|_{L^2((0,1)\times(0,T))} \leqslant \frac{C}{\delta^{1/2}}\big(\|\mathbf{f}\|_{L^2(0,T;(L_r^2(D_\varepsilon))^2)} + \|g\|_{L^2((0,1)\times(0,T))}\big),$$

$$\varepsilon^{1/2}\left\|\frac{\partial^3 d}{\partial z^2 \partial t}\right\|_{L^\infty(0,T;L^2(0,1))} \leqslant \frac{C}{\delta^{1/2}}\big(\|\mathbf{f}\|_{L^2(0,T;(L_r^2(D_\varepsilon))^2)} + \|g\|_{L^2((0,1)\times(0,T))}\big).$$

4 Asymptotic Analysis of the Problem

In order to approximate the solution $(\mathbf{u}, p, d)$ of (2.6) with more regular functions which can be determined, we suppose that the small parameters characterizing our problem satisfy the relation $\delta = \varepsilon^\gamma$, with $\gamma \in \mathbb{N}$.

For obtaining the desired regularity of the asymptotic solution and some initial conditions, we suppose that the data $\mathbf{f}$ and g have the following properties:

$$\begin{aligned}
&\mathbf{f} = f_z(z,t)\mathbf{e}_z,\ f_z \in C^\infty([0,1]\times[0,T]), f_z \text{ 1-periodic in } z,\\
&g \in C^\infty([0,1]\times[0,T]), g \text{ 1-periodic in } z, \langle g\rangle(t) = 0 \quad t\in[0,T],\\
&\exists\ 0 < t^* < T : f_z(z,t) = g(z,t) = 0 \quad \forall\ (z,t) \in [0,1]\times[0,t^*),
\end{aligned} \tag{4.1}$$

where

$$\langle \alpha\rangle(t) = \int_0^1 \alpha(z,t)\,\mathrm{d}z.$$

We consider the same asymptotic expansions as in the two-dimensional periodic case (cf. [12]), i.e.,

$$\begin{aligned}
u_z^{(K)}(z,r,t) &= \sum_{j=0}^{K} \varepsilon^{j+2} u_{z,j}\big(z,\frac{r}{\varepsilon},t\big),\\
u_r^{(K)}(z,r,t) &= \sum_{j=0}^{K} \varepsilon^{j+3} u_{r,j}\big(z,\frac{r}{\varepsilon},t\big),\\
p^{(K)}(z,r,t) &= \sum_{j=0}^{K} \varepsilon^{j+1} p_j\big(z,\frac{r}{\varepsilon},t\big) + \sum_{j=0}^{K} \varepsilon^{j} q_j\big(z,t\big),\\
d^{(K)}(z,t) &= \sum_{j=0}^{K} \varepsilon^{j+\gamma} d_j\big(z,t\big),
\end{aligned} \tag{4.2}$$

where $\mathbf{u}_j$, p_j, q_j, d_j are smooth functions, 1-periodic in z.

A standard scaling (cf. [12]) leads to the following problem for $u_{z,j}$, $u_{r,j}$, p_j, q_j, d_j, for $s = r/\varepsilon$:

$$\begin{aligned}
&-\mu\Big(\frac{\partial^2 u_{z,j}}{\partial s^2} + \frac{1}{s}\frac{\partial u_{z,j}}{\partial s}\Big) + \frac{\partial q_j}{\partial z} = f_z\delta_{j0} - \frac{\partial p_{j-1}}{\partial z}\\
&\qquad\qquad + \mu\frac{\partial^2 u_{z,j-2}}{\partial z^2} - \rho_f\frac{\partial u_{z,j-2}}{\partial t},\\
&\frac{\partial p_j}{\partial s} = -\rho_f\frac{\partial u_{r,j-3}}{\partial t} + \mu\Big(\frac{\partial^2 u_{r,j-3}}{\partial z^2} + \frac{\partial^2 u_{r,j-1}}{\partial s^2}\\
&\qquad\qquad + \frac{1}{s}\frac{\partial u_{r,j-1}}{\partial s} - \frac{1}{s^2}u_{r,j-1}\Big),\\
&\frac{\partial u_{z,j}}{\partial z} + \frac{\partial u_{r,j}}{\partial s} + \frac{1}{s}u_{r,j} = 0,\\
&b\frac{\partial^4 d_j}{\partial z^4} - c\frac{\partial^2 d_j}{\partial z^2} + f d_j - q_j = g\delta_{j0} - a\frac{\partial^2 d_{j-\gamma}}{\partial t^2}\\
&\qquad\qquad - \nu\frac{\partial^5 d_{j-\gamma}}{\partial z^4\partial t} + p_{j-1}/_{s=1} + 2\mu u_{r,j-2}/_{s=1},\\
&\mathbf{u}_j, p_j, q_j, d_j \text{ 1-periodic in } z,
\end{aligned} \tag{4.3}$$

$$
\begin{aligned}
& u_{r,j}(z,0,t) = 0, \\
& u_{z,j}(z,1,t) = 0, \ u_{r,j}(z,1,t) = \frac{\partial d_{j-\gamma+3}}{\partial t}, \\
& \langle d_j \rangle (t) = 0, \\
& \langle q_j \rangle (t) = -\langle p_{j-1}/_{s=1} \rangle (t) - 2\mu \langle u_{r,j-2}/_{s=1} \rangle (t).
\end{aligned}
$$

The last relation is a consequence of $(4.3)_4$, $(4.3)_5$ and $(4.3)_8$. The next lemma gives expressions of the functions $\mathbf{u}_j$, p_j, q_j, d_j.

Lemma 4.1. *The unknowns of the system* (4.3) *are given by*

$$
\begin{aligned}
u_{z,j} &= \frac{1-s^2}{4\mu}\Big(f_z\delta_{j0} - \frac{\partial q_j}{\partial z}\Big) - \frac{1}{\mu}\int_s^1 \frac{1}{\tau}\int_0^\tau \sigma A_{j-1}(z,\sigma,t)\mathrm{d}\sigma\mathrm{d}\tau, \\
u_{r,j} &= \frac{s(1-s^2/2)}{8\mu}\Big(-\frac{\partial f_z}{\partial z}\delta_{j0} + \frac{\partial^2 q_j}{\partial z^2}\Big) \\
&\quad + \frac{1}{\mu s}\int_0^s \lambda \int_\lambda^1 \frac{1}{\tau}\int_0^\tau \sigma \frac{\partial A_{j-1}}{\partial z}(z,\sigma,t)\mathrm{d}\sigma\mathrm{d}\tau\mathrm{d}\lambda, \\
p_j &= \int_0^s \Big(-\rho_f\frac{\partial u_{r,j-3}}{\partial t} + \mu\Big(\frac{\partial^2 u_{r,j-3}}{\partial z^2} + \frac{\partial^2 u_{r,j-1}}{\partial s^2} \\
&\quad + \frac{1}{s}\frac{\partial u_{r,j-1}}{\partial s} - \frac{1}{s^2}u_{r,j-1}\Big)\Big)\mathrm{d}\sigma, \qquad (4.4) \\
&\frac{1}{16\mu}\Big(\frac{\partial^2 q_j}{\partial z^2} - \frac{\partial f_z}{\partial z}\delta_{j0}\Big) \\
&\quad + \frac{1}{\mu}\int_0^1 \lambda\int_\lambda^1 \frac{1}{\tau}\int_0^\tau \sigma\frac{\partial A_{j-1}}{\partial z}(z,\sigma,t)\mathrm{d}\sigma\mathrm{d}\tau\mathrm{d}\lambda = \frac{\partial d_{j+3-\gamma}}{\partial t}, \\
&b\frac{\partial^4 d_j}{\partial z^4} - c\frac{\partial^2 d_j}{\partial z^2} + f d_j - q_j = g\delta_{j0} - a\frac{\partial^2 d_{j-\gamma}}{\partial t^2} \\
&\quad - \nu\frac{\partial^5 d_{j-\gamma}}{\partial z^4\partial t} + p_{j-1}/_{s=1} + 2\mu u_{r,j-2}/_{s=1}, \\
&\langle d_j\rangle(t) = 0, \\
&\langle q_j\rangle(t) = -\langle p_{j-1}/_{s=1}\rangle(t), \\
&\mathbf{u}_j, p_j, q_j, d_j \ \text{1-periodic in } z,
\end{aligned}
$$

where

$$
A_{j-1} = \frac{\partial p_{j-1}}{\partial z} - \mu\frac{\partial^2 u_{z,j-2}}{\partial z^2} + \rho_f\frac{\partial u_{z,j-2}}{\partial t}.
$$

Proof. The relation $(4.4)_1$ is as follows. We multiply $(4.3)_1$ by s and integrate from 0 to s. Then we integrate the obtained relation from s to 1 and use $(4.3)_7$. To find $u_{r,j}$, we substitute $(4.4)_1$ in $(4.3)_3$, multiply by s, and integrate from 0 to s. Thus, we obtain $(4.4)_2$. The expression $(4.4)_3$ for p_j is obtained by integrating $(4.3)_2$ from 0 to s (it is assumed that all the functions depending only on z and t are contained in q_j). Finally, $(4.4)_7$ is obtained from $(4.4)_2$ and the 1-periodicity in z. The expression $(4.4)_4$ follows from $(4.4)_2$ and $(4.3)_7$. □

Remark 4.1. The choice of the unknown function in $(4.4)_4$ depends on the value of γ. In the sequel, we consider three cases: $\gamma > 3$, $\gamma = 3$, and $\gamma < 3$. In each case, we solve the system (4.4) and analyze the leading terms.

Case $\gamma > 3$ (*high rigidity of the wall*). In this case, the unknown of $(4.4)_4$ is q_j. For $j = 0$, the system (4.4) leads to an asymptotic solution of order K given b

$$\begin{aligned}
u_z^{(K)}(z,r,t) &= \varepsilon^2 \frac{\langle f_z\rangle(t)}{4\mu}\Big(1-\frac{r^2}{\varepsilon^2}\Big) + \mathcal{O}(\varepsilon^3),\\
u_r^{(K)}(z,r,t) &= \mathcal{O}(\varepsilon^4),\\
p^{(K)}(z,r,t) &= \int_0^z \widetilde{f}_z(\zeta,t)\mathrm{d}\zeta - \Big\langle \int_0^z \widetilde{f}_z(\zeta,t)\mathrm{d}\zeta\Big\rangle + \mathcal{O}(\varepsilon),\\
d^{(K)}(z,t) &= \varepsilon^\gamma d_0 + \mathcal{O}(\varepsilon^{\gamma+1}),
\end{aligned} \tag{4.5}$$

where $\widetilde{f}_z(z,t) = f_z(z,t) - \langle f_z\rangle(t)$ and d_0 is the unique 1-periodic in z solution of the problem

$$\begin{aligned}
&b\frac{\partial^4 d_0}{\partial z^4} - c\frac{\partial^2 d_0}{\partial z^2} + f d_0 = g + \int_0^z \widetilde{f}_z(\zeta,t)\mathrm{d}\zeta - \Big\langle \int_0^z \widetilde{f}_z(\zeta,t)\mathrm{d}\zeta\Big\rangle,\\
&\langle d_0\rangle(t) = 0.
\end{aligned} \tag{4.6}$$

Case $\gamma < 3$ (*low rigidity of the wall*). For $\gamma \in \{1,2\}$ the unknown of $(4.4)_4$ is $d_{j+3-\gamma}$, while q_j is defined from the previous approximations. The asymptotic solution of order K has the following expression:

$$\begin{aligned}
u_z^{(K)}(z,r,t) &= \varepsilon^2\Big(\frac{\partial g}{\partial z} + f_z\Big)\frac{1-r^2/\varepsilon^2}{4\mu} + \mathcal{O}(\varepsilon^3),\\
u_r^{(K)}(z,r,t) &= -\varepsilon^3\Big(\frac{\partial^2 g}{\partial z^2} + \frac{\partial f_z}{\partial z}\Big)\frac{1}{8\mu}\frac{r}{\varepsilon}\Big(1-\frac{r^2}{2\varepsilon^2}\Big) + \mathcal{O}(\varepsilon^4),\\
p^{(K)}(z,r,t) &= -g(z,t) + \mathcal{O}(\varepsilon),\\
d^{(K)}(z,t) &= -\frac{\varepsilon^3}{16\mu}\int_0^t \Big(\frac{\partial^2 g}{\partial z^2} + \frac{\partial f_z}{\partial z}\Big)(z,\tau)\mathrm{d}\tau + \mathcal{O}(\varepsilon^4).
\end{aligned} \tag{4.7}$$

Case $\gamma = 3$ (*transitory case*). In this case, Equation $(4.4)_4$ contains two unknowns q_j and d_j. By $(4.4)_{4,5}$, d_j satisfies the parabolic problem

$$
\begin{aligned}
16\mu\frac{\partial d_j}{\partial t} - b\frac{\partial^6 d_j}{\partial z^6} + c\frac{\partial^4 d_j}{\partial z^4} - f\frac{\partial^2 d_j}{\partial z^2} &= -\Big(\frac{\partial^2 g}{\partial z^2} + \frac{\partial f_z}{\partial z}\Big)\delta_{j0} \\
&\quad + a\frac{\partial^4 d_{j-3}}{\partial z^2 \partial t^2} + \nu\frac{\partial^7 d_{j-3}}{\partial z^6 \partial t} - \frac{\partial^2 p_{j-1}}{\partial z^2}/_{s=1} - 2\mu\frac{\partial^2 u_{r,j-2}}{\partial z^2}/_{s=1} \\
&+ 16\int_0^1 s\int_s^1 \frac{1}{\tau}\int_0^\tau \sigma\frac{\partial A_{j-1}}{\partial z}\mathrm{d}\sigma\mathrm{d}\tau\mathrm{d}s, \qquad (4.8) \\
&d_j \text{ 1-periodic in } z, \quad \langle d_j\rangle(t) = 0, \quad d_j(z,0) = 0.
\end{aligned}
$$

The asymptotic solution of order K is given by

$$
\begin{aligned}
u_z^{(K)}(z,r,t) &= \frac{\varepsilon^2 - r^2}{4\mu}\Big(f_z - \frac{\partial q_0}{\partial z}\Big) + \mathcal{O}(\varepsilon^3), \\
u_r^{(K)}(z,r,t) &= \frac{r(\varepsilon^2 - r^2)}{8\mu}\Big(\frac{\partial^2 q_0}{\partial z^2} - \frac{\partial f_z}{\partial z}\Big) + \mathcal{O}(\varepsilon^4), \qquad (4.9) \\
p^{(K)}(z,r,t) &= q_0(z,t) + \mathcal{O}(\varepsilon), \\
d^{(K)}(z,t) &= \varepsilon^3 d_0(z,t) + \mathcal{O}(\varepsilon^4),
\end{aligned}
$$

where d_0 is a unique solution of the problem

$$
\begin{aligned}
&16\mu\frac{\partial d_0}{\partial t} - b\frac{\partial^6 d_0}{\partial z^6} + c\frac{\partial^4 d_0}{\partial z^4} - f\frac{\partial^2 d_0}{\partial z^2} = -\Big(\frac{\partial^2 g}{\partial z^2} + \frac{\partial f_z}{\partial z}\Big)\delta_{j0}, \qquad (4.10) \\
&d_0 \text{ 1-periodic in } z, \quad \langle d_0\rangle(t) = 0, \quad d_0(z,0) = 0.
\end{aligned}
$$

Remark 4.2. Comparing (4.5), (4.7), and (4.9) with the asymptotic solution in the two-dimensional case (cf. [12]), we note that the leading terms obtained in this paper have the same expressions as in the 2D problem.

To conclude the section, we show that the approximations of the asymptotic solution possess the same property as the exact solution.

Proposition 4.1. *The functions $u_{z,j}$ and $u_{r,j}$ satisfy the following relations:*

$$
\begin{aligned}
&\lim_{s\to 0}\frac{\partial u_{z,j}}{\partial s} = 0, \\
&\lim_{s\to 0}\Big(\frac{1}{s}\frac{\partial u_{r,j}}{\partial s} - \frac{1}{s^2}u_{r,j}\Big) = \textit{finite}, \qquad (4.11) \\
&\lim_{s\to 0}\frac{u_{r,j}}{s} = \textit{finite}.
\end{aligned}
$$

Proof. We obtain $(4.11)_1$ as follows:

$$
\begin{aligned}
\lim_{s\to 0}\frac{\partial u_{z,j}}{\partial s} &= \lim_{s\to 0}\Big(-\frac{s}{2\mu}\Big(f_z\delta_{j0} - \frac{\partial q_j}{\partial z}\Big) + \frac{1}{\mu s}\int_0^s \sigma A_{j-1}(z,\sigma,t)\mathrm{d}\sigma\Big) \\
&= \lim_{s\to 0}\frac{1}{\mu}sA_{j-1}(z,s,t) = 0.
\end{aligned}
$$

The other two relations of (4.11) are an obvious consequence of the l'Hospital theorem. □

5 The Justification of the Asymptotic Solution

The purpose of this section is to justify the asymptotic solution (4.2) by obtaining a small error between the exact solution of (2.6) and (4.2). We obtain, for the asymptotic solution of order K, a problem of the same type as (2.6), but with different right-hand sides in $(2.6)_{1,2,4,7}$. In order to write the problem satisfied by the asymptotic solution, we introduce the notation:

$$R_{z,K}=\varepsilon^{K+1}\Big(\Big(-\rho_f\frac{\partial u_{z,K-1}}{\partial t}+\mu\frac{\partial^2 u_{z,K-1}}{\partial z^2}-\frac{\partial p_K}{\partial z}\Big)+\varepsilon\Big(-\rho_f\frac{\partial u_{z,K}}{\partial t}+\mu\frac{\partial^2 u_{z,K}}{\partial z^2}\Big)\Big), \tag{5.1}$$

$$R_{r,K}=\varepsilon^{K+1}\Big(\Big(-\rho_f\frac{\partial u_{r,K-2}}{\partial t}+\mu\Big(\frac{\partial^2 u_{r,K-2}}{\partial z^2}+\frac{\partial^2 u_{r,K}}{\partial s^2}+\frac{1}{s}\frac{\partial u_{r,K}}{\partial s}-\frac{1}{s^2}u_{r,K}\Big)\Big)+\varepsilon\Big(-\rho_f\frac{\partial u_{r,K-1}}{\partial t}+\mu\frac{\partial^2 u_{r,K-1}}{\partial z^2}\Big)+\varepsilon^2\Big(-\rho_f\frac{\partial u_{r,K}}{\partial t}+\mu\frac{\partial^2 u_{r,K}}{\partial z^2}\Big)\Big), \tag{5.2}$$

$$R_{d,K}=\begin{cases}\varepsilon^{K+1}\Big(\Big(-a\dfrac{\partial^2 d_K}{\partial t^2}+p_K/_{s=1}+2\mu u_{r,K-1}/_{s=1}\Big)+2\mu\varepsilon u_{r,K}\Big), & \gamma=1,\\[2ex] \varepsilon^{K+1}\Big(\Big(-a\dfrac{\partial^2 d_{K-1}}{\partial t^2}+p_K/_{s=1}+2\mu u_{r,K-1}/_{s=1}\Big)+\varepsilon\Big(-a\dfrac{\partial^2 d_K}{\partial t^2}+2\mu u_{r,K}/_{s=1}\Big)\Big), & \gamma=2,\\[2ex] \varepsilon^{K+1}\Big(\Big(-a\dfrac{\partial^2 d_{K+1-\gamma}}{\partial t^2}+p_K/_{s=1}+2\mu u_{r,K-1}\Big)+\varepsilon\Big(-a\dfrac{\partial^2 d_{K+2-\gamma}}{\partial t^2}+2\mu u_{r,K}/_{s=1}\Big)+\varepsilon^2\Big(-a\dfrac{\partial^2 d_{K+3-\gamma}}{\partial t^2}\Big)+\ldots+\varepsilon^{\gamma-1}\Big(-a\dfrac{\partial^2 d_K}{\partial t^2}\Big)\Big), & \gamma\geqslant 3,\end{cases} \tag{5.3}$$

$$R_{\Gamma,K} = \begin{cases} \varepsilon^{K+\gamma+1}(u_{r,K+\gamma-2}/_{s=1} + \ldots + \varepsilon^{2-\gamma} u_{r,K}/_{s=1}), & \gamma < 3, \\ 0, & \gamma = 3, \\ -\varepsilon^{K+4}\Big(\dfrac{\partial d_{K+4-\gamma}}{\partial t} + \ldots + \varepsilon^{\gamma-4}\dfrac{\partial d_K}{\partial t}\Big), & \gamma > 3. \end{cases} \tag{5.4}$$

Obvious computations lead to the following problem for the asymptotic solution of order K:

$$\begin{aligned}
&\rho_f \frac{\partial u_z^{(K)}}{\partial t} - \mu\Big(\frac{\partial^2 u_z^{(K)}}{\partial z^2} + \frac{\partial^2 u_z^{(K)}}{\partial r^2} + \frac{1}{r}\frac{\partial u_z^{(K)}}{\partial r}\Big) \\
&\qquad\qquad + \frac{\partial p^{(K)}}{\partial z} = f_z + R_{z,K}, \\
&\rho_f \frac{\partial u_r^{(K)}}{\partial t} - \mu\Big(\frac{\partial^2 u_r^{(K)}}{\partial z^2} + \frac{\partial^2 u_r^{(K)}}{\partial r^2} + \frac{1}{r}\frac{\partial u_r^{(K)}}{\partial r} - \frac{1}{r^2}u_r^{(K)}\Big) \\
&\qquad\qquad + \frac{\partial p^{(K)}}{\partial r} = f_r + R_{r,K}, \\
&\frac{\partial u_z^{(K)}}{\partial z} + \frac{\partial u_r^{(K)}}{\partial r} + \frac{1}{r}u_r^{(K)} = 0, \\
&a\frac{\partial^2 d^{(K)}}{\partial t^2} + \frac{b}{\varepsilon^\gamma}\frac{\partial^4 d^{(K)}}{\partial z^4} - \frac{c}{\varepsilon^\gamma}\frac{\partial^2 d^{(K)}}{\partial z^2} + \frac{f}{\varepsilon^\gamma}d^{(K)} + \nu\frac{\partial^5 d^{(K)}}{\partial z^4 \partial t} \\
&\qquad\qquad = \Big(p^{(K)} + \frac{2\mu}{\varepsilon}u_r^{(K)}\Big)/_{r=\varepsilon} + g + R_{d,K}, \\
&u_z^{(K)}, u_r^{(K)}, p^{(K)}, d^{(K)} \text{ 1-periodic in } z, \\
&u_r^{(K)}(z,0,t) = 0, \\
&\mathbf{u}^{(\mathbf{K})}(z,\varepsilon,t) = \frac{\partial d^{(K)}}{\partial t}(z,t)\mathbf{e}_r + R_{\Gamma,K}\mathbf{e}_r, \\
&\mathbf{u}^{(\mathbf{K})}(z,r,0) = \mathbf{0}, \\
&d^{(K)}(z,0) = \frac{\partial d^{(K)}}{\partial t}(z,0) = 0, \\
&\int_0^1 d^{(K)}(z,t)\mathrm{d}z = 0.
\end{aligned} \tag{5.5}$$

The first result which gives the error between the exact and the asymptotic solution of order K is obtained by means of the *a priori* estimates (3.22), (3.23), and (3.20).

Remark 5.1. The estimates (3.8), (3.12), and (3.20) cannot be applied directly since the boundary conditions $(2.6)_7$ and $(5.5)_7$ are different. So, we have to define a modified asymptotic solution, $(\mathbf{u}^{(\mathbf{K})}, p^{(K)}, D^{(K)})$, which satisfies

$$\mathbf{u}^{(K)}(z,\varepsilon,t) = \frac{\partial D^{(K)}}{\partial t}(z,t)\mathbf{e}_r. \tag{5.6}$$

The problem satisfied by the modified asymptotic solution is given by the following obvious lemma.

Lemma 5.1. *Let* $D^{(K)} : [0,1) \times [0,T] \mapsto \mathbb{R}$ *be defined by*

$$D^{(K)}(z,t) = d^{(K)}(z,t) + \int_0^t R_{\Gamma,K}(z,\tau)\mathrm{d}\tau. \tag{5.7}$$

Then $(\mathbf{u}^{(K)}, p^{(K)}, D^{(K)})$ *is a solution of the problem*

$$\begin{aligned}
&\rho_f \frac{\partial u_z^{(K)}}{\partial t} - \mu\Big(\frac{\partial^2 u_z^{(K)}}{\partial z^2} + \frac{\partial^2 u_z^{(K)}}{\partial r^2} + \frac{1}{r}\frac{\partial u_z^{(K)}}{\partial r}\Big) \\
&\qquad\qquad\qquad + \frac{\partial p^{(K)}}{\partial z} = f_z + R_{z,K}, \\
&\rho_f \frac{\partial u_r^{(K)}}{\partial t} - \mu\Big(\frac{\partial^2 u_r^{(K)}}{\partial z^2} + \frac{\partial^2 u_r^{(K)}}{\partial r^2} + \frac{1}{r}\frac{\partial u_r^{(K)}}{\partial r} - \frac{1}{r^2}u_r^{(K)}\Big) \\
&\qquad\qquad\qquad + \frac{\partial p^{(K)}}{\partial r} = f_r + R_{r,K}, \\
&\frac{\partial u_z^{(K)}}{\partial z} + \frac{\partial u_r^{(K)}}{\partial r} + \frac{1}{r}u_r^{(K)} = 0, \\
&a\frac{\partial^2 D^{(K)}}{\partial t^2} + \frac{b}{\varepsilon^\gamma}\frac{\partial^4 D^{(K)}}{\partial z^4} - \frac{c}{\varepsilon^\gamma}\frac{\partial^2 D^{(K)}}{\partial z^2} + \frac{f}{\varepsilon^\gamma}D^{(K)} + \nu\frac{\partial^5 D^{(K)}}{\partial z^4 \partial t} \\
&\qquad - \big(p^{(K)} + \frac{2\mu}{\varepsilon}u_r^{(K)}\big)/_{r=\varepsilon} = g + R_{d,K} + a\frac{\partial R_{\Gamma,K}}{\partial t} \\
&\qquad + \frac{b}{\varepsilon^\gamma}\int_0^t \frac{\partial^4 R_{\Gamma,K}}{\partial z^4}(z,\tau)\mathrm{d}\tau - \frac{c}{\varepsilon^\gamma}\int_0^t \frac{\partial^2 R_{\Gamma,K}}{\partial z^2}(z,\tau)\mathrm{d}\tau \\
&\qquad + \frac{f}{\varepsilon^\gamma}\int_0^t R_{\Gamma,K}(z,\tau)\mathrm{d}\tau + \nu\frac{\partial^4 R_{\Gamma,K}}{\partial z^4}, \\
&u_z^{(K)}, u_r^{(K)}, p^{(K)}, D^{(K)} \quad \text{1-periodic in } z, \\
&u_r^{(K)}(z,0,t) = 0, \\
&\mathbf{u}^{(K)}(z,\varepsilon,t) = \frac{\partial D^{(K)}}{\partial t}(z,t)\mathbf{e}_r, \\
&\mathbf{u}^{(K)}(z,r,0) = \mathbf{0}, \\
&D^{(K)}(z,0) = \frac{\partial D^{(K)}}{\partial t}(z,0) = 0, \\
&\int_0^1 D^{(K)}(z,t)\mathrm{d}z = 0.
\end{aligned} \tag{5.8}$$

The first error estimates between the exact and the asymptotic solutions are established by means of (3.22), (3.23), and (3.20).

Lemma 5.2. *Let* $(\mathbf{u}^{(K)}, p^{(K)}, d^{(K)})$ *be the asymptotic solution given by* (4.2) *and* $(\mathbf{u}, p, d)$ *the unique solution of* (2.6). *Then the following estimates hold:*

$$\|\mathbf{u} - \mathbf{u}^{(K)}\|_{L^\infty(0,T;(L^2_r(D_\varepsilon))^2)} = \mathcal{O}(\varepsilon^{\min(K+1,K+4-\gamma)}),$$

$$\Bigg(\left\|\frac{\partial}{\partial z}(u_z - u_z^{(K)})\right\|^2_{L^2(0,T;L^2_r(D_\varepsilon))} + \left\|\frac{\partial}{\partial r}(u_z - u_z^{(K)})\right\|^2_{L^2(0,T;L^2_r(D_\varepsilon))}$$
$$+ \left\|\frac{\partial}{\partial z}(u_r - u_r^{(K)})\right\|^2_{L^2(0,T;L^2_r(D_\varepsilon))} + \left\|\frac{\partial}{\partial r}(u_r - u_r^{(K)})\right\|^2_{L^2(0,T;L^2_r(D_\varepsilon))}$$
$$+ \left\|\frac{(u_r - u_r^{(K)})}{r}\right\|^2_{L^2(0,T;L^2_r(D_\varepsilon))}\Bigg)^{1/2} = \mathcal{O}(\varepsilon^{\min(K+1,K+4-\gamma)}),$$

$$\left\|\frac{\partial}{\partial t}(d - d^{(K)})\right\|_{L^\infty(0,T;L^2(0,1))} = \mathcal{O}(\varepsilon^{\min(K+1,K+4-\gamma)-1/2}), \tag{5.9}$$

$$\left\|\frac{\partial^2}{\partial t^2}(d - d^{(K)})\right\|_{L^2((0,1)\times(0,T))} = \mathcal{O}(\varepsilon^{\min(K+1,K+4-\gamma)-(\gamma+1)/2}),$$

$$\left\|\frac{\partial^2}{\partial z^2}(d - d^{(K)})\right\|_{L^\infty(0,T;L^2(0,1))} = \mathcal{O}(\varepsilon^{\min(K+1,K+4-\gamma)+(\gamma-1)/2}),$$

$$\|\nabla(p - p^{(K)})\|_{L^2(0,T;(L^2_r(D_\varepsilon))^2)} = \mathcal{O}(\varepsilon^{\min(K+4-\gamma,K+1)-\gamma/2}).$$

Proof. Taking into account the right-hand sides of $(2.6)_{1,2,4}$ and $(5.8)_{1,2,4}$ respectively, and denoting their difference by $\widehat{\mathbf{f}}$ and $\widehat{g}$, we get

$$\|\widehat{\mathbf{f}}\|^2_{L^2(0,T;(L^2_r(D_\varepsilon))^2)} + \|\widehat{g}\|^2_{L^2((0,1)\times(0,T))}$$
$$= \|R_{z,K}\|^2_{L^2(0,T;L^2_r(D_\varepsilon))} + \|R_{r,K}\|^2_{L^2(0,T;L^2_r(D_\varepsilon))} + \|R_{d,K}\|^2_{L^2((0,1)\times(0,T))}$$
$$+ \Bigg\| a\frac{\partial R_{\Gamma,K}}{\partial t} + \frac{b}{\varepsilon^\gamma}\int_0^t \frac{\partial^4 R_{\Gamma,K}}{\partial z^4}(z,\tau)\mathrm{d}\tau - \frac{c}{\varepsilon^\gamma}\int_0^t \frac{\partial^2 R_{\Gamma,K}}{\partial z^2}(z,\tau)\mathrm{d}\tau$$
$$+ \frac{f}{\varepsilon^\gamma}\int_0^t R_{\Gamma,K}(z,\tau)\mathrm{d}\tau + \nu\frac{\partial^4 R_{\Gamma,K}}{\partial z^4}\Bigg\|^2_{L^2((0,1)\times(0,T))}.$$

From the definitions (5.1)–(5.4) we get

$$\|\widehat{\mathbf{f}}\|_{L^2(0,T;(L^2_r(D_\varepsilon))^2)} + \|\widehat{g}\|_{L^2((0,1)\times(0,T))} = \mathcal{O}(\varepsilon^{\min(K+1,K+4-\gamma)}).$$

Hence the relations (5.9) are a consequence of (3.22), (3.23), and (3.20) and the proof is complete. □

In the last theorem of the paper, we improve the previous estimates.

Theorem 5.1. *The error between the exact solution of* (2.6) *and the asymptotic solution defined by* (4.2) *is given by*

$$\|\mathbf{u}-\mathbf{u}^{(L)}\|_{L^\infty(0,T;(L^2_r(D_\varepsilon))^2)} = \mathcal{O}(\varepsilon^{L+7/2}),$$

$$\Bigg(\Big\|\frac{\partial}{\partial z}(u_z-u_z^{(L)})\Big\|_{L^2(0,T;L^2_r(D_\varepsilon))} + \Big\|\frac{\partial}{\partial r}(u_z-u_z^{(K)})\Big\|^2_{L^2(0,T;L^2_r(D_\varepsilon))} + \Big\|\frac{\partial}{\partial z}(u_r-u_r^{(K)})\Big\|^2_{L^2(0,T;L^2_r(D_\varepsilon))} + \Big\|\frac{\partial}{\partial r}(u_r-u_r^{(L)})\Big\|^2_{L^2(0,T;L^2_r(D_\varepsilon))} + \Big\|\frac{(u_r-u_r^{(K)})}{r}\Big\|^2_{L^2(0,T;L^2_r(D_\varepsilon))}\Bigg)^{1/2} = \mathcal{O}(\varepsilon^{L+5/2}),$$

$$\Big\|\frac{\partial}{\partial t}(d-d^{(L)})\Big\|_{L^\infty(0,T;L^2(0,1))} = \mathcal{O}(\varepsilon^{L+\gamma+1}), \tag{5.10}$$

$$\Big\|\frac{\partial^2}{\partial t^2}(d-d^{(L)})\Big\|_{L^2((0,1)\times(0,T))} = \mathcal{O}(\varepsilon^{L+\gamma+1}),$$

$$\Big\|\frac{\partial^2}{\partial z^2}(d-d^{(L)})\Big\|_{L^\infty(0,T;L^2(0,1))} = \mathcal{O}(\varepsilon^{L+\gamma+1}),$$

$$\|\nabla(p-p^{(L)})\|_{L^2(0,T;(L^2_r(D_\varepsilon))^2)} = \mathcal{O}(\varepsilon^{L+1}).$$

Proof. Let L be a fixed integer, and let $K \gg L$. Then

$$\begin{aligned}&\|\mathbf{u}-\mathbf{u}^{(L)}\|_{L^\infty(0,T;(L^2_r(D_\varepsilon))^2)}\\ &\quad\leqslant \|\mathbf{u}-\mathbf{u}^{(K)}\|_{L^\infty(0,T;(L^2_r(D_\varepsilon))^2)} + \|\mathbf{u}^{(L)}-\mathbf{u}^{(K)}\|_{L^\infty(0,T;(L^2_r(D_\varepsilon))^2)}\\ &\quad= \mathcal{O}(\varepsilon^{\min(K+1,K+4-\gamma)}+\mathcal{O}(\varepsilon^{L+7/2}) = \mathcal{O}(\varepsilon^{L+7/2}).\end{aligned}$$

One of the estimates for the displacement is obtained as follows:

$$\begin{aligned}&\Big\|\frac{\partial^2}{\partial t^2}(d-d^{(L)})\Big\|_{L^2((0,1)\times(0,T))}\\ &\leqslant \Big\|\frac{\partial^2}{\partial t^2}(d-d^{(K)})\Big\|_{L^2((0,1)\times(0,T))} + \Big\|\frac{\partial^2}{\partial t^2}(d^{(K)}-d^{(L)})\Big\|_{L^2((0,1)\times(0,T))}\\ &= \mathcal{O}(\varepsilon^{\min(K+1,K+4-\gamma)-(\gamma+1)/2}+\mathcal{O}(\varepsilon^{L+\gamma+1}) = \mathcal{O}(\varepsilon^{L+\gamma+1}).\end{aligned}$$

The other estimates of (5.10) are obtained in a similar way. □

Acknowledgments. The authors are grateful to Professor E. Sanchez-Palencia for his useful suggestions during the preparation of this paper.

The first author is supported by the grant of the French Ministery of Research and Education ALLIANA, by the GDRE "Regular and chaotic hydrodynamics" and by the ANR project MECAMERGE.

The work is done during the stay of the second author in Saint-Etienne University.

References

1. Čanić, S., Mikelić, A.: Effective equations describing the flow of a viscous incompressible fluid through a long elastic tube. C. R. Acad. Sci. Paris, Série IIb **330**, 661–666 (2002)
2. Desjardins, B., Esteban, M.J.: Existence of weak solutions for the motion of rigid bodies in a viscous fluid. Arch. Ration. Mech. Anal. **146**, no. 1, 59–71 (1999)
3. Desjardins, B., Esteban, M.J., Grandmont, C., le Talec, P.: Weak solutions for a fluid-structure interaction model. Rev. Mat. Comput. **14**, no. 2, 523–538 (2001)
4. Galdi, G.P.: An Introduction to the Mathematical Theory of the Navier-Stokes Equations. Vol. I. Linearized Steady Problems. Springer, New York (1994)
5. Grandmont, C., Maday, Y.: Existence for an unsteady fluid-structure interaction problem. M^2AN Math. Model. Numer. Anal. **34**, no. 3, 609–636 (2000)
6. Kozlov, V., Maz'ya, V., Movchan, A.: Asymptotic Analysis of Fields in Multi-Structures Oxford Univ. Press, New York (1999)
7. Maz'ya, V.G.: Sobolev Spaces. Springer, Berlin etc. (1985)
8. Maz'ya, V., Nazarov, S., Plamenevskii, B.: Asymptotic Theory of Elliptic Boundary Value Problems in Singularly Perturbed Domains. I. II. Birkhäuser, Basel (2000)
9. Maz'ya, V., Slutskii, A.: Asymptotic analysis of the Navier–Stokes system in a plane domain with thin channels. Asymptotic Anal. **23**, no. 1, 59–89 (2000)
10. Panasenko, G.: Asymptotic expansion for the solution of the Navier–Stokes equation in a tube structure. CRAS, Ser. IIb **326**, 867–872 (1998)
11. Panasenko, G.P., Sirakov, Y., Stavre, R.: Asymptotic and numerical modelling of a flow in a thin channel with visco-elastic wall. Int. J. Multiscale Comput. Engng. **5**, no. 6, 473–482 (2007)
12. Panasenko, G.P., Stavre, R.: Asymptotic analysis of a periodic flow in a thin channel with visco-elastic wall. J. Math. Pures Appl. **(9)85**, no. 4, 558–579 (2006)
13. Panasenko, G.P., Stavre, R.: Asymptotic analysis of a non-periodic flow in a thin channel with visco-elastic wall. Networks and Heterogeneous Media, **3**, no. 3, 651–673 (2008)
14. Panasenko, G.P., Stavre, R.: Asymptotic analysis of a non-periodic flow with variable viscosity in a thin elastic channel. [Submitted]
15. Temam, R.: Navier-Stokes Equations. North-Holland, Amsterdam (1984)

On Solvability of Integral Equations for Harmonic Single Layer Potential on the Boundary of a Domain with Cusp

Sergei V. Poborchi

Abstract We present solvability theorems for boundary integral equations of the Dirichlet and the Neumann problems on a multi-dimensional domain with cusp which were established in recent joint papers by V.G. Maz'ya and the author.

1 Introduction

During more than a hundred years boundary integral equations induced by elliptic boundary value problems were intensively studied (for the history, see [5, 2]). By now, a comprehensive classical layer potential theory for domains with piecewise smooth and Lipschitz boundaries has been developed. The works by Maz'ya made an important contribution to the theory.

In 1967, Burago and Maz'ya [1] examined harmonic single and double layer potentials in the space C for a wide class of nonsmooth surfaces. In 1981, Maz'ya [4] suggested an approach to study boundary integral equations with the aid of some auxiliary boundary value problem. In this way, he succeeded to prove solvability theorems for classical boundary integral equations on piecewise smooth surfaces and to find the asymptotics of their solutions near boundary singularities. Later, Maz'ya and Shaposhnikova [11] generalized this method to Lipschitz graph domains. Maz'ya and Solov'ev were the first to consider integral equations on a planar contour with zero angles and 2π angles. Using the above method with an auxiliary boundary value problem [12]–[14], they developed the theory of logarithmetics potentials suitable for

Sergei V. Poborchi
Mathematical Department, St. Petersburg State University, 28, Universitetskii pr., Petrodvorets, St. Petersburg 198504, Russia
e-mail: poborchi@mail.ru

A. Laptev (ed.), *Around the Research of Vladimir Maz'ya II: Partial Differential Equations*, 303
International Mathematical Series 12, DOI 10.1007/978-1-4419-1343-2_14,

elasticity problems on planar domains with outward and inward cusps on the boundary.

In this article, I review recent works [7]–[10] written jointly by V. G. Maz'ya and myself. They mostly concern properties of the harmonic single layer potential on the boundary of a multi-dimensional domain having an isolated cusp. Our aim was to find out if solutions of the Dirichlet and the Neumann problems for the Laplace equation can be represented as the single layer potential with density in a sufficiently wide class of distributions. It turns out that the answer is affirmative for domains with "inward cusp" and generally negative for domains with "outward cusp."

2 Domains and Function Spaces

We now describe a surface with cusp which we deal with in what follows. Let $\Omega \subset \mathbb{R}^n$, $n > 2$, be a bounded simply connected domain whose boundary contains the origin and some neighborhood of the origin intersects Ω or $\mathbb{R}^n \setminus \overline{\Omega}$ by the set

$$\{x = (y, z) \in \mathbb{R}^n : z \in (0,1), y/\varphi(z) \in \omega\}, \tag{2.1}$$

where φ is a continuous increasing function on $[0,1]$ such that $\varphi(0) = 0$ and ω is a bounded domain in $\mathbb{R}^{n-1}$. For the simplicity of presentation, we do not try to attain maximal generality and suppose that $\partial\omega \in C^2$, $\varphi \in C^2(0,1] \cap C^1[0,1]$. Furthermore, $\varphi'(0) = 0$ and $\partial\Omega \setminus \{O\} \in C^2$. If the set (2.1) lies in Ω, we say that Ω has an *outward cusp* with vertex O. If the set (2.1) lies in $\mathbb{R}^n \setminus \overline{\Omega}$, then, by definition, O is the vertex of an *inward cusp* for Ω.

We put $\Gamma = \partial\Omega$. For all $x \in \Gamma\setminus\{O\}$ there exists a normal to Γ at x. A unit normal vector, directed into the exterior of Ω, is denoted by $\nu(x)$. In what follows, $\Omega^+ = \Omega$ and $\Omega^- = \mathbb{R}^n \setminus \overline{\Omega}$. Note that Ω^+ has an outward (inward) cusp if Ω^- has an inward (outward) cusp. The symbol $B_r(x)$ designates an open ball in $\mathbb{R}^n$ with radius r and center x, $B_r = B_r(0)$. Let G be a domain in $\mathbb{R}^n$. Then $C_0^\infty(G)$ is the set of infinitely differentiable functions compactly supported in G, and $H(G)$ is the space of harmonic functions on G with finite Dirichlet integral over G. We equip $H(G)$ with (generally speaking) seminorm $\|\nabla(\cdot)\|_{L_2(G)}$, where ∇u is the gradient of u. By $L_2^1(G)$ we mean the space of functions in $L_{2,\text{loc}}(G)$ whose gradient is in $L_2(G)$.

We also introduce the space $H(\Gamma)$ of functions on $\mathbb{R}^n$ with finite Dirichlet integral which are harmonic in $\mathbb{R}^n\setminus\Gamma$. The space $H(\Gamma)$ is endowed with norm $\|\nabla(\cdot)\|_{L_2(\mathbb{R}^n)}$.

To state the Dirichlet or the Neumann problem in $\Omega^\pm$, we need trace spaces on Γ for functions defined on $\Omega^\pm$. Let $L_2^1(\Omega^-)$ denote the closure with respect to the norm $\|\nabla(\cdot)\|_{L_2(\Omega^-)}$ of the set of functions in $C^\infty(\Omega^-)\cap L_2^1(\Omega^-)$ having bounded supports in Ω^-. Let $Tr^-(\Gamma)$ be the trace space $\{u|_\Gamma : u \in L_2^1(\Omega^-)\}$

with norm

$$\|f\|_{Tr^-(\Gamma)} = \inf\{\|\nabla u\|_{L_2(\Omega^-)} : u \in L_2^1(\Omega^-), \quad u|_\Gamma = f\}.$$

For an internal domain Ω^+ we introduce the trace space $Tr^+(\Gamma) = \{u|_\Gamma : u \in L_2^1(\Omega^+)\}$ equipped with seminorm

$$\|f\|_{Tr^+(\Gamma)} = \inf\{\|\nabla u\|_{L_2(\Omega^+)} : u \in L_2^1(\Omega^+), u|_\Gamma = f\}.$$

Note that $\|f\|_{Tr^+(\Gamma)} = 0$ is equivalent to $f = \text{const}$. We also define

$$Tr(\Gamma) = Tr^+(\Gamma) \cap Tr^-(\Gamma)$$

and put

$$\|f\|_{Tr(\Gamma)} = \left(\|f\|^2_{Tr^+(\Gamma)} + \|f\|^2_{Tr^-(\Gamma)}\right)^{1/2}.$$

Since the minimum of the Dirichlet integral over a domain is attained on a harmonic function, it follows that the maps

$$H(\Omega^\pm) \ni u \mapsto u|_\Gamma \in Tr^\pm(\Gamma), \quad H(\Gamma) \ni u \mapsto u|_\Gamma \in Tr(\Gamma)$$

are isometric isomorphisms.

Let $\widehat{\Omega}$ be that domain Ω^+ or Ω^- which has an inward cusp. Then there exists a linear continuous extension operator $E : L_2^1(\widehat{\Omega}) \to L_2^1(\mathbb{R}^n)$, i.e., $Eu|_{\widehat{\Omega}} = u$ for all $u \in L_2^1(\widehat{\Omega})$ (cf. [3]). By this extension theorem, one can obtain the following relations between the spaces $Tr^\pm(\Gamma)$ and $Tr(\Gamma)$ (cf. [10]).

Lemma 2.1. (i) *If Ω^+ has an outward cusp, then*

$$Tr(\Gamma) = Tr^-(\Gamma)$$

with equivalence of norms.

(ii) *If Ω^+ has an inward cusp, then the following equivalence relation holds:*

$$\|f\|_{Tr(\Gamma)} \sim \|f\|_{Tr^+(\Gamma)} + \|f\|_{L_2(\Gamma)}.$$

We also introduce the dual spaces $Tr(^\pm\Gamma)^*$ and $Tr(\Gamma)^*$ of the corresponding trace spaces.

3 Description of the Spaces $Tr(\Gamma)$ and $Tr(\Gamma)^*$

The spaces $Tr^\pm(\Gamma)$ and $Tr(\Gamma)$, introduced above, admit an explicit characterization [6, Chapter 7]. We state here the assertion concerning the last space.

Theorem 3.1. *Let $\varepsilon > 0$ be so small that $\overline{B}_\varepsilon$ lies in the neighborhood of the origin which intersects Ω or $\mathbb{R}^n \setminus \overline{\Omega}$ by the set* (2.1). *Let $f \in L_{2,\mathrm{loc}}(\Gamma \setminus \{O\})$. Then f belongs to $Tr(\Gamma)$ if and only if*

$$\|f\|^2_{L_2(\Gamma)} + \int_{\Gamma \cap B_\varepsilon} f(x)^2 \frac{d\Gamma(x)}{\varphi(z)} + \iint_{\Gamma \times \Gamma} |f(x) - f(\xi)|^2 \frac{d\Gamma(x) d\Gamma(\xi)}{|x-\xi|^n} < \infty \quad (3.1)$$

for $n > 3$, and

$$\int_{\Gamma \cap B_\varepsilon} f(x)^2 \frac{d\Gamma(x)}{(\varphi(z)\log(z/\varphi(z)))} + \iint_{\Gamma \times \Gamma} |f(x) - f(\xi)|^2 \frac{d\Gamma(x) d\Gamma(\xi)}{r^3}$$

$$+ \|f\|^2_{L_2(\Gamma)} + \iint_{\{x,\xi \in \Gamma \cap B_\varepsilon : r > M(z,\zeta)\}} |f(x) - f(\xi)|^2 \frac{M(z,\zeta)^{-2} d\Gamma(x) d\Gamma(\xi)}{r\big(\log(1 + r/M(z,\zeta))\big)^2} < \infty \quad (3.2)$$

for $n = 3$. Here, $r = |x - \xi|$, $x = (y, z)$, $\xi = (\eta, \zeta)$, and $M(z,\zeta) = \max\{\varphi(z), \varphi(\zeta)\}$. For $n = 3$ the additional assumption $\varphi'(z) = O(\varphi(z)z^{-1})$ is required. Furthermore, the expression on the left-hand side of (3.1) *and* (3.2) *is equivalent to $\|f\|^2_{Tr(\Gamma)}$.*

In order to state an assertion describing the space $Tr(\Gamma)^*$, we need to construct some special partition of unity for $\Gamma \setminus \{O\}$.

Let $\{z_k\}$ be defined by

$$z_0 \in (0,1), \quad z_{k+1} + \varphi(z_{k+1}) = z_k, \quad k = 0, 1, \ldots$$

It is clear that $\{z_k\}$ is decreasing and

$$z_k \to 0, \quad z_{k+1}^{-1} z_k \to 1, \quad \varphi(z_{k+1})^{-1}\varphi(z_k) \to 1.$$

Consider a smooth partition of unity $\{\mu_k\}_{k \geqslant 1}$ for interval $(0, z_1]$, subordinate to the covering by intervals $\Delta_k = (z_{k+1}, z_{k-1})$, i.e., $\mu_k \in C_0^\infty(\Delta_k)$,

$$0 \leqslant \mu_k \leqslant 1, \quad \sum_{k \geqslant 1} \mu_k(z) = 1, \quad z \in (0, z_1].$$

We can assume that

$$\mathrm{dist}(\mathrm{supp}\,\mu_k, \mathbb{R}^1 \setminus \Delta_k) \geqslant \mathrm{const} \cdot \varphi(z_k), \quad |\mu_k'| \leqslant \mathrm{const} \cdot \varphi(z_k)^{-1}$$

with constants depending only on φ, and the equality

$$\sum_{k \geqslant 1} \mu_k(z) = 1$$

is valid for $z \in (0, \delta]$ with some $\delta > z_1$.

Let $\{\lambda_k\}_{k\geqslant 1}$ be a set of functions such that

$$0 \leqslant \lambda_k \leqslant 1, \quad \lambda_k \in C_0^\infty(\Delta_k), \quad \lambda_k|_{\text{supp}\,\mu_k} = 1.$$

Then $\lambda_k \mu_k = \mu_k$ for all $k \geqslant 1$. Define $\mu_0(z) = 0$ for $0 < z < z_1$ and $\mu_0(z) = 1 - \mu_1(z)$ for $z \geqslant z_1$. It is clear that $\sum_{k\geqslant 0} \mu_k(z) = 1$ for all $z \in (0,1]$. The partition of unity for $(0,1]$ just constructed and the set $\{\lambda_k\}$ depend only on z_0 and φ. Next, we put

$$\Gamma' = \{x = (y,z) \in \mathbb{R}^n : z \in (0,1), y/\varphi(z) \in \partial\omega\},$$

$$\Gamma_k = \{(y,z) \in \Gamma' : z \in \Delta_k\}, \quad \Delta_k = (z_{k+1}, z_{k-1}), \quad k = 1,2\ldots$$

and

$$\Gamma_0 = \Gamma \setminus \{x \in \overline{\Gamma'} : z \leqslant z_1\}.$$

We observe that the partition of unity $\{\mu_k\}$ for $(0,1]$ generates a partition of unity for $\Gamma\setminus\{O\}$ if we define $\mu_0 = 1$ on $\Gamma_0\setminus\Gamma'$, $\mu_k(x) = \mu_k(z)$ for $x \in \Gamma_k \cap \Gamma'$, $k \geqslant 0$, and $\mu_k = 0$ on $\Gamma \setminus \Gamma'$ for $k \geqslant 1$. This partition of unity is subordinate to the covering $\{\Gamma_k\}_{k\geqslant 0}$ in the sense that $\text{dist}(\text{supp}\,\mu_k, \Gamma \setminus \Gamma_k) > 0$, $k \geqslant 0$.

Let $F \in Tr(\Gamma)^*$ and $v \in Tr(\Gamma)$. By $\langle F, v\rangle$ we mean the value of F of an element v. If $F \in Tr(\Gamma)^*$ and $\lambda \in C^{0,1}(\Gamma)$, we define

$$\langle \lambda F, v\rangle = \langle F, \lambda v\rangle, \quad v \in Tr(\Gamma).$$

Let v be defined on Γ'. Then its mean value on the section of Γ' by hyperplane $z = \text{const}$ is

$$\overline{v}(z) = \frac{1}{|\gamma|}\int_\gamma v(\varphi(z)y, z)d\gamma(y), \quad \gamma = \partial\omega, \quad z \in (0,1),$$

where $|\gamma|$ is the $(n-2)$-dimensional area of γ.

If $F \in Tr(\Gamma)^*$, $v \in Tr(\Gamma)$, and $\text{supp}\, v \subset \Gamma'$, we put

$$\langle \overline{F}, v\rangle = \langle F, \overline{v}\rangle.$$

We say that the support of a functional F lies in Γ_k if $v|_{\Gamma_k} = 0$ implies $\langle F, v\rangle = 0$.

Let S be a Lipschitz surface. The space $W_2^{1/2}(S)$ consists of those functions in $L_2(S)$ for which the norm

$$\|v\|_{W_2^{1/2}(S)} = \left(\|v\|^2_{L_2(S)} + \iint\limits_{S\times S} |v(x) - v(\xi)|^2 \frac{ds_x ds_\xi}{|x-\xi|^n}\right)^{1/2}$$

is finite. Here, ds_x and ds_ξ are the area elements on S. For the same S we introduce $W_2^{-1/2}(S)$ as the space of continuous linear functionals on $W_2^{1/2}(S)$

equipped with norm

$$\|F\|_{W_2^{-1/2}(S)} = \sup\{|\langle F, v\rangle| : v \in W_2^{1/2}(S), \|v\| \leqslant 1\}.$$

For $n = 3, 4, \ldots$ we consider the space of functions on $(0,1)$ with finite norm

$$\|u\|_{H_n(0,1)} = \Bigg(\int\limits_0^1 |u(z)|^2 \frac{dz}{\varphi(z)^{3-n}} + \iint\limits_{\{z,\zeta\in(0,1):|z-\zeta|<M(z,\zeta)\}} |u(z)-u(\zeta)|^2 \frac{M(z,\zeta)^{n-2}}{|z-\zeta|^2} dz d\zeta\Bigg)^{\frac{1}{2}}$$

if $n > 3$. Here, the same notation is used as in Theorem 3.1. The norm in $H_3(0,1)$ is given by

$$\|u\|_{H_3(0,1)} = \Bigg(\int\limits_0^1 \frac{|u(z)|^2 dz}{\log(z/\varphi(z))} + \int\limits_0^1\int\limits_0^1 \frac{|u(z)-u(\zeta)|^2}{|z-\zeta|} \sigma(z,\zeta) dz d\zeta\Bigg)^{1/2},$$

where

$$\sigma(z,\zeta) = \chi_{(1/2,2)}(z/\zeta) Q\left(|z-\zeta|(M(z,\zeta))^{-1}\right),$$

$\chi_{(1/2,2)}$ is the characteristic function of the interval $(1/2, 2)$ and

$$Q(t) = \begin{cases} t^{-1}, & t \in (0,1), \\ (\log(et))^{-2}, & t > 1. \end{cases}$$

Now, we are ready to state a theorem describing the space $Tr(\Gamma)^*$ (cf. [7, 8]).

Theorem 3.2. *Let $\{\mu_k\}_{k\geqslant 0}$ be the partition of unity constructed above.*

(i) *Any functional $F \in Tr(\Gamma)^*$ can be represented by the sum*

$$F = \mu_0 F + (1-\mu_0)\overline{F} + (1-\mu_0)(F - \overline{F}) = F^{(1)} + F^{(2)} + F^{(3)},$$

where each term belongs to the same space. Furthermore, $F^{(1)}$ has support in Γ_0 and belongs to $W_2^{-1/2}(\Gamma_0)$, whereas $F^{(2)}$ is supported in $\{x \in \Gamma' : z \leqslant z_0\}$ and belongs to the space $H_n(0,1)^*$ in the sense that the following estimate holds:

$$|\langle F^{(2)}, v\rangle| \leqslant \text{const} \cdot \|\overline{v}\|_{H_n(0,1)}.$$

The third term can be written as the sum

$$F^{(3)} = \sum_{k\geqslant 1} \mu_k (F - \overline{F})$$

and the following estimate holds:

$$\left(\sum_{k\geqslant 1}\|\mu_k(F-\overline{F})\|^2_{W_2^{-1/2}(\Gamma_k)}\right)^{1/2} \leqslant c\,\|(1-\mu_0)(F-\overline{F})\|_{Tr(\Gamma)^*}.$$

(ii) *Let $F_k \in W_2^{-1/2}(\Gamma_k)$ for $k \geqslant 1$. Suppose that* $\operatorname{supp} F_k \subset \Gamma_k$ *and* $\langle F_k, v\rangle = 0$ *if $v \in W_2^{1/2}(\Gamma_k)$ and $v(y,z)$ depends only on z. Assume that*

$$\sum_{k\geqslant 1}\|\lambda_k F_k\|^2_{W_2^{-1/2}(\Gamma_k)} < \infty.$$

If g and h belong to the spaces $H_n(0,1)^*$ and $W_2^{-1/2}(\Gamma_0)$ respectively, then each functional $F^{(1)} = \mu_0 h$,

$$Tr(\Gamma) \ni v \mapsto \langle F^{(2)}, v\rangle = \langle g, (1-\mu_0)\overline{v}\rangle,$$

$$Tr(\Gamma) \ni v \mapsto \langle F^{(3)}, v\rangle = \sum_{k\geqslant 1}\langle \lambda_k F_k, v\rangle$$

is continuous in $Tr(\Gamma)$. Furthermore, $F^{(1)} \in W_2^{-1/2}(\Gamma_0)$, and the norm of $F^{(3)}$ can be bound as follows

$$\|F^{(3)}\|_{Tr(\Gamma)^*} \leqslant c\left(\sum_{k\geqslant 1}\|\lambda_k F_k\|^2_{W_2^{-1/2}(\Gamma_k)}\right)^{1/2}.$$

4 The Single Layer Potential for Γ

For $x, y \in \mathbb{R}^n$ let $E(x,y)$ be the fundamental solution of the Poisson equation on $\mathbb{R}^n$, i.e.,

$$E(x,y) = ((2-n)|S^{n-1}||x-y|^{n-2})^{-1},$$

where $|S^{n-1}|$ is the area of the unit sphere in $\mathbb{R}^n$.

The single layer potential with density ϱ is defined by

$$(V\varrho)(x) = \int_\Gamma \varrho(y)E(x,y)d\Gamma(y), \quad x \in \mathbb{R}^n. \tag{4.1}$$

From the above assumptions imposed on Γ it follows that for $x \in \Gamma\setminus\{O\}$ there exists a ball $B_\delta(x)$ such that

$$\left|\frac{\partial E(\xi,\eta)}{\partial \nu_\xi}\right| \leqslant c(x,\Gamma)|\xi-\eta|^{2-n}$$

for all $\xi, \eta \in B_\delta(x)\cap\Gamma$ with $c(x,\Gamma) > 0$ depending only on x, Γ. Hence the integral on the right-hand side of (4.2) below makes sense if $\varrho \in L_\infty(\Gamma)$ and

$x \in \Gamma \setminus \{O\}$, so that the quantity

$$(\overline{V}\varrho)(x) = \int_{\Gamma} \varrho(\xi) \frac{\partial E(x,\xi)}{\partial \nu_x} d\Gamma(\xi) \tag{4.2}$$

is well defined. Moreover, it is readily verified that the function $\overline{V}\varrho$ is continuous on $\Gamma \setminus \{O\}$ for $\varrho \in L_\infty(\Gamma)$.

Let u be a function on $\mathbb{R}^n \setminus \Gamma$, and let $x \in \Gamma \setminus \{O\}$. The values $u^\pm(x)$ are defined as one-sided limits (if they exist)

$$u^\pm(x) = \lim_{t \to \mp 0} u(x + t\nu(x)).$$

For $\varrho \in C(\Gamma)$ we have

$$\frac{\partial (V\varrho)^\pm}{\partial \nu}(x) = \overline{V}\varrho(x) \mp \varrho(x)/2, \quad x \in \Gamma \setminus \{O\}. \tag{4.3}$$

Formula (4.3) is well known if Γ is a smooth surface (cf., for example, [15, Chapter 14, Section 7]). One should repeat this argument to establish (4.3) for our surface Γ. Because of the continuity of the function on the left-hand side of (4.2), we conclude that for $\varrho \in C(\Gamma)$ the limit values $\partial(V\varrho)^\pm/\partial\nu$ are continuous on $\Gamma \setminus \{O\}$.

We now mention some properties of the single layer potential (4.1) as an operator acting in some function spaces on Γ. One can show that for $\lambda \in (0,1]$ the integral

$$\int_{\Gamma} |\xi - x|^{1-n+\lambda} d\Gamma(\xi)$$

is bounded uniformly in $x \in \mathbb{R}^n$. This fact enables us to obtain the following result (cf. [9]).

Lemma 4.1. *The single layer potential* (4.1) *is continuous as an operator* $V : L_2(\Gamma) \to H(\Gamma)$.

We have already seen earlier that $H(\Gamma)$ is isometrically isomorphic to the space $Tr(\Gamma)$, so that $H(\Gamma)$ can be replaced by $Tr(\Gamma)$ in Lemma 4.1. However, it is possible to state a stronger result for the space $Tr(\Gamma)$ (cf. [9]).

Theorem 4.1. *Let $C_0(\Gamma)$ denote the set of continuous functions on Γ vanishing in the vicinity of the origin, and let V be given by* (4.1)*. Then the map*

$$C_0(\Gamma) \ni \varrho \mapsto V\varrho \in Tr(\Gamma)$$

can be uniquely extended to an isometric isomorphism between the spaces $Tr(\Gamma)^*$ and $Tr(\Gamma)$ if $\varrho \in C_0(\Gamma)$ is identified with the functional

$$\langle \varrho, g \rangle = \int_{\Gamma} \varrho g d\Gamma, \quad g \in Tr(\Gamma).$$

This theorem implies a direct consequence.

Corollary 4.1. *Let u belong to one of the spaces $H(\Omega^-)$ or $H(\Omega^+)$. Then $u = V\varrho$ for some $\varrho \in Tr(\Gamma)^*$ if and only if $u|_\Gamma \in Tr(\Gamma)$ and hence $u \in H(\Gamma)$.*

By Theorem 3.1, we can state one more consequence of Theorem 4.1.

Corollary 4.2. *A necessary and sufficient condition for the unique solvability of the equation*

$$f(x) = \int_\Gamma \varrho(y)E(x,y)d\Gamma(y), \quad f \in Tr(\Gamma),$$

in $Tr(\Gamma)^*$ is the inequality (3.1) *for $n > 3$ and the inequality* (3.2) *for $n = 3$.*

5 The Dirichlet Problem for the Laplace Equation

Consider the Dirichlet problem

$$\Delta u = 0 \ \text{ in } \ \Omega^\pm, \quad u|_\Gamma = f, \tag{5.1}$$

where $f \in Tr^\pm(\Gamma)$. Its solution is that function $u \in L_2^1(\Omega^+)$ (or $u \in L_2^1(\Omega^-)$ for which $u|_\Gamma = f$ and

$$\int_{\Omega^\pm} \nabla u \nabla v dx = 0$$

for all $v \in C_0^\infty(\Omega^\pm)$. It is well known that the problem (5.1) is uniquely solvable and its solution belongs to $H(\Omega^\pm)$. We say that the problem (5.1) for Ω^+ is an *internal* Dirichlet problem and for Ω^- is an *external* one. The following assertion is a consequence of Theorem 4.1 and Corollary 4.1.

Corollary 5.1. *The solution of the problem* (5.1) *can be written as the single layer potential* (4.1) *for some density $\varrho \in Tr(\Gamma)^*$ if and only if the two-sided Dirichlet problem*

$$u \in H(\Gamma), \quad u|_\Gamma = f$$

is solvable.

Combining Lemma 2.1, Theorem 4.1, and the preceding assertion, we arrive at the following statement.

Theorem 5.1. (i) *If Ω^+ has an inward cusp, then the solution of the internal Dirichlet problem is represented as the single layer potential* (4.1) *where $\varrho \in Tr(\Gamma)^*$ satisfies the uniquely solvable equation $V\varrho = f$ for any $f \in Tr^+(\Gamma)$.*

(ii) *If Ω^- has an inward cusp, then the solution of the external Dirichlet problem can be written as the single layer potential* (4.1) *with density uniquely determined by the equation $V\varrho = f$ for any $f \in Tr^-(\Gamma)$.*

(iii) *If $\Omega^{\pm}$ has an outward cusp, then there exist Dirichlet data $f \in Tr^{\pm}(\Gamma)$ such that the solution of the problem* (5.1) *cannot be written in the form $V\varrho$ with $\varrho \in Tr(\Gamma)^*$.*

Note that statement (iii) of Theorem 5.1 follows from Corollary 4.1 and the fact that the set $Tr^{\pm}(\Gamma) \setminus Tr(\Gamma)$ is nonempty if $\Omega^{\pm}$ has an outward cusp (cf. [6, Chapter 7]).

6 The Neumann Problem for the Laplace Equation

First we consider an external Neumann problem

$$\Delta u = 0 \text{ in } \Omega^-, \quad \partial u/\partial \nu|_{\Gamma} = \psi^-, \tag{6.1}$$

where $\psi^- \in Tr^-(\Gamma)^*$. Its solution is that function $u \in L_2^1(\Omega^-)$ for which the identity

$$\int_{\Omega^-} \nabla u \nabla w dx = -\langle \psi^-, w \rangle$$

holds for all $w \in L_2^1(\Omega^-)$.

It is well known that the problem (6.1) is uniquely solvable for all $\psi^- \in Tr^-(\Gamma)^*$. Moreover, the solution belongs to $H(\Omega^-)$ and the estimate

$$\|\nabla u\|_{L_2(\Omega^-)} \leqslant \|\psi^-\|_{Tr^-(\Gamma)^*}$$

is valid. The following assertion concerns the possibility to write the solution of the problem (6.1) in the form of a single layer potential (cf. [10]).

Theorem 6.1. *If Ω^+ has an outward cusp (and hence Ω^- has an inward cusp), the solution of the problem* (6.1) *can be represented as the single layer potential $u = V\varrho$ with density ϱ determined by a uniquely solvable equation $A\varrho = \psi^-$, where $A : Tr^-(\Gamma)^* \to Tr^-(\Gamma)^*$ is an extension by continuity of the operator*

$$C_0(\Gamma) \ni \varrho \mapsto A\varrho = \overline{V}\varrho + \varrho/2 \in Tr^-(\Gamma)^*$$

and $\overline{V}\varrho$ is defined by (4.2).

We now turn to the internal Neumann problem

$$\Delta u = 0 \text{ in } \Omega^+, \quad \partial u/\partial \nu|_{\Gamma} = \psi^+, \tag{6.2}$$

where $\psi^+ \in Tr^+(\Gamma)^*$. A function $u \in L_2^1(\Omega^+)$ is called the solution of the problem (6.2) if

$$\int_{\Omega^+} \nabla u \nabla w dx = \langle \psi^+, w \rangle$$

for all $w \in L_2^1(\Omega^+)$.

It is clear that the condition $\langle\psi^+, 1\rangle = 0$ is necessary for the solvability of (6.2). Let $Tr^+(\Gamma)^* \ominus 1$ denote the subspace of functionals in $Tr^+(\Gamma)^*$ orthogonal to 1. It is known that for $\psi^+ \in Tr^+(\Gamma)^* \ominus 1$ the problem (6.2) has a solution, unique up to a constant summand, and this solution belongs to $H(\Omega^+)$. The following assertion is established in [10].

Theorem 6.2. *Let Ω^+ be a domain with inward cusp, and let $\psi^+ \in Tr^+(\Gamma)^* \ominus 1$. Then any solution u of the problem* (6.2) *can be written in the form $u = V\varrho$, where $V\varrho$ is given by* (4.1) *and ϱ is an element of $Tr(\Gamma)^*$ satisfying the boundary integral equation $A\varrho = \psi^+$. Here, $A : Tr(\Gamma)^* \to Tr^+(\Gamma)^* \ominus 1$ is an extension by continuity of the operator*

$$C_0(\Gamma) \ni \varrho \mapsto A\varrho = \overline{V}\varrho - \varrho/2 \in Tr^+(\Gamma)^* \ominus 1$$

and $\overline{V}\varrho$ is defined in (4.2). *Furthermore, for any $\psi^+ \in Tr^+(\Gamma)^* \ominus 1$, ϱ is determined by the equation $A\varrho = \psi^+$ uniquely up to a term $const \cdot V^{-1}(1)$.*

In conclusion, we show that Theorems 6.1 and 6.2 fail for domains with outward cusp, i.e., if Ω^- (or Ω^+) has an outward cusp, then there exist functionals $\psi^- \in Tr^-(\Gamma)^*$ (or $\psi^+ \in Tr^+(\Gamma)^* \ominus 1$) such that the Neumann problem (6.1) (or (6.2)) for this domain has solutions which cannot be represented as a harmonic single layer potential with density in $Tr(\Gamma)^*$. In view of Corollary 4.1, it suffices to check that there exist solutions of the Neumann problem lying in $H(\Omega^-)$ (or in $H(\Omega^+)$, but not lying in $H(\Gamma)$.

Consider, for example, Ω^+ and apply Lemma 2.1 (i). Since $Tr^+(\Gamma) \neq Tr^-(\Gamma)$ [6, Sections 7.3 and 7.4], it follows that the set $Tr^+(\Gamma) \setminus Tr(\Gamma)$ is nonempty. Let f belong to this set, and let $u^+ \in H(\Omega^+)$ satisfy $u^+|_\Gamma = f$. Define ψ^+ by

$$L_2^1(\Omega^+) \ni v \mapsto \langle\psi^+, v\rangle = \int_{\Omega^+} \nabla u^+ \nabla v dx.$$

Then $\psi^+ \in Tr^+(\Gamma)^* \ominus 1$ and the problem (6.2) has the solution $u^+ + \text{const}$ whose trace on Γ does not belong to $Tr(\Gamma)$. Hence this solution cannot be written in the form (4.1). An argument for domain Ω^- is similar.

References

1. Burago, Yu.D., Maz'ya, V.G.: Certain questions of potential theory and function theory for regions with irregular boundaries (Russian). Zap. Nauchn. Semin. LOMI. **3** (1967); English transl.: Consultants Bureau, New York (1969)
2. Kenig, C.E.: Harmonic Analysis Technics for Second Order Elliptic Boundary Value Problems. Am. Math. Soc., Providence RI (1994)
3. Jones, P.W.: Quasiconformal mappings and extendability of functions in Sobolev spaces. Acta Math. **147**, 71–88 (1981)

4. Maz'ya, V.G.: The integral equations of potential theory in domains with piecewise smooth boundary (Russian). Usp. Mat. Nauk. **36**, no. 4, 229–230 (1981)
5. Maz'ya, V.G.: Boundary integral equations (Russian). In: Current Problems in Mathematics. Fundamental Directions. Itogi Nauki i Tekhniki, Akad. Nauk SSSR, VINITI, Moscow **27**, 131–228 (1988); English transl.: Analysis IV. Encyclop. Math. Sci. **27**, pp. 127–222. Springer, Berlin (1991)
6. Maz'ya, V.G., Poborchi, S.V.: Differentiable Functions on Bad Domains. World Scientific, Singapore etc. (1997)
7. Maz'ya, V.G., Poborchi, S.V.: On solvability of the Neumann problem in energy space for a domain with peak. Georgian Math. J. **14**, no. 3, 499–518 (2007)
8. Maz'ya, V.G., Poborchi S.V.: On solvability of the Neumann problem for a domain with cusp (Russian). Algebra Anal. **20**, no. 5, 108–153 (2008); English transl.: St. Petersburg Math. J. [To appear]
9. Maz'ya, V.G., Poborchi, S.V.: Unique solvability of integral equation for harmonic single layer potential on the boundary of a domain with cusp (Russian). Vestnik St. Petersburg State Univ. no. 2, 63–73 (2009)
10. Maz'ya, V.G., Poborchi, S.V.: On representation of solution to the Neumann problem for a domain with cusp by harmonic single layer potential (Russian). Vestnik St. Petersburg State Univ. (2009) [To appear]
11. Maz'ya, V.G., Shaposhnikova, T.O.: Higher regularity in the classical layer potential theory for Lipschitz domains. Indiana Univ. Math. J. **54**, no. 1, 99–142 (2005)
12. Maz'ya, V.G., Solov'ev, A.A.: On integral equation of the Dirichlet problem in a plane domain with cusps on the boundary (Russian). Mat. Sb. **180**, no. 9, 1211–1233, 1296 (1989); English transl.: Math. USSR-Sb. **68**, no. 1, 61–83 (1991)
13. Maz'ya, V.G., Solov'ev, A.A.: On boundary integral equation for the Neumann problem in a domain with cusp (Russian). Tr. Leningrad. Mat. O-va **1**, 109–134 (1990)
14. Maz'ya, V.G., Solov'ev, A.A.: Integral equations of logarithmetics potential theory for contours with cusps in Hölder spaces (Russian). Algebra Anal. **10** no. 5, 85–142 (1998); English transl.: St. Petersburg Math. J. **10**, no. 5. 791–832 (1999)
15. Mikhlin, S.G.: Linear Partial Differential Equations (Russian). Vyssh. Shkola, Moscow (1977)

Hölder Estimates for Green's Matrix of the Stokes System in Convex Polyhedra

Jürgen Roßmann

Abstract The paper deals with Green's matrix of the Dirichlet problem for the Stokes system in a polyhedron. In particular, Hölder estimates for the derivatives of the elements of this matrix are obtained in the case where the polyhedron is convex.

1 Introduction

This paper is closely related to V. Maz'ya's results in the theory of elliptic boundary value problems for domains with nonsmooth boundaries. The development of this theory was significantly influenced by the works of Maz'ya who began his research in this area in the 1960s. One of his first works in this direction was that with Verzhbinskii in 1967 which appeared at the same time as Kondrat'ev's fundamental work on elliptic boundary value problems in domains with angular and conical points. In [17] (the complete proofs are given in [18]), Maz'ya and Verzhbinskii created an extensive asymptotic theory of the Dirichlet problem for the Laplacian in n-dimensional domains with various types of point boundary singularities. The comparison principle obtained here leads directly to an estimate of the Green function. In particular, for a cone $\mathcal{K}$ with vertex at the origin, this estimate takes the form

$$G(x,\xi) \leqslant c\,|\xi|^{\lambda}\,|x|^{2-n-\lambda} \quad \text{for } |\xi| \leqslant |x|/2,$$

where λ is a positive number such that $\lambda(\lambda+n-2)$ is the first eigenvalue of the Dirichlet problem for the Beltrami operator $-\delta$ on $\mathcal{K} \cap S^{n-2}$.

Jürgen Roßmann
University of Rostock, Institute of Mathematics, D-18051 Rostock, Germany e-mail: juergen.rossmann@uni-rostock.de

A. Laptev (ed.), *Around the Research of Vladimir Maz'ya II: Partial Differential Equations*, 315
International Mathematical Series 12, DOI 10.1007/978-1-4419-1343-2_15,

In 1979, Maz'ya and Plamenevskii [7] proved asymptotic formulas for the Green function of general elliptic boundary value problems in domains with conical points. Shortly after that, they obtained point estimates for the Green functions of elliptic boundary value problems in a dihedral angle (cf. [5]). Estimates for the Green matrix of the Dirichlet problem for the Stokes and the Lamé systems in three-dimensional domains of polyhedral type were proved by the same authors in [9]. In particular, they obtained the following estimate for the elements of the Green matrix if x and ξ lie in a neighborhood of the vertex $x^{(\nu)}$ and $|x-x^{(\nu)}| > 2\,|\xi - x^{(\nu)}|$:

$$\big|\partial_x^\alpha \partial_\xi^\gamma G_{i,j}(x,\xi)\big| \leqslant c\,\rho_\nu(x)^{-1-\Lambda_\nu^+-\delta_{i,4}-|\alpha|+\varepsilon}\,\rho_\nu(\xi)^{\Lambda_\nu^+-\delta_{j,4}-|\gamma|-\varepsilon} \\ \times \prod_{k\in I_\nu}\Big(\frac{r_k(x)}{\rho_\nu(x)}\Big)^{\delta_k^+-\delta_{i,4}-|\alpha|-\varepsilon} \prod_{k\in I_\nu}\Big(\frac{r_k(\xi)}{\rho_\nu(\xi)}\Big)^{\delta_k^+-\delta_{j,4}-|\gamma|-\varepsilon}.$$

Here, $\rho_\nu(x) = |x-x^{(\nu)}|$, $r_k(x)$ denotes the distance to the edge M_k, I_ν is the set of all indices k such that $\overline{M}_k \supset x^{(\nu)}$, δ_k^+ and Λ_ν^+ are the real parts of eigenvalues of certain operator pencils (cf. Section 3), and ε is an arbitrarily small positive number. Similar estimates are given in [9] for the cases $|\xi - x^{(\nu)}| > 2\,|x-x^{(\nu)}|$ and $|x-x^{(\nu)}|/2 < |\xi - x^{(\nu)}| < 2|x-x^{(\nu)}|$. Maz'ya and Rossmann [11] obtained analogous formulas for the Green function of the Dirichlet problem for higher order elliptic equations in domains of polyhedral type. The Green matrices of Neumann and mixed boundary value problems for second order elliptic systems in polyhedral domains were studied by Maz'ya and Rossmann in [12], a mixed boundary value problem for the Stokes system was considered in [13].

Point estimates of the Green functions are important for many application. For example, Maz'ya and Plamenevskii [5, 6] established a solvability and regularity theory for elliptic boundary value problems in weighted Hölder spaces by means of such estimates. In [9], they proved maximum modulus estimates for solutions of the Stokes and Lamé systems in domains of polyhedral type using estimates for the Green matrix. The Miranda–Agmon maximum principle for higher order elliptic equations proved in Maz'ya's and Rossmann's paper [11] is also based on such estimates. Of course, many applications require additional information on the spectra of the operator pencils which are generated by the boundary value problem at the singular boundary points. The spectral properties of these operator pencils are studied in a multitude of papers by Maz'ya and his collaborators. These results are collected in the book [3]. I mention here only the papers [8, 9] by Maz'ya and Plamenevskii and the paper [4] by Kozlov, Maz'ya, and Schwab which deal with the Dirichlet problem for the Stokes system. The results of [4, 8, 9] are used in the present paper.

Here is an overview of the main results of the present paper. In the case of the Dirichlet problem for the Stokes system, the estimates for the Green matrix obtained in [13] improve the above given result in [9] if the edge angles are less than π. However, both estimates given in [13] and [9] are not sharp for the Dirichlet problem for the Stokes system in convex polyhedra. The goal of this paper is to obtain more precise estimates in the case of a convex polyhedron which allow to derive Hölder estimates for the derivatives of the elements of the Green matrix in the whole domain. In particular, it is shown here that

$$\left|\partial_x^\alpha \partial_\xi^\gamma G_{i,j}(x,\xi)\right| \leqslant c\,|x-\xi|^{-1-\delta_{i,4}-\delta_{j,4}-|\alpha|-|\gamma|} \tag{1.1}$$

for $|\alpha| \leqslant 1-\delta_{i,4}$, $|\gamma| \leqslant 1-\delta_{j,4}$ if the polyhedron $\mathcal{G}$ is convex. Moreover, we prove the Hölder estimates

$$\begin{aligned}&\frac{\left|\partial_x^\alpha \partial_\xi^\gamma G_{i,j}(x,\xi) - \partial_y^\alpha \partial_\xi^\gamma G_{i,j}(y,\xi)\right|}{|x-y|^\sigma} \\ &\leqslant c\left(|x-\xi|^{-2-\sigma-\delta_{j,4}-|\gamma|} + |y-\xi|^{-2-\sigma-\delta_{j,4}-|\gamma|}\right)\end{aligned} \tag{1.2}$$

for $x \neq y$, $i = 1,2,3$, $|\alpha| = 1$, $|\gamma| \leqslant 1-\delta_{j,4}$, and

$$\begin{aligned}&\frac{\left|\partial_x^\alpha \partial_\xi^\gamma G_{i,j}(x,\xi) - \partial_x^\alpha \partial_\xi^\gamma G_{i,j}(x,\eta)\right|}{|\xi-\eta|^\sigma} \\ &\leqslant c\left(|x-\xi|^{-2-\sigma-\delta_{i,4}-|\alpha|} + |x-\eta|^{-2-\sigma-\delta_{i,4}-|\alpha|}\right)\end{aligned} \tag{1.3}$$

for $j = 1,2,3$, $|\alpha| \leqslant 1-\delta_{i,4}$, $|\gamma| = 1$. Here σ is a sufficiently small positive number which is specified below (cf. Theorem 5.1).

Note that, as a consequence of (1.1), the estimate

$$\begin{aligned}&\frac{\left|\partial_\xi^\gamma G_{i,j}(x,\xi) - \partial_\xi^\gamma G_{i,j}(y,\xi)\right|}{|x-y|^\sigma} \\ &\leqslant c\left(|x-\xi|^{-1-\sigma-\delta_{j,4}-|\gamma|} + |y-\xi|^{-1-\sigma-\delta_{j,4}-|\gamma|}\right)\end{aligned} \tag{1.4}$$

with arbitrary $\sigma \in (0,1)$ holds for $x,y,\xi \in \mathcal{G}$, $x \neq y$, $i = 1,2,3$, $|\gamma| \leqslant 1-\delta_{j,4}$. Analogously,

$$\begin{aligned}&\frac{\left|\partial_x^\alpha G_{i,j}(x,\xi) - \partial_x^\alpha G_{i,j}(x,\eta)\right|}{|\xi-\eta|^\sigma} \\ &\leqslant c\left(|x-\xi|^{-1-\sigma-\delta_{i,4}-|\alpha|} + |x-\eta|^{-1-\sigma-\delta_{i,4}-|\alpha|}\right)\end{aligned} \tag{1.5}$$

for $j = 1,2,3$, $|\alpha| \leqslant 1-\delta_{i,4}$.

Finally, I mention that the Hölder estimate

$$\frac{\left|\partial_{x_i}\partial_{\xi_j} G(x,\xi) - \partial_{y_i}\partial_{\xi_j} G(y,\xi)\right|}{|x-y|^\sigma} \leqslant c\left(|x-\xi|^{-3-\sigma} + |y-\xi|^{-3-\sigma}\right)$$

with a certain positive $\sigma < 1$ was derived in [2] for the Green matrix of the Dirichlet problem for the Laplacian in a convex polyhedral domain. This coincides with (1.2) for $i,j=1,2,3$ and $|\alpha|=|\gamma|=1$. The result was applied in [2] to estimate the error in finite element methods.

2 The Green Matrix for the Stokes System

We assume that $\mathcal{G}$ is a bounded polyhedron in $\mathbb{R}^3$, the boundary $\partial\mathcal{G}$ of which consists of the plane faces Γ_j, $j=1,\dots,N$, the edges M_k, $k=1,\dots,N'$, and the vertices $x^{(1)},\dots,x^{(d)}$. It is well known that the boundary value problem

$$-\Delta u+\nabla p=f,\quad -\nabla\cdot u=g \ \text{ in } \mathcal{G}, \tag{2.1}$$

$$u=0 \quad \text{on } \partial\mathcal{G} \tag{2.2}$$

is solvable in $W^{1,2}(\mathcal{G})^3\times L_2(\mathcal{G})$ for arbitrary $f\in W^{-1,2}(\mathcal{G})^3$ and $g\in L_2(\mathcal{G})$ satisfying the condition $g\perp 1$, i.e.,

$$\int_{\mathcal{G}} g(x)\,dx=0.$$

The solution (u,p) is unique up to vectors $(0,c)$, where c is a constant.

Let ϕ be an infinitely differentiable function in $\mathcal{G}$ which vanishes in a neighborhood of the edges such that

$$\int_{\mathcal{G}} \phi(x)\,dx=1.$$

The matrix

$$G(x,\xi)=\big(G_{i,j}(x,\xi)\big)_{i,j=1,2,3,4}$$

is called the *Green matrix* for the problem (2.1), (2.2) if the vector functions $\mathbf{G}_j=(G_{1,j},G_{2,j},G_{3,j})^t$ and the function $G_{4,j}$ are solutions of the problems

$$\begin{aligned}
&-\Delta_x\mathbf{G}_j(x,\xi)+\nabla_x G_{4,j}(x,\xi)=\delta(x-\xi)\,(\delta_{1,j},\delta_{2,j},\delta_{3,j})^t \quad \text{for } x,\xi\in\mathcal{G},\\
&-\nabla_x\cdot\mathbf{G}_j(x,\xi)=\big(\delta(x-\xi)-\phi(x)\big)\,\delta_{4,j} \quad \text{for } x,\xi\in\mathcal{G},\\
&\mathbf{G}_j(x,\xi)=0 \ \text{ for } x\in\partial\mathcal{G},\ \xi\in\mathcal{G}
\end{aligned}$$

and $G_{4,j}$ satisfies the condition

$$\int_{\mathcal{G}} G_{4,j}(x,\xi)\,\phi(x)\,dx=0 \text{ for } \xi\in\mathcal{G},\ j=1,2,3,4.$$

As was shown in [9] (cf. also [13, Theorem 4.5]), there exists a uniquely determined Green matrix $G(x,\xi)$ such that the vector functions

$$x\to\zeta(x,\xi)\,\big(\mathbf{G}_j(x,\xi),G_{4,j}(x,\xi)\big)$$

belong to the space $\overset{\circ}{W}^{1,2}(\mathcal{G})^3 \times L_2(\mathcal{G})$ for each $\xi \in \mathcal{G}$ and for every infinitely differentiable function $\zeta(\cdot,\xi)$ equal to zero in a neighborhood of the point $x = \xi$. Note that

$$G_{i,j}(x,\xi) = G_{j,i}(\xi,x) \quad \text{for } x,\xi \in \mathcal{G},\ i,j = 1,2,3,4.$$

Furthermore, if $f \in W^{-1,2}(\mathcal{G})^3$, $g \in L_2(\mathcal{G})$, $g \perp 1$, and $(u,p) \in \overset{\circ}{W}{}^{1,2}(\mathcal{G})^3 \times L_2(\mathcal{G})$ is the uniquely determined solution of the Stokes system (2.1) satisfying the condition

$$\int_{\mathcal{G}} p(x)\,\phi(x)\,dx = 0,$$

then the components of (u,p) admit the representations

$$u_i(x) = \int_{\mathcal{G}} f(\xi)\cdot \mathbf{G}_i(\xi,x)\,d\xi + \int_{\mathcal{G}} g(\xi)\,G_{4,i}(\xi,x)\,d\xi,\ i = 1,2,3,$$

$$p(x) = \int_{\mathcal{G}} f(\xi)\cdot \mathbf{G}_4(\xi,x)\,d\xi + \int_{\mathcal{G}} g(\xi)\,G_{4,4}(x,\xi)\,d\xi.$$

3 A Local Estimate for Solutions of the Stokes System

We introduce the following positive real numbers μ_k and Λ_ν.

1) Let θ_k denote the angle at the edge M_k. If $\theta_k < \pi$, then we define $\delta_k^+ = 1$ and $\mu_k = \pi/\theta_k$. In the case $\theta_k > \pi$, the numbers μ_k and δ_k^+ are defined as the smallest positive solution of the equation

$$\sin(\mu\theta_k) + \mu \sin\theta_k = 0.$$

Note that δ_k^+ is the smallest positive eigenvalue of the pencil $A_k(\lambda)$ introduced in [13, Section 4.2]. In the case $\theta_k < \pi$, the number μ_k is the smallest eigenvalue greater than $\delta_k^+ = 1$. Furthermore, let $\mu_k' = \min(2,\mu_k)$.

2) Let $\mathcal{K}_\nu$ be the infinite cone with vertex at $x^{(\nu)}$ which coincides with $\mathcal{G}$ in a neighborhood of $x^{(\nu)}$, i.e.,

$$\mathcal{K}_\nu = \{x \in \mathbb{R}^3 :\ (x - x^{(\nu)})/|x - x^{(\nu)}| \in \Omega_\nu\}.$$

Here Ω_ν is a domain of polygonal type on the sphere $|x - x^{(\nu)}| = 1$. We denote the distance from x to the vertex $x^{(\nu)}$ by $\rho_\nu(x)$, the distance to the edge M_k by $r_k(x)$, and consider the operator

$$\overset{\circ}{W}{}_2^1(\Omega_\nu)^3 \times L_2(\Omega_\nu) \ni \binom{u}{p} \to \mathfrak{A}_\nu(\lambda)\binom{u}{p}$$
$$= \begin{pmatrix} \rho_\nu^{2-\lambda}\big(-\Delta(\rho_\nu^\lambda u) + \nabla(\rho_\nu^{\lambda-1}p)\big) \\ \rho_\nu^{1-\lambda}\nabla\cdot(\rho_\nu^\lambda u)\end{pmatrix} \in W_2^{-1}(\Omega_\nu)^3 \times L_2(\Omega_\nu).$$

The operator $\mathfrak{A}_\nu(\lambda)$ depends quadratically on the complex parameter λ. As was shown in [8] (cf. also [3, Theorem 5.5.6]), the strip $-1 \leqslant \operatorname{Re}\lambda \leqslant 0$ is free of eigenvalues of the pencil $\mathfrak{A}_\nu(\lambda)$. We denote by Λ_ν^+ the greatest positive real number such that the strip $0 < \operatorname{Re}\lambda < \Lambda_\nu^+$ is free of eigenvalues of the pencil $\mathfrak{A}_\nu(\lambda)$. Note that $\lambda = 1$ is always an eigenvalue of this pencil. If the cone $\mathcal{K}_\nu$ is convex, then the strip $-1/2 \leqslant \operatorname{Re}\lambda \leqslant 1$ does not contain other eigenvalues (cf. [1, 4] and [3, Theorem 5.5.6]). We define Λ_ν as the smallest positive number such that the strip $0 < \operatorname{Re}\lambda < \Lambda_\nu$ contains at most the eigenvalue $\lambda = 1$ of the pencil $\mathfrak{A}_\nu(\lambda)$.

We introduce the weighted Sobolev spaces $V_{\beta,\delta}^{l,s}(\mathcal{K}_\nu)$ and $W_{\beta,\delta}^{l,s}(\mathcal{K}_\nu)$, where l is an integer, $l \geqslant 0$, s, β are real numbers, $s > 1$, and $\delta = (\delta_k)_{k\in I_\nu}$ is a tuple of real numbers. The space $V_{\beta,\delta}^{l,s}(\mathcal{K}_\nu)$ is defined as the closure of the set $C_0^\infty(\overline{\mathcal{K}}_\nu\backslash\mathcal{S}_\nu)$ with respect to the norm

$$\|u\|_{V_{\beta,\delta}^{l,s}(\mathcal{K}_\nu)} = \Big(\int_{\mathcal{K}_\nu} \sum_{|\alpha|=0}^{l} \rho_\nu^{s(\beta-l+|\alpha|}\prod_{k\in I_\nu}\Big(\frac{r_k}{\rho_\nu}\Big)^{s(\delta_k-l+|\alpha|)} \big|\partial_x^\alpha u(x)\big|^s\,dx\Big)^{1/s}.$$

Here, $\mathcal{S}_\nu$ is the set of all edge points of the cone $\mathcal{K}_\nu$, and $C_0^\infty(\overline{\mathcal{K}}_\nu\backslash\mathcal{S}_\nu)$ is the set of all infinitely differentiable functions with compact support in $\overline{\mathcal{K}}_\nu\backslash\mathcal{S}_\nu$. If $\delta_k > -2/s$ for all $k \in I_\nu$, then $W_{\beta,\delta}^{l,s}(\mathcal{K}_\nu)$ denotes the weighted Sobolev space with the norm

$$\|u\|_{W_{\beta,\delta}^{l,s}(\mathcal{K}_\nu)} = \Big(\int_{\mathcal{K}_\nu} \sum_{|\alpha|=0}^{l} \rho_\nu^{s(\beta-l+|\alpha|}\prod_{k\in I_\nu}\Big(\frac{r_k}{\rho_\nu}\Big)^{s\delta_k} \big|\partial_x^\alpha u(x)\big|^s\,dx\Big)^{1/s}.$$

More precisely, $W_{\beta,\delta}^{l,s}(\mathcal{K}_\nu)$ is the closure of the set $C_0^\infty(\overline{\mathcal{K}}_\nu\backslash\{x^{(\nu)}\})$ with respect to the above norm. Using the inequalities

$$c_1\, r(x) \leqslant \rho_\nu(x) \prod_{k\in I_\nu} \frac{r_k(x)}{\rho_\nu(x)} \leqslant c_2\, r(x) \quad \text{for all } x \in \mathcal{K}_\nu\,, \tag{3.1}$$

where $r(x) = \min\limits_{k\in I_\nu} r_k(x)$ is the distance to the set $\mathcal{S}_\nu$ and c_1, c_2 are certain positive constants, one can obtain various equivalent norms in the spaces $V_{\beta,\delta}^{l,s}(\mathcal{K}_\nu)$ and $W_{\beta,\delta}^{l,s}(\mathcal{K}_\nu)$.

The next lemma follows immediately from [15, Lemmas 3.2 and 3.4].

Lemma 3.1. 1) *Let $u \in W_{\beta,\delta}^{l,s}(\mathcal{K}_\nu)$, $l > 3/s$ and $\delta_k < l - 3/s$ for all k. Then*

$$\big\|\rho_\nu^{\beta-l+3/s}\,u\big\|_{L_\infty(\mathcal{K}_\nu)} \leqslant c\,\|u\|_{W_{\beta,\delta}^{l,s}(\mathcal{K}_\nu)}$$

with a constant c independent of u.

2) *Let $u \in V_{\beta,\delta}^{l,s}(\mathcal{K}_\nu)$ and $l > 3/s$. Then*

$$\Big\| \rho_\nu^{\beta-l+3/s} \prod_{k\in I_\nu} \Big(\frac{r_k}{\rho_\nu}\Big)^{\delta_k-l+3/s} u \Big\|_{L_\infty(\mathcal{K}_\nu)} \leqslant c\, \|u\|_{V^{l,s}_{\beta,\delta}(\mathcal{K}_\nu)}$$

with a constant c independent of u.

We consider the problem

$$-\Delta u + \nabla p = f, \quad -\nabla\cdot u = g \ \text{ in } \mathcal{K}_\nu, \quad u|_{\partial\mathcal{K}_\nu} = 0. \tag{3.2}$$

Let ζ and η be smooth functions on $[0,\infty)$ such that $\zeta = 1$ on $[a,b]$, $\eta = 1$ in a neighborhood of $\operatorname{supp}\zeta$ and $\eta = 0$ on $[b',\infty)$, where a, b, b' are certain constants, $0 \leqslant a < b < b'$. Furthermore, we put

$$\zeta_R(x) = \zeta(R^{-1}|x - x^{(\nu)}|), \quad \eta_R(x) = \eta(R^{-1}|x - x^{(\nu)}|).$$

Lemma 3.2. *Let (u,p) be a solution of the boundary value problem (3.2), where $\eta_R(u,p) \in \overset{\circ}{W}{}^{1,2}(\mathcal{K}_\nu)^3 \times L_2(\mathcal{K}_\nu)$, $\eta_R f \in V^{l-2,s}_{\beta,\delta}(\mathcal{K}_\nu)^3$, $\eta_R g \in V^{l-1,s}_{\beta,\delta}(\mathcal{K}_\nu)$, l is an arbitrary integer, $l \geqslant 2$, $0 < l-\beta-3/s < \Lambda_\nu$ and $l-\mu'_k < \delta_k+2/s < l$ for $k \in I_\nu$. Then*

$$\rho_\nu(x)^{\beta-l+|\alpha|+3/s} \prod_{k\in I_\nu} \Big(\frac{r_k(x)}{\rho_\nu(x)}\Big)^{\delta_k-l+|\alpha|+3/s} \big|\partial_x^\alpha u(x)\big| \leqslant c\, A \tag{3.3}$$

for $|\alpha| < l-3/s$, $a \leqslant \rho_\nu(x) \leqslant b$ and

$$\rho_\nu(x)^{\beta-l+1+|\alpha|+3/s} \prod_{k\in I_\nu} \Big(\frac{r_k(x)}{\rho_\nu(x)}\Big)^{\delta_k-l+1+|\alpha|+3/s} \big|\partial_x^\alpha p(x)\big| \leqslant c\, A \tag{3.4}$$

for $1 \leqslant |\alpha| < l-1-3/s$, $a \leqslant \rho_\nu(x) \leqslant b$. Here, A denotes the expression

$$A = \|\eta_R f\|_{V^{l-2,s}_{\beta,\delta}(\mathcal{K}_\nu)^3} + \|\eta_R g\|_{V^{l-1,s}_{\beta,\delta}(\mathcal{K}_\nu)} + R^{\beta-l+(6-s)/(2s)} \Big(\|\eta_R\|_{\overset{\circ}{W}{}^{1,2}(\mathcal{K}_\nu)^3} + \|\eta_R p\|_{L_2(\mathcal{K}_\nu)} \Big)$$

and c is a constant independent of u, p, R and x. Furthermore,

$$\big|p(x)\big| \leqslant c\, R^{l-1-\beta-3/s}\, A \tag{3.5}$$

for $a \leqslant \rho_\nu(x) \leqslant b$ if $l-1 > 3/s$ and $\delta_k + 3/s < l-1$.

Proof. Without loss of generality, we may assume that $x^{(\nu)}$ is the origin. Then we set

$$v(x) = u(Rx), \quad q(x) = Rp(Rx), \quad F(x) = R^2 f(Rx), \quad G(x) = Rg(Rx).$$

It is obvious that

$$-\Delta v + \nabla q = F \quad \text{and} \quad -\nabla \cdot v = G$$

in $\mathcal{K}_\nu$. Let ζ_R, η_R be the above introduced functions. In particular, $\zeta_1(x) = \zeta(|x|)$ and $\eta_1(x) = \eta(|x|)$. By [14, Corollary 4.11], there exists a constant c_0 such that

$$\|\zeta_1 v\|_{W^{l,s}_{\beta,\delta}(\mathcal{K}_\nu)^3} + \|\zeta_1(q-c_0)\|_{W^{l-1,s}_{\beta,\delta}(\mathcal{K}_\nu)} + |c_0|$$
$$\leqslant c\Big(\|\eta_1 F\|_{V^{l-2,s}_{\beta,\delta}(\mathcal{K}_\nu)^3} + \|\eta_1 G\|_{V^{l-1,s}_{\beta,\delta}(\mathcal{K}_\nu)} + \|\eta_1 v\|_{\overset{\circ}{W}{}^{1,2}(\mathcal{K}_\nu)^3} + \|\eta_1 q\|_{L_2(\mathcal{K}_\nu)}\Big).$$

The right-hand side of the last inequality is equal to $c\,R^{l-\beta-3/s}\,A$. Since $\zeta_1 v \in W^{l,s}_{\beta,\delta}(\mathcal{K}_\nu)^3$ and $\delta_k + 2/s < l-1$, the trace of $\zeta_1 v$ and of the derivatives $\partial_{x_j}(\zeta_1 v)$ on the edges of the cone $\mathcal{K}_\nu$ exist. The Dirichlet condition $v = 0$ on $\partial\mathcal{K}_\nu$ implies $\zeta_1 v = \partial_{x_j}(\zeta_1 v) = 0$ on M_k for $k \in I_\nu$, $j = 1, 2, 3$. Since moreover $\delta_k + 2/s > l-2$, it follows that $\zeta_1 v \in V^{l,s}_{\beta,\delta}(\mathcal{K}_\nu)^3$ and

$$\|\zeta_1 v\|_{V^{l,s}_{\beta,\delta}(\mathcal{K}_\nu)^3} \leqslant c\,\|\zeta_1 v\|_{W^{l,s}_{\beta,\delta}(\mathcal{K}_\nu)^3}$$

(cf. [10, Lemmas 1.1 and 1.3] and [16, Theorem 4, Corollary 2]). Furthermore, $V^{l-2,s}_{\beta,\delta}(\mathcal{K}_\nu) = W^{l-2,s}_{\beta,\delta}(\mathcal{K}_\nu)$ if $\delta_k + 2/s > l-2$ for all $k \in I_\nu$. Thus, we obtain

$$\|\zeta_1 v\|_{V^{l,s}_{\beta,\delta}(\mathcal{K}_\nu)^3} + \|\zeta_1 \nabla q\|_{V^{l-2,s}_{\beta,\delta}(\mathcal{K}_\nu)^3} \leqslant c\,R^{l-\beta-3/s}\,A$$

which implies

$$\|\zeta_R u\|_{V^{l,s}_{\beta,\delta}(\mathcal{K}_\nu)^3} + \|\zeta_R \nabla p\|_{V^{l-2,s}_{\beta,\delta}(\mathcal{K}_\nu)^3} \leqslant c\,A.$$

Applying the second part of Lemma 3.1, we get (3.3) and (3.4). Furthermore, the inequality

$$\|\zeta_1(q-c_0)\|_{W^{l-1,s}_{\beta,\delta}(\mathcal{K}_\nu)} + |c_0| \leqslant c\,R^{l-\beta-3/s}\,A$$

yields

$$\|\zeta_R(p-c_0/R)\|_{W^{l-1,s}_{\beta,\delta}(\mathcal{K}_\nu)} + R^{\beta-l+3/s}\,|c_0| \leqslant c\,A.$$

If $l-1 > 3/s$ and $\delta_k + 3/s < l-1$, then the inequality

$$\|\rho_\nu^{\beta-l+1+3/s}\zeta_R(p-c_0/R)\|_{L_\infty(\mathcal{K}_\nu)} + R^{\beta-l+3/s}\,|c_0| \leqslant c\,A$$

holds by means of the first part of Lemma 3.1. Since $l-1-\beta-3/s > 0$ and $\rho_\nu(x) < bR$ on the support of ζ_R, it follows that

$$\|\zeta_R p\|_{L_\infty(\mathcal{K}_\nu)} \leqslant c\,R^{l-1-\beta-3/s}\,A.$$

The proof is complete. □

4 Point Estimates of the Green Matrix

In the sequel, $\mathcal{U}_\nu$ denotes a neighborhood of the vertex $x^{(\nu)}$ which has a positive distance to all edges M_k, $k \notin I_\nu$. In the following two theorems, we obtain estimates for the elements of the matrix $G(x,\xi)$ when x and ξ lie in a neighborhood $\mathcal{U}_\nu$ of the same vertex $x^{(\nu)}$.

Theorem 4.1. *Suppose that $x,\xi \in \mathcal{G} \cap \mathcal{U}_\nu$ and $\rho_\nu(\xi) < \rho_\nu(x)/2$. Then*

$$\big|\partial_x^\alpha \partial_\xi^\gamma G_{i,j}(x,\xi)\big| \leqslant c\,\rho_\nu(x)^{-1-\Lambda_\nu-\delta_{i,4}-|\alpha|+\varepsilon}\,\rho_\nu(\xi)^{\Lambda_\nu-\delta_{j,4}-|\gamma|-\varepsilon} \times \prod_{k\in I_\nu}\Big(\frac{r_k(x)}{\rho_\nu(x)}\Big)^{\mu_k'-\delta_{i,4}-|\alpha|-\varepsilon} \prod_{k\in I_\nu}\Big(\frac{r_k(\xi)}{\rho_\nu(\xi)}\Big)^{\mu_k'-\delta_{j,4}-|\gamma|-\varepsilon}$$

for $|\alpha| \geqslant \delta_{i,4}$, $|\gamma| \geqslant \delta_{j,4}$,

$$\big|\partial_x^\alpha G_{i,4}(x,\xi)\big| \leqslant c\,\rho_\nu(x)^{-2-\delta_{i,4}-|\alpha|} \prod_{k\in I_\nu}\Big(\frac{r_k(x)}{\rho_\nu(x)}\Big)^{\mu_k'-\delta_{i,4}-|\alpha|-\varepsilon}$$

for $|\alpha| \geqslant \delta_{i,4}$,

$$\big|\partial_\xi^\gamma G_{4,j}(x,\xi)\big| \leqslant c\,\rho_\nu(x)^{-2-\Lambda_\nu+\varepsilon}\,\rho_\nu(\xi)^{\Lambda_\nu-\delta_{j,4}-|\gamma|-\varepsilon} \prod_{k\in I_\nu}\Big(\frac{r_k(\xi)}{\rho_\nu(\xi)}\Big)^{\mu_k'-\delta_{j,4}-|\gamma|-\varepsilon}$$

for $|\gamma| \geqslant \delta_{j,4}$, and

$$\big|G_{4,4}(x,\xi)\big| \leqslant c\,\rho_\nu(x)^{-3}.$$

Here, ε is an arbitrarily small positive number.

Proof. Let $x \in \mathcal{G} \cap \mathcal{U}_\nu$. Furthermore, let η be a smooth cut-off function such that $\eta(\xi) = 1$ for $\rho_\nu(\xi) \leqslant 5\rho_\nu(x)/8$ and $\eta(\xi) = 0$ for $\rho_\nu(\xi) \geqslant 3\rho_\nu(x)/4$. We consider the vector functions $\mathbf{H}_i = (G_{i,1}, G_{i,2}, G_{i,3})^t$ and the functions $G_{i,4}$. Since

$$-\Delta_\xi \mathbf{H}_i(x,\xi) + \nabla_\xi G_{i,4}(x,\xi) = \delta(x-\xi)\,(\delta_{i,1},\delta_{i,2},\delta_{i,3})^t \quad \text{for } x,\xi \in \mathcal{G},$$
$$-\nabla_\xi \cdot \mathbf{H}_i(x,\xi) = \big(\delta(x-\xi) - \phi(\xi)\big)\,\delta_{i,4} \quad \text{for } x,\xi \in \mathcal{G},$$
$$\mathbf{H}_i(x,\xi) = 0 \quad \text{for } x \in \mathcal{G}, \quad xi \in \partial\mathcal{G},$$

it follows that

$$\eta(\xi)\,\big(-\Delta_\xi \partial_x^\alpha \mathbf{H}_i(x,\xi) + \nabla_\xi \partial_x^\alpha G_{i,4}(x,\xi)\big) = 0 \quad \text{for } \xi \in \mathcal{G}.$$

Furthermore,

$$\eta(\xi)\,\nabla_\xi \cdot \mathbf{H}_i(x,\xi) = \eta(\xi)\,\phi(\xi)\,\delta_{i,4} \quad \text{and} \quad \eta(\xi)\,\nabla_\xi \cdot \partial_x^\alpha \mathbf{H}_i(x,\xi) = 0 \ \text{ for } |\alpha| \geqslant 1.$$

Let l be sufficiently large, and let β and δ_k be real numbers satisfying the inequalities

$$1 < l - \beta - 3/s < \Lambda_\nu, \quad l - \mu_k' < \delta_k + 2/s < l - 1 - 1/s. \tag{4.1}$$

Then, by Lemma 3.2,

$$\rho_\nu(\xi)^{\beta-l+|\gamma|+3/s} \prod_{k\in I_\nu} \Big(\frac{r_k(\xi)}{\rho_\nu(\xi)}\Big)^{\delta_k-l+|\gamma|+3/s} \big|\partial_x^\alpha \partial_\xi^\gamma \mathbf{H}_i(x,\xi)\big| \leqslant c\, A_i \tag{4.2}$$

for $\rho_\nu(\xi) < \rho_\nu(x)/2$, where

$$A_i = \rho_\nu(x)^{\beta-l+(6-s)/(2s)} \Big(\|\eta(\cdot)\partial_x^\alpha \mathbf{H}_i(x,\cdot)\|_{\overset{\circ}{W}{}^{1,2}(\mathcal{G})^3} + \|\eta(\cdot)\partial_x^\alpha G_{i,4}(x,\cdot)\|_{L_2(\mathcal{G})}\Big).$$

Furthermore,

$$\rho_\nu(\xi)^{\beta-l+1+|\gamma|+3/s} \prod_{k\in I_\nu} \Big(\frac{r_k(\xi)}{\rho_\nu(\xi)}\Big)^{\delta_k-l+1+|\gamma|+3/s} \big|\partial_x^\alpha \partial_\xi^\gamma G_{i,4}(x,\xi)\big| \leqslant c\, A_i \tag{4.3}$$

for $\rho_\nu(\xi) < \rho_\nu(x)/2$, $|\gamma| \geqslant 1$, and

$$\begin{aligned} \big|\partial_x^\alpha G_{i,4}(x,\xi)\big| \leqslant c\, \rho_\nu(x)^{-3/2} \Big(&\|\eta(\cdot)\,\partial_x^\alpha \mathbf{H}_i(x,\cdot)\|_{\overset{\circ}{W}{}^{1,2}(\mathcal{K}_\nu)^3} \\ &+ \|\eta(\cdot)\,\partial_x^\alpha G_{i,4}(x,\cdot)\|_{L_2(\mathcal{K}_\nu)}\Big). \end{aligned} \tag{4.4}$$

Let $f \in W^{-1,2}(\mathcal{G})^3$, $g \in L_2(\mathcal{G})$, and let ψ be an infinitely differentiable function such that $\psi(y) = 1$ for $|\rho_\nu(x) - \rho_\nu(y)| < \rho_\nu(x)/8$ and $\psi(y) = 0$ for $|\rho_\nu(x) - \rho_\nu(y)| > \rho_\nu(x)/4$. We consider the vector function (u,p) with the components

$$u_i(x) = \int_{\mathcal{G}} \eta(y)\, f(y)\cdot \mathbf{H}_i(x,y)\, dy + \int_{\mathcal{G}} \eta(y)\, g(y)\, G_{i,4}(x,y)\, dy, \tag{4.5}$$

$$p(x) = \int_{\mathcal{G}} \eta(y)\, f(y)\cdot \mathbf{H}_4(x,y)\, dy + \int_{\mathcal{G}} \eta(y)\, g(y)\, G_{4,4}(x,y)\, dy, \tag{4.6}$$

which satisfies the equations

$$-\Delta u + \nabla p = \eta f, \qquad -\nabla\cdot u = \eta g - C\phi, \quad \text{where } C = \int_{\mathcal{G}} \eta(y)\, g(y)\, dy.$$

Since $\psi(-\Delta u + \nabla p) = 0$ and $\psi\nabla\cdot u = C\psi\phi$, it follows from Lemma 3.2 that

$$\begin{aligned} &\rho_\nu(x)^{\beta-l+|\alpha|+3/s} \prod_{k\in I_\nu} \Big(\frac{r_k(x)}{\rho_\nu(x)}\Big)^{\delta_k-l+|\alpha|+3/s} \big|\partial_x^\alpha u(x)\big| \\ &\qquad \leqslant c\, \rho_\nu(x)^{\beta-l+(6-s)/(2s)} \Big(\|\psi u\|_{\overset{\circ}{W}{}^{1,2}(\mathcal{G})^3} + \|\psi p\|_{L_2(\mathcal{G})}\Big). \end{aligned}$$

Consequently,

$$\left|\partial_x^\alpha u(x)\right| \leqslant c\,\rho_\nu(x)^{-|\alpha|-1/2} \prod_{k\in I_\nu} \Big(\frac{r_k(x)}{\rho_\nu(x)}\Big)^{l-\delta_k-|\alpha|-3/s} \|(f,g)\|.$$

Here by $\|(f,g)\|$, we mean the norm of (f,g) in $W^{-1,2}(\mathcal{G})^3 \times L_2(\mathcal{G})$. Furthermore,

$$\left|\partial_x^\alpha p(x)\right| \leqslant c\,\rho_\nu(x)^{-|\alpha|-3/2} \prod_{k\in I_\nu} \Big(\frac{r_k(x)}{\rho_\nu(x)}\Big)^{l-1-\delta_k-|\alpha|-3/s} \|(f,g)\|$$

for $|\alpha| \geqslant 1$ and

$$\left|p(x)\right| \leqslant c\,\rho_\nu(x)^{-3/2} \|(f,g)\|.$$

Thus,

$$\left\|\eta(\cdot)\,\partial_x^\alpha \mathbf{H}_i(x,\cdot)\right\|_{\overset{\circ}{W}{}^{1,2}(\mathcal{G})^3} \leqslant c\,\rho_\nu(x)^{-|\alpha|-\delta_{i,4}-1/2} \prod_{k\in I_\nu} \Big(\frac{r_k(x)}{\rho_\nu(x)}\Big)^{l-\delta_{i,4}-\delta_k-|\alpha|-3/s}$$

for $|\alpha| \geqslant \delta_{i,4}$, while

$$\left\|\eta(\cdot)\,\mathbf{H}_4(x,\cdot)\right\|_{\overset{\circ}{W}{}^{1,2}(\mathcal{G})^3} \leqslant c\,\rho_\nu(x)^{-3/2}.$$

Analogously,

$$\left\|\eta(\cdot)\,\partial_x^\alpha G_{i,4}(x,\cdot)\right\|_{L_2(\mathcal{G})} \leqslant c\,\rho_\nu(x)^{-|\alpha|-\delta_{i,4}-1/2} \prod_{k\in I_\nu} \Big(\frac{r_k(x)}{\rho_\nu(x)}\Big)^{l-\delta_{i,4}-\delta_k-|\alpha|-3/s}$$

for $|\alpha| \geqslant \delta_{i,4}$ and

$$\left\|\eta(\cdot)\,G_{4,4}(x,\cdot)\right\|_{L_2(\mathcal{G})^3} \leqslant c\,\rho_\nu(x)^{-3/2}.$$

We can choose l, s, β and δ_k such that

$$l-\beta-3/s = \Lambda_\nu - \varepsilon, \quad l-\delta_k-3/s = \mu'_k - \varepsilon, \quad 1/s < \varepsilon.$$

Combining then the last four inequalities with (4.2)–(4.4), we obtain the desired estimates for the elements of the matrix $G(x,\xi)$. □

Using the equality $G_{i,j}(x,\xi) = G_{j,i}(\xi,x)$, one can deduce analogous estimates for the elements of the Green matrix in the case $\rho_\nu(\xi) > 2\rho_\nu(x)$.

Next, we consider the case where x and ξ lie in a neighborhood $\mathcal{U}_\nu$ of the vertex $x^{(\nu)}$ and $\rho_\nu(x)/2 < \rho_\nu(\xi) < 2\rho_\nu(x)$. In the sequel, μ' denotes the minimum of the numbers μ'_k.

Theorem 4.2. 1) *If* $x,\xi \in \mathcal{G}\cap\mathcal{U}_\nu$ *and* $|x-\xi| < \min(r(x), r(\xi))$, *then*

$$\left|\partial_x^\alpha \partial_\xi^\gamma G_{i,j}(x,\xi)\right| \leqslant c_{\alpha,\gamma}\, |x-\xi|^{-1-\delta_{i,4}-\delta_{j,4}-|\alpha|-|\gamma|}.$$

The constant $c_{\alpha,\gamma}$ is independent of x and ξ.

2) *Suppose that $x,\xi \in \mathcal{G}\cap\mathcal{U}_\nu$, $\rho_\nu(x)/2 < \rho_\nu(\xi) < 2\rho_\nu(x)$ and $|x-\xi| > \min(r(x), r(\xi))$. Then*

$$\left|\partial_x^\alpha \partial_\xi^\gamma G_{i,j}(x,\xi)\right| \leqslant c_{\alpha,\gamma} |x-\xi|^{-1-\delta_{i,4}-\delta_{j,4}-|\alpha|-|\gamma|} \Big(\frac{r(x)}{|x-\xi|}\Big)^{\mu'-\delta_{i,4}-|\alpha|-\varepsilon} \times \Big(\frac{r(\xi)}{|x-\xi|}\Big)^{\mu'-\delta_{j,4}-|\gamma|-\varepsilon} \tag{4.7}$$

for $|\alpha| \geqslant \delta_{i,4}$, $|\gamma| \geqslant \delta_{j,4}$,

$$\left|\partial_x^\alpha G_{i,4}(x,\xi)\right| \leqslant c_{\alpha,\gamma} |x-\xi|^{-2-\delta_{i,4}-|\alpha|} \Big(\frac{r(x)}{|x-\xi|}\Big)^{\mu'-\delta_{i,4}-|\alpha|-\varepsilon} \tag{4.8}$$

for $|\alpha| \geqslant \delta_{i,4}$,

$$\left|\partial_\xi^\gamma G_{4,j}(x,\xi)\right| \leqslant c_{\alpha,\gamma} |x-\xi|^{-2-\delta_{j,4}-|\gamma|} \Big(\frac{r(\xi)}{|x-\xi|}\Big)^{\mu'-\delta_{j,4}-|\gamma|-\varepsilon} \tag{4.9}$$

for $|\gamma| \geqslant \delta_{j,4}$, and $|G_{4,4}(x,\xi)| \leqslant c\,|x-\xi|^{-3}$. Here, ε is an arbitrarily small positive number.

Proof. The first estimate is shown in [13, Theorem 4.11]. Furthermore, it follows from [13, Theorems 4.11 and 4.14] that

$$\left|\partial_x^\alpha \partial_\xi^\gamma G_{i,j}(x,\xi)\right| \leqslant c\, |x-\xi|^{-1-\delta_{i,4}-\delta_{j,4}-|\alpha|-|\gamma|} \Big(\frac{r(x)}{|x-\xi|}\Big)^{\min(0,\mu'-\delta_{i,4}-|\alpha|-\varepsilon)} \times \Big(\frac{r(\xi)}{|x-\xi|}\Big)^{\min(0,\mu'-\delta_{j,4}-|\gamma|-\varepsilon)} \tag{4.10}$$

for $\rho_\nu(\xi)/8 < \rho_\nu(x) < 8\rho_\nu(\xi)$ and $4|x-\xi| > 3\min(r(x), r(\xi))$. Thus, in particular, the estimate (4.7) is valid for $|\alpha| + \delta_{i,4} \geqslant 2$ and $|\gamma| + \delta_{j,4} \geqslant 2$, the estimate (4.8) is valid for $|\alpha| + \delta_{i,4} \geqslant 2$, while (4.9) holds for $|\gamma| + \delta_{j,4} \geqslant 2$.

We prove (4.7) for $i,j = 1,2,3$, $|\alpha| = 1$ and $|\gamma| \geqslant 2$. If $|x-\xi| < 4r(x)$, then the estimate

$$\left|\partial_x^\alpha \partial_\xi^\gamma G_{i,j}(x,\xi)\right| \leqslant c\, |x-\xi|^{-2-|\gamma|} \Big(\frac{r(x)}{|x-\xi|}\Big)^{\mu'-1-\varepsilon} \Big(\frac{r(\xi)}{|x-\xi|}\Big)^{\mu'-|\gamma|-\varepsilon}$$

follows immediately from (4.10). Let $|x-\xi| > 4r(x)$. Then we denote by y the nearest point to x on the set

$$S_\nu = \bigcup_{k\in I_\nu} \overline{M}_k\,.$$

It can b e easily verified that the inequalities

$$\rho_\nu(\xi)/8 < \rho_\nu(z) < 8\rho_\nu(\xi) \quad \text{and} \quad 3|x-\xi| < 4|z-\xi| < 5|x-\xi|$$

are satisfied for the point $z = y + t(x-y)$, where $t \in (0,1)$. Moreover, $r(z) = t|x-y| = t\, r(x) > 3\min(r(z), r(\xi))/4$. Consequently by (4.10),

$$\begin{aligned} \big|(\nabla_x \partial_x^\alpha \partial_\xi^\gamma G_{i,j})(z,\xi)\big| &\leqslant c\,|z-\xi|^{-3-|\gamma|}\Big(\frac{r(z)}{|z-\xi|}\Big)^{\mu'-2-\varepsilon}\Big(\frac{r(\xi)}{|z-\xi|}\Big)^{\mu'-|\gamma|-\varepsilon} \\ &\leqslant c\,|x-\xi|^{-3-|\gamma|}\Big(\frac{t\, r(x)}{|x-\xi|}\Big)^{\mu'-2-\varepsilon}\Big(\frac{r(\xi)}{|x-\xi|}\Big)^{\mu'-|\gamma|-\varepsilon}. \end{aligned}$$

Since $G_{i,j}(x,\xi) = 0$ for $x \in \partial\mathcal{G}$ and $i \neq 4$, we have

$$G_{i,j}(x,\xi) = 0 \quad \text{and} \quad \nabla_x G_{i,j}(x,\xi) = 0 \quad \text{for } x \in \mathcal{S},\ i \neq 4.$$

This implies

$$\begin{aligned} \big|\partial_x^\alpha \partial_\xi^\gamma G_{i,j}(x,\xi)\big| &= \big|(\partial_x^\alpha \partial_\xi^\gamma G_{i,j})(x,\xi) - (\partial_x^\alpha \partial_\xi^\gamma G_{i,j})(y,\xi)\big| \\ &= \Big|\int_0^1 \frac{d}{dt}(\partial_x^\alpha \partial_\xi^\gamma G_{i,j})(y+t(x-y),\xi)\,dt\Big| \\ &\leqslant |x-y| \int_0^1 \big|(\nabla_x \partial_x^\alpha \partial_\xi^\gamma G_{i,j})(y+t(x-y),\xi)\big|\,dt \\ &\leqslant c\,|x-\xi|^{-2-|\gamma|}\Big(\frac{r(x)}{|x-\xi|}\Big)^{\mu'-1-\varepsilon}\Big(\frac{r(\xi)}{|x-\xi|}\Big)^{\mu'-|\gamma|-\varepsilon}. \end{aligned}$$

Thus, (4.7) is proved for $i,j = 1,2,3$, $|\alpha| = 1$, $|\gamma| \geqslant 2$. Repeating this argument, we get

$$\big|\partial_\xi^\gamma G_{i,j}(x,\xi)\big| \leqslant |x-\xi|^{-1-|\gamma|}\Big(\frac{r(x)}{|x-\xi|}\Big)^{\mu'-\varepsilon}\Big(\frac{r(\xi)}{|x-\xi|}\Big)^{\mu'-|\gamma|-\varepsilon}$$

for $i,j = 1,2,3$, $|\gamma| \geqslant 2$. This proves (4.7) for $i,j = 1,2,3$, $|\gamma| \geqslant 2$ and arbitrary α. Analogously, the inequality (4.7) for $i = 1,2,3$, $j = 4$, $|\gamma| \geqslant 1$ and the inequality (4.8) for $i = 1,2,3$ hold.

Using the fact that

$$G_{i,j}(x,\xi) = 0 \quad \text{and} \quad \nabla_\xi G_{i,j}(x,\xi) = 0 \quad \text{for } \xi \in \mathcal{S},\ j \neq 4.$$

we obtain the inequalities (4.7) and (4.9) for $j = 1,2,3$ and $|\gamma| \leqslant 1$ by the same arguments as above. This completes the proof. □

Finally, we consider the case where x and ξ lie in neighborhoods $\mathcal{U}_\mu$ and $\mathcal{U}_\nu$ of different vertices $x^{(\mu)}$ and $x^{(\nu)}$, respectively.

Theorem 4.3. *Suppose that $x \in \mathcal{U}_\mu$ and $\xi \in \mathcal{U}_\nu$, where $\mathcal{U}_\mu$ and $\mathcal{U}_\nu$ are neighborhoods of the vertices $x^{(\mu)}$ and $x^{(\nu)}$, respectively, which have a positive distance. Then*

$$\big|\partial_x^\alpha \partial_\xi^\gamma G_{i,j}(x,\xi)\big| \leqslant c\,\rho_\mu(x)^{\Lambda_\mu-\delta_{i,4}-|\alpha|-\varepsilon}\,\rho_\nu(\xi)^{\Lambda_\nu-\delta_{j,4}-|\gamma|-\varepsilon} \\ \times \prod_{k\in I_\mu}\Big(\frac{r_k(x)}{\rho_\mu(x)}\Big)^{\mu_k'-\delta_{i,4}-|\alpha|-\varepsilon} \prod_{k\in I_\nu}\Big(\frac{r_k(\xi)}{\rho_\nu(\xi)}\Big)^{\mu_k'-\delta_{j,4}-|\gamma|-\varepsilon}.$$

for $|\alpha| \geqslant \delta_{i,4}$, $|\gamma| \geqslant \delta_{j,4}$,

$$\big|\partial_x^\alpha G_{i,4}(x,\xi)\big| \leqslant c\,\rho_\mu(x)^{\Lambda_\mu-\delta_{i,4}-|\alpha|-\varepsilon} \prod_{k\in I_\mu}\Big(\frac{r_k(x)}{\rho_\mu(x)}\Big)^{\mu_k'-\delta_{i,4}-|\alpha|-\varepsilon}$$

for $|\alpha| \geqslant \delta_{i,4}$,

$$\big|\partial_\xi^\gamma G_{4,j}(x,\xi)\big| \leqslant c\,\rho_\nu(\xi)^{\Lambda_\nu-\delta_{j,4}-|\gamma|-\varepsilon} \prod_{k\in I_\nu}\Big(\frac{r_k(\xi)}{\rho_\nu(\xi)}\Big)^{\mu_k'-\delta_{j,4}-|\gamma|-\varepsilon}$$

for $|\gamma| \geqslant \delta_{j,4}$, and $\big|G_{4,4}(x,\xi)\big| \leqslant c$. Here, c is a constant independent of x and ξ, and ε is an arbitrarily small positive number.

Proof. Let ζ and η be smooth cut-off functions, $\zeta = 1$ on $\mathcal{U}_\mu$, $\eta = 1$ on $\mathcal{U}_\nu$, $\zeta = 0$ near all edges M_k, $k \notin I_\mu$, $\eta = 0$ near all edges M_k, $k \notin I_\nu$. We assume that the distance between the supports of ζ and η is positive. Let A_i denote the expression

$$A_i = \big\|\eta(\cdot)\,\partial_x^\alpha \mathbf{H}_i(x,\cdot)\big\|_{\overset{\circ}{W}{}^{1,2}(\mathcal{G})^3} + \big\|\eta(\cdot)\,\partial_x^\alpha G_{i,4}(x,\cdot)\big\|_{L_2(\mathcal{G})}.$$

Here again $\mathbf{H}_i$ is the vector with the components $G_{i,1}, G_{i,2}, G_{i,3}$. Since

$$\eta(z)\,\big(-\Delta_z \mathbf{H}_i(x,z) + \nabla_z G_{i,4}(x,z)\big) = 0$$

and

$$\eta(z)\,\nabla_z \cdot \mathbf{H}_i(x,z) = \eta(z)\,\varphi(z)\,\delta_{i,4}$$

for $z \in \mathcal{G}$, it follows from Lemma 3.2 that

$$\big|\partial_x^\alpha \partial_\xi^\gamma \mathbf{H}_i(x,\xi)\big| \leqslant c\,\rho_\nu(\xi)^{l-\beta-|\gamma|-3/s} \prod_{k\in I_\nu}\Big(\frac{r_k(\xi)}{\rho_\nu(\xi)}\Big)^{l-\delta_k-|\gamma|-3/s} A_i$$

for all γ,

$$\big|\partial_x^\alpha \partial_\xi^\gamma G_{i,4}(x,\xi)\big| \leqslant c\,\rho_\nu(\xi)^{l-1-\beta-|\gamma|-3/s} \prod_{k\in I_\nu}\Big(\frac{r_k(\xi)}{\rho_\nu(\xi)}\Big)^{l-1-\delta_k-|\gamma|-3/s} A_i$$

for $|\gamma| \geqslant 1$ and $|\partial_x^\alpha G_{i,4}(x,\xi)| \leqslant c\, A_i$, where c is a constant independent of x and ξ, l is a sufficiently large integer, β and δ_k are real numbers satisfying the inequalities (4.1).

Let $f \in W^{-1,2}(\mathcal{G})$ and $g \in L_2(\mathcal{G})$. We consider the vector function with the components (4.5) and (4.6). Since

$$\zeta(-\Delta u + \nabla p) = 0 \quad\text{and}\quad \zeta\nabla\cdot u = C\zeta\varphi, \quad\text{where } C = \int_{\mathcal{G}} \eta(y)\, g(y)\, dy,$$

Lemma 3.2 yields

$$|\partial_x^\alpha u(x)| \leqslant c\,\rho_\mu(x)^{l-\beta'-|\alpha|-3/s} \prod_{k\in I_\mu} \Big(\frac{r_k(x)}{\rho_\mu(x)}\Big)^{l-\delta_k-|\alpha|-3/s} \|(f,g)\|$$

for all α,

$$|\partial_x^\alpha p(x)| \leqslant c\,\rho_\mu(x)^{l-1-\beta'-|\alpha|-3/s} \prod_{k\in I_\mu} \Big(\frac{r_k(x)}{\rho_\mu(x)}\Big)^{l-1-\delta_k-|\alpha|-3/s} \|(f,g)\|$$

for $|\alpha| \geqslant 1$, and

$$|p(x)| \leqslant c\,\|(f,g)\|.$$

Here by $\|(f,g)\|$, we mean the norm of (f,g) in the space $W^{-1,2}(\mathcal{G})^3 \times L_2(\mathcal{G})$, and β' is an arbitrary real number such that $1 < l - \beta' - 3/s < \Lambda_\mu$. Thus, we obtain the following estimates for the expressions A_i:

$$A_i \leqslant c\,\rho_\mu(x)^{l-\beta'-\delta_{i,4}-|\alpha|-3/s} \prod_{k\in I_\mu} \Big(\frac{r_k(x)}{\rho_\mu(x)}\Big)^{l-\delta_k-\delta_{i,4}-|\alpha|-3/s} \quad \text{for } |\alpha| \geqslant \delta_{i,4}$$

and $A_4 \leqslant c$. We can choose l, s, β, β' and δ_k such that

$$l - \beta - 3/s = \Lambda_\nu - \varepsilon, \quad l - \beta' - 3/s = \Lambda_\mu - \varepsilon, \quad l - \delta_k - 3/s = \mu_k' - \varepsilon,$$

where $\varepsilon > 1/s$. Combining then the last two inequalities for A_i with the estimates for $\partial_x^\alpha \partial_\xi^\gamma \mathbf{H}_i(x,\xi)$ and $\partial_x^\alpha \partial_\xi^\gamma G_{i,4}(x,\xi)$ in the first part of the proof, we obtain the assertion of the theorem. □

As a consequence of the estimates in Theorems 4.1–4.3, we obtain the following result. Here, we use the inequalities

$$\mu_k' > 1 \quad\text{and}\quad \Lambda_\nu > 1$$

which are valid in the case of a convex polyhedron.

Theorem 4.4. *Let the polyhedron $\mathcal{G}$ be convex. Then the elements of the matrix $G(x,\xi)$ satisfy the estimate* (1.1) *for* $|\alpha| \leqslant 1 - \delta_{i,4}$, $|\gamma| \leqslant 1 - \delta_{j,4}$.

Proof. If x and ξ lie in (sufficiently small) neighborhoods of different vertices, then the assertion of the lemma follows immediately from the estimates in Theorem 4.3. We assume that x and ξ lie in a neighborhood $\mathcal{U}_\nu$ of the same vertex $x^{(\nu)}$. Then, by Theorem 4.1,

$$\left|\partial_x^\alpha \partial_\xi^\gamma G_{i,j}(x,\xi)\right| \leqslant c\,\rho_\nu(x)^{-1-\delta_{i,4}-\delta_{j,4}-|\alpha|-|\gamma|}$$

for $\rho_\nu(\xi) < \rho_\nu(x)/2$, $|\alpha| + \delta_{i,4} \leqslant 1$, $|\gamma| + \delta_{j,4} \leqslant 1$. Using the inequality $|x-\xi| < 3\rho_\nu(x)/2$ for $\rho_\nu(\xi) < \rho_\nu(x)/2$, we obtain (1.1). The same estimate holds for $\rho_\nu(x) < \rho_\nu(\xi)/2$ by means of the equality $G_{i,j}(x,\xi) = G_{j,i}(\xi,x)$ and Theorem 4.1. In the case $\rho_\nu(x)/2 < \rho_\nu(\xi) < 2\rho_\nu(x)$, the inequality (1.1) follows from Theorem 4.2. □

5 Hölder Estimates for the Elements of the Green Matrix

The following Hölder estimate for the elements $\partial_\xi^\gamma G_{i,j}(x,\xi)$, $i = 1,2,3$ can be easily deduced from the last theorem.

Corollary 5.1. *Let the polyhedron $\mathcal{G}$ be convex. Then the estimate* (1.4) *with arbitrary $\sigma \in (0,1)$ holds for $x, y, \xi \in \mathcal{G}$, $x \neq y$, $i = 1,2,3$, $|\gamma| \leqslant 1 - \delta_{j,4}$. Analogously, the estimate* (1.5) *is satisfied for $x, y, \xi \in \mathcal{G}$, $x \neq y$, $j = 1,2,3$, $|\alpha| \leqslant 1 - \delta_{i,4}$.*

Proof. If $|x-\xi| < 2|x-y|$, then $|y-\xi| < 3|x-y|$ and, consequently,

$$\begin{aligned}\left|\partial_\xi^\gamma G_{i,j}(x,\xi) - \partial_\xi^\gamma G_{i,j}(y,\xi)\right| &\leqslant c\left(|x-\xi|^{-1-\delta_{j,4}-|\gamma|} + |y-\xi|^{-1-\delta_{j,4}-|\gamma|}\right)\\ &\leqslant c\,3^\sigma\left(|x-\xi|^{-1-\sigma-\delta_{j,4}-|\gamma|} + |y-\xi|^{-1-\sigma-\delta_{j,4}-|\gamma|}\right)|x-y|^\sigma.\end{aligned}$$

Suppose that $|x-\xi| > 2|x-y|$. Then, by the mean value theorem,

$$\left|\partial_\xi^\gamma G_{i,j}(x,\xi) - \partial_\xi^\gamma G_{i,j}(y,\xi)\right| \leqslant |x-y|\,\left|\nabla_z \partial_\xi^\gamma G_{i,j}(z,\xi)\right|,$$

where $z = x + t(y-x)$, $0 < t < 1$. Since $|z-\xi| > |x-\xi|/2 > |x-y|$, Theorem 4.4 yields

$$\begin{aligned}\left|\partial_\xi^\gamma G_{i,j}(x,\xi) - \partial_\xi^\gamma G_{i,j}(y,\xi)\right| &\leqslant c\,|x-y|\;|z-\xi|^{-2-\delta_{j,4}-|\gamma|}\\ &\leqslant c\,|x-y|^\sigma\,|x-\xi|^{-1-\sigma-\delta_{j,4}-|\gamma|}.\end{aligned}$$

This proves (1.4). The estimate (1.5) follows from (1.4) and from the equality $G_{i,j}(x,\xi) = G_{j,i}(\xi,x)$. □

We prove an analogous estimate for the derivatives $\partial_{x_k}\partial_\xi^\gamma G_{i,j}(x,\xi)$, $i = 1,2,3$ and $\partial_x^\alpha \partial_{\xi_k}^\gamma G_{i,j}(x,\xi)$, $j = 1,2,3$, respectively.

Theorem 5.1. *Suppose that the polyhedron $\mathcal{G}$ is convex and that σ is a positive number, $\sigma < 1$, $\sigma < \mu_k - 1$ for all k, and $\sigma < \Lambda_\nu - 1$ for all ν. Then the estimate* (1.2) *holds for $i = 1,2,3$, $|\alpha| = 1$, $|\gamma| \leqslant 1 - \delta_{j,4}$, while* (1.3) *is satisfied for $j = 1,2,3$, $|\alpha| \leqslant 1 - \delta_{i,4}$, $|\gamma| = 1$.*

6 Proof of Theorem 5.1

Since $G_{i,j}(x,\xi) = G_{j,i}(\xi,x)$, it suffices to prove (1.2). The validity of this estimate follows from the subsequent Lemmas 6.1–6.4, where we consider the cases $|x-\xi| < m|x-y|$, $|x-\xi| > m|x-y| > r(x)$ and $m|x-y| < \min(|x-\xi|, r(x))$ separately. Here, m is an arbitrary fixed (sufficiently large) positive number.

Lemma 6.1. *Let $\mathcal{G}$ be a convex polyhedron, and let m be an arbitrary positive number. Then the estimate* (1.2) *is satisfied for $i = 1,2,3$, $|\alpha| = 1$, $|\gamma| \leqslant 1 - \delta_{j,4}$, $x,y,\xi \in \mathcal{G}$, $x \neq y$, $|x-\xi| < m|x-y|$.*

Proof. By Theorem 4.4,

$$\frac{\big|\partial_x^\alpha \partial_\xi^\gamma G_{i,j}(x,\xi)\big|}{|x-y|^\sigma} \leqslant c\,\frac{|x-\xi|^{-2-\delta_{j,4}-|\gamma|}}{|x-y|^\sigma} \leqslant c\,m^\sigma\,|x-\xi|^{-2-\sigma-\delta_{j,4}-|\gamma|}$$

for $|x-\xi| < m|x-y|$. Analogously,

$$\frac{\big|\partial_x^\alpha \partial_\xi^\gamma G_{i,j}(x,\xi)\big|}{|x-y|^\sigma} \leqslant c\,(m+1)^\sigma\,|x-\xi|^{-2-\sigma-\delta_{j,4}-|\gamma|}$$

since $|y-\xi| < (m+1)|x-y|$. This proves the lemma. □

Next, we consider the case $|x-\xi| > m|x-y|$. In the following lemma, we assume that x and ξ lie in neighborhoods $\mathcal{U}_\mu$ and $\mathcal{U}_\nu$ of different vertices.

Lemma 6.2. *Let $\mathcal{G}$ be a convex polyhedron, and let σ be a positive number, $\sigma < 1$, $\sigma < \Lambda_\nu - 1$ for all ν, and $\sigma < \mu_k - 1$ for all k. Furthermore, let δ be an arbitrary fixed positive number, and let m be sufficiently large. Then* (1.2) *is satisfied for $i = 1,2,3$, $|\alpha| = 1$, $|\gamma| \leqslant 1 - \delta_{j,4}$, $x,y \in \mathcal{G} \cap \mathcal{U}_\mu$, $\xi \in \mathcal{G} \cap \mathcal{U}_\nu$, $x \neq y$, $|x-\xi| > m|x-y|$, $|x-\xi| > \delta$.*

Proof. First note that the inequalities $|x-\xi| > m|x-y|$ and $|x-\xi| > \delta$ imply $|y-\xi| > (m-1)|x-y|$ and $|y-\xi| > (m-1)\delta/m$. Suppose first that $r(x) < m|x-y|$. Then $r(y) < (m+1)|x-y|$, and Theorem 4.3 together with (3.1) yield

$$\begin{aligned}\big|\partial_x^\alpha \partial_\xi^\gamma G_{i,j}(x,\xi)\big| &\leqslant c\,\rho_\mu(x)^{\Lambda_\mu-1-\varepsilon} \prod_{k\in I_\mu} \Big(\frac{r_k(x)}{\rho_\mu(x)}\Big)^{\mu_k'-1-\varepsilon} \\ &\leqslant c\,r(x)^\sigma \leqslant c\,m^\sigma |x-y|^\sigma\end{aligned}$$

and, analogously,

$$\left|\partial_y^\alpha \partial_\xi^\gamma G_{i,j}(y,\xi)\right| \leqslant c\,(m+1)^\sigma\,|x-y|^\sigma$$

for $i \neq 4$, $|\alpha| = 1$, $|\gamma| \leqslant 1 - \delta_{j,4}$. This implies (1.2).

If $r(x) > m|x-y|$, then there exists a point $z = x + t(y-x)$, $0 < t < 1$, such that

$$\begin{aligned}\left|\partial_x^\alpha \partial_\xi^\gamma G_{i,j}(x,\xi) - \partial_y^\alpha \partial_\xi^\gamma G_{i,j}(y,\xi)\right| &\leqslant |x-y|\,\left|\nabla_z \partial_z^\alpha \partial_\xi^\gamma G_{i,j}(z,\xi)\right| \\ &\leqslant c\,|x-y|\,\rho_\mu(z)^{\Lambda_\mu-2-\varepsilon} \prod_{k\in I_\mu} \Big(\frac{r_k(x)}{\rho_\mu(x)}\Big)^{\mu_k'-2-\varepsilon} \\ &\leqslant c\,|x-y|\,r(z)^{\sigma-1} \leqslant c\,(m-1)^{\sigma-1}|x-y|^\sigma.\end{aligned}$$

This proves the lemma. □

The last lemma allows us to restrict ourselves in the proof of the following two lemmas to the case where x and ξ lie in a neighborhood of the same vertex $x^{(\nu)}$.

Lemma 6.3. *Let $\mathcal{G}$ be a convex polyhedron, and let σ be a positive number, $\sigma < 1$, $\sigma < \Lambda_\nu - 1$ for all ν, and $\sigma < \mu_k - 1$ for all k. Then the estimate (1.2) is satisfied for $i = 1,2,3$, $|\alpha| = 1$, $|\gamma| \leqslant 1 - \delta_{j,4}$, $x,y,\xi \in \mathcal{G}$, $x \neq y$, $|x-\xi| > m|x-y| > r(x)$.*

Proof. Since $|x-\xi| > m|x-y|$, where m is sufficiently large, we may suppose that x and y lie in a neighborhood $\mathcal{U}_\nu$ of the same vertex $x^{(\nu)}$. We assume that ξ lies in the same neighborhood and consider the following cases:

1. $\xi \in \mathcal{U}_\nu$, $\rho_\nu(\xi) < \rho_\nu(x)/2$,
2. $\xi \in \mathcal{U}_\nu$, $\rho_\nu(\xi) > 2\rho_\nu(x)$,
3. $\xi \in \mathcal{U}_\nu$, $\rho_\nu(x)/2 < \rho_\nu(\xi) < 2\rho_\nu(x)$.

We start with Case 1. Then obviously $|x-\xi| < \rho_\nu(x) + \rho_\nu(\xi) < 3\rho_\nu(x)/2$, and Theorem 4.1 together with (3.1) yield

$$\begin{aligned}\left|\partial_x^\alpha \partial_\xi^\gamma G_{i,j}(x,\xi)\right| &\leqslant c\,\rho_\nu(x)^{-2-\delta_{j,4}-|\gamma|} \Big(\frac{r(x)}{\rho_\nu(x)}\Big)^\sigma \\ &\leqslant c\,|x-\xi|^{-2-\sigma-\delta_{j,4}-|\gamma|}\,|x-y|^\sigma\end{aligned}$$

for $i = 1,2,3$, $|\alpha| = 1$, $|\gamma| + \delta_{j,4} \leqslant 1$. Since

$$\rho_\nu(y) > \rho_\nu(x) - |x-y| > \rho_\nu(x) - \frac{|x-\xi|}{m} > \Big(1 - \frac{3}{2m}\Big)\rho_\nu(x) > \Big(2 - \frac{3}{m}\Big)\rho_\nu(\xi),$$

we obtain analogously

$$\left|\partial_y^\alpha \partial_\xi^\gamma G_{i,j}(y,\xi)\right| \leqslant c\,|y-\xi|^{-2-\sigma-\delta_{j,4}-|\gamma|}\,|x-y|^\sigma.$$

This proves (1.2) for Case 1.

Case 2. In this case, $|x-\xi| < 3\rho_\nu(\xi)/2$ and

$$\rho_\nu(y) < \rho_\nu(x) + |x-y| < \rho_\nu(x) + \frac{|x-\xi|}{m} < \Big(\frac{1}{2} + \frac{3}{2m}\Big)\,\rho_\nu(\xi).$$

Using Theorem 4.1 and the inequality $G_{i,j}(x,\xi) = G_{j,i}(\xi,x)$, we obtain

$$\begin{aligned}\big|\partial_x^\alpha \partial_\xi^\gamma G_{i,j}(x,\xi)\big| &\leqslant c\,\rho_\nu(x)^{\Lambda_\nu - 1-\varepsilon}\,\rho_\nu(\xi)^{-1-\Lambda_\nu-\delta_{j,4}-|\gamma|+\varepsilon}\Big(\frac{r(x)}{\rho_\nu(x)}\Big)^\sigma \\ &\leqslant c\,\rho_\nu(\xi)^{-2-\sigma-\delta_{j,4}-|\gamma|}\,r(x)^\sigma \leqslant c\,|x-\xi|^{-2-\sigma-\delta_{j,4}-|\gamma|}\,|x-y|^\sigma\end{aligned}$$

and, analogously,

$$\big|\partial_y^\alpha \partial_\xi^\gamma G_{i,j}(y,\xi)\big| \leqslant c\,|y-\xi|^{-2-\sigma-\delta_{j,4}-|\gamma|}\,|x-y|^\sigma$$

for $i = 1,2,3$, $|\alpha| = 1$, $|\gamma| + \delta_{j,4} \leqslant 1$. Thus, (1.2) holds for Case 1.

Case 3. Then $|x-\xi| < 3\rho_\nu(\xi)$, and from the conditions of the lemma it follows that

$$\Big(\frac{1}{2} - \frac{3}{m}\Big)\,\rho_\nu(\xi) < \rho_\nu(y) < \Big(2 + \frac{3}{m}\Big)\,\rho_\nu(\xi).$$

Furthermore, $r(x) < |x-\xi|$ and

$$r(y) < (m+1)\,|x-y| < \frac{m+1}{m}\,|x-\xi| < \frac{m+1}{m-1}\,|y-\xi|.$$

Applying Theorem 4.2, we obtain

$$\begin{aligned}\big|\partial_x^\alpha \partial_\xi^\gamma G_{i,j}(x,\xi)\big| &\leqslant c\,|x-\xi|^{-2-\delta_{j,4}-|\gamma|}\Big(\frac{r(x)}{|x-\xi|}\Big)^\sigma \\ &\leqslant c\,|x-\xi|^{-2-\sigma-\delta_{j,4}-|\gamma|}\,|x-y|^\sigma\end{aligned}$$

and analogously

$$\big|\partial_y^\alpha \partial_\xi^\gamma G_{i,j}(y,\xi)\big| \leqslant c\,|y-\xi|^{-2-\sigma-\delta_{j,4}-|\gamma|}\,|x-y|^\sigma$$

for $|\alpha| = 1$, $|\gamma| + \delta_{j,4} \leqslant 1$. Consequently, the inequality (1.2) is also valid in Case 3. This completes the proof. □

It remains to prove (1.2) for the case $m|x-y| < \min(|x-\xi|, r(x))$.

Lemma 6.4. *Let $\mathcal{G}$ be a convex polyhedron, and let σ be a positive number, $\sigma < 1$, $\sigma < \Lambda_\nu - 1$ for all ν, and $\sigma < \mu_k - 1$ for all k. Then the estimate (1.2) is satisfied for $i = 1,2,3$, $|\alpha| = 1$, $|\gamma| \leqslant 1 - \delta_{j,4}$, $x, y, \xi \in \mathcal{G}$, $x \neq y$, $|x-\xi| > m|x-y|$, $r(x) > m|x-y|$.*

Proof. By the mean value theorem,

$$\left|\partial_x^\alpha \partial_\xi^\gamma G_{i,j}(x,\xi) - \partial_y^\alpha \partial_\xi^\gamma G_{i,j}(y,\xi)\right| \leqslant |x-y|\, \left|\nabla_z \partial_z^\alpha \partial_\xi^\gamma G_{i,j}(z,\xi)\right|, \tag{6.1}$$

where $z = x+t(y-x)$, $0<t<1$. Since $m|x-y| < |x-\xi|$ and m is assumed to be sufficiently large, we may suppose that x, ξ and y lie in a neighborhood $\mathcal{U}_\nu$ of the same vertex $x^{(\nu)}$. We estimate the right-hand side of (6.1) for $|\alpha|=1$ and $|\gamma| \leqslant 1-\delta_{j,4}$. To this end, we consider the same cases as in the proof of Lemma 6.3.

Case 1. $\xi \in \mathcal{U}_\nu$, $\rho_\nu(\xi) < \rho_\nu(x)/2$. Since $|x-\xi| < 3\rho_\nu(x)/2$ and

$$\begin{aligned}\rho_\nu(z) &> \rho_\nu(x) - |x-y| > \rho_\nu(x) - \frac{|x-\xi|}{m} > \Big(1-\frac{3}{2m}\Big)\rho_\nu(x)\\ &> \Big(2-\frac{3}{m}\Big)\rho_\nu(\xi),\end{aligned}$$

Theorem 4.1 yields

$$\left|\nabla_z \partial_z^\alpha \partial_\xi^\gamma G_{i,j}(z,\xi)\right| \leqslant c\,\rho_\nu(z)^{-3-\delta_{j,4}-|\gamma|} \Big(\frac{r(z)}{\rho_\nu(z)}\Big)^{\sigma-1}.$$

for $i=1,2,3$, $|\alpha|=1$, $|\gamma|+\delta_{j,4} \leqslant 1$. Using the inequalities $r(z) > (m-1)|x-y|$ and

$$\rho_\nu(z) > \rho_\nu(x) - |x-y| > \Big(\frac{2}{3} - \frac{1}{m}\Big)\, |x-\xi|,$$

we obtain

$$\left|\nabla_z \partial_z^\alpha \partial_\xi^\gamma G_{i,j}(z,\xi)\right| \leqslant c\,|x-\xi|^{-2-\sigma-\delta_{j,4}-|\gamma|}\, |x-y|^{\sigma-1}.$$

This together with (6.1 implies (1.2).

Case 2. $\xi \in \mathcal{U}_\nu$, $\rho_\nu(x) < \rho_\nu(\xi)/2$. Then $|x-\xi| < 3\rho_\nu(\xi)/2$ and

$$\rho_\nu(z) < \rho_\nu(x) + |x-y| < \rho_\nu(x) + \frac{|x-\xi|}{m} < \Big(\frac{1}{2} + \frac{3}{2m}\Big)\rho_\nu(\xi).$$

Theorem 4.1 and the equality $G_{i,j}(x,\xi) = G_{j,i}(\xi,x)$ imply

$$\left|\nabla_z \partial_z^\alpha \partial_\xi^\gamma G_{i,j}(z,\xi)\right| \leqslant c\,\rho_\nu(z)^{\Lambda_\nu-2-\varepsilon}\, \rho_\nu(\xi)^{-1-\Lambda_\nu-\delta_{j,4}-|\gamma|+\varepsilon} \Big(\frac{r(z)}{\rho_\nu(z)}\Big)^{\sigma-1}$$

for $i=1,2,3$, $|\alpha|=1$, $|\gamma|+\delta_{j,4} \leqslant 1$. The number ε can be chosen such that $\Lambda_\nu - 1 - \sigma - \varepsilon \geqslant 0$. Then we get

$$\begin{aligned}\left|\nabla_z \partial_z^\alpha \partial_\xi^\gamma G_{i,j}(z,\xi)\right| &\leqslant c\,\rho_\nu(\xi)^{-2-\sigma-\delta_{j,4}-|\gamma|}\, r(z)^{\sigma-1}\\ &\leqslant c\,|x-\xi|^{-2-\sigma-\delta_{j,4}-|\gamma|}\, |x-y|^{\sigma-1}.\end{aligned}$$

This proves (1.2) for the case $\xi \in \mathcal{U}_\nu$, $\rho_\nu(x) < \rho_\nu(\xi)/2$.

Case 3. $\xi \in \mathcal{U}_\nu$, $\rho_\nu(x)/2 < \rho_\nu(\xi) < 2\rho_\nu(x)$. In this case,

$$\left(\frac{1}{2} - \frac{3}{m}\right) \rho_\nu(\xi) < \rho_\nu(z) < \left(2 + \frac{3}{m}\right) \rho_\nu(\xi).$$

Furthermore, by the conditions of the lemma,

$$\left(1 - \frac{1}{m}\right) |x - \xi| < |z - \xi| < \left(1 + \frac{1}{m}\right) |x - \xi|$$

and

$$\left(1 - \frac{1}{m}\right) r(x) < r(z) < \left(1 + \frac{1}{m}\right) r(x).$$

Let $i \neq 4$, $|\alpha| = 1$ and $|\gamma| + \delta_{j,4} \leqslant 1$. Using Theorem 4.2, we get

$$\begin{aligned} \left|\nabla_z \partial_z^\alpha \partial_\xi^\gamma G_{i,j}(z,\xi)\right| &\leqslant c\,|z-\xi|^{-3-\delta_{j,4}-|\gamma|} \left(\frac{r(z)}{|z-\xi|}\right)^{\sigma-1} \\ &\leqslant c\,|x-\xi|^{-2-\sigma-\delta_{j,4}-|\gamma|}\, r(x)^{\sigma-1} \\ &\leqslant c\,|x-\xi|^{-2-\sigma-\delta_{j,4}-|\gamma|}\, |x-y|^{\sigma-1} \end{aligned}$$

for $|z - \xi| > \min(r(z), r(\xi))$ and

$$\begin{aligned} \left|\nabla_z \partial_z^\alpha \partial_\xi^\gamma G_{i,j}(z,\xi)\right| &\leqslant c\,|z-\xi|^{-3-\delta_{j,4}-|\gamma|} \\ &\leqslant c\,|x-\xi|^{-2-\sigma-\delta_{j,4}-|\gamma|}\, |x-y|^{\sigma-1} \end{aligned}$$

for $|z - \xi| < \min(r(z), r(\xi))$. This, together with (6.1), implies (1.2). The proof is complete. □

References

1. Dauge, M.: Stationary Stokes and Navier-Stokes systems on two- and three-dimensional with corners. Part 1: Linearized equations. SIAM J. Math. Anal. **20**, 74–97 (1989)
2. Guzmán, J., Leykekhman, D., Rossmann, J., Schatz, A.-H.: Hölder estimates for Green's functions on convex polyhedral domains and their applications to finite element methods. Numer. Math. **112**, 221–243 (2009)
3. Kozlov, V.A., Maz'ya, V.G., Rossmann, J.: Spectral Problems Associated with Corner Singularities of Solutions to Elliptic Equations. Am. Math. Soc., Providence, RI (2001)
4. Kozlov, V.A., Maz'ya, V.G., Schwab, C.: On singularities of solutions to the Dirichlet problem of hydrodynamics near the vertex of a cone. J. Rein. Angew. Math. **456**, 65–97 (1994)
5. Maz'ya, V.G., Plamenevskii, B.A.: Estimates of Green's functions and Schauder estimates for solutions of elliptic boundary value problems in a dihedral angle (Russian). Sib. Mat. Zh. **19**, no. 5, 1064–1082 (1978); English transl.: Sib. Math. J. **19**, 752–764 (1979)
6. Maz'ya, V.G., Plamenevskii, B.A.: Schauder estimations of solutions of elliptic boundary value problems in domains with edges on the boundary (Russian).

Tr. Semin. S.L. Soboleva 1978, No. 2, 69–102 (1978); English transl.: Am. Math. Soc. Transl. **123**, 141–169 (1984)

7. Maz'ya, V.G., Plamenevskii, B.A.: On the asymptotics of the fundamental solutions of elliptic boundary value problems in regions with conical points (Russian). Probl. Mat. Anal. **7**, 100–145 (1979); English transl.: Sel. Math. Sov. **4** 363–397 (1985)
8. Maz'ya, V.G., Plamenevskii, B.A.: On properties of solutions of three-dimensional problems of elasticity theory and hydrodynamics in domains with isolated singular points (Russian). Dinam. Sploshn. Sred. **50**, 99–120 (1981); English transl.: Am. Math. Soc. Transl. (2) **123**, 109–123 (1984) ,
9. Maz'ya, V.G., Plamenevskii, B.A.: The first boundary value problem for classical equations of mathematical physics in domains with piecewise smooth boundaries. Part 1: Z. Anal. Anwend. **2**, no. 4, 335–359 (1983); Part 2. ibid **2**, no. 6, 523–551 (1983)
10. Maz'ya, V.G., Roßmann (Rossmann), J.: Über die Asymptotik der Lösungen elliptischer Randwertaufgaben in der Umgebung von Kanten. Math. Nachr. **138**, 27–53 (1988)
11. Maz'ya, V.G., Roßmann (Rossmann), J.: On the Agmon–Miranda maximum principle for solutions of elliptic equations in polyhedral and polygonal domains. Ann. Global Anal. Geom. **9**, 253–303 (1991)
12. Maz'ya, V.G., Rossmann, J.: Point estimates for Green's matrix to boundary value problems for second order elliptic systems in a polyhedral cone. Z. Angew. Math. Mech. **82**, no. 5, 291–316 (2002)
13. Maz'ya, V.G., Rossmann, J.: Pointwise estimates for Green's kernel of a mixed boundary value problem to the Stokes system in a polyhedral cone. Math. Nachr. **278**, no. 15, 1766–1810 (2005)
14. Maz'ya, V.G., Rossmann, J.: L_p estimates of solutions to mixed boundary value problems for the Stokes system in polyhedral domains. Math. Nachr. **280**, no. 7, 751–793 (2007)
15. Maz'ya, V.G., Rossmann, J.: Mixed boundary value problems for the Navier-Stokes system in polyhedral domains. Arch. Ration. Mech. Anal. [To appear]
16. Rossmann, J.: On two classes of weighted Sobolev-Slobodetskiĭ spaces in a dihedral angle. Partial Differential Equations, Banach Center Publications **27**, 399–424 (1992)
17. Verzhbinskii, G.M., Maz'ya, V.G.: On the asymptotics of solutions of the Dirichlet problem near a nonregular boundary. Dokl. Akad. Nauk SSSR **176**, no. 3, 498–501 (1967)
18. Verzhbinskii, G.M., Maz'ya, V.G.: Asymptotic behaviour of solutions of elliptic equations of second order near the boundary I, II (Russian). Sib. Mat. Zh. **12**, no. 6, 1217–1249, (1971); ibid **13**, no. 6, 1239–1271 (1972)

Boundary Integral Methods for Periodic Scattering Problems

Gunther Schmidt

Abstract The paper is devoted to the scattering of a plane wave obliquely illuminating a periodic surface. Integral equation methods lead to a system of singular integral equations over the profile. Using boundary integral techniques, we study the equivalence of these equations to the electromagnetic formulation, the existence and uniqueness of solutions under general assumptions on the permittivity and permeability of the materials. In particular, new results for materials with negative permittivity or permeability are established.

1 Introduction

We study the scattering of time-harmonic plane waves by a surface, which in Cartesian coordinates (x, y, z) is periodic in x- and invariant in z-direction and separates two different materials. This is the simplest model of diffraction gratings, which have found several applications in micro-optics, where tools from the semiconductor industry are used to fabricate optical devices with complicated structural features within the length-scale of optical waves. Such diffractive elements have many technological advantages and can be designed to perform functions unattainable with traditional optical elements.

The electromagnetic formulation of the scattering by gratings, which are modeled as infinite periodic structures, can be reduced to a system of Helmholtz equations for the z-components of the electric and magnetic fields in $\mathbf{R}^2$, where the solutions have to be quasiperiodic in one variable, subject to radiation conditions with respect to the other and satisfy certain jump con-

Gunther Schmidt
Weierstrass Institute of Applied Analysis and Stochastics, Mohrenstr. 39, 10117 Berlin, Germany
e-mail: schmidt@wias-berlin.de

A. Laptev (ed.), *Around the Research of Vladimir Maz'ya II: Partial Differential Equations*, International Mathematical Series 12, DOI 10.1007/978-1-4419-1343-2_16,

ditions at the interfaces between different materials of the diffraction grating. In the case of classical diffraction, where the incident wave is orthogonal to the z-direction, the system degenerates to independent transmission problems for the two basic polarizations of the incident wave, whereas for the so called conical diffraction the boundary values of the z-components as well as their normal and tangential derivatives at the interfaces are coupled.

The electromagnetic theory of gratings has been studied since Rayleigh's time. For an introduction to this problem along with some numerical methods see the collection of articles in [14]. By far the largest number of papers in the literature has come from the optics and engineering community, whereas rigorous mathematical results have been obtained only during the last 15 years.

In the case of classical diffraction, existence and uniqueness results are based on the observation that the weak form of corresponding boundary value problems in a periodic cell satisfies a Gårding inequality if the argument of the in general complex permittivity ε of the non-magnetic grating materials satisfies $0 \leqslant \arg \varepsilon < \pi$ (cf. [3, 4] and the references contained therein). Here the radiation condition is reformulated as a nonlocal boundary condition imposed on one part of the boundary of the periodic cell. The reduction of conical diffraction to a transmission problem for the system of Helmholtz equations in $\mathbf{R}^2$ goes back to [15] (in the case of one interface) and [2], where results, similar to classical diffraction, have been established by extending the variational approach.

The variational formulation is also used for the numerical solution of periodic diffraction problems with the Finite-Element-Method, which is now accepted also in the optics community. But the most popular numerical methods for grating problems are methods based on Rayleigh or eigenmode expansions, differential and integral methods, which have been developed since 1970. Especially integral equation methods are very efficient for solving the classical diffraction problems in certain scenarios with large ratio period over wavelength, profile curves with corners and gratings with thin coated layers. Various integral formulations have been proposed and implemented, for example, [12, 16, 17, 5], but a rigorous mathematical and numerical analysis of these methods can not be found in the literature. The mathematical papers dealing with integral formulations of grating problems are mainly concerned with perfectly reflecting gratings or the study of the fundamental solution and radiation conditions.

Recently in [19] an integral equation approach from [12, 16] was extended to conical diffraction, resulting in a system of integral equations over the interface, which contains besides the single and double layer potentials of periodic diffraction also the integral operator with the tangential derivative of the fundamental solution as kernel. It was shown, in particular, that this system of singular integral equations generates a strongly elliptic operator if the materials satisfy the assumptions of the variational approach. For the analysis the interface is conformly mapped to a close curve, such that the transformed integral operators are compact perturbations of boundary integral operators for the Laplacian on that curve.

This allows to apply techniques from the method of boundary integral equations which has been established as a powerful tool for the analysis and numerical solution of boundary value problems for partial differential equations and which was always on the list of the mathematical interests of V. Maz'ya. Among his important contributions to the theory of boundary integral equations are deep results for potential operators on non-smooth surfaces, the solvability of equations and the asymptotics of their solutions, integral operators on contours with peak and the Fredholm radius of double layer potentials (cf., for example, [7, 8, 10, 11]). We mention also Maz'ya's encyclopedia article [9], which is a fine representation of the fundamental theoretical aspects of the application of the potential method in various problems in mathematical physics and mechanics.

In this paper, we apply ideas of [9, 18] to study conical diffraction with quite general assumptions on the permittivity and permeability of the materials. Motivated by recent proposals for the design of optical metamaterials we allow magnetic materials with complex permeability μ, $\arg \mu \in [0, \pi)$, and consider also the case that either ε or μ are negative. Then the variational formulation does not satisfy a Gårding type inequality and strong ellipticity principles do not work. However, the integral formulation can be analyzed by using some more or less standard techniques from singular and second kind integral equations with double layer potentials. We find conditions for the existence and uniqueness of solutions of the integral equation system and for its equivalence to the transmission problem for the Helmholtz equations.

To give an example. Let the profile of the surface in the (x, y)-plane be given by a smooth periodic function $y = f(x)$ and denote the permittivities and permeabilities of the materials above and below the surface by $\varepsilon_{\pm}$ and $\mu_{\pm}$, respectively. Then the integral formulation is solvable if $\varepsilon_- \neq -\varepsilon_+$ and $\mu_- \neq -\mu_+$, and its solution generates a solution of the conical diffraction problem. Moreover, the solution is unique except for certain real ε_- and μ_-, where resonances or so called trapped modes can occur.

In the case of profiles with corners, the existence of solutions can be guaranteed if the ratios

$$\frac{\varepsilon_+ + \varepsilon_-}{\varepsilon_+ - \varepsilon_-} \quad \text{and} \quad \frac{\mu_+ + \mu_-}{\mu_+ - \mu_-}$$

belong to the essential spectrum of the double layer logarithmic potential in the Sobolev space $H^{1/2}$ on the associated closed curve.

The outline of the paper is as follows. Section 2 is devoted to the conical diffraction by periodic structures and the formulation by partial differential equations. Quasiperiodic potentials for Helmholtz equations and integral operators of periodic diffraction are discussed in Section 3. In Section 4, we derive the system of singular integral equations for conical diffraction and study its equivalence to the differential equations. Conditions for the existence and uniqueness of solutions are obtained in the final Section 5.

2 Conical Diffraction

Let Σ be a non self-intersecting curve in the (x, y)-plane which is d-periodic in x-direction. The surface $\Sigma \times \mathbf{R}$ separates two regions $G_\pm \times \mathbf{R} \subset \mathbf{R}^3$ filled with materials of constant electric permittivity $\varepsilon_\pm$ and magnetic permeability $\mu_\pm$ (cf. Fig. 1).

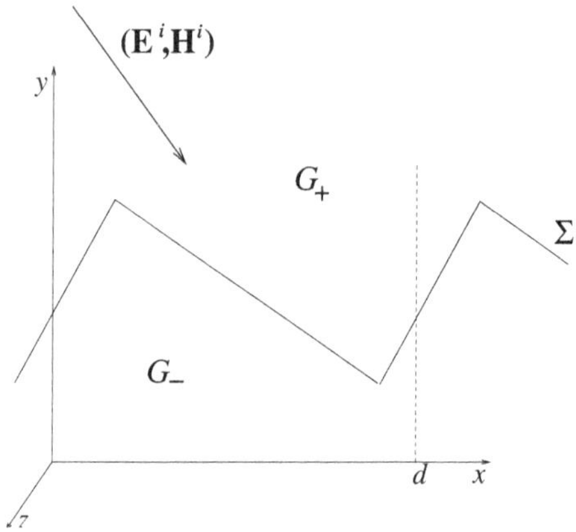

Fig. 1 Schematic presentation of a simple grating.

The surface is illuminated from $G_+ \times \mathbf{R}$, where $\varepsilon_+, \mu_+ > 0$, by a electromagnetic plane wave at oblique incidence

$$\mathbf{E}^i = \mathbf{p}\,e^{i(\alpha x - \beta y + \gamma z)} e^{-i\omega t}\,, \quad \mathbf{H}^i = \mathbf{s}\,e^{i(\alpha x - \beta y + \gamma z)} e^{-i\omega t}\,, \tag{2.1}$$

which due to the periodicity of Σ is scattered into a finite number of plane waves in $G_+ \times \mathbf{R}$ and possibly in $G_- \times \mathbf{R}$. The wave vectors of these outgoing modes lie on the surface of a cone whose axis is parallel to the z–axis. Therefore, in optics one speaks of conical diffraction, which occurs in a variety of technological applications.

2.1 Maxwell equations

The wave $(\mathbf{E}^i, \mathbf{H}^i)$ is scattered by the surface, and the total fields will be given by

$$\begin{aligned} &\mathbf{E}_+ = \mathbf{E}^i + \mathbf{E}^{refl}\,,\ \mathbf{H}_+ = \mathbf{H}^i + \mathbf{H}^{refl} && \text{in the region } G_+ \times \mathbf{R}\,, \\ &\mathbf{E}_- = \mathbf{E}^{tran}\,,\ \mathbf{H}_- = \mathbf{H}^{tran} && \text{in the region } G_- \times \mathbf{R}\,. \end{aligned}$$

Dropping the factor $e^{-i\omega t}$, the incident, diffracted, and total fields satisfy the time–harmonic Maxwell equations

$$\nabla \times \mathbf{E} = i\omega\mu\mathbf{H} \quad \text{and} \quad \nabla \times \mathbf{H} = -i\omega\varepsilon\mathbf{E}\,, \tag{2.2}$$

with piecewise constant functions $\varepsilon(x,y) = \varepsilon_\pm$, $\mu(x,y) = \mu_\pm$ for $(x,y) \in G_\pm$. The components of the wave vector $\mathbf{k}_+ = (\alpha, -\beta, \gamma)$ of the incoming field satisfy

$$\beta > 0 \quad \text{and} \quad \alpha^2 + \beta^2 + \gamma^2 = \omega^2\varepsilon_+\mu_+\,,$$

and they are expressed using the incidence angles $|\theta|, |\phi| < \pi/2$

$$(\alpha, -\beta, \gamma) = \omega\sqrt{\varepsilon_+\mu_+}\,(\sin\theta\cos\phi, -\cos\theta\cos\phi, \sin\phi)\,.$$

If the angle $\phi = 0$, then one speaks of classical periodic diffraction, whereas $\phi \neq 0$ characterizes conical diffraction. To be a solution of the Maxwell system above the surface the coefficient vectors $\mathbf{p}, \mathbf{s}$, which determine the polarization of the incident light (2.1), and the wave vector $\mathbf{k}_+$ are connected by certain compatibility relations.

When crossing the surface the tangential components of the total fields are continuous

$$\mathbf{n} \times (\mathbf{E}_+ - \mathbf{E}_-) = 0 \quad \text{and} \quad \mathbf{n} \times (\mathbf{H}_+ - \mathbf{H}_-) = 0 \quad \text{on } \Sigma \times \mathbf{R}\,, \tag{2.3}$$

where $\mathbf{n}$ is the unit normal to the interface $\Sigma \times \mathbf{R}$. Taking the divergence of (2.2) leads to

$$\nabla \cdot (\varepsilon\mathbf{E}) = 0 \quad \text{and} \quad \nabla \cdot (\mu\mathbf{H}) = 0\,. \tag{2.4}$$

We look for vector fields $\mathbf{E}, \mathbf{H}$ satisfying (2.2) and (2.3) such that

$$\mathbf{E}\,,\ \mathbf{H}\,,\ \nabla \times \mathbf{E}\,,\ \nabla \times \mathbf{H} \in \left(L^2_{loc}(\mathbf{R}^3)\right)^3\,, \tag{2.5}$$

i.e., possessing locally a finite energy.

2.2 Helmholtz equations

Since the geometry is invariant with respect to any translation parallel to the z–axis, we make the ansatz for the total field

$$(\mathbf{E}, \mathbf{H})(x, y, z) = (E, H)(x, y)\,\mathrm{e}^{i\gamma z} \tag{2.6}$$

with $E, H : \mathbf{R}^2 \to \mathbb{C}^3$. This transforms the equations in $\mathbf{R}^3$ into a two-dimensional problem. The Maxwell equations (2.2) provide

$$
\begin{aligned}
E &= (E_x, E_y, E_z) = \frac{i}{\omega\varepsilon}\big(\partial_y H_z - i\gamma H_y, i\gamma H_x - \partial_x H_z, \partial_x H_y - \partial_y H_x\big), \\
H &= (H_x, H_y, H_z) = \frac{1}{i\omega\mu}\big(\partial_y E_z - i\gamma E_y, i\gamma E_x - \partial_x E_z, \partial_x E_y - \partial_y E_x\big).
\end{aligned}
\tag{2.7}
$$

Hence if (2.6) holds, then the condition of locally finite energy (2.5) is satisfied only if the z-components of E, H are H^1- regular, since

$$
\begin{aligned}
\partial_x E_z &= i\gamma E_x - i\omega\mu H_y\,, \quad & \partial_y E_z &= i\gamma E_y + i\omega\mu H_x\,, \\
\partial_x H_z &= i\gamma H_x + i\omega\varepsilon E_y\,, \quad & \partial_y H_z &= i\gamma H_y - i\omega\varepsilon E_x\,.
\end{aligned}
$$

Moreover, from (2.7) we derive

$$
E_x = \frac{i}{\omega\varepsilon}\big(\partial_y H_z - i\gamma H_y\big) = \frac{i}{\omega\varepsilon}\partial_y H_z + \frac{i\gamma}{\omega^2\varepsilon\mu}\partial_x E_z + \frac{\gamma^2}{\omega^2\varepsilon\mu}E_x\,,
$$

implying

$$
\frac{\omega^2\varepsilon\mu - \gamma^2}{\omega^2\varepsilon\mu}\,E_x = \frac{i}{\omega\varepsilon}\,\partial_y H_z + \frac{i\gamma}{\omega^2\varepsilon\mu}\,\partial_x E_z, \tag{2.8}
$$

and similar relations for E_y, H_x and H_y. Noting $\gamma = \omega\sqrt{\varepsilon_+\mu_+}\,\sin\phi$, we introduce the piecewise constant function

$$
\varkappa(x,y) = \begin{cases} \sqrt{\varepsilon_+\mu_+ - \varepsilon_+\mu_+\sin^2\phi} = \varkappa_+\,, & (x,y)\in G_+\,, \\ \sqrt{\varepsilon_-\mu_- - \varepsilon_+\mu_+\sin^2\phi} = \varkappa_-\,, & (x,y)\in G_-\,, \end{cases} \tag{2.9}
$$

where we choose the square root $\sqrt{z} = \sqrt{r}\,e^{i\varphi/2}$ for $z = r\,e^{i\varphi}$, $0 \leqslant \varphi < 2\pi$. Thus, (2.8) and the other relations give

$$
\begin{aligned}
\omega^2\varkappa^2 E_x &= i\gamma\partial_x E_z + i\omega\mu\partial_y H_z, \quad & \omega^2\varkappa^2 E_y &= i\gamma\partial_y E_z - i\omega\mu\partial_x H_z, \\
\omega^2\varkappa^2 H_x &= i\gamma\partial_x H_z - i\omega\varepsilon\partial_y E_z, \quad & \omega^2\varkappa^2 H_y &= i\gamma\partial_y H_z + i\omega\varepsilon\partial_x E_z,
\end{aligned}
\tag{2.10}
$$

implying that under the condition $\varkappa \neq 0$, which will be assumed throughout, the components E_z, H_z of the electric and the magnetic field determine the other components, which in general belong only to L^2.

It follows from (2.7) and (2.4) that the Maxwell equations (2.2) can be replaced by Helmholtz equations

$$
(\Delta + \omega^2\varkappa^2)\,E_z = (\Delta + \omega^2\varkappa^2)\,H_z = 0 \tag{2.11}
$$

in $G_\pm$. The continuity conditions (2.3) on the surface take the form

$$
\big[(n,0)\times E\big]_{\Sigma\times\mathbf{R}} = \big[(n,0)\times H\big]_{\Sigma\times\mathbf{R}} = 0\,,
$$

where $(n, 0) = (n_x, n_y, 0)$ is the normal vector and $\big[(n,0) \times E\big]_{\Sigma \times \mathbf{R}}$ denotes the jump of the function $(n, 0) \times E$ across the interface $\Sigma \times \mathbf{R}$. Since

$$(n,0) \times E = \big(n_y E_z, -n_x E_z, n_x E_y - n_y E_x\big)$$

we conclude that

$$\big[E_z\big]_\Sigma = \big[H_z\big]_\Sigma = 0\,.$$

Furthermore, because of $\varkappa \neq 0$ relations (2.10) give

$$n_x E_y - n_y E_x = \frac{1}{\omega^2 \varkappa^2}\Big(i\gamma(n_x \partial_y E_z - n_y \partial_x E_z) - i\omega\mu(n_x \partial_x H_z + n_y \partial_y H_z)\Big),$$

$$n_x H_y - n_y H_x = \frac{1}{\omega^2 \varkappa^2}\Big(i\gamma(n_x \partial_y H_z - n_y \partial_x H_z) + i\omega\varepsilon(n_x \partial_x E_z + n_y \partial_y E_z)\Big),$$

which implies the jump conditions

$$\Big[\frac{\gamma}{\omega^2\varkappa^2}\,\partial_t H_z + \frac{\omega\varepsilon}{\omega^2\varkappa^2}\,\partial_n E_z\Big]_\Sigma = \Big[\frac{\gamma}{\omega^2\varkappa^2}\,\partial_t E_z - \frac{\omega\mu}{\omega^2\varkappa^2}\,\partial_n H_z\Big]_\Sigma = 0\,.$$

Here $\partial_n = n_x\partial_x + n_y\partial_y$ and $\partial_t = -n_y\partial_x + n_x\partial_y$ are the normal resp. tangential derivatives on Σ. Introduce $B_z = \sqrt{\mu_+/\varepsilon_+}\, H_z$ and use $\gamma = \omega\sqrt{\varepsilon_+\mu_+}\,\sin\phi$ to rewrite the jump conditions in the form

$$\Big[\frac{\varepsilon\,\partial_n E_z}{\varkappa^2}\Big]_\Sigma = -\varepsilon_+ \sin\phi\Big[\frac{\partial_t B_z}{\varkappa^2}\Big]_\Sigma,\quad \Big[\frac{\mu\,\partial_n B_z}{\varkappa^2}\Big]_\Sigma = \mu_+ \sin\phi\Big[\frac{\partial_t E_z}{\varkappa^2}\Big]_\Sigma. \tag{2.12}$$

In addition, the z-components of the incoming field

$$E_z^i(x,y) = p_z \mathrm{e}^{i(\alpha x - \beta y)}\,,\quad B_z^i(x,y) = q_z \mathrm{e}^{i(\alpha x - \beta y)}\,, \tag{2.13}$$

are α-quasiperiodic in x of period d, i.e., satisfy the relation

$$u(x+d, y) = \mathrm{e}^{id\alpha} u(x,y)\,.$$

The periodicity of ε and μ, together with the form of the incident wave, motivates to seek for solutions E_z, B_z which are α-quasiperiodic, too. Because the domain is unbounded, a radiation condition on the scattering problem must be imposed at infinity, namely that the diffracted fields remain bounded. This implies the so called outgoing wave condition

$$\begin{aligned} (E_z, B_z)(x,y) &= (E_z^i, B_z^i) + \sum_{n\in\mathbf{Z}} (E_n^+, B_n^+)\mathrm{e}^{\,i(\alpha_n x + \beta_n^+ y)}, & y \geqslant H,\\ (E_z, B_z)(x,y) &= \sum_{n\in\mathbf{Z}} (E_n^-, B_n^-)\mathrm{e}^{\,i(\alpha_n x - \beta_n^- y)}, & y \leqslant -H, \end{aligned} \tag{2.14}$$

where $\Sigma \subset \{(x,y) : |y| < H\}$, and α_n, $\beta_n^\pm$ are given by

$$\alpha_n = \alpha + \frac{2\pi n}{d}, \quad \beta_n^\pm = \sqrt{\omega^2 \varkappa_\pm^2 - \alpha_n^2} \quad \text{with } 0 \leqslant \arg \beta_n^\pm < \pi . \tag{2.15}$$

The Rayleigh coefficients $E_n^\pm, B_n^\pm \in \mathbb{C}$ are the main characteristics of diffraction gratings. In particular, if $\beta_n^\pm \in \mathbf{R}$ (which is possible only for a finite number of indices), then the Rayleigh coefficients indicate the energy and the phase shift of the propagating modes, i.e., of the outgoing plane waves with the wave vectors

$$(\alpha_n, \beta_n^+, \gamma) \text{ in } G_+ \text{ and } (\alpha_n, -\beta_n^-, \gamma) \text{ in } G_- .$$

In view of (2.15), we have to specify the assumptions for the material parameters ε_- and μ_-. In the following it is always assumed that

$$\operatorname{Im} \varepsilon_-, \operatorname{Im} \mu_- \geqslant 0 \quad \text{unless} \quad \varepsilon_- \text{ and } \mu_- < 0 , \tag{2.16}$$

which holds for all existing optical (meta)materials.

We will not consider the case $\varepsilon_-, \mu_- < 0$, which corresponds to negative refraction index materials, proposed in [20]. Then $\omega^2 \varkappa_-^2 - \alpha_n^2 = \omega^2 \varepsilon_- \mu_- - \gamma^2 - \alpha_n^2$ can be positive and one has to choose $\beta_n^- = -\sqrt{\omega^2 \varkappa_-^2 - \alpha_n^2} < 0$.

3 Potential Methods

Here we describe some potential-theoretic methods for quasiperiodic Helmholtz equations in $\mathbf{R}^2$ and the mapping properties of the resulting integral operators which have been studied in [19]. They are consequences of well-known properties of the classical logarithmic potentials on closed curves.

3.1 Potentials of periodic diffraction

Assume that Σ is non self-intersecting and is given by a piecewise C^2 parametrization

$$\sigma(t) = (X(t), Y(t)), \quad X(t+1) = X(t) + d,\; Y(t+1) = Y(t) , \quad t \in \mathbf{R},$$

i.e., the continuous functions X, Y are piecewise C^2 with

$$|\sigma'(t)| = \sqrt{(X'(t))^2 + (Y'(t))^2} > 0 ,$$

and $\sigma(t_1) \neq \sigma(t_2)$ if $t_1 \neq t_2$. Suppose additionally that if Σ has corners, then the angles between adjacent tangents at the corners are strictly between 0 and 2π.

The potentials which provide α-quasiperiodic solutions of the Helmholtz equation

$$\Delta u + k^2 u = 0 \tag{3.1}$$

are based on the quasiperiodic fundamental solution of period d

$$\Psi_{k,\alpha}(P) = \frac{i}{4} \sum_{n\in\mathbf{Z}} H_0^{(1)}\Big(k\sqrt{(X-nd)^2+Y^2}\Big)\mathrm{e}^{ind\alpha}, \quad P=(X,Y), \tag{3.2}$$

with the Hankel function of the first kind $H_0^{(1)}$ for $\arg k \in (-\pi,\pi)$. The single and double layer potentials are defined by

$$\begin{aligned} \mathcal{S}_{k,\alpha}\varphi(P) &= 2\int_\Gamma \varphi(Q)\,\Psi_{k,\alpha}(P-Q)\,d\sigma_Q\,, \\ \mathcal{D}_{k,\alpha}\varphi(P) &= 2\int_\Gamma \varphi(Q)\,\partial_{n(Q)}\Psi_{k,\alpha}(P-Q)\,d\sigma_Q\,, \end{aligned} \tag{3.3}$$

where Γ is one period of the interface Σ, i.e., $\Gamma = \{\sigma(t) : t \in [t_0, t_0+1]\}$ for some t_0. In (3.3), $d\sigma_Q$ denotes the integration with respect to the arc length and $n(Q)$, $Q \in \Sigma$, is the normal to Σ pointing into G_-. Obviously, for α-quasiperiodic densities φ on Σ the value of the potentials does not depend on the choice of Γ.

The series (3.2) converges uniformly over compact sets in $\mathbf{R}^2 \setminus \bigcup_{n\in\mathbf{Z}} \{(nd,0)\}$ if

$$k^2 \neq \alpha_n^2 = \left(\alpha + \frac{2\pi n}{d}\right)^2 \quad \text{for all} \quad n \in \mathbf{Z}. \tag{3.4}$$

Moreover, setting $\beta_n = \sqrt{k^2 - \alpha_n^2}$ (recall that $\operatorname{Im}\beta_n \geqslant 0$), the Poisson summation formula leads to the representation

$$\Psi_{k,\alpha}(P) = \lim_{N\to\infty} \frac{i}{2d} \sum_{n=-N}^{N} \frac{\mathrm{e}^{i\alpha_n X + i\beta_n|Y|}}{\beta_n}. \tag{3.5}$$

Define the function spaces

$$H_\alpha^s(\Gamma) = \left\{\mathrm{e}^{i\alpha X}\varphi\circ\sigma : \varphi\circ\sigma \in H_p^s(0,1)\right\}, \tag{3.6}$$

where $H_p^s(0,1)$, $s \in \mathbf{R}$, denotes the Sobolev space of 1-periodic functions on the real line and suppose (3.4). For $\varphi \in H_\alpha^{-1/2}(\Gamma)$ and $\psi \in H_\alpha^{1/2}(\Gamma)$ the potentials $u = V_\Gamma\varphi(P)$ resp. $u = K_\Gamma\psi(P)$, $P \notin \Sigma$, are H^1-regular and α-quasiperiodic solutions of the Helmholtz equation (3.1) which satisfy the radiation condition

$$u(x,y) = \sum_{n=-\infty}^{\infty} u_n e^{i\alpha_n x + i\beta_n |y|}, \quad |y| \geqslant H. \tag{3.7}$$

The potentials provide also the usual representation formulas. Suppose that the α-quasiperiodic function u given in G_+ is locally H^1 such that $\Delta u \in L^2_{loc}(G_+)$, satisfies the Helmholtz equation (3.1) almost everywhere and the radiation condition (3.7). Then

$$\frac{1}{2}\left(\mathcal{S}_{k,\alpha}\partial_n u - \mathcal{D}_{k,\alpha}u\right) = \begin{cases} u \text{ in } G_+, \\ 0 \text{ in } G_-, \end{cases} \tag{3.8}$$

where the normal n points into G_-. Under the same assumptions for a function u in G_- the following representation holds:

$$\frac{1}{2}\left(\mathcal{D}_{k,\alpha}u - \mathcal{S}_{k,\alpha}\partial_n u\right) = \begin{cases} 0 \text{ in } G_+, \\ u \text{ in } G_-. \end{cases} \tag{3.9}$$

3.2 Boundary integrals for periodic diffraction

Boundary integral operators are restriction of $\mathcal{S}_{k,\alpha}$ and $\mathcal{D}_{k,\alpha}$ to the profile curve Σ. The potentials provide the usual jump relations of classical potential theory [8]. The single layer potential is continuous across Σ

$$(\mathcal{S}_{k,\alpha}\varphi)^+(P) = (\mathcal{S}_{k,\alpha}\varphi)^-(P) = V_{k,\alpha}\varphi(P),$$

where the upper sign $+$ resp. $-$ denotes the limits of the potentials for points in $G_\pm$ tending in non-tangential direction to $P \in \Sigma$, and

$$V_{k,\alpha}\varphi(P) = 2\int_\Gamma \Psi_{k,\alpha}(P-Q)\,\varphi(Q)\,d\sigma_Q, \quad P \in \Sigma. \tag{3.10}$$

The double layer potential has a jump if crossing Γ:

$$\left(\mathcal{D}_{k,\alpha}\varphi\right)^+ = K_{k,\alpha}\varphi - \varphi, \quad \left(\mathcal{D}_{k,\alpha}\varphi\right)^- = K_{k,\alpha}\varphi + \varphi \tag{3.11}$$

with the boundary double layer potential

$$K_{k,\alpha}\varphi(P) = 2\int_\Gamma \varphi(Q)\,\partial_{n(Q)}\Psi_{k,\alpha}(P-Q)\,d\sigma_Q + (\delta(P)-1)\varphi(P). \tag{3.12}$$

Here $\delta(P) \in (0,2)$, $P \in \Sigma$, denotes the ratio of the angle in G_+ at P and π, i.e., $\delta(P) = 1$ outside corner points of Σ. The normal derivative of $\mathcal{S}_{k,\alpha}\varphi$ at Σ exists outside corners and has the limits

$$\left(\partial_n \mathcal{S}_{k,\alpha}\varphi\right)^+ = L_{k,\alpha}\varphi + \varphi, \quad \left(\partial_n \mathcal{S}_{k,\alpha}\varphi\right)^- = L_{k,\alpha}\varphi - \varphi, \tag{3.13}$$

where we denote

$$L_{k,\alpha}\varphi(P) = 2\int_\Gamma \varphi(Q)\,\partial_{n(P)}\Psi_{k,\alpha}(P-Q)\,d\sigma_Q\,, \quad P \in \Sigma\,. \tag{3.14}$$

In the following we consider also operators of the form

$$2\int_\Gamma \varphi(Q)\,\partial_{t(Q)}\Psi_{k,\alpha}(P-Q)\,d\sigma_Q = -2\int_\Gamma \Psi_{k,\alpha}(P-Q)\,\partial_t\varphi(Q)\,d\sigma_Q\,, \tag{3.15}$$

where φ has an α-quasiperiodic extension to Σ. If $P \notin \Sigma$, then equality (3.15) follows from integration by parts and the quasi-periodicity

$$\varphi(\sigma(t_0+1)) = e^{id\alpha}\varphi(\sigma(t_0))\,,\ \Psi_{k,\alpha}(P-\sigma(t_0+1)) = e^{-id\alpha}\Psi_{k,\alpha}(P-\sigma(t_0))$$

at the end points of Γ. If $P \in \Sigma$, then the integral on the left of (3.15) is defined as the principal value integral

$$H_{k,\alpha}\varphi(P) = 2\lim_{\delta\to 0}\int_{\Gamma\setminus\Gamma(P,\delta)} \varphi(Q)\,\partial_{t(Q)}\Psi_{k,\alpha}(P-Q)\,d\sigma_Q\,, \tag{3.16}$$

where $\Gamma(P,\delta)$ is the subarc of Γ of length 2δ with the mid point P. Let us denote the integral operator

$$\mathcal{H}_{k,\alpha}\varphi(P) = 2\int_\Gamma \varphi(Q)\,\partial_{t(Q)}\Psi_{k,\alpha}(P-Q)\,d\sigma_Q = -\mathcal{S}_{k,\alpha}\big(\partial_t\varphi\big)(P)\,, \quad P \notin \Sigma\,,$$

which satisfies for $P \in \Sigma$ the relation

$$(\mathcal{H}_{k,\alpha}\varphi)^+(P) = (\mathcal{H}_{k,\alpha}\varphi)^-(P) = H_{k,\alpha}\varphi(P) = -V_{k,\alpha}\big(\partial_t\varphi\big)(P)\,, \tag{3.17}$$

with the singular integral operator $H_{k,\alpha}$ defined by (3.16). Finally, for $P \in \Sigma$ we also define the singular integral operator

$$J_{k,\alpha}\varphi(P) = 2\int_\Gamma \varphi(Q)\,\partial_{t(P)}\Psi_{k,\alpha}(P-Q)\,d\sigma_Q = \partial_t\big(V_{k,\alpha}\varphi\big)(P)\,. \tag{3.18}$$

Mapping properties of the boundary integral operators for the quasiperiodic Helmholtz equation in the function spaces $H^s_\alpha(\Gamma)$ have been studied in [19]. In particular, the operators

$$V_{k,\alpha} : H_\alpha^{-1/2}(\Gamma) \to H_\alpha^{1/2}(\Gamma),\ H_{k,\alpha},\, K_{k,\alpha} : H_\alpha^{1/2}(\Gamma) \to H_\alpha^{1/2}(\Gamma),$$
$$J_{k,\alpha},\, L_{k,\alpha} : H_\alpha^{-1/2}(\Gamma) \to H_\alpha^{-1/2}(\Gamma)$$

are bounded, and $V_{k,\alpha}$, $H_{k,\alpha}$ and $J_{k,\alpha}$ are Fredholm operators with index 0.

The single layer potential operator $V_{k,\alpha}$ is invertible if and only if the homogeneous Dirichlet problems in the domains G_+ and G_-

$$\Delta u + k^2 u = 0\,, \quad u|_\Sigma = 0 \text{ and } u \text{ satisfies (3.7)}\,, \tag{3.19}$$

have only the trivial solution.

Remark 3.1. Well-known sufficient conditions for the unique solvability of (3.19) in G_+ (and consequently in G_-) are
- $\operatorname{Im} k^2 > 0$ or $\operatorname{Re} k^2 < 0$;
- the profile curve Σ is non-overhanging, i.e., $n_y(Q) \leqslant 0$ for all $Q \in \Sigma$.

In the following we consider also equations with transposed operators. For the physical interpretation we need that their kernel functions satisfy a radiation condition similar to (3.7). To this end we introduce the bilinear form

$$[\varphi, \psi]_\Gamma = \int_\Gamma \varphi\, \psi\, d\sigma\,, \tag{3.20}$$

which extends to a duality between the spaces $H^s_\alpha(\Gamma)$ and $H^{-s}_{-\alpha}(\Gamma)$, see (3.6). Then, for bounded $A : H^s_\alpha(\Gamma) \to H^t_\alpha(\Gamma)$ the transposed A' with respect to (3.20) maps $H^{-t}_{-\alpha}(\Gamma)$ into $H^{-s}_{-\alpha}(\Gamma)$. From the relation

$$\Psi_{k,-\alpha}(P) = \Psi_{k,\alpha}(-P) \quad \text{for all } P \in \mathbf{R}^2$$

one easily concludes that the integral operators associated with $\Psi_{k,\alpha}$ and $\Psi_{k,-\alpha}$ are connected by

$$\big(V_{k,\alpha}\big)' = V_{k,-\alpha}\,, \ \big(K_{k,\alpha}\big)' = L_{k,-\alpha}\,, \ \big(H_{k,\alpha}\big)' = J_{k,-\alpha}\,. \tag{3.21}$$

3.3 Boundary integrals for the Laplacian

The mapping properties mentioned above follow from the close connection of the boundary integral operators on Γ with corresponding operators for the Laplacian on the simple closed curve

$$\widetilde{\Gamma} = \big\{ \mathrm{e}^{-Y(t)} \big(\cos X(t), \sin X(t) \big) : t \in [0,1] \big\}\,, \tag{3.22}$$

which is the image of Γ under the conformal mapping $\mathrm{e}^{i\zeta}$, $\zeta \in \mathbb{C}$. Note that $\widetilde{\Gamma}$ has the same smoothness as Γ and additionally, if Σ has corners, then the angles at corner points of Σ in G_+ and interior angles at the corresponding corner points of $\widetilde{\Gamma}$ coincide.

On $\widetilde{\Gamma}$ we consider boundary integral operators corresponding to the fundamental solution of the Laplacian $\gamma(x) = -\log|x|/2\pi$. The operators $\widetilde{V}$, $\widetilde{K}$,

$\widetilde{L}$, $\widetilde{H}$, and $\widetilde{J}$ are defined as in (3.10), (3.12), (3.14) (3.16), and (3.18) with Γ and $\Psi_{k,\alpha}$ replaced by $\widetilde{\Gamma}$ and γ, and n is the exterior normal to $\widetilde{\Gamma}$.

For completeness we give some properties of the operators in the energy spaces $H^{\pm 1/2}(\widetilde{\Gamma})$, where $H^s(\widetilde{\Gamma})$ denotes the usual Sobolev space over the close curve $\widetilde{\Gamma}$ (cf. [6]): The operators $\widetilde{V} : H^{-1/2}(\widetilde{\Gamma}) \to H^{1/2}(\widetilde{\Gamma})$, and $\widetilde{K}, \widetilde{H} : H^{1/2}(\widetilde{\Gamma}) \to H^{1/2}(\widetilde{\Gamma})$ are bounded. With respect to the L_2-duality the operators $\widetilde{L}$ and $\widetilde{J}$ are the adjoints of $\widetilde{K}$ and $\widetilde{H}$, respectively, i.e., $\widetilde{L} = \widetilde{K}'$, $\widetilde{J} = \widetilde{H}'$, whereas $\widetilde{V}$ is symmetric. Denote by $\mathcal{C}$ the set of constant functions and by $\widetilde{I}$ the identity operator on $\widetilde{\Gamma}$. Then $\ker(\widetilde{I}+\widetilde{K}) = \ker H = \mathcal{C}$. Moreover, the operator $\widetilde{I} - \widetilde{K}$ is invertible, $\widetilde{V}$ and $\widetilde{H} = -\widetilde{V}\partial_t$ are Fredholm operators with index 0. Among the interesting relations between the integral operators we mention

$$\begin{aligned} \widetilde{V}\widetilde{K}' &= \widetilde{K}\widetilde{V}\,, & \widetilde{H}\widetilde{V} &= -\widetilde{V}\widetilde{H}'\,, \\ \widetilde{H}\widetilde{K} &= -\widetilde{K}\widetilde{H}\,, & \widetilde{K}^2 - \widetilde{H}^2 &= \widetilde{I}\,. \end{aligned} \tag{3.23}$$

The last two equations follow from the representation

$$\frac{1}{\pi i}\int_{\widetilde{\Gamma}} \frac{\varphi(t)dt}{t - (x+iy)} = -\widetilde{K}\varphi(x,y) + i\widetilde{H}\varphi(x,y) \quad \text{for } (x,y) \in \widetilde{\Gamma}\,,$$

of the Cauchy singular integral (cf. [13]).

Defining the isomorphisms $M : H^s_\alpha(\Gamma) \to H^s_\alpha(\Gamma)$ and $\vartheta^*_\alpha : H^s(\widetilde{\Gamma}) \to H^s_\alpha(\Gamma)$ by

$$M\varphi(P) = \mathrm{e}^{Y}\varphi(P)\,, \quad \vartheta^*_\alpha\varphi(P) = \mathrm{e}^{i\alpha X}\varphi(\vartheta(P))\,, \quad P = (X,Y) \in \Gamma$$

where $\vartheta : \Gamma \ni P = (X,Y) \to \mathrm{e}^{-Y}(\cos X, \sin X) \in \widetilde{\Gamma}$, one can show by using the asymptotics of the fundamental solution $\Psi_{k,\alpha}$ that the differences

$$\begin{aligned} &V_{k,\alpha} - \vartheta^*_\alpha\widetilde{V}(\vartheta^*_\alpha)^{-1}M : H^{-1/2}_\alpha(\Gamma) \to H^{1/2}_\alpha(\Gamma)\,, \\ &X_{k,\alpha} - \vartheta^*_\alpha\widetilde{X}(\vartheta^*_\alpha)^{-1} : \; H^{1/2}_\alpha(\Gamma) \to H^{1/2}_\alpha(\Gamma)\,, \\ &Y_{k,\alpha} - M^{-1}\vartheta^*_\alpha\widetilde{Y}(\vartheta^*_\alpha)^{-1}M : H^{-1/2}_\alpha(\Gamma) \to H^{-1/2}_\alpha(\Gamma)\,, \end{aligned} \tag{3.24}$$

are compact operators, where X stands for K or H and Y for L or J.

4 Integral Formulation

Here we derive the system of integral equations for conical diffraction in the case of one surface and study the equivalence to the electromagnetic formulation.

4.1 Integral equation

Denoting the components of the total fields

$$E_z = \begin{cases} u_+ + E_z^i, \\ u_-, \end{cases} \quad B_z = \begin{cases} v_+ + B_z^i & \text{in } G_+, \\ v_- & \text{in } G_- . \end{cases}$$

the transmission problem described in Subsection 2.2 can be formulated as follows. We seek H^1-regular α-quasiperiodic functions $u_\pm, v_\pm$ such that

$$\Delta u_\pm + \omega^2 \varkappa_\pm^2 u_\pm = \Delta v_\pm + \omega^2 \varkappa_\pm^2 v_\pm = 0 \quad \text{in } G_\pm, \tag{4.1}$$

subject to the transmission conditions on Σ

$$\begin{aligned} u_- &= u_+ + E_z^i, \quad \frac{\varepsilon_- \partial_n u_-}{\varkappa_-^2} - \frac{\varepsilon_+ \partial_n (u_+ + E_z^i)}{\varkappa_+^2} = \varepsilon_+ \sin\phi \Big(\frac{1}{\varkappa_+^2} - \frac{1}{\varkappa_-^2} \Big) \partial_t v_-, \\ v_- &= v_+ + B_z^i, \quad \frac{\mu_- \partial_n v_-}{\varkappa_-^2} - \frac{\mu_+ \partial_n (v_+ + B_z^i)}{\varkappa_+^2} = -\mu_+ \sin\phi \Big(\frac{1}{\varkappa_+^2} - \frac{1}{\varkappa_-^2} \Big) \partial_t u_-, \end{aligned} \tag{4.2}$$

and satisfying the outgoing wave condition

$$\begin{aligned} (u_+, v_+)(x,y) &= \sum_{n=-\infty}^{\infty} (u_n^+, v_n^+) \mathrm{e}^{i(\alpha_n x + \beta_n^+ y)} \quad \text{for } y \geqslant H, \\ (u_-, v_-)(x,y) &= \sum_{n=-\infty}^{\infty} (u_n^-, v_n^-) \mathrm{e}^{i(\alpha_n x - \beta_n^- y)} \quad \text{for } y \leqslant -H, \end{aligned} \tag{4.3}$$

with α_n and $\beta_n^\pm$ given in (2.15) and $u_n^\pm, v_n^\pm \in \mathbb{C}$.

There exist different ways to transform (4.1)–(4.3) to integral equations. We combine here the direct and indirect approach as proposed in [12, 16] for the case of classical diffraction ($\phi = 0$).

In order to represent $u_\pm$ and $v_\pm$ as layer potentials, we assume in what follows that the parameters are such that $\beta_n^\pm \neq 0$ for all n. Since $\arg \varkappa_- \in [0, \pi)$ (see (2.16)) the boundary integral operators corresponding to the fundamental solution $\Psi_{\omega\varkappa_\pm,\alpha}$ are well defined and by (3.8), (3.9) we can write

$$\begin{aligned} u_+ &= \frac{1}{2} \big(\mathcal{S}_\alpha^+ \partial_n u_+ - \mathcal{D}_\alpha^+ u_+ \big), \qquad & v_+ &= \frac{1}{2} \big(\mathcal{S}_\alpha^+ \partial_n v_+ - \mathcal{D}_\alpha^+ v_+ \big) \qquad \text{in } G_+, \\ E_z^i &= \frac{1}{2} \big(\mathcal{D}_\alpha^+ E_z^i - \mathcal{S}_\alpha^+ \partial_n E_z^i \big), \qquad & B_z^i &= \frac{1}{2} \big(\mathcal{D}_\alpha^+ B_z^i - \mathcal{S}_\alpha^+ \partial_n B_z^i \big) \qquad \text{in } G_- . \end{aligned}$$

Here we denote by $\mathcal{S}_\alpha^\pm$ the single layer potential defined on Γ with the fundamental solution $\Psi_{\omega\varkappa_\pm,\alpha}$. Correspondingly $\mathcal{D}_\alpha^\pm$ is the double layer potential over Γ with the normal derivative of $\Psi_{\omega\varkappa_\pm,\alpha}$ as kernel function. Taking the limits on Σ the jump relations (3.11) lead to

$$\begin{aligned} V_\alpha^+ \partial_n(u_+ + E_z^i) - \left(I + K_\alpha^+\right)(u_+ + E_z^i) &= 2E_z^i|_\Sigma \,, \\ V_\alpha^+ \partial_n(v_+ + B_z^i) - \left(I + K_\alpha^+\right)(v_+ + B_z^i) &= 2B_z^i|_\Sigma \,, \end{aligned} \tag{4.4}$$

where $V_\alpha^\pm$ denote the single layer potential operators

$$V_\alpha^\pm \varphi(P) = 2\int_\Gamma \varphi(Q)\, \Psi_{\omega\varkappa_\pm,\alpha}(P-Q)\, d\sigma_Q \,, \quad P \in \Sigma \,, \tag{4.5}$$

and the operators $K_\alpha^\pm$ and $L_\alpha^\pm$ are defined analogously.

The solutions in G_- are sought as single layer potentials

$$u_- = \mathcal{S}_\alpha^- w \,, \quad v_- = \mathcal{S}_\alpha^- \tau \tag{4.6}$$

with certain auxiliary densities $w, \tau \in H_\alpha^{-1/2}(\Gamma)$. Since by (3.13)

$$u_-|_\Sigma = V_\alpha^- w, \ \partial_n u_-|_\Sigma = (L_\alpha^- - I)w, \ v_-|_\Sigma = V_\alpha^- \tau, \ \partial_n v_-|_\Sigma = (L_\alpha^- - I)\tau,$$

we see from (4.4) that the transmission conditions (4.2) are valid, when the unknowns w, τ satisfy the equations

$$\begin{aligned} \frac{\varepsilon_- \varkappa_+^2}{\varepsilon_+ \varkappa_-^2} V_\alpha^+ (L_\alpha^- - I) w - \left(I + K_\alpha^+\right) V_\alpha^- w - \sin\phi\Big(1 - \frac{\varkappa_+^2}{\varkappa_-^2}\Big) V_\alpha^+ \partial_t V_\alpha^- \tau &= 2E_z^i, \\ \frac{\mu_- \varkappa_+^2}{\mu_+ \varkappa_-^2} V_\alpha^+ (L_\alpha^- - I)\tau - \left(I + K_\alpha^+\right) V_\alpha^- \tau + \sin\phi\Big(1 - \frac{\varkappa_+^2}{\varkappa_-^2}\Big) V_\alpha^+ \partial_t V_\alpha^- w &= 2B_z^i. \end{aligned}$$

Noting $V_\alpha^+ \partial_t = -H_\alpha^+$ (see (3.17)) and introducing the coefficients

$$a = \frac{\varepsilon_- \varkappa_+^2}{\varepsilon_+ \varkappa_-^2} \,, \ b = \frac{\mu_- \varkappa_+^2}{\mu_+ \varkappa_-^2} \,, \ c = \sin\phi\Big(1 - \frac{\varkappa_+^2}{\varkappa_-^2}\Big), \tag{4.7}$$

we obtain the system of singular integral equations on Γ

$$\mathcal{A}\begin{pmatrix} w \\ \tau \end{pmatrix} = -2\begin{pmatrix} E_z^i \\ B_z^i \end{pmatrix} \tag{4.8}$$

with the operator matrix

$$\mathcal{A} = \begin{pmatrix} (I + K_\alpha^+)V_\alpha^- + aV_\alpha^+(I - L_\alpha^-) & -c\,H_\alpha^+ V_\alpha^- \\ c\,H_\alpha^+ V_\alpha^- & (I + K_\alpha^+)V_\alpha^- + b\,V_\alpha^+(I - L_\alpha^-) \end{pmatrix}. \tag{4.9}$$

Recall that we have assumed (2.16), $\varkappa_-^2 \neq 0$ and $\omega^2\varkappa_\pm^2 - \alpha_n^2 \neq 0$ for all n, which implies that $\mathcal{A}$ maps $\left(H_\alpha^{-1/2}(\Gamma)\right)^2$ boundedly into $\left(H_\alpha^{1/2}(\Gamma)\right)^2$.

4.2 Equivalence

It is evident from (4.6) that any solution of (4.1)–(4.3) provides a solution of the integral equations (4.8) if the operator V_α^- is invertible.

Lemma 4.1. Let $w, \tau \in H_\alpha^{-1/2}(\Gamma)$ be a solution of (4.8) and assume $\ker V_\alpha^+ = \{0\}$. Then the functions

$$\begin{aligned} u_+ &= \frac{1}{2}\left(a\,\mathcal{S}_\alpha^+(L_\alpha^- - I)w - \mathcal{D}_\alpha^+ V_\alpha^- w + c\,\mathcal{H}_\alpha^+ V_\alpha^- \tau\right), \\ v_+ &= -\frac{1}{2}\left(c\,\mathcal{H}_\alpha^+ V_\alpha^- w + b\,\mathcal{S}_\alpha^+(L_\alpha^- - I)\tau + \mathcal{D}_\alpha^+ V_\alpha^- \tau\right), \end{aligned} \tag{4.10}$$

with the coefficients a, b, c given by (4.7) and

$$u_- = \mathcal{S}_\alpha^- w\,, \; v_- = \mathcal{S}_\alpha^- \tau\,, \tag{4.11}$$

are a solution of the transmission problem (4.1)–(4.3).

Proof. For any densities $w, \tau \in H_\alpha^{-1/2}(\Gamma)$ the single layer potentials u_-, v_- are quasi-periodic solutions of $\Delta u + \omega^2 \varkappa_-^2 u = 0$ in G_- and satisfy the outgoing wave condition (4.3). Moreover, since $u_-|_\Gamma, v_-|_\Gamma \in H_\alpha^{1/2}(\Gamma)$, $\partial_n u_-, \partial_n v_- \in H_\alpha^{-1/2}(\Gamma)$, the functions

$$\begin{aligned} u_+ &= \frac{1}{2}\left(\mathcal{S}_\alpha^+\left(a\,\partial_n u_- - c\,\partial_t v_-\right) - \mathcal{D}_\alpha^+ u_-\right), \\ v_+ &= \frac{1}{2}\left(\mathcal{S}_\alpha^+\left(b\,\partial_n v_- + c\,\partial_t u_-\right) - \mathcal{D}_\alpha^+ v_-\right) \end{aligned} \tag{4.12}$$

are H^1 regular solutions of $\Delta u + \omega^2 \varkappa_+^2 u = 0$ in G_+, satisfy (4.3) and have the boundary values

$$\begin{aligned} u_+|_\Gamma &= \frac{1}{2}\left(V_\alpha^+\left(a\,\partial_n u_- - c\,\partial_t v_-\right) + (I - K_\alpha^+)u_-\right), \\ v_+|_\Gamma &= \frac{1}{2}\left(V_\alpha^+\left(b\,\partial_n v_- + c\,\partial_t u_-\right) + (I - K_\alpha^+)v_-\right). \end{aligned}$$

Since $\partial_n u_-|_\Gamma = (L_\alpha^+ - I)w$, $H_\alpha^+ V_\alpha^- w = -V_\alpha^+\,\partial_t u_-$, and w, τ satisfy (4.8),

$$\begin{aligned} a\,V_\alpha^+ \partial_n u_- + (I - K_\alpha^+)u_- - c\,V_\alpha^+ \partial_t v_- &= 2(u_- - E_z^i)|_\Gamma\,, \\ b\,V_\alpha^+ \partial_n v_- - (I - K_\alpha^+)v_- + c\,V_\alpha^+ \partial_t u_- &= 2(v_- - B_z^i)|_\Gamma\,. \end{aligned}$$

This gives $u_+ + E_z^i = u_-$ and $v_+ + B_z^i = v_-$ on Σ. Since by (3.8)

$$\mathcal{D}_\alpha^+ E_z^i = \mathcal{S}_\alpha^+ \partial_n E_z^i\,, \quad \mathcal{D}_\alpha^+ B_z^i = \mathcal{S}_\alpha^+ \partial_n B_z^i \quad \text{in } G_+\,,$$

formulas (4.12) transform to

$$u_+ = \frac{1}{2}\left(\mathcal{S}_\alpha^+\left(a\,\partial_n u_- - c\,\partial_t v_-\right) - \mathcal{D}_\alpha^+ u_+ - \mathcal{S}_\alpha^+ \partial_n E_z^i\right),$$
$$v_+ = \frac{1}{2}\left(\mathcal{S}_\alpha^+\left(b\,\partial_n v_- + c\,\partial_t u_-\right) - \mathcal{D}_\alpha^+ v_+ - \mathcal{S}_\alpha^+ \partial_n B_z^i\right).$$

Again by (3.8) we find that in G_+

$$\mathcal{S}_\alpha^+\left(a\,\partial_n u_- - c\,\partial_t v_-\right) = \mathcal{S}_\alpha^+ \partial_n (u_+ + E_z^i),$$
$$\mathcal{S}_\alpha^+\left(b\,\partial_n v_- + c\,\partial_t u_-\right) = \mathcal{S}_\alpha^+ \partial_n (v_+ + B_z^i),$$

which shows that conditions (4.2) are satisfied if $\ker V_\alpha^+ = \{0\}$. □

5 Existence and Uniqueness of Solutions

Here we obtain conditions that the operator matrix $\mathcal{A}$ defined by (4.9) is a Fredholm mapping with index 0. Then we show that the system (4.8) is always solvable and describe cases where the solution is unique. For the following we denote by $\Phi_0(X)$ the set of bounded Fredholm operators of index 0 in the space X.

5.1 Fredholmness

Theorem 5.1. The matrix $\mathcal{A} : \left(H_\alpha^{-1/2}(\Gamma)\right)^2 \to \left(H_\alpha^{1/2}(\Gamma)\right)^2$ is a Fredholm operator with index 0 if

$$(\varepsilon_+ + \varepsilon_-)\widetilde{I} + (\varepsilon_+ - \varepsilon_-)\widetilde{K}\,,\ (\mu_+ + \mu_-)\widetilde{I} + (\mu_+ - \mu_-)\widetilde{K} \in \Phi_0\big(H^{1/2}(\widetilde{\Gamma})\big). \tag{5.1}$$

Here $\widetilde{K}$ is the double layer logarithmic potential on the closed curve $\widetilde{\Gamma}$ introduced in 3.3. Note that for sufficiently smooth $\widetilde{\Gamma}$, for example C^2, the operator $\widetilde{K}$ is compact in $H^{1/2}(\widetilde{\Gamma})$. Hence, if the profile curve Σ is sufficiently smooth, then the operator matrix $\mathcal{A}$ is Fredholm with index 0 if $\varepsilon_- \neq -\varepsilon_+$ and $\mu_- \neq -\mu_+$.

The study of Fredholm properties of the operator $\lambda\widetilde{I} + \widetilde{K}$ on nonsmooth curves $\widetilde{\Gamma}$ has a long history. A excellent overview is given by Maz'ya in [9, Section 4.1], where also higher dimensional cases and double layer potentials of other equations are discussed (cf. also [7]). Unfortunately, the results on the essential spectrum and Fredholm domain of double layer potentials cannot be applied directly since they were obtained mainly for (weighted) spaces of continuous and Hölder-continuous functions and L_p spaces. But due to the relations (3.23) it is simple to get conditions in the "energy" space $H^{1/2}(\widetilde{\Gamma})$, which are sufficient for the Fredholmness of $\mathcal{A}$.

Lemma 5.1. For any $\lambda \notin (-1,1)$ the operator $\lambda\widetilde{I} + \widetilde{K} \in \Phi_0\big(H^{1/2}(\widetilde{\Gamma})\big)$.

This result for space dimension $n \geqslant 3$ and Lipschitz domains was proved in [1]. For $n = 2$ one needs a slight modification of the proof due to the fact that the gradient of the single layer logarithmic potential does not belong to $L_2(\mathbf{R}^2)$. This holds however for the gradient of the double layer logarithmic potential, which gives together with the application of Theorem 12 in [9, Chapter 1] that in the quotient space over the constants $H^{1/2}(\widetilde{\Gamma})\,/\,\mathcal{C}$ the induced operator $\lambda\widetilde{I} + \widetilde{K}$ is invertible if $\lambda \notin [-1,1]$.

Thus, if the grating profile Σ has corners, then by Lemma 5.1 the matrix $\mathcal{A}$ is Fredholm with index 0 for $\varepsilon_-, \mu_- \notin (-\infty, 0]$. It should be noted however, that for piecewise C^2 curves one could expect the existence of $\rho < 1$ depending on the angles of $\widetilde{\Gamma}$, such that $\lambda\widetilde{I} + \widetilde{K} \in \Phi_0\big(H^{1/2}(\widetilde{\Gamma})\big)$ if $\lambda \notin (-\rho, \rho)$. For example, in the space $C(\widetilde{\Gamma})$ the parameter ρ is equal to $\max|\pi - \alpha_s|/\pi$, where the maximum is taken over all interior angles α_s of $\widetilde{\Gamma}$, but for Sobolev spaces the answer is unknown.

The proof of Theorem 5.1 follows from Lemmas 5.2, 5.3 and 5.4 given below. As in 3.3 we associate to $\mathcal{A}$ boundary integral operators for the Laplacian, more precisely, we consider the 2×2 matrix

$$\widetilde{\mathcal{A}} = \begin{pmatrix} (\widetilde{I} + \widetilde{K})\widetilde{V} + a\widetilde{V}(\widetilde{I} - \widetilde{L}) & -c\,\widetilde{H}\widetilde{V} \\ c\,\widetilde{H}\widetilde{V} & (\widetilde{I} + \widetilde{K})\widetilde{V} + b\,\widetilde{V}(\widetilde{I} - \widetilde{L}) \end{pmatrix}$$

with the coefficients a, b and c given by (4.7). From (3.24) and (4.9) it follows immediately that the difference

$$\mathcal{A} - \begin{pmatrix} \vartheta_\alpha^* & 0 \\ 0 & \vartheta_\alpha^* \end{pmatrix} \widetilde{\mathcal{A}} \begin{pmatrix} (\vartheta_\alpha^*)^{-1}M & 0 \\ 0 & (\vartheta_\alpha^*)^{-1}M \end{pmatrix} : \big(H_\alpha^{-1/2}(\Gamma)\big)^2 \to \big(H_\alpha^{1/2}(\Gamma)\big)^2$$

is compact, which provides

Lemma 5.2. $\mathcal{A} : \big(H_\alpha^{-1/2}(\Gamma)\big)^2 \to \big(H_\alpha^{1/2}(\Gamma)\big)^2$ is Fredholm if and only if $\widetilde{\mathcal{A}} : \big(H^{-1/2}(\widetilde{\Gamma})\big)^2 \to \big(H^{1/2}(\widetilde{\Gamma})\big)^2$ is Fredholm and $\operatorname{ind}\mathcal{A} = \operatorname{ind}\widetilde{\mathcal{A}}$.

Using the relation $\widetilde{K}\widetilde{V} = \widetilde{V}\widetilde{L}$, we can write

$$\widetilde{\mathcal{A}} = \widetilde{\mathcal{A}}_0 \begin{pmatrix} \widetilde{V} & 0 \\ 0 & \widetilde{V} \end{pmatrix} \quad \text{with} \quad \widetilde{\mathcal{A}}_0 = \begin{pmatrix} (1+a)\widetilde{I} + (1-a)\widetilde{K} & -c\,\widetilde{H} \\ c\,\widetilde{H} & (1+b)\widetilde{I} + (1-b)\widetilde{K} \end{pmatrix}.$$

The single layer logarithmic potential $\widetilde{V} : H^{-1/2}(\widetilde{\Gamma}) \to H^{1/2}(\widetilde{\Gamma})$ is Fredholm with index 0, hence it remains to study Fredholm properties of $\widetilde{\mathcal{A}}_0$.

Lemma 5.3. Let $c = \sin\phi\,(1 - \varkappa_+^2/\varkappa_-^2) = 0$. Then $\widetilde{\mathcal{A}}_0 \in \Phi_0\big(\big(H^{1/2}(\widetilde{\Gamma})\big)^2\big)$ if condition (5.1) holds.

Proof. $\widetilde{\mathcal{A}}_0$ is diagonal and therefore Fredholm with index 0 if and only if

$$(1+a)\widetilde{I}+(1-a)\widetilde{K}\,,\ (1+b)\widetilde{I}+(1-b)\widetilde{K}\in\Phi_0\big(H^{1/2}(\widetilde{\Gamma})\big)\,.$$

This is (5.1) when $\varkappa_+^2=\varkappa_-^2$. If otherwise $\phi=0$, then

$$a=\frac{\mu_+}{\mu_-}\,,\ b=\frac{\varepsilon_+}{\varepsilon_-}\,,$$

and (5.1) follows from the simple observation that $\widetilde{I}+\lambda\widetilde{K}\in\Phi_0\big(H^{1/2}(\widetilde{\Gamma})\big)$ implies $\widetilde{I}-\lambda\widetilde{K}\in\Phi_0\big(H^{1/2}(\widetilde{\Gamma})\big)$, which is due to $\widetilde{K}\widetilde{H}=-\widetilde{H}\widetilde{K}$, see (3.23). □

Lemma 5.4. The assertion of Lemma 5.3 is also valid in the case $c\neq 0$.

Proof. Since the relation $A\in\Phi_0(X)$ is equivalent to the existence of a compact perturbation T such that $A+T$ is invertible in X, we can apply a method to check the invertibility of operator matrices.

It is easy to see that the operator $\widetilde{H}_1=\widetilde{H}+\mathrm{j}$ with the rank 1 operator

$$\mathrm{j}u=(u,e)_{L_2(\widetilde{\Gamma})}\,e\,,\qquad e=1\in\mathcal{C}\,,$$

is invertible in $H^{1/2}(\widetilde{\Gamma})$. Instead of $\widetilde{\mathcal{A}}_0$ we consider the perturbed matrix

$$\widetilde{\mathcal{A}}_1=\begin{pmatrix}(1+a)\widetilde{I}+(1-a)\widetilde{K} & -c\widetilde{H}_1\\ c\widetilde{H}_1 & (1+b)\widetilde{I}+(1-b)\widetilde{K}\end{pmatrix}.$$

with invertible off-diagonal elements. Using the abbreviation

$$A_\pm=(1+a)\widetilde{I}\pm(1-a)\widetilde{K}\,,\quad B_\pm=(1+b)\widetilde{I}\pm(1-b)\widetilde{K}\,,$$

we transform

$$\widetilde{\mathcal{A}}_1=\begin{pmatrix}A_+ & -c\widetilde{H}_1\\ c\widetilde{H}_1 & B_+\end{pmatrix}\begin{pmatrix}-(c\widetilde{H}_1)^{-1}B_+ & \widetilde{I}\\ \widetilde{I} & 0\end{pmatrix}\begin{pmatrix}0 & \widetilde{I}\\ I & (c\widetilde{H}_1)^{-1}B_+\end{pmatrix}$$

$$=\begin{pmatrix}-A_+(c\widetilde{H}_1)^{-1}B_+-c\widetilde{H}_1 & A_+\\ 0 & c\widetilde{H}_1\end{pmatrix}\begin{pmatrix}0 & \widetilde{I}\\ \widetilde{I} & (c\widetilde{H}_1)^{-1}B_+\end{pmatrix}.$$

Now the relation $\widetilde{H}\widetilde{K}=-\widetilde{K}\widetilde{H}$ implies

$$\widetilde{H}_1A_+=A_-\widetilde{H}_1+(1-a)\,\mathrm{j}\,(\widetilde{K}-\widetilde{I})\,,$$

and therefore we get

$$\widetilde{\mathcal{A}}_1 = \begin{pmatrix} -(c\widetilde{H}_1)^{-1}(A_-B_+ + (c\widetilde{H}_1)^2) + \mathrm{j}_1 & A_+ \\ 0 & c\widetilde{H}_1 \end{pmatrix} \begin{pmatrix} 0 & \widetilde{I} \\ \widetilde{I} & (c\widetilde{H}_1)^{-1}B_+ \end{pmatrix}$$

with another rank 1 operator j_1. Hence $\widetilde{\mathcal{A}}_1$ is Fredholm with index 0 if this is true for $A_-B_+ + (c\widetilde{H}_1)^2$, and consequently for

$$A_-B_+ + c^2\widetilde{H}^2 = ((1+a)(1+b) - c^2)\widetilde{I} + 2(a-b)\widetilde{K} - ((1-a)(1-b) - c^2)\widetilde{K}^2,$$

where we make use of $\widetilde{H}^2 = \widetilde{K}^2 - \widetilde{I}$. Now (4.7), (2.9) and simple computations give

$$\begin{aligned}
(1+a)(1+b) - c^2 &= \frac{1}{\varepsilon_+\mu_+\varkappa_-^4}\Big(\varkappa_+^2\varkappa_-^2(\varepsilon_+\mu_- + \varepsilon_-\mu_+ + 2\varepsilon_+\mu_+\sin^2\phi) \\
&\quad + (\varepsilon_+\mu_+ - \varepsilon_+\mu_+\sin^2\phi)\varkappa_-^4 + (\varepsilon_-\mu_- - \varepsilon_+\mu_+\sin^2\phi)\varkappa_+^4\Big) \\
&= \frac{\varkappa_+^2}{\varepsilon_+\mu_+\varkappa_-^2}(\varepsilon_+ + \varepsilon_-)(\mu_+ + \mu_-), \\
(1-a)(1-b) - c^2 &= \frac{\varkappa_+^2}{\varepsilon_+\mu_+\varkappa_-^2}(\varepsilon_+ - \varepsilon_-)(\mu_+ - \mu_-), \\
a - b &= \frac{\varkappa_+^2}{\varepsilon_+\mu_+\varkappa_-^2}(\varepsilon_-\mu_+ - \varepsilon_+\mu_-).
\end{aligned}$$

Thus, we get the explicit representation

$$A_-B_+ + c^2\widetilde{H}^2 = \frac{\varkappa_+^2}{\varepsilon_+\mu_+\varkappa_-^2}((\varepsilon_+ + \varepsilon_-)\widetilde{I} - (\varepsilon_+ - \varepsilon_-)\widetilde{K})((\mu_+ + \mu_-)\widetilde{I} + (\mu_+ - \mu_+)\widetilde{K}),$$

which completes the proof. □

5.2 Uniqueness

Theorem 5.2. Assume $\operatorname{Im}\varepsilon_- \geqslant 0$ and $\operatorname{Im}\mu_- \geqslant 0$ with $\operatorname{Im}(\varepsilon_- + \mu_-) > 0$, which implies that $\arg\varkappa_-^2 \in (0, 2\pi)$. If $\ker V_\alpha^+ = \{0\}$ and $\arg\varkappa_-^2 \in (0, 3\pi/2)$, then the system (4.8) has at most one solution. The assertion is valid also in the case $\arg\varkappa_-^2 \in [3\pi/2, 2\pi)$ if additionally $\ker V_\alpha^- = \{0\}$.

Proof. Let w, τ be a solution of (4.8) with the right-hand side $E_z^i = B_z^i = 0$. Then in view of Lemma 4.1 the functions $u = u_\pm|_{G_\pm}$ and $v = v_\pm|_{G_\pm}$, given by (4.10) and (4.11), satisfy (4.1), (4.3) and the transmission condition

$$\Big[\frac{\varepsilon}{\varepsilon_+\varkappa^2}\,\partial_n u + \frac{\sin\phi}{\varkappa^2}\,\partial_t v\Big]_\Sigma = \Big[\frac{\mu}{\mu_+\varkappa^2}\,\partial_n v - \frac{\sin\phi}{\varkappa^2}\,\partial_t u\Big]_\Sigma = 0\,. \qquad (5.2)$$

Our aim is to obtain a variational equality for u and v in a periodic cell Ω_H, which has in x-direction the width d, is bounded by the straight lines $\{y = \pm H\}$ and contains Γ. We multiply the Helmholtz equations (4.1) respectively with

$$\frac{\varepsilon}{\varepsilon_+ \varkappa^2}\,\overline{u} \quad \text{and} \quad \frac{\mu}{\mu_+ \varkappa^2}\,\overline{v}\,,$$

and apply the Green formula in the subdomains $\Omega_H \cap G_\pm$. Then by using (5.2)

$$\int_{\Omega_H} \frac{\varepsilon}{\varepsilon_+}\Big(\frac{1}{\varkappa^2}|\nabla u|^2 - \omega^2\,|u|^2\Big) + \sin\phi\Big(\frac{1}{\varkappa_+^2} - \frac{1}{\varkappa_-^2}\Big)\int_\Gamma \partial_t v\,\overline{u} \\ - \frac{1}{\varkappa_+^2}\int_{\Gamma(H)} \partial_n u\,\overline{u} - \frac{\varepsilon_-}{\varepsilon_+ \varkappa_-^2}\int_{\Gamma(-H)} \partial_n u\,\overline{u} = 0\,, \tag{5.3}$$

$$\int_{\Omega_H} \frac{\mu}{\mu_+}\Big(\frac{1}{\varkappa^2}|\nabla v|^2 - \omega^2\,|v|^2\Big) - \sin\phi\Big(\frac{1}{\varkappa_+^2} - \frac{1}{\varkappa_-^2}\Big)\int_\Gamma \partial_t u\,\overline{v} \\ - \frac{1}{\varkappa_+^2}\int_{\Gamma(H)} \partial_n v\,\overline{v} - \frac{\mu_-}{\mu_+ \varkappa_-^2}\int_{\Gamma(-H)} \partial_n v\,\overline{v} = 0\,, \tag{5.4}$$

where $\Gamma(\pm H)$ denotes the upper resp. lower straight boundary of Ω_H. By the notation $\nabla^\perp = (\partial_y, -\partial_x)$ and the Green formula, the integral over Γ equals

$$\int_\Gamma \partial_t v\,\overline{u} = \int_{\Gamma(\pm H)} \partial_x v\,\overline{u} \mp \int_{\Omega_H \cap G_\pm} \nabla v \cdot \nabla^\perp \overline{u}\,,$$

and (4.3) gives

$$\int_{\Gamma(\pm H)} \partial_n u\,\overline{u} = i \sum_{n\in\mathbf{Z}} \beta_n^\pm |u_n^\pm|^2\,\mathrm{e}^{-2H\,\mathrm{Im}\,\beta_n^\pm}\,,$$

$$\int_{\Gamma(\pm H)} \partial_x v\,\overline{u} = i \sum_{n\in\mathbf{Z}} \alpha_n v_n^\pm\,\overline{u_n^\pm}\,\mathrm{e}^{-2H\,\mathrm{Im}\,\beta_n^\pm}\,.$$

Hence (5.3) and (5.4) can be rewritten in the form

$$\int_{\Omega_H} \Big(\frac{\varepsilon}{\varepsilon_+ \varkappa^2}\,|\nabla u|^2 - \frac{\sin\phi}{\varkappa^2}\nabla v \cdot \nabla^\perp \overline{u} - \frac{\omega^2 \varepsilon}{\varepsilon_+}\,|u|^2\Big)$$

$$= \frac{i}{\varkappa_+^2} \sum_{n\in\mathbf{Z}} \Big(\beta_n^+ u_n^+ - \alpha_n \sin\phi\, v_n^+\Big)\overline{u_n^+}\,\mathrm{e}^{-2H\,\mathrm{Im}\,\beta_n^+} \\ + \frac{i}{\varkappa_-^2} \sum_{n\in\mathbf{Z}} \Big(\frac{\varepsilon_- \beta_n^-}{\varepsilon_+}\,u_n^- - \alpha_n \sin\phi\, v_n^-\Big)\overline{u_n^-}\,\mathrm{e}^{-2H\,\mathrm{Im}\,\beta_n^-}\,, \tag{5.5}$$

$$\int_{\Omega_H} \Big(\frac{\mu}{\mu_+ \varkappa^2} |\nabla v|^2 + \frac{\sin\phi}{\varkappa^2} \nabla u \cdot \nabla^\perp \overline{v} - \frac{\omega^2 \mu}{\mu_+} |v|^2 \Big)$$
$$= \frac{i}{\varkappa_+^2} \sum_{n\in\mathbf{Z}} \Big(\beta_n^+ v_n^+ + \alpha_n \sin\phi \, u_n^+ \Big) \overline{v_n^+} \, \mathrm{e}^{-2H \,\mathrm{Im}\, \beta_n^+}$$
$$+ \frac{i}{\varkappa_-^2} \sum_{n\in\mathbf{Z}} \Big(\frac{\mu_- \beta_n^-}{\mu_+} v_n^- + \alpha_n \sin\phi \, u_n^- \Big) \overline{v_n^-} \, \mathrm{e}^{-2H \,\mathrm{Im}\, \beta_n^-} .$$

To write the quadratic forms in (5.5) more compactly we introduce the 4×4 matrix B and the vector U

$$B = \frac{1}{\varkappa^2} \begin{pmatrix} \varepsilon/\varepsilon_+ & 0 & 0 & -\sin\phi \\ 0 & \mu/\mu_+ & \sin\phi & 0 \\ 0 & \sin\phi & \varepsilon/\varepsilon_+ & 0 \\ -\sin\phi & 0 & 0 & \mu/\mu_+ \end{pmatrix}, \quad U = \begin{pmatrix} \partial_x u \\ \partial_x v \\ \partial_y u \\ \partial_y v \end{pmatrix},$$

which allow to write the left of (5.5) in the form

$$\int_{\Omega_H} \Big(BU \cdot \overline{U} - \frac{\omega^2 \varepsilon}{\varepsilon_+} |u|^2 - \frac{\omega^2 \mu}{\mu_+} |v|^2 \Big).$$

Noting that $\mathrm{Im}\, \beta_n^- > 0$ for all n and $\mathrm{Im}\, \beta_n^+ > 0$ for almost all n we see that, if $H \to \infty$, then the right-hand side of (5.5) tends to

$$\sum_{\beta_n^+ \geqslant 0} M_n \begin{pmatrix} u_n^+ \\ v_n^+ \end{pmatrix} \cdot \overline{\begin{pmatrix} u_n^+ \\ v_n^+ \end{pmatrix}}, \quad \text{where } M_n = \frac{i}{\varkappa_+^2} \begin{pmatrix} \beta_n^+ & -\alpha_n \sin\phi \\ \alpha_n \sin\phi & \beta_n^+ \end{pmatrix}.$$

Hence (5.5) states that

$$\int_{\Omega_H} \Big(BU \cdot \overline{U} - \frac{\omega^2 \varepsilon}{\varepsilon_+} |u|^2 - \frac{\omega^2 \mu}{\mu_+} |v|^2 \Big) \longrightarrow \sum_{\beta_n^+ \geqslant 0} M_n \begin{pmatrix} u_n^+ \\ v_n^+ \end{pmatrix} \cdot \overline{\begin{pmatrix} u_n^+ \\ v_n^+ \end{pmatrix}}$$

when $H \to \infty$. Obviously, if $\beta_n^+ \geqslant 0$, then $\mathrm{Re}(iM_n) \leqslant 0$. On the other side, the assumption $\mathrm{Im}\, \varepsilon_-, \mathrm{Im}\, \mu_- \geqslant 0$ implies

$$\mathrm{Re} \Big(-i \int_{\Omega_H} \Big(\frac{\omega^2 \varepsilon}{\varepsilon_+} |u|^2 + \frac{\omega^2 \mu}{\mu_+} |v|^2 \Big) \Big) = \omega^2 \int_{\Omega_H \cap G_-} \Big(\frac{\mathrm{Im}\, \varepsilon_-}{\varepsilon_+} |u|^2 + \frac{\mathrm{Im}\, \mu_-}{\mu_+} |v|^2 \Big) \geqslant 0,$$

and in addition

$$\mathrm{Re} \int_{\Omega_H} i \, BU \cdot \overline{U} \geqslant 0, \tag{5.6}$$

which can be shown similar to [2]. Taking the unitary matrix

$$\mathcal{U} = \frac{1}{\sqrt{2}} \begin{pmatrix} I & iI \\ iI & I \end{pmatrix} \quad \text{with} \quad \mathcal{U}^* = \mathcal{U}^{-1} = \frac{1}{\sqrt{2}} \begin{pmatrix} I & -iI \\ -iI & I \end{pmatrix},$$

where I denotes the 2×2 identity matrix, we obtain

$$i\mathcal{U}^{-1} B \mathcal{U} = \begin{pmatrix} N^+ & 0 \\ 0 & N^- \end{pmatrix}, \quad \text{where} \quad N^\pm = \frac{1}{\varkappa^2} \begin{pmatrix} i\varepsilon/\varepsilon_+ & \pm \sin\phi \\ \mp \sin\phi & i\mu/\mu_+ \end{pmatrix}.$$

Introducing the differential operators

$$\partial^+ = \frac{1}{\sqrt{2}} (\partial_x - i\partial_y), \quad \partial^- = \frac{1}{\sqrt{2}} (\partial_y - i\partial_x),$$

we get

$$\int_{\Omega_H} i\, BU \cdot \overline{U} = \int_{\Omega_H} \left(N^+ \partial^+ \begin{pmatrix} u \\ v \end{pmatrix} \cdot \overline{\partial^+ \begin{pmatrix} u \\ v \end{pmatrix}} + N^- \partial^- \begin{pmatrix} u \\ v \end{pmatrix} \cdot \overline{\partial^- \begin{pmatrix} u \\ v \end{pmatrix}} \right).$$

Note that $\operatorname{Re} N^\pm = 0$ in $\Omega_H \cap G_+$. Thus, it remains to consider the real part of the matrices in G_-

$$\operatorname{Re} N^\pm = \begin{pmatrix} -\operatorname{Im} \dfrac{\varepsilon_-}{\varepsilon_+ \varkappa_-^2} & \pm i \operatorname{Im} \dfrac{\sin\phi}{\varkappa_-^2} \\ \mp i \operatorname{Im} \dfrac{\sin\phi}{\varkappa_-^2} & -\operatorname{Im} \dfrac{\mu_-}{\mu_+ \varkappa_-^2} \end{pmatrix},$$

which is nonnegative if and only if the inequalities

$$-\operatorname{Im} \frac{\varepsilon_-}{\varkappa_-^2} \geqslant 0 \quad \text{and} \quad \operatorname{Im} \frac{\varepsilon_-}{\varkappa_-^2} \operatorname{Im} \frac{\mu_-}{\varkappa_-^2} - \varepsilon_+ \mu_+ \sin^2\phi \Big(\operatorname{Im} \frac{1}{\varkappa_-^2} \Big)^2 \geqslant 0 \qquad (5.7)$$

are valid. Let us denote $\phi_\varepsilon = \arg \varepsilon_-$, $\phi_\mu = \arg \mu_-$, $\phi_\varkappa = \arg \varkappa_-^2$. The assumptions

$$\phi_\varepsilon, \phi_\mu \in [0, \pi] \quad \text{and} \quad \phi_\varkappa \in (0, 2\pi),$$

together with $\varkappa_-^2 = \varepsilon_- \mu_- - \varepsilon_+ \mu_+ \sin^2\phi$ lead to $0 < \phi_\varkappa - \phi_\varepsilon, \phi_\varkappa - \phi_\mu \leqslant \pi$, which gives

$$-\operatorname{Im} \frac{\varepsilon_-}{\varkappa_-^2} = \left| \frac{\varepsilon_-}{\varkappa_-^2} \right| \sin(\phi_\varkappa - \phi_\varepsilon) \geqslant 0.$$

Since

$$\operatorname{Im} \frac{\varepsilon_+ \mu_+ \sin^2\phi}{\varkappa_-^2} = \operatorname{Im} \frac{\varepsilon_- \mu_-}{\varkappa_-^2},$$

the second inequality in (5.7) is equivalent to

$$\sin(\phi_\varepsilon - \phi_\varkappa) \sin(\phi_\mu - \phi_\varkappa) + \sin(\phi_\varepsilon + \phi_\mu - \phi_\varkappa) \sin\phi_\varkappa = \sin\phi_\varepsilon \sin\phi_\mu \geqslant 0,$$

which establishes (5.6).

Hence the solution u, v of the homogeneous problem satisfies

$$\int_{\Omega_H \cap G_-} \Big(\frac{\operatorname{Im} \varepsilon_-}{\varepsilon_+} |u|^2 + \frac{\operatorname{Im} \mu_-}{\mu_+} |v|^2 \Big) = 0 \,,$$

$$\int_{\Omega_H \cap G_-} \left(\operatorname{Re} N^+ \partial^+ \binom{u}{v} \cdot \overline{\partial^+ \binom{u}{v}} + \operatorname{Re} N^- \partial^- \binom{u}{v} \cdot \overline{\partial^- \binom{u}{v}} \right) = 0.$$

If $\operatorname{Im} \varepsilon_-, \operatorname{Im} \mu_- > 0$, then $u_- = v_- = 0$ in G_-. If otherwise, for example, $\operatorname{Im} \varepsilon_- = 0$, then $\sin \phi_\varkappa \neq 0$ and $v_- = 0$. Hence

$$\int_{\Omega_H \cap G_-} \left(\operatorname{Re} N^+ \partial^+ \binom{u}{0} \cdot \overline{\partial^+ \binom{u}{0}} + \operatorname{Re} N^- \partial^- \binom{u}{0} \cdot \overline{\partial^- \binom{u}{0}} \right)$$
$$= -2 \operatorname{Im} \frac{\varepsilon_-}{\varepsilon_+ \varkappa_-^2} \int_{\Omega_H \cap G_-} |\nabla u|^2 = \frac{2 \varepsilon_- \sin \phi_\varkappa}{\varepsilon_+ |\varkappa_-^2|} \int_{\Omega_H \cap G_-} |\nabla u|^2 = 0 \,,$$

and together with $\Delta u_- + \varkappa_-^2 u_- = 0$ this implies $u_- = 0$. By (4.11), we get $w = \tau = 0$ if the single layer potential V_α^- with the fundamental solution $\Psi_{\omega \varkappa_-, \alpha}$ is invertible. By Remark 3.1, this is always true if $\arg \varkappa_-^2 \in (0, 3\pi/2)$. □

5.3 Existence of solutions

It follows from Theorem 5.1 that under (5.1) and the conditions of Theorem 5.2 the integral equation system (4.8) has a unique solution $w, \tau \in H_\alpha^{-1/2}(\Gamma)$. Moreover, due to Lemma 4.1 the functions $u_\pm$ and $v_\pm$ from (4.10), (4.11) are a solution of the diffraction problem (4.1)–(4.3), which is unique if $V_\alpha^- = \{0\}$ is invertible.

Let us consider the remaining case of real ε_- and μ_-, where $\mathcal{A}$ can possess a nontrivial kernel. To show that the right-hand side of (4.8) is in the range of $\mathcal{A}$ we define in accordance with (3.20) the bilinear form

$$[W, \Phi] = [w, \varphi]_\Gamma + [\tau, \psi]_\Gamma \tag{5.8}$$

for $W = (w, \tau) \in \big(H_\alpha^s(\Gamma)\big)^2$, $\Phi = (\varphi, \psi) \in \big(H_{-\alpha}^{-s}(\Gamma)\big)^2$. In view of (3.21), the operator matrix transposed to $\mathcal{A}$ is given by

$$\mathcal{A}' = \begin{pmatrix} V_{-\alpha}^-(I + L_{-\alpha}^+) + a(I - K_{-\alpha}^-)V_{-\alpha}^+ & c V_{-\alpha}^- J_{-\alpha}^+ \\ -c V_{-\alpha}^- J_{-\alpha}^+ & V_{-\alpha}^-(I + L_{-\alpha}^+) + b\,(I - K_{-\alpha}^-)V_{-\alpha}^+ \end{pmatrix}$$

(see (4.5) for the definition of the integral operators, corresponding now to the fundamental solution $\Psi_{\omega \varkappa_\pm, -\alpha}$). Note that the range of $\mathcal{A}$ is orthogonal to the kernel of $\mathcal{A}'$ with respect to (5.8).

Theorem 5.3. Suppose (5.1) and let the material parameters ε_- and μ_- be real and at least one of them be positive. If V_α^- is invertible, then there exists a solution $w, \tau \in H_\alpha^{-1/2}(\Gamma)$ of the system (4.8).

Proof. The range of $\mathcal{A}$ is closed, hence it suffices to show that

$$[E_z^i, \varphi]_\Gamma + [B_z^i, \psi]_\Gamma = 0 \quad \text{for all } \Phi = (\varphi, \psi) \in \left(H_{-\alpha}^{-1/2}(\Gamma)\right)^2 \quad \text{with } \mathcal{A}'\Phi = 0\,.$$

The functions $u_+ = \mathcal{S}_{-\alpha}^+\varphi$, $v_+ = \mathcal{S}_{-\alpha}^+\psi$ in G_+ and

$$\begin{aligned} u_- &= -\frac{1}{2}\left(\mathcal{S}_{-\alpha}^-\Big(\frac{1}{a}\,(I + L_{-\alpha}^+)\varphi + \frac{c}{a}\,J_{-\alpha}^+\psi\Big) - \mathcal{D}_{-\alpha}^- V_{-\alpha}^+\varphi\right), \\ v_- &= -\frac{1}{2}\left(\mathcal{S}_{-\alpha}^-\Big(\frac{1}{b}\,(I + L_{-\alpha}^+)\psi - \frac{c}{b}\,J_{-\alpha}^+\varphi\Big) - \mathcal{D}_{-\alpha}^- V_{-\alpha}^+\psi\right) \end{aligned} \tag{5.9}$$

in G_- are $-\alpha$-quasiperiodic of period d, satisfy the Helmholtz equations (4.1) and the outgoing wave condition (4.3) with α replaced by $-\alpha$:

$$\begin{aligned} (u_+, v_+)(x,y) &= \sum_{n=-\infty}^{\infty} (u_n^+, v_n^+)\mathrm{e}^{\,i(-\alpha_n x + \beta_n^+ y)} \quad \text{for } y \geqslant H, \\ (u_-, v_-)(x,y) &= \sum_{n=-\infty}^{\infty} (u_n^-, v_n^-)\mathrm{e}^{\,i(-\alpha_n x - \beta_n^- y)} \quad \text{for } y \leqslant -H, \end{aligned} \tag{5.10}$$

the numbers α_n and $\beta_n^\pm$ are given in (2.15). The functions u_-, v_- have the boundary values

$$\begin{aligned} u_-|_\Gamma &= -\frac{1}{2}\left(V_{-\alpha}^-\Big(\frac{1}{a}\,(I + L_{-\alpha}^+)\varphi + \frac{c}{a}\,J_{-\alpha}^+\psi\Big) - (I + K_{-\alpha}^-)V_{-\alpha}^+\varphi\right), \\ v_-|_\Gamma &= -\frac{1}{2}\left(V_{-\alpha}^-\Big(\frac{1}{b}\,(I + L_{-\alpha}^+)\psi - \frac{c}{b}\,J_{-\alpha}^+\varphi\Big) - (I + K_{-\alpha}^-)V_{-\alpha}^+\psi\right), \end{aligned}$$

by using $\mathcal{A}'\Phi = 0$ we get

$$u_-|_\Gamma = V_{-\alpha}^+\varphi = u_+|_\Gamma\,, \quad v_-|_\Gamma = V_{-\alpha}^+\psi = v_+|_\Gamma\,.$$

Now (5.9) and (3.8) imply that

$$\begin{aligned} \mathcal{S}_{-\alpha}^-\Big(\frac{1}{a}\,(I + L_{-\alpha}^+)\varphi + \frac{c}{a}\,J_{-\alpha}^+\psi\Big) &= \mathcal{S}_{-\alpha}^-\partial_n u_-\,, \\ \mathcal{S}_{-\alpha}^-\Big(\frac{1}{b}\,(I + L_{-\alpha}^+)\psi - \frac{c}{b}\,J_{-\alpha}^+\varphi\Big) &= \mathcal{S}_{-\alpha}^-\partial_n v_-\,, \end{aligned}$$

and since $V_{-\alpha}^- = (V_\alpha^-)'$ is invertible, this gives the jump relations

$$\Big[\frac{\varepsilon}{\varkappa^2}\partial_n u+\frac{\varepsilon_+\sin\phi}{\varkappa^2}\,\partial_t v\Big]_\Sigma=\Big[\frac{\mu}{\varkappa^2}\partial_n v-\frac{\mu_+\sin\phi}{\varkappa^2}\,\partial_t u\Big]_\Sigma=0\,. \tag{5.11}$$

We proceed as in the proof of Lemma 5.2 and obtain the same variational equalities (5.3), (5.4) over the periodic cell Ω_H, but now for the $-\alpha$-quasiperiodic functions u and v. Since $\varepsilon_-,\mu_-\in\mathbf{R}$ and $\mathrm{Re}\,\beta_n^\pm=0$ for almost all n, the imaginary parts of (5.3), (5.4) are equal to

$$\sin\phi\Big(\frac{1}{\varkappa_+^2}-\frac{1}{\varkappa_-^2}\Big)\,\mathrm{Im}\int_\Gamma\partial_t v\,\overline{u}-\frac{1}{\varkappa_+^2}\sum_{\beta_n^+>0}\beta_n^+|u_n^+|^2-\frac{\varepsilon_-}{\varepsilon_+\varkappa_-^2}\sum_{\beta_n^->0}\beta_n^-|u_n^-|^2=0,$$

$$-\sin\phi\Big(\frac{1}{\varkappa_+^2}-\frac{1}{\varkappa_-^2}\Big)\,\mathrm{Im}\int_\Gamma\partial_t u\,\overline{v}-\frac{1}{\varkappa_+^2}\sum_{\beta_n^+>0}\beta_n^+|v_n^+|^2-\frac{\mu_-}{\mu_+\varkappa_-^2}\sum_{\beta_n^->0}\beta_n^-|v_n^-|^2=0.$$

Note that if ε_- and μ_- have different signs, then $\varkappa_-^2<0$ and $\mathrm{Re}\,\beta_n^-=0$ for all n. Because of

$$\mathrm{Im}\int_\Gamma\partial_t v\,\overline{u}=\mathrm{Im}\int_\Gamma\partial_t u\,\overline{v}$$

we derive $u_n^+=v_n^+=0$ if $\beta_n^+>0$ and $u_n^-=v_n^-=0$ if $\beta_n^->0$. The Rayleigh coefficients u_n^+,v_n^+ can be computed by the formula

$$\begin{aligned}u_n^+&=\frac{1}{d}\int_0^d u_+(X,H)\,\mathrm{e}^{i\alpha_n X-i\beta_n^+H}dX\\&=\frac{2}{d}\int_\Gamma\varphi(Q)\,d\sigma_Q\int_0^d\Psi_{\omega\varkappa_+,-\alpha}((X,H)-Q)\,\mathrm{e}^{\,i\alpha_n X-i\beta_n^+H}\,dX\,.\end{aligned}$$

From (3.5) we obtain for $Q=(x,y)$

$$\int_0^d\Psi_{\omega\varkappa_+,-\alpha}((X,H)-Q)\,\mathrm{e}^{i\alpha_n X-i\beta_n^+H}\,dX=\frac{i}{2\beta_n^+}\,\mathrm{e}^{i\alpha_n x-i\beta_n^+y}\,,$$

which gives for $n=0$

$$u_0^+=\frac{i}{d\beta}\int_\Gamma\varphi(Q)\,\mathrm{e}^{\,i\alpha x-\beta y}d\sigma_Q\,,\quad v_0^+=\frac{i}{d\beta}\int_\Gamma\psi(Q)\,\mathrm{e}^{\,i\alpha x-\beta y}d\sigma_Q\,.$$

By (2.13), the components of the incoming field E_z^i and B_z^i are multiples of $\mathrm{e}^{\,i\alpha x-\beta y}$, hence

$$\big[E_z^i,\varphi\big]_\Gamma=\big[B_z^i,\psi\big]_\Gamma=0$$

because of $u_0^+=v_0^+=0$. □

References

1. Elschner, J.: The double layer potential operator over polyhedral domains. I. Solvability in weighted Sobolev spaces. Appl. Anal. **45**, 117–134 (1992)
2. Elschner, J., Hinder, R., Penzel, F., Schmidt, G.: Existence, uniqueness and regularity for solutions of the conical diffraction problem. Math. Mod. Meth. Appl. Sci. **10**, 317–341 (2000)
3. Elschner, J., Schmidt, G.: Diffraction in Periodic structures and optimal design of binary gratings I. Direct problems and gradient formulas. Math. Meth. Appl. Sci. **21**, 1297–1342 (1998)
4. Friedman, A.: Mathematics in Industrial Problems, Part 7. Springer, Berlin (1995)
5. Goray, L.I., Sadov, S.Yu.: Numerical modelling of coated gratings in sensitive cases. OSA Diffractive Optics Micro-Optics **75**, 365–379 (2002)
6. Hsiao, G.C., Wendland, W.L.: Boundary Integral Equations. Springer, Berlin (2008)
7. Kresin, G.I., Mazya, V.G.: The norm and the essential norm of the double layer elastic and hydrodynamic potentials in the space of continuous functions. Math. Meth. Appl. Sci. **18**, 1095–1131 (1995)
8. Mazya, V.G.: The integral equations of potential theory in domains with piecewise smooth boundary (Russian). Usp. Mat. Nauk **68**, 229–230 (1981)
9. Maz'ya, V.G.: Boundary integral equations (Russian). In: Current Problems in Mathematics. Fundamental Directions. Itogi Nauki i Tekhniki, Akad. Nauk SSSR, VINITI, Moscow **27**, 131–228 (1988); English transl.: Analysis IV. Encyclop. Math. Sci. **27**, pp. 127–222. Springer, Berlin (1991)
10. Maz'ya, V., Shaposhnikova, T.: Higher regularity in the layer potential theory for Lipschitz domains. Indiana Univ. Math. J. **54**, 99–142 (2005)
11. Maz'ya, V. Solov'ev, A.: L_p-theory of boundary integral equations on a contour with outward peak. Int. Equ. Oper. Th. **32**, 75–100 (1998)
12. Maystre, D.: Integral methods. In: [14], 63–100
13. Muskhelishvili, N.I.: Singular Integral Equations, P. Noordhoff, Groningen, (1953)
14. Petit, R. (Ed.): Electromagnetic theory of gratings. Topics in Current Physics, **22**. Springer, Berlin (1980)
15. Petit R., Zolla, F.: The method of fictitious source as applied to the electromagnetic diffraction of a plane wave by a grating in conical diffraction mounts. PIE Proc. **2532**, 374–385 (1997)
16. Pomp, A.: The integral method for coated gratings: computational cost. J. Mod. Optics **38**, 109–120 (1991)
17. Popov, E., Bozhkov, B., Maystre, D., Hoose J.: Integral methods for echelles covered with lossless or absorbing thin dielectric layers. Appl. Optics **38**, 47–55 (1999)
18. Prössdorf, S.: Linear integral equations. Analysis, IV. Encyclop. Math. Sci. **27**, pp. 1–125. Springer, Berlin (1991)
19. Schmidt, G.: Integral equations for conical diffraction by coated gratings. WIAS Preprint No. 1296 (2008) [To appear in J. Int. Equ. Appl.]
20. Veselago, V.G.: The electrodynamics of substances with simultaneously negative values of ε and μ (Russian). Usp. Fiz. Nauk. **92**, 517–526 (1967)

Boundary Coerciveness and the Neumann Problem for 4th Order Linear Partial Differential Operators

Gregory C. Verchota

Abstract The relationship between the classical interior coercive estimate over $W^{2,2}(\Omega)$ and a required boundary coercive estimate for solutions to the $L^2(\partial\Omega)$ Neumann problem is discussed. A conditional lemma in which boundary coerciveness implies interior is proved. Hilbert's theorem that elliptic operators need not be sums of squares of differential operators and therefore cannot, in general, have formally positive integro-differential forms is discussed. An elliptic operator that is a sum of squares yet has no formally positive coercive form is displayed. The existence of coercive forms for elliptic operators far away from sums of squares is questioned.

Let $\Omega \subset \mathbb{R}^n$ denote a Lipschitz domain. Denote by $D = (D_1, \dots, D_n)$ the gradient operator, where $D_i = \dfrac{\partial}{\partial X_i}$ for $X = (X_i, \dots, X_n) \in \mathbb{R}^n$. Let $\gamma = (\gamma_1, \dots, \gamma_n) \in \mathbb{N}_0^n$ denote a multiindex so that $D^\gamma = D_1^{\gamma_1} \dots D_n^{\gamma_n}$, and define $|\gamma| = \gamma_1 + \dots + \gamma_n$.

A homogeneous 4th order differential operator with real constant coefficients

$$L = L(D) = \sum_{|\gamma|=4} a_\gamma D^\gamma$$

can be associated, upon integration by parts, with a *Dirichlet form*, also with constant coefficients,

$$A[u,v] = \sum_{|\beta|=|\gamma|=2} a_{\beta\gamma} \int_\Omega D^\gamma u \, D^\beta v \, dX, \tag{1}$$

Gregory C. Verchota
Syracuse University, 215 Carnegie, Syracuse NY 13244, USA
e-mail: gverchot@syr.edu

A. Laptev (ed.), *Around the Research of Vladimir Maz'ya II: Partial Differential Equations*, International Mathematical Series 12, DOI 10.1007/978-1-4419-1343-2_17,

when u and v are smooth enough up to the boundary by

$$\int_\Omega Lu\ vdX = A[u,v] + \sum_{|\beta|=|\gamma|=2} a_{\beta\gamma} \int_{\partial\Omega} N^{\beta'} D^{\beta''} D^\gamma u\ v - D^\gamma u\ N^{\beta''} D^{\beta'} v\ ds. \quad (2)$$

Here, $|\beta'| = |\beta''| = 1$, $\beta = \beta' + \beta''$, ds denotes the surface (Lebesgue) measure on the Lipschitz boundary, and $N = (N_1, \dots, N_n)$ denotes the *outer* unit normal vector to the boundary defined *a.e.*(ds). The normal derivative of a function u will be written $\dfrac{\partial u}{\partial N}$.

The operator L has of necessity been rewritten, here as

$$L = \sum_{|\beta|=|\gamma|=2} a_{\beta\gamma} D^\beta D^\gamma.$$

Such rewrites are not unique, nor are their associated Dirichlet forms. Given any four multiindices of order 2 satisfying $\beta + \gamma = \widetilde{\beta} + \widetilde{\gamma}$, a Dirichlet form associated with L may be modified by the addition of any multiple of $D^\gamma u D^\beta v - D^{\widetilde{\beta}} u D^{\widetilde{\gamma}} v$ and the associated boundary integral simultaneously modified by the addition of certain terms that involve a tangential differentiation. The left-hand side of (2) remains unchanged. For example,

$$0 = \int_\Omega D_1 D_2 u D_3 D_4 v - D_1 D_3 u D_2 D_4 v\ dX - \int_{\partial\Omega} \frac{\partial}{\partial T_{32}} D_1 u\ D_4 v\ ds, \quad (3)$$

where we introduce the notation

$$\frac{\partial}{\partial T_{ij}} = N_i D_j - N_j D_i = -\frac{\partial}{\partial T_{ji}}, \quad 1 \leqslant i, j \leqslant n.$$

In boundary value theory based on classical Hilbert space methods, the boundary derivatives of u in (2) are said to give rise to *unstable* or *natural* boundary operators, while those of v give rise to *stable* operators. (See [2, p. 6]. The distinction is tied to the order of the boundary operator.)

This terminology may be justified as follows. No matter which form is associated with L, the boundary integral in (2) can be rewritten in terms of $(v, -\frac{\partial v}{\partial N})$, the Dirichlet data of v, without alteration of the form. To illustrate this, and with *summation convention* in k (and slightly extending the notation for tangential derivatives), write

$$D^{\beta'} v = N_k \frac{\partial v}{\partial T_{k\beta'}} + N^{\beta'} \frac{\partial v}{\partial N}.$$

Using this formula in (2), tangential differentiations may be transferred to u-terms. This results in two boundary operators (formally) applied to u of orders 3 and 2 respectively

$$K = K_A = \sum_{|\beta|=|\gamma|=2} a_{\beta\gamma} \left(N^{\beta'} D^{\beta''} D^{\gamma} + \frac{\partial}{\partial T_{k\beta'}} \left[N_k N^{\beta''} D^{\gamma} \right] \right); \tag{4}$$

$$M = M_A = \sum_{|\beta|=|\gamma|=2} a_{\beta\gamma} N^{\beta} D^{\gamma} \tag{5}$$

that pair with the Dirichlet data of v:

$$\int_{\Omega} Lu \; v \; dX = A[u, v] + \int_{\partial\Omega} K_A u \; v - M_A u \; \frac{\partial v}{\partial N} \; ds. \tag{6}$$

Unlike the Dirichlet boundary operators, the operators K and M depend on the particular form associated with L. The Dirichlet operators are stable, while K and M are not.

Additionally, the form $A[u, v]$ is unaltered if the roles of β' and β'' are interchanged in (2). This interchange carries over to (4). However, a calculation shows that the operator K is, in fact, not changed. This illustrates the theorem of Aronszajn and Milgram [5], to which Agmon refers in p. 6, essentially stating that boundary value problems are completely determined by a form together with the presence of any stable boundary operators. The absence of stable operators on the boundary means that the boundary value problem is determined by the form only. These problems, with boundary operators defined as above from integration by parts and without stable operators, will be called *Neumann problems*. Given a choice of form A for a real constant coefficient *elliptic* operator

$$L = L(D) = \sum_{|\beta|=|\gamma|=2} a_{\beta\gamma} D^{\beta} D^{\gamma},$$

the pair (K_A, M_A) will be termed a *Neumann operator* for L. Ellipticity here means that $L(\xi) > 0$ for all nonzero $\xi \in \mathbb{R}^n$.

Because of work in singular integral theory [6, 11, 9] it is possible to seek real analytic solutions to the homogeneous equation

$$Lu = 0 \tag{7}$$

from the class of functions defined in Ω that have *nontangential maximal function* of the second order derivatives in the Lebesgue space $L^p(\partial\Omega)$ (defined with respect to surface measure). A solution u to (7) from this class is said to solve the Neumann problem with *Neumann data* $(\Lambda, f) \in W^{-1,p}(\partial\Omega) \times L^p(\partial\Omega)$ if

$$K_A u = \Lambda \tag{8}$$

in a *sense of distributions* defined by way of parallel boundaries, and if

$$M_A u = f \tag{9}$$

in the sense of *nontangential convergence a.e.*(ds). Let $p + p' = pp'$ denote dual exponents. The Banach space $W^{-1,p}(\partial\Omega)$ is defined to be dual to the Sobolev space $W^{1,p'}(\partial\Omega)$ of $L^{p'}(\partial\Omega)$ functions with weak first derivatives (tangential derivatives) in $L^{p'}(\partial\Omega)$ (cf. [28] for complete definitions in the case $L = \triangle^2$, the bi-Laplacian).

Given $1 < p < \infty$, layer potential solutions u formed from the fundamental solution for L [18] can be defined that belong to the above L^p nontangential maximal function classes. As in [28], Coifman, McIntosh and Meyer's theorems on the boundedness of singular integrals on Lipschitz boundaries [9] can then be used to show that the Neumann data (Ku, Mu) belongs to the corresponding boundary space $W^{-1,p} \times L^p$ and is taken on in the required distributional and pointwise senses.

In order to show the existence of a potential solution that solves boundary equations (8), (9) for given data (Λ, f), an a priori estimate on *all* second derivatives of a solution by the data is needed. In the case of second order equations and systems [16, 25, 10, 12, 21], this has been supplied by certain integral identities on Lipschitz boundaries known generally as *Rellich identities*. Identities (also termed Rellich), not quite as complete as those just mentioned, were used by the author and Pipher to study the Dirichlet problem for higher order equations and systems of all orders in [22, 27]. Recently a genuine Rellich identity for the *biharmonic* equation, leading to a solution of both Neumann and Dirichlet regularity problems for biharmonic functions, was studied by the author [28].

Let α denote a smooth vector field in $\mathbb{R}^n$ that is transverse to $\partial\Omega$, i.e., there is a constant $c_\alpha > 0$ depending on α and on the Lipschitz geometry of Ω so that the inner product

$$\alpha \cdot N = \alpha(Q) \cdot N_Q \geqslant c_\alpha > 0$$

for almost every $Q \in \partial\Omega$. By choosing a *symmetric form* A for L, i.e., one for which $a_{\beta\gamma} = a_{\gamma\beta}$, and considering solutions u to (7), the directional derivative $2\alpha \cdot Du$ can be substituted for v in (6) and the Gauss divergence theorem applied to get the Rellich identity

$$\begin{aligned}\sum_{|\beta|=|\gamma|=2} a_{\beta\gamma} \int_{\partial\Omega} D^\gamma u \, D^\beta u \, \alpha \cdot N \, ds \\ \approx 2\int_{\partial\Omega} M_A u \, \frac{\partial(\alpha \cdot Du)}{\partial N} - K_A u \, \alpha \cdot Du \, ds,\end{aligned} \tag{10}$$

where $\approx$ indicates *equality* modulo lower order terms involving differentiations of α. The Dirichlet form has been placed on the boundary as a *quadratic form* and the pairing of Neumann and Dirichlet data realized now as a pairing of dual spaces $W^{-1,2} \times L^2$ and $W^{1,2} \times L^2$. The Neumann operator, however, has been restricted to those that arise from symmetric forms.

It is not apparent from (10) that the quadratic form on the left bounds from above all second derivatives of a solution restricted to the boundary. It would suffice to prove this modulo Neumann data. In other words, a *boundary coercive estimate* is wanted for a Neumann problem derived from a symmetric form for L. For example,

$$\sum_{|\beta|=|\gamma|=2} a_{\beta\gamma} \int_{\partial\Omega} D^\gamma u \, D^\beta u \, \alpha \cdot N \, ds$$
$$\geqslant c \int_{\partial\Omega} |DDu|^2 |\alpha| ds - C \left(\|K_A u\|^2_{-1,2} + \|M_A u\|^2_2 \right) \qquad (11)$$

for some $c > 0$ and C depending only on A and the Lipschitz geometry of Ω. The norms are for $W^{-1,2}(\partial\Omega; |\alpha| ds)$ and $L^2(\partial\Omega; |\alpha| ds)$ respectively and DDu denotes all second derivatives of u.

Again we are willing to restrict the set of Neumann operators in order to secure the coercive estimate (11). For example, if $L = D_1^4 + D_2^4 + \ldots + D_n^4$, then a coercive estimate (a *strong* coercive estimate since one will be able to take $C = 0$) follows by choosing quadratic forms with integrands

$$(1-\theta) \sum_{i=1}^{n} (D_i^2 u)^2 + \frac{\theta}{2n} \sum_{i,j=1}^{n} \left[(D_i^2 u - D_j^2 u)^2 + 2(D_i D_j u)^2 \right], \qquad (12)$$

when the parameter θ is restricted to the open interval $(0\ ,1)$. The evident quadratic form with terms

$$(D_1^2 u)^2 + \ldots + (D_n^2 u)^2 \qquad (13)$$

has been modified, in the manner discussed before (3), by adding the terms

$$\frac{\theta}{n} \left((D_i D_j u)^2 - D_i^2 u D_j^2 u \right). \qquad (14)$$

Because $0 < \frac{\theta}{n}$ is small enough, the newly introduced terms that lack a definite sign are bounded in a pointwise manner by the terms (13). Here, it is useful that the elliptic operator L is itself a *sum of squares* of differential operators, leading to (13) and thence to (12). Forms that are themselves sums of squares like (12) and (13) are said to be *formally positive* [4], [1, p. 184] regardless of any coercive properties they might or might not own. Similarly, the bi-Laplacian $\triangle^2$ is a sum of squares, and $(\triangle u)^2$ can be modified to be coercive on the boundary by adding the terms (14) for $0 < \theta \leqslant (n-1)^{-1} n^2$, as shown in [28].

Hilbert [15], [14, p. 56] proved that every constant coefficient 4th order homogeneous elliptic operator in $\mathbb{R}^3$ can be written as a sum of squares of second order operators. That it is every such operator makes possible in $\mathbb{R}^3$ the above argument for pointwise coerciveness. For given an elliptic L, the operator $L - \epsilon \sum_{i,j}(D_i D_j)^2$ would be elliptic and therefore a sum of squares when $\epsilon > 0$ is small enough. Consequently, L would have a symmetric form that was the sum of a formally positive form plus the coercive form $\epsilon \sum_{i,j}(D_i D_j u)^2$.

But Hilbert also proved that in all higher dimensions 4th order elliptic operators existed that could not be written as sums of squares. Thus a coerciveness problem.

Perhaps, the most elegant example of an operator that cannot be written as a sum of squares comes from a polynomial of Motzkin type

$$\mathcal{Q}(x, y, z, w) = w^4 + (xy)^2 + (yz)^2 + (zx)^2 - 4xyzw$$

proved by Choi and Lam [7] not to be a sum of squares. A homogeneous polynomial (also called a *form*) is *positive definite* if it is nonnegative and vanishes only at the origin, i.e., satisfies the ellipticity condition. The arithmetic-geometric mean inequality and inspection show that Choi and Lam's polynomial is only *positive semidefinite* However, as Robinson argues [23, pp. 267-268], any positive semidefinite form that is not a sum of squares can be perturbed to be a positive definite form that is still not a sum of squares. Thus, for $0 < \eta$ small enough the operators

$$\mathcal{Q}_\eta(D) = \eta(D_1^4 + D_2^4 + D_3^4) + \mathcal{Q}(D) \tag{15}$$

are elliptic in $\mathbb{R}^4$ and not sums of squares. The evident quadratic forms all contain the indefinite term $-4D_1D_2uD_3D_4u$ (or other choices or sums of such choices) making it not clear how the boundary coercive estimate (11) could be obtained in this case.

A fortiori it is not clear how the stronger estimate for *solutions*

$$\sum_{|\beta|=|\gamma|=2} a_{\beta\gamma} \int_{\partial\Omega} D^\gamma u \; D^\beta u \; \alpha \cdot N \; ds \geqslant c \int_{\partial\Omega} |DDu|^2 |\alpha| ds \tag{16}$$

could be obtained for (15). Such a scale invariant estimate is not true in the bounded domain case. For example, the quadratic form $(D_1^2 u - D_2^2 u)^2 + 4(D_1 D_2 u)^2$ associated with the bi-Laplacian operator $\triangle^2$ is known to yield the classical coercive estimate over the domain Sobelev space $W^{2,2}(\Omega)$ when $\Omega \subset \mathbb{R}^2$ is a bounded domain [2]. However, if h is *any* harmonic function defined in a neighborhood of the *origin*, then $u_\epsilon(x, y) := (x^2 + y^2)h(\epsilon x, \epsilon y)$ are solutions to the biharmonic equation with the property that the left-hand side of

$$\int_{\partial\Omega} \Big(\frac{\partial^2 u_\epsilon}{\partial x^2} - \frac{\partial^2 u_\epsilon}{\partial y^2}\Big)^2 + 4\Big(\frac{\partial^2 u_\epsilon}{\partial x \partial y}\Big)^2 \alpha \cdot N \; ds$$

$$\geqslant c \int_{\partial\Omega} \left(\frac{\partial^2 u_\epsilon}{\partial x^2}\right)^2 + \left(\frac{\partial^2 u_\epsilon}{\partial x \partial y}\right)^2 + \left(\frac{\partial^2 u_\epsilon}{\partial y^2}\right)^2 |\alpha| ds$$

vanishes as ϵ vanishes, while the right-hand side converges to

$$c8h^2(0,0) \int_{\partial\Omega} |\alpha| ds$$

(cf. [28, Lemma 7.3]).

On the other hand, when $\Omega = \{(x,y) : y > \phi(x), x \in \mathbb{R}^{n-1}\}$ is the domain above the graph of a compactly supported Lipschitz function $\phi : \mathbb{R}^{n-1} \to \mathbb{R}$, the strong boundary coercive estimate for solutions (16) depending only on $a_{\beta\gamma}$ and the Lipschitz constant $\|\nabla\phi\|_\infty$, when true, implies the coercive estimate over the domain Ω for *solutions*

$$\sum_{|\beta|=|\gamma|=2} a_{\beta\gamma} \int_\Omega D^\gamma u \; D^\beta u \; dX \geqslant c_1 \int_\Omega |DDu|^2 dX \tag{17}$$

by integrating (16) formulated on parallel boundaries $\partial\Omega_t = \{(x, t + \phi(x)) : x \in \mathbb{R}^{n-1}\}$, $t > 0$. The left-hand sides of (16) and (17), integrands that can take *negative* values, are related by the identity

$$\int_0^\infty dt \sum_{\beta,\gamma} a_{\beta\gamma} \int_{\partial\Omega_t} D^\gamma u \; D^\beta u |N_n| ds = \sum_{\beta,\gamma} a_{\beta\gamma} \int_\Omega D^\gamma u \; D^\beta u \; dX.$$

Now, the transverse vector field α has been chosen to be the constant basis-vector field $-e_n$ and c_1 depends only on the Lipschitz constant and $a_{\beta\gamma}$.

Below it will be shown that (17) *then implies*

$$\sum_{\beta,\gamma} a_{\beta\gamma} \int_\Omega D^\gamma v \; D^\beta v \; dX \geqslant c \int_\Omega |DDv|^2 dX \tag{18}$$

over all $v \in W^{2,2}(\Omega)$ *with* c *depending only on the Lipschitz constant and* $a_{\beta\gamma}$.

As Agmon [1, p. 201] argues in the half-space, (18) then implies the classical coercive estimate

$$\sum_{\beta,\gamma} a_{\beta\gamma} \int_\Omega D^\gamma v \; D^\beta v \; dX$$
$$\geqslant c \int_\Omega |DDv|^2 dX + c \int_\Omega |Dv|^2 dX - c_0 \int_\Omega |v|^2 dX \tag{19}$$

as a consequence of the Gagliardo–Nirenberg inequality.

If the constant in (18) *depends only on* $\|\nabla\phi\|$ *and* $a_{\beta\gamma}$ *so will the constants in* (19). To prove this define for any function f on Ω and $M \geqslant 0$ the partial scalings $f_M(x,y) = f(x, y + M(y - \phi(x)))$, $(x,y) \in \Omega$. Then by the Fubini theorem, the change of variable $y' = y + M(y - \phi(x))$ for each $x \in \mathbb{R}^{n-1}$, and $dX = dxdy$

$$\int_\Omega f_M dX = \frac{1}{M+1}\int_\Omega f dX. \tag{20}$$

Because $f_M = f$ on the boundary, the divergence theorem yields

$$\int_\Omega (Dv)_M \cdot Dv dX = \int_{\partial\Omega} \frac{\partial v}{\partial N} v ds - \int_\Omega \frac{\partial}{\partial x_j} D_j v(x, y + M(y - \phi(x)))v(x,y) + \frac{\partial}{\partial y} D_n v(x, y + M(y - \phi(x)))v(x,y) dxdy.$$

Subtracting the $M > 0$ cases from the $M = 0$ case yields

$$\int_\Omega |Dv|^2 dX = \int_\Omega (Dv)_M \cdot Dv dX - \int_\Omega \triangle v v dX + \int_\Omega (\triangle' v)_M v - M(D' D_n v)_M \cdot \nabla\phi(x) v + (M+1)(D_n^2 v)_M v dX,$$

where $D' = (D_1, \ldots, D_{n-1}, 0)$, and $\triangle$ and $\triangle'$ are the respective Laplace operators. Choosing $M = 3$, and using the Schwarz inequality and (20) on the first integral allows the first integral to be hidden on the left. Using the Young inequality

$$\int fg \leqslant \epsilon \int f + \frac{1}{4\epsilon}\int g$$

and (20) on the others, it follows that

$$\int_\Omega |Dv|^2 dX \leqslant A\left(\epsilon \int_\Omega |DDv|^2 dX + (1 + \|\nabla\phi\|_\infty^2)\epsilon^{-1}\int_\Omega |v|^2 dX\right),$$

where A is a universal constant. Now, ϵ can be chosen to depend only on the constant in (18) to obtain (19).

Conversely, another scaling argument Agmon uses in the half-space can also be applied for the Lipschitz graph domain Ω, as long as the constant in (19) depends only on the Lipschitz constant (and $a_{\beta\gamma}$), in order to show that (19) implies (18). Thus, (19) and (18) are equivalent in the scale invariant graph setting. To do this, we define $\Omega_1 = \Omega$ and $\Omega_M = \{(x,y) : y > M^{-1}\phi(Mx)\}$. Then $X \in \Omega_M$ if and only if $MX \in \Omega_1$. Given $v \in W^{2,2}(\Omega)$, define v_M in Ω_M by $v_M(X) = v(MX)$ so that $D^\beta v_M(X) = M^{|\beta|} D^\beta v(MX)$. Since $M^{-1}\phi(Mx)$ and $\phi(x)$ have the same Lipschitz norm, letting $M \to \infty$ and arguing exactly as in p. 201 of [1], shows that (19) implies (18). Just as well, given that the Rellich compactness theorem is not valid in unbounded graph domains, precluding the standard application of the estimate (19) (cf., for example, [13, pp. 249-250]).

Consequently, it is reasonable to look for (18) in graph domains, and this estimate would be a consequence of the strong boundary coercive statement (16) for solutions.

However, as far as the author knows, given an elliptic operator

$$L = \sum_{|\gamma|=4} a_\gamma D^\gamma$$

it is not known whether there exists a quadratic form for it,

$$L = \sum_{|\beta|=|\gamma|=2} a_{\beta\gamma} D^\beta D^\gamma,$$

that makes (18) true in arbitrary half-spaces!

Indeed, even in the case of an operator L that is a sum of squares, it is not immediately apparent that it possesses a coercive quadratic form in the classical setting, i.e., over $W^{2,2}(\Omega)$. To see this, recall Aronszajn's characterization [4, 3] of coercive formally positive integro-differential quadratic forms which here may be stated

Let $L(D) = \sum_j p_j^2(D)$ be a homogeneous elliptic operator of order $2m$ with constant coefficients and $\Omega \subset \mathbb{R}^n$ a bounded Lipschitz domain. Then a necessary and sufficient condition for the coercive estimate over $W^{m,2}(\Omega)$

$$\sum_j \int_\Omega |p_j(D)v|^2 dX \geqslant c \int_\Omega |D^m v|^2 dX - c_0 \int_\Omega |v|^2 dX$$

is that there be no solution $z \in \mathbb{C}^n \setminus \{0\}$ to the system of algebraic equations

$$p_1(z) = p_2(z) = \ldots = 0.$$

Allowing the parameter η in (15) to increase until the first sum of squares elliptic operator is obtained, one can arrive at a family of such operators

$$\begin{aligned} L_\rho(D) = (D_4^2 - \rho(D_1^2 + D_2^2 + D_3^2))^2 + (D_4 D_1 - D_2 D_3)^2 \\ + (D_4 D_2 - D_3 D_1)^2 + (D_4 D_3 - D_1 D_2)^2 \end{aligned}$$

for $0 < \rho < 1/3$ that will satisfy Aronszajn's algebraic condition and thus each will have a coercive formally positive form. However, this form can be proved to be the *only* coercive formally positive form that each possesses by using Gram matrix methods for modeling sums of squares polynomials [8]. By introducing two more real variables in the following way *each of the elliptic operators*

$$(D_6^2 + D_5^2 + D_5 D_4)^2 + L_\rho(D)$$

can also be shown to have a unique formally positive integro-differential form. However, now Aronszajn's condition is violated by $z = (0,0,0,0,1,i)$. Consequently, *these elliptic operators possess no formally positive coercive forms.* These assertions are proved in [29].

There remains the possibility that all sums of squares operators do have coercive forms over $W^{m,2}(\Omega)$ but not, in general, of formally positive type. The author claims this is the case at least for $W^{2,2}(\Omega)$ and proof will appear in [30]. However, if an elliptic operator is far away from the sum of squares operators, as with (15) when η is small, the existence of a coercive form seems to be unknown.

Now, we prove the following assertion.

Conditional lemma. *If the boundary coercive estimate* (16) *for solutions* u, *with nontangential maximal function* $(DDu)^* \in L^2(\partial\Omega)$, *holds in a domain* $\Omega \in \mathbb{R}^n$ *above the graph of a compactly supported Lipschitz function, then the coercive estimate* (18) *over all* $v \in W^{2,2}(\Omega)$ *must hold.*

It suffices to consider $v \in C_0^\infty(\mathbb{R}^n)$. Solutions with Dirichlet data equal to that of v on $\partial\Omega$ may be constructed as follows.

By the Gårding inequality and the Lax–Milgram theorem, there are unique $v_j^0 \in W_0^{2,2}(\Omega_j)$ satisfying $Lv_j^0 = Lv$ in each Ω_j. By the Weyl lemma, the solutions $u_j = v - v_j^0$ are C^∞ in Ω_j and have, by the scale invariant estimates in [22, Theorem 2, p. 5] for the *regularity problem*, the nontangential maximal functions of DDu_j controlled in $L^2(\partial\Omega_j)$ by the first tangential derivatives of the gradient $\dfrac{\partial}{\partial T}Dv_{|\partial\Omega}$ *uniformly* in j. The nontangential maximal functions of the Du_j are likewise uniformly controlled in $L^2(\partial\Omega_j)$ by the gradient $Dv_{|\partial\Omega}$. Denote these maximal functions by DDu_j^* and Du_j^* respectively with the understanding that they are taken over the (bounded) nontangential approach regions of Ω_j. As was shown in [22], the u_j and all derivatives DDu_j and Du_j have pointwise limits *a.e.* on $\partial\Omega_j$, the limits of the Du_j agreeing with Dv.

Let $K \subset \Omega$ be the compact closure of a subdomain of Ω. Averaging straightforward applications of the fundamental theorem of calculus yields

$$|u_j(X)|^2 \leqslant C_K\Big(\int_{\partial\Omega} v^2 ds + \int_{\partial\Omega} Du_j^{*2} ds\Big)$$

uniformly in $X \in K$, where C_K depends on $\operatorname{dist}(K, \partial\Omega)$. Also uniformly in j by the uniform bound on the Du_j^*. Similarly, by the uniform bound on the DDu_j^*, the u_j form an equicontinuous sequence on K. By interior estimates [18, p. 155], derivatives of any order form equicontinuous sequences. Thus, there is a uniformly convergent subsequence to a *solution* on K. By considering a further sequence of compact $K_k \uparrow \Omega$ and a diagonalization argument, a subsequence of the u_j converges to a solution $Lu = 0$ in Ω.

Fix a K again and denote by $(DDu_j^*)_K$ the nontangential maximal functions formed over the part of the approach regions intersecting K. Then, by uniform convergence on K and the uniform nontangential bound in j,

$$\int_{\partial\Omega} (DDu^*)_K^2 ds = \lim_{j\to\infty} \int_{\partial\Omega} (DDu_j^*)_K^2 ds \leqslant C \int_{\partial\Omega} \Big(\frac{\partial}{\partial T} Dv\Big)^2 ds.$$

By monotone convergence as $K \uparrow \Omega$,

$$\int_{\partial\Omega} (DDu^*)^2 ds \leqslant C \int_{\partial\Omega} \Big(\frac{\partial}{\partial T} Dv\Big)^2 ds, \tag{21}$$

where the maximal function is now formed over the unbounded approach regions for Ω. By the same arguments and the uniform bound on the Du_j^*, we have

$$\int_{\partial\Omega} (Du^*)^2 ds \leqslant C \int_{\partial\Omega} (Dv)^2 ds \tag{22}$$

(cf. [17] for the prototype of this argument).

Because of (21) DDu is square integrable over the strip $\Omega\backslash\{(x, t+\phi(x)) : t > M\}$ for any M. Because of (22) the interior estimates [18] over balls of radius half the distance to the boundary, and the geometry of the approach regions

$$|DDu(x, t+\phi(x))| \leqslant Ct^{-1} Du^*(x, \phi(x)), \quad t > M,$$

where C depends only on $a_{\beta\gamma}$. Consequently,

$$DDu \in L^2(\Omega). \tag{23}$$

To show that u and its derivatives have nontangential limits on $\partial\Omega$ we do so inside one of the bounded domains $\Omega_0 = \Omega_{j_0}$. First the Dirichlet data is shown to exist pointwise nontangentially. Again consider a compact K. For $k, j > j_0$ the integrals

$$\int_{\partial\Omega_0\backslash\partial\Omega\backslash K} |Du_k - Du_j|^2 ds$$

behave like $\operatorname{dist}(K, \partial\Omega)$ by the fundamental theorem of calculus and the uniform bound on the maximal functions. The same integrands over $\partial\Omega_0 \cap K$ are small uniformly as $j, k \to \infty$. Thus, for j, k large enough the Dirichlet data measured in $W^{1,2}(\partial\Omega_0)$ for solutions $u_k - u_j$ in Ω_0 is as small as desired. By estimates in [22, Theorem 1], so too are the corresponding maximal functions for $D(u_k - u_j)$ and then $D(u - u_j)$ in Ω_0. Now, the standard lim sup – lim inf argument [24, p. 8] can be employed to yield nontangential limits for u andDu on $\partial\Omega$ *a.e.* and equal to v and Dv respectively on $\partial\Omega \cap \partial\Omega_0$. By (21), this Dirichlet data for u has tangential derivatives in $L^2(\partial\Omega_0)$ and thus supplies data in the Whitney array space $WA^{2,2}(\partial\Omega_0)$ for the *regularity problem* [26, 22]. The unique solution to this problem has nontangential limits *a.e.* for its 2nd derivatives. Since it also solves the same Dirichlet problem as does u it is identical to u. In this way all derivatives up to order 2 of u are seen to have nontangential limits *a.e.* on $\partial\Omega$.

Now, it is possible to give the orthogonality argument that yields (18) from (17). Given $v \in C_0^\infty(\mathbb{R}^n)$ and the solution u, as constructed, with Dirichlet boundary values on $\partial\Omega$ those of v, put $v_0 = v - u$. Then v_0 satisfies (21), (22) and (23) also and has vanishing Dirichlet data on $\partial\Omega$. Consequently, by the Lebesgue dominated convergence,

$$\begin{aligned} A[u, v_0] = \sum_{\beta,\gamma} a_{\beta\gamma} \int_\Omega D^\gamma u \, D^\beta v_0 \, dX &= \lim_j \sum_{\beta,\gamma} a_{\beta\gamma} \int_{\Omega_j} D^\gamma u \, D^\beta v_0 \, dX \\ &= -\lim_j \int_{\partial\Omega_j \setminus \partial\Omega} K_A u \, v_0 - M_A u \, \frac{\partial v_0}{\partial N} \, ds, \end{aligned}$$

where the vanishing on $\partial\Omega$ of the term with 3rd derivatives on u in the resulting boundary integral is justified by the interior estimates for $D^3 u$ and the fundamental theorem for v_0, taken close to the boundary, together with (21) and (22). The geometry of nontangential approach regions and these justifications again show that the boundary integrals are controlled by

$$int_{\partial\Omega \setminus \partial\Omega_{j/2}} DDu^* \; Dv_0^* ds$$

which vanishes in the limit. Thus, $A[u, v_0] = 0$.

By the Plancherel identity,

$$\int_\Omega (DDv_0)^2 dX \leqslant c_2 A[v_0, v_0].$$

Recalling the constant in (17),

$$\begin{aligned} \int_\Omega (DDv)^2 dX &\leqslant 2 \int_\Omega (DDv_0)^2 dX + 2 \int_\Omega (DDu)^2 dX \\ &\leqslant 2c_2 A[v_0, v_0] + 2c_1^{-1} A[u, u] \\ &\leqslant 2(c_2 + c_1^{-1}) A[v, v], \end{aligned}$$

where the last inequality is obtained by the orthogonality of u and v_0.

Thus, (18) follows from (16).

If it happens that (18) is false in the graph case for operators far away form the sum of squares, then a weak boundary coercive estimate like (11) is the only remaining possibility. However, it too is problematic. An example of constant coefficient elliptic operator that has no coercive form is not known. But the only forms (1) for the bi-Laplacian that lead to solutions for the L^2 Neumann problem in planar Lipschitz domains are those that are formally positive. This is proved by counterexamples in [28]. It seems unlikely that an operator far from the sum of squares operators could then have the weak estimate (11) on Lipschitz boundaries.

Strong corroboration for this point of view comes from work of Kozlov and Maz'ya [19]. They establish in thin cones Ω that biharmonic singular solutions, with vanishing Neumann data and 2nd derivatives in $weak-L^2(\Omega)$, exist for the Neumann problems that arise from those quadratic forms for the bi-Laplacian that are *not* formally positive but are coercive in smooth domains (cf. [20, pp. 397–398]).

References

1. Agmon, S.: The coerciveness problem for integro-differential forms. J. Anal. Math. **6**, 183–223 (1958)
2. Agmon, S.: Remarks on self-adjoint and semi-bounded elliptic boundary value problems. In: Proc. Internat. Sympos. Linear Spaces (Jerusalem, 1960), pp. 1–13. Jerusalem Acad. Press, Jerusalem (1960)
3. Agmon, S.: Lectures on Elliptic Boundary Value Problems. Van Nostrand, Princeton, NJ (1965)
4. Aronszajn, N.: On coercive integro-differential quadratic forms. In: Conference on Partial Differential Equations. Univ. Kansas, Technical Report No. 14, 94–106 (1954)
5. Aronszajn, N., Milgram, A.N.: Differential operators on Riemannian manifolds. Rend. Circ. Mat. Palermo (2) **2**, 266–325 (1953)
6. Calderón, A.P.: Cauchy integrals on Lipschitz curves and related operators, Proc. Nat. Acad. Sci. USA **74**, no. 4, 1324–1327 (1977)
7. Choi M. D., Lam, T.Y.: Extremal positive semidefinite forms. Math. Ann. **231**, no. 1, 1–18 (1977/78)
8. Choi M. D., Lam, T.Y., Reznick, B.: Sums of squares of real polynomials. Proc. Sympos. Pure Math. **58**, 103–126 (1995)
9. Coifman, R.R., McIntosh, A., Meyer, Y.: L'intégrale de Cauchy définit un opérateur borné sur L^2 pur les courbes lipschiziennes, Ann. Math. (2) **116**, no. 2, 361–387 (1982)
10. Dahlberg, B.E.J., Kenig, C.E., frankVerchota, G.C.: Boundary value problems for the systems of elastostatics in Lipschitz domains. Duke Math. J. **57**, no. 3, 795–818 (1988)
11. Fabes, E.B., Jodeit, M. Jr., Rivière, N.M.: Potential techniques for boundary value problems on C^1-domains. Acta Math. **141**, no. 3–4, 165–186 (1978)
12. Fabes, E.B., Kenig, C.E., Verchota, G.C.: The Dirichlet problem for the Stokes system on Lipschitz domains. Duke Math. J. **57** no. 3, 769–793 (1988)
13. Folland, G.B.: Introduction to Partial Differential Equations. Princeton Univ. Press, Princeton, NJ (1995)
14. Hardy, G.H., Littlewood, J.E., Pólya, G.: Inequalities. Cambridge Univ. Press, Cambridge (1952)
15. Hilbert, D.: Uber die Darstellung definiter Formen als Summe von Formenquadraten. Math. Ann. **32** 342–350 (1888)
16. Jerison, D.S., Kenig, C.E.: The Neumann problem on Lipschitz domains. Bull. Am. Math. Soc. (N.S.) **4**, no. 2, 203–207 (1981)
17. Jerison D.S., Kenig, C.E.: Boundary value problems on Lipschitz domains. In: Studies in Partial Differential Equations, MAA Stud. Math. **23**, pp. 1–68. MAA, Washington, DC (1982)

18. John, F.: Plane Waves and Spherical Means Applied to Partial Differential Equations. Intersci. Publ., New York etc. (1955)
19. Kozlov, V.A., Maz'ya, V.G.: On the spectrum of an operator pencil generated by the Neumann problem in a cone (Russian). Algebra Anal. **3**, no. 2, 111–131 (1991); English transl.: St. Petersburg Math. J. **3**, no. 2, 333–353 (1992)
20. Kozlov, V.A., Maz'ya, V.G., Rossmann, J.: Spectral Problems Associated with Corner Singularities of Solutions to Elliptic Equations. Am. Math. Soc., Providence, RI (2001)
21. Mitrea, M.: The method of layer potentials in electromagnetic scattering theory on nonsmooth domains. Duke Math. J. **77**, no. 1, 111–133 (1995)
22. Pipher, J., Verchota, G.C.: Dilation invariant estimates and the boundary Gårding inequality for higher order elliptic operators. Ann. Math. (2) **142**, no. 1, 1–38 (1995)
23. Robinson, R.M.: Some definite polynomials which are not sums of squares of real polynomials (Russian). In: Selected Questions of Algebra and Logic (Collection Dedicated to the Memory of A. I. Malcev), pp. 264–282. Nauka, Novosibirsk (1973)
24. Stein, E.M.: Singular Integrals and Differentiability Properties of Functions. Princeton Univ. Press, Princeton, NJ (1970)
25. Verchota, G.C.: Layer potentials and regularity for the Dirichlet problem for Laplace's equation in Lipschitz domains. J. Funct. Anal. **59**, no. 3, 572–611 (1984)
26. Verchota, G.C.: The Dirichlet problem for the polyharmonic equation in Lipschitz domains. Indiana Univ. Math. J. **39**, no. 3, 671–702 (1990)
27. Verchota, G.C.: Potentials for the Dirichlet problem in Lipschitz domains. In: Potential Theory (Kouty, 1994), pp. 167–187. Walter de Gruyter, Berlin (1996)
28. Verchota, G.C.: The biharmonic Neumann problem in Lipschitz domains. Acta Math. **194**, no. 2, 217–279 (2005)
29. Verchota, G.C.: Noncoercive sums of squares in $\mathbb{R}[x_1, \ldots, x_n]$. J. Pure Appl. Algebra (2009) doi:10.1016/j.jpaa.2009.05.012
30. Verchota, G.C.: On the necessity of Agmon's coerciveness for the analysis of operators with formally positive integro-differential forms [In preparation]

Index

Bergman polynomial 219
Bergman shift 221

Capacity 205, 220
— harmonic 65
— logarithmic 220
capacity test 207
cusp 304
— inward 304
— outward 304

Diffraction
— conical 340
— periodic 344
Douglis–Nirenberg ellipticity 166

Equation
— biharmonic 368
Euclidean capacity 210
expansion
— matched asymptotic 97
— multiscale 97

Hardy constant
— global 5
Hardy inequality 2
Hessenberg matrix 221
homogenization 63

Fredholm alternative 261
Fredholm operator 170, 353

Gagliardo–Nirenberg inequality 371
Green matrix 318, 318

Helmholtz equations 341
Helmholtz operator 123
Hersch isoperimetric
inequality 233

Kato class 240
Keller–Osserman bound 10
Koiter equation 277
Koiter shell system 172, 161

Laplace–Beltrami operator 136
Laplace–Dirichlet problem 99

Maxwell equations 341

Navier–Stokes
equations 277, 135
— linearized 89
Neumann eigenvalue 231
Neumann problem 367
Newtonian capacity 241

Operator pencil 83

Perturbation 95
— selfsimilar 95
potential
— double layer 345
— single layer 309, 345
point, regular 241
problem
— sensitive 181, 182
— super-variational 97

Rayleigh quotient 231
Rellich identity 368
regularity index 171

Shapiro–Lopatinskii condition 166
sistem
— membrane 172
— — tension 172
— rigidity 172
Sobolev space 162, 204, 345
— subelliptic 148
— weighted 102, 320
solution
— layer potential 368
Stokes equations 277, 318
sub-harmonic 4
— global 4
— local 4
sub-solution 7
super-harmonic 4
— global 4
super-harmonic 4
— local 4
super-solution 7

Trace 305

Volume test 214

Whitney distance 10

p-Laplacian 146

References to Maz'ya's Publications Made in Volume II

Monographs

- Maz'ya, V.G., Nazarov, S.A., Plamenevskii, B.A.: Asymptotics of Solutions to Elliptic Boundary Value Problems under Singular Pertubation of Domains (Russian). Tbilisi Univ. Press, Tbilisi (1981)

  61, 93 Chechkin [33]
- Maz'ya, V.G.: Sobolev Spaces. Springer, Berlin etc. (1985)

  5, 22 Bandle–Moroz-Reichel [9]
  145, 157 Domokos–Manfredi [14]
  210, 215, 218 Grigor'yan–Kondratiev [9]
  224, 228 Khavinson–Stylianopoulos [13]
  232, 238 Kokarev–Nadirashvil [7]
  277, 301 Panasenko–Stavre [7]
  cf. also Vols. I and III
- Maz'ya, V.G., Shaposhnikova, T.O.: Theory of Multipliers in Spaces of Differentiable Functions. Pitman, Boston etc. (1985) Russian edition: Leningrad. Univ. Press, Leningrad (1986)

  145, 157 Domokos–Manfredi [16]
  cf. also Vol. I
- Kozlov, V.A., Maz'ya, V.G., Rossmann, J.: Elliptic Boundary Value Problems in Domains with Point Singularities. Am. Math. Soc., Providence, RI (1997)

  96, 133 Dauge–Tordeux–Vial [13]
  cf. also Vols. I, III
- Maz'ya, V.G., Poborchi, S.V.: Differentiable Functions on Bad Domains. World Scientific, Singapore (1997)

  305, 312, 314 Poborchi [6]
  cf. also Vol. I
- Kozlov, V., Maz'ya, V., Movchan, A.: Asymptotic Analysis of Fields in Multi-structures. Clarendon Press, Oxford Univ. Press, New York (1999)

  25, 60 Barbatis–Burenkov–Lamberti [15]
  160, 202 Egorov–Meunier–Sanchez-Palencia [11]
  277, 301 Panasenko–Stavre [6]
- Maz'ya, V., Nazarov, S., Plamenevskii, B.: Asymptotic Theory of Elliptic Boundary Value Problems in Singularly Perturbed Domains. I. II. Birkhäuser, Basel (2000)

  25, 60 Barbatis–Burenkov–Lamberti [18]
  61, 93 Chechkin [35]
  97, 101, 123, 125, 127, 133 Dauge–Tordeux–Vial [19]
  160, 202 Egorov–Meunier–Sanchez-Palencia [12]
  277, 301 Panasenko–Stavre [8]
- Kozlov, V.A., Maz'ya, V.G., Rossmann, J.: Spectral Problems Associated with Corner Singularities of Solutions to Elliptic Equations. Am. Math. Soc., Providence, RI (2001)

  316, 320, 320, 335 Roßmann [3]
  377, 378 Verchota [20]
  cf. also Vol. I
- Kuznetsov, N., Maz'ya, V., Vainberg, B.: Linear Water Waves: A Mathematical Approach. Cambridge Univ. Press, Cambridge (2002)

  254–261, 265, 271–273 Motygin–Kuznetsov [8]

Research papers

- Burago, Yu.D., Maz'ya, V.G.: Certain questions of potential theory and function theory for regions with irregular boundaries (Russian). Zap. Nauchn. Semin. LOMI. **3** (1967); English transl.: Consultants Bureau, New York (1969)
  303, 313 Poborchi [1]
  cf. also Vol. I
- Verzhbinskii, G.M., Maz'ya, V.G.: On the asymptotics of solutions of the Dirichlet problem near a nonregular boundary. Dokl. Akad. Nauk SSSR **176**, no. 3, 498-501 (1967)
  315, 336 Roßmann [17]
- Maz'ya, V.: On Neumann's problem in domains with nonregular boundaries (Russian). Sib. Mat. Zh. **9**, 1322–1350 (1968); English transl.: Sib. Math. J. **9**, 990–1012 (1968)
  230, 238 Kokarev–Nadirashvili [6]
- Maz'ya, V.G.: The continuity at a boundary point of solutions of quasilinear equations (Russian). Vestnik Leningrad. Univ. **25**, no. 13, 42–55 (1970). Correction, ibid **27:1**, 160 (1972); English transl.: Vestnik Leningrad Univ. Math. **3**, 225–242 (1976)
  145, 157 Domokos–Manfred [13]
  239, 252 Liskevich–Skrypnik [12]
  cf. also Vol. I
- Verzhbinskii, G.M., Maz'ya, V.G.: Asymptotic behaviour of solutions of elliptic equations of second order near the boundary I, II (Russian). Sib. Mat. Zh. **12**, no. 6, 1217-1249, (1971); ibid **13**, no. 6, 1239-1271 (1972)
  315, 336 Roßmann [18]
- Maz'ya, V.G., Khavin, V.P.: Nonlinear potential theory (Russian). Usp. Mat. Nauk **27**, 67–138 (1972); English transl.: Russ. Math. Surv. **27**, 71–148 (1973)
  145, 157 Domokos–Manfredi [15]
  cf. also Vol. I
- Vainberg, B.R., Maz'ya, V.G.: On the plane problem of the motion of a body immersed in a fluid (Russian). Tr. Mosk. Mat. Ob. **28**, 35–56 (1973); English transl.: Trans. Moscow Math. Soc. **28**, 33–55 (1973)
  253, 265, 274 Motygin–Kuznetsov [16]
- Maz'ya, V.G., Plamenevskii, B.A.: Estimates of Green's functions and Schauder estimates for solutions of elliptic boundary value problems in a dihedral angle (Russian). Sib. Mat. Zh. **19**, no. 5, 1064–1082 (1978); English transl.: Sib. Math. J. **19**, 752–764 (1979)
  316, 335 Roßmann [5]
- Maz'ya, V.G., Plamenevskii, B.A.: Schauder estimations of solutions of elliptic boundary value problems in domains with edges on the boundary (Russian). Tr. Semin. S.L. Soboleva 1978, No. 2, 69–102 (1978); English transl.: Am. Math. Soc. Transl. **123**, 141–169 (1984)
  316, 336 Roßmann [6]
- Maz'ya, V.G., Nazarov, S.A., Plamenevskii, B.A.: Asymptotic solutions of elliptic boundary value problems when the domain is varied close to conical points (Russian). Dokl. Akad. Nauk SSSR **249**, no. 1, 94–96 (1979); English transl.: Sov. Phys. Dokl. **24**, 904-905 (1979)
  97, 133 Dauge–Tordeux–Vial [17]
- Maz'ya, V.G., Plamenevskii, B.A.: On the asymptotics of the fundamental solutions of elliptic boundary value problems in regions with conical points (Russian). Probl. Mat. Anal. **7**, 100–145 (1979); English transl.: Sel. Math. Sov. **4** 363–397 (1985)
  316, 336 Roßmann [7]

- Maz'ya, V.G., Nazarov, S.A., Plamenevskii, B.A.: On the asymptotic behavior of solutions of elliptic boundary value problems with irregular perturbations of the domain (Russian). Probl. Mat. Anal. **8**, 72–153 (1981)
  97, 133 Dauge–Tordeux–Vial [18]
- Mazya, V.G.: The integral equations of potential theory in domains with piecewise smooth boundary (Russian). Usp. Mat. Nauk **68**, 229–230 (1981)
  303, 314 Poborchi [4]
  339, 346, 363 Schmidt [8]
- Maz'ya, V.G., Plamenevskii, B.A.: On properties of solutions of three-dimensional problems of elasticity theory and hydrodynamics in domains with isolated singular points (Russian). Dinam. Sploshn. Sred. **50**, 99-120 (1981); English transl.: Am. Math. Soc. Transl. (2) **123**, pp. 109–123. Am. Mat. Soc., Providence, RI (1984)
  316, 320, 336 Roßmann [8]
- Maz'ya, V.G., Plamenevskii, B.A.: The first boundary value problem for classical equations of mathematical physics in domains with piecewise smooth boundaries (Russian). Part 1: Z. Anal. Anwend. **2**, no. 4, 335–359 (1983); Part 2: ibid **2**, no. 6, 523–551 (1983)
  316, 318, 336 Roßmann [9]
- Maz'ya, V.G., Nazarov, S.A., Plamenevskii, B.A.: Asymptotic expansions of the eigenvalues of boundary value problems for the Laplace operator in domains with small holes (Russian). Izv. Akad. Nauk SSSR, Ser. Mat. **48**, no. 2, 347–371 (1984); English transl.: Math. USSR, Izv. **24**, 321–345 (1985)
  71, 93 Chechkin [34]
- Maz'ya, V.G., Plamenevskii, B.A.: Estimates in L^p and in Hölder classes and the Miranda–Agmon maximum principle for solutions of elliptic boundary value problems in domains with singular points on the boundary. Transl., Ser. 2, Am. Math. Soc. **123**, 1–56 (1984)
  96, 104, 133 Dauge–Tordeux–Vial [20]
- Maz'ya, V.G., Plamenevskii, B.A.: On the coefficients in the asymptotic of solutions of the elliptic boundary problem in domains with conical points. Transl., Ser. 2, Am. Math. Soc. **123**, 57–88 (1984)
  104, 105, 133 Dauge–Tordeux–Vial [21]
- Maz'ya, V.G., Slutskii, A.S.: Homogenization of a differential operator on a fine periodic curvilinear mesh. Math. Nachr. **133**, 107–133 (1986)
  61, 93 Chechkin [36]
- Maz'ya, V.G., Nazarov, S.A.: Paradoxes of limit passage in solutions of boundary value problems involving the approximation of smooth domains by polygonal domains (Russian). Izv. Akad. Nauk SSSR Ser. Mat. **50**, no. 6, 1156–1177, 1343 (1986); English transl.: Math. USSR, Izv. **29**, 511–533 (1987)
  61, 93 Chechkin [32]
- Maz'ya, V.G., Roßmann (Rossmann), J.: Über die Asymptotik der Lösungen elliptischer Randwertaufgaben in der Umgebung von Kanten. Math. Nachr. **138**, 27–53 (1988)
  322, 336 Roßmann [10]
- Maz'ya, V.G. Boundary integral equations (Russian). In: Current Problems in Mathematics. Fundamental Directions. Itogi Nauki i Tekhniki, Akad. Nauk SSSR, VINITI, Moscow **27**, pp. 131–228 (1988); English transl.: Analysis IV. Encyclop. Math. Sci. **27**, pp. 127-222, Springer, Berlin (1991)
  303, 314 Poborchi [5]
  339, 353, 363 Schmidt [9]

- Kuznetsov, N.G., Maz'ya, V.G.: On unique solvability of the plane Neumann–Kelvin problem (Russian). Mat. Sb. **135**, 440–462 (1988); English transl.: Math. USSR Sb. **63**, 425–446 (1989)

  253, 273 Motygin–Kuznetsov [7]
- Maz'ya, V.G., Solov'ev, A.A.: On integral equation of the Dirichlet problem in a plane domain with cusps on the boundary (Russian). Mat. Sb. **180**, no. 9, 1211–1233, 1296 (1989); English transl.: Math. USSR-Sb. **68**, no. 1, 61–83 (1991)

  303, 314 Poborchi [12]
- Maz'ya, V.G., Solov'ev, A.A.: On boundary integral equation for the Neumann problem in a domain with cusp (Russian). Tr. Leningrad. Mat. O-va **1**, 109–134 (1990)

  303, 314 Poborchi [13]
- Kozlov, V.A., Maz'ya, V.G.: On the spectrum of an operator pencil generated by the Neumann problem in a cone (Russian). Algebra Anal. **3**, no. 2, 111–131 (1991); English transl.: St. Petersburg Math. J. **3**, no. 2, 333–353 (1992)

  377, 378 Verchota [19]
- Maz'ya, V.G., Roßmann (Rossmann), J.: On the Agmon–Miranda maximum principle for solutions of elliptic equations in polyhedral and polygonal domains. Ann. Global Anal. Geom. **9**, 253–303 (1991)

  316, 336 Roßmann [11]
- Maz'ya, V.G., Hänler, M.: Approximation of solutions to the Neumann problem in disintegrating domains. Math. Nachr. **162**, 261–278 (1993)

  61, 93 Chechkin [30]
- Kozlov, V.A., Maz'ya, V.G., Schwab, C.: On singularities of solutions to the Dirichlet problem of hydrodynamics near the vertex of a cone. J. Rein. Angew. Math. **456**, 65–97 (1994)

  316, 320, 335 Roßmann [4]
- Kresin, G.I., Mazya, V.G.: The norm and the essential norm of the double layer elastic and hydrodynamic potentials in the space of continuous functions. Math. Meth. Appl. Sci. **18**, 1095–1131 (1995)

  339, 353, 363 Schmidt [7]
- Maz'ya, V.G.: Unsolved problems connected with the Wiener criterion. In: The Legacy of Norbert Wiener: A Centennial Symposium (Cambridge, Mass., 1994). Proc. Sympos. Pure Math. **60**, pp. 199–208. Am. Math. Soc., Providence, RI, (1997)

  239, 252 Liskevich–Skrypnik [13]
- Maz'ya, V. Solov'ev, A.: L_p-theory of boundary integral equations on a contour with outward peak. Int. Equ. Oper. Theor. **32**, 75–100 (1998)

  339, 363 Schmidt [11]
- Maz'ya, V.G., Solov'ev, A.A.: Integral equations of logarithmetics potential theory for contours with cusps in Hölder spaces (Russian). Algebra Anal. **10** no. 5, 85–142 (1998); English transl.: St. Petersburg Math. J. **10**, no. 5. 791–832 (1999)

  303, 314 Poborchi [14]
- Maz'ya, V., Slutskii, A.: Asymptotic analysis of the Navier–Stokes system in a plane domain with thin channels. Asymptotic Anal. **23**, no. 1, 59–89 (2000)

  160, 202 Egorov–Meunier–Sanchez-Palencia [13]
  277, 301 Panasenko–Stavre [9]
- Maz'ya, V.G.: The Wiener test for higher order elliptic equations. Duke Math. J. **115**, 479–512 (2002)

  239, 252 Liskevich–Skrypnik [14]

- Maz'ya, V.G., Rossmann, J.: Point estimates for Green's matrix to boundary value problems for second order elliptic systems in a polyhedral cone. Z. Angew. Math. Mech. **82**, no. 5, 291–316 (2002)
  316, 336 Roßmann [12]
- Maz'ya, V., Rossmann, J.: Pointwise estimates for Green's kernel of mixed boundary value problem to the Stokes system in polyhedral cone. Math. Nachr. **278**, no. 15, 1766–1810 (2005)
  135, 144 Dindoš [16]
  316, 318, 319, 326, 336 Roßmann [13]
- Maz'ya, V.G., Shaposhnikova, T.O.: Higher regularity in the classical layer potential theory for Lipschitz domains. Indiana Univ. Math. J. **54**, no. 1, 99–142 (2005)
  303, 314 Poborchi [11]
  339, 363 Schmidt [10]
- Maz'ya, V., Rossmann, J.: Schauder estimates for solution to a mixed boundary value problem for the Stokes system in polyhedral domains. Math. Meth. Appl. Sci. **29**, 965–1017 (2006)
  135, 144 Dindoš [17]
- Maz'ya, V.G., Poborchi, S.V.: On solvability of the Neumann problem in energy space for a domain with peak. Georgian Math. J. **14**, no. 3, 499–518 (2007)
  304, 308, 314 Poborchi [7]
- Maz'ya, V.G., Rossmann, J.: L_p estimates of solutions to mixed boundary value problems for the Stokes system in polyhedral domains. Math. Nachr. **280**, no. 7, 751–793 (2007)
  322, 336 Roßmann [14]
- Filippas, S., Maz'ya, V., Tertikas A.: Critical Hardy–Sobolev inequalities. J. Math. Pures Appl. **87**, 37–56 (2007)
  5, 22 Bandle–Moroz-Reichel [6]
  cf. also Vol. I
- Maz'ya, V.G., Poborchi S.V.: On solvability of the Neumann problem for a domain with cusp (Russian). Algebra Anal. **20**, no. 5, 108-153 (2008); English transl.: St. Petersburg Math. J. [To appear]
  304, 308, 314 Poborchi [8]
- Maz'ya, V.G., Poborchi, S.V.: Unique solvability of integral equation for harmonic single layer potential on the boundary of a domain with cusp (Russian). Vestnik St. Petersburg State Univ. no. 2, 63–73 (2009)
  304, 310, 314 Poborchi [9]
- Maz'ya, V.G., Poborchi, S.V.: On representation of solution to the Neumann problem for a domain with cusp by harmonic single layer potential (Russian). Vestnik St. Petersburg State Univ. (2009) [To appear]
  304, 305, 312, 314 Poborchi [10]
- Maz'ya, V.G., Rossmann, J.: Mixed boundary value problems for the Navier-Stokes system in polyhedral domains. Arch. Rational Mech. Anal. [To appear]
  320, 336 Roßmann [15]
- Maz'ya, V.G., Movchan, A.S.: Asymptotic treatment of perforated domains without homogenization. Math. Nachr. [To appear]
  61, 93 Chechkin [31]
- Maz'ya, V., Rossmann, J.: A Maximum Modules Estimate for Solutions of the Navier–Stokes System in Domains of Polyhedral Type. Preprint.
  135, 136, 137, 144 Dindoš [18]

FSC
www.fsc.org
MIX
Papier | Fördert gute Waldnutzung
FSC® C083411

Kolibri 360 GmbH
Ferdinand-Jühlke-Straße 7
99095 Erfurt, Deutschland
produktsicherheit@kolibri360.de